Progress in HPLC-HPCE Volume 6

Progress in HPLC-HPCE

Series Editors: H. Parvez and S. Parvez

Previously published in this series:

Volume 1
Gel Permeation and Ion-Exchange Chromatography of Proteins and Peptides

Volume 2
Electrochemical Detection in Medicine and Chemistry

Volume 3
Flow Through Radioactivity Detection in HPLC

Volume 4
Supercritical Fluid Chromatography and Micro-HPLC

Volume 5
Capillary Electrophoresis in Biotechnology and Environmental Analysis

Progress in HPLC-HPCE Volume 6

Coulometric electrode array detectors for HPLC

Editors

I.N. Acworth, M. Naoi,
H. Parvez and S. Parvez

///VSP///

Utrecht, The Netherlands, 1997

VSP BV
P.O. Box 346
3700 AH Zeist
The Netherlands

First published in 1997

ISBN 90-6764-232-0

Printed in The Netherlands by Ridderprint bv, Ridderkerk.

CONTENTS

This book is dedicated to Kim, Pat, Paul, Edwin, and Guadalupe

Preface

The sixth volume in the well received series ***Progress in HPLC-HPCE*** examines the power of high-performance liquid chromatography with coulometric array detection. This analytical technique markedly extends the utility of electrochemical detection by offering improvements in sensitivity and selectivity over the amperometric "thin-layer" type of electrochemical detector (ECD). Furthermore, by using the coulometric electrode array to examine the voltammetric behavior of analytes, a third chromatographic dimension is produced. Now compounds can be identified and qualified by both retention time and characteristic voltammetric redox behavior. The coulometric electrode array, therefore, offers the resolution of the photodiode array with the extreme sensitivity of an ECD.

When it comes to HPLC, not all ECDs are the same and, in fact, the efficiency of an electrode plays an important role. In 1978, Dr. Wayne Matson was trying to measure the levels of essential neurotransmitters and their metabolites in tissues obtained from children exposed to heavy metals. Unfortunately, the commercially available detectors at that time were neither sufficiently sensitive nor stable enough for routine measurement of these analytes. Matson realized that by forcing the eluent from the column through a "filter electrode", instead of over the electrode in the case of the thin-layer design, analyte conversion efficiencies could approach 100%. Such highly efficient electrodes he termed "coulometric". These flow-through graphite electrodes proved not only to be sensitive, but also incredibly resilient and reliable. In 1981, based on the studies and efforts of Matson and his colleague Dr. Eric Zink, the first coulometrically efficient electrodes became commercially available. In 1988 the powerful coulometric electrode array detector was introduced. In 1995, the modular CoulArray® detector was released.

This volume examines the use of the coulometric electrode array detector. The editors wish to acknowledge the achievements of Dr. Matson and his continued quest to improve electrochemical detectors for use in HPLC. It is hoped that this collection of reviews will provide a valuable reference source to all workers in the field of HPLC and will help to stimulate further research in all aspects of electrochemical detection.

The Editors and Publisher wish to express their gratitude to all contributors for the effort and care with which they prepared their manuscripts. Finally, the Editors wish to thank Dr. Kim Acworth and Ms. Debbie Aldrich for their help in the preparation of this volume.

Progress in HPLC-HPCE, Vol. 6, pp. 3–50
I.N. Acworth *et al.* (Eds)

AN INTRODUCTION TO HPLC-BASED ELECTROCHEMICAL DETECTION: FROM SINGLE ELECTRODE TO MULTI-ELECTRODE ARRAYS

IAN N. ACWORTH and MARK BOWERS

ESA, Inc., 22 Alpha Road, Chelmsford, MA 01824-4171

INTRODUCTION

High performance liquid chromatography (HPLC) is one of the most popular analytical techniques today. The growth of this technique is a direct consequence of the demands of the analytical chemist forcing continued research into the development of more efficient separation processes and detectors with greater sensitivity. Although today's chromatograph is superior in many ways to those that came before and column technology has been vastly improved, it is still usually the detector that is the limiting component in many HPLC analyses. Why the electrochemical detector (ECD) was originally developed and the reason for its popularity is perhaps best described by McClintock and Purdy when they state "many compounds function *in vivo* as a result of electron transfer processes; it is therefore only a logical extension to assume that these same processes can be monitored by utilizing the 'electron transfer' process, i.e. measurement by electrochemical means" (McClintock and Purdy, 1987). To the analyst, the ECD offers the advantages of superior sensitivity and selectivity over other HPLC detector.

Many different electrochemical detectors have appeared over the years and have been used successfully by the analytical chemist, including conductometric, potentiometric, oscillometric, tensammetric, and voltammetric detectors (Barth *et al.*, 1986). However, whether rightfully or not, many published articles using HPLC-ECD refer to the use of one of three voltammetric detectors: the amperometric, the coulometric, or the polarographic detector. It has been more than forty years since Wiktor Kemula first published his design for a dropping mercury polarographic detector for liquid chromatography (Kemula, 1952). Kemula's technique and its possible importance languished for more than ten years because it was impractical (the detector had to be hand-made, and suffered from poor peak shape and limits of detection) and was prone to difficulties ("spent" mercury had to be removed, mercury drops caused recorder oscillations or noise, the electrode had a limited anodic potential range, and oxygen had to be removed to prevent interference with reductive processes of interest).

In the early 1970's, electrochemical analysis underwent a major revolution. The combination of efficient HPLC separation with a "low volume" amperometric electrode by Adams and his group resulted in the first working HPLC-ECD. They also reported routine quantitation of catecholamines at the nanogram level (Kissinger *et al.*, 1973). The success of Adam's work overshadowed another report, that one describing the first wall-jet amperometric detector using a carbon working electrode (Varadi *et al.*, 1974). See below for further details on wall-jet electrodes.

Neuroscientists[1] initially exploited amperometric HPLC-ECD, with its advantages of simple instrumentation, inherent sensitivity, and selectivity, to measure labile neurotransmitters such as catecholamines and indole-alkylamines in brain tissue (Kissinger, 1977, 1989; Kissinger *et al.*, 1973; Horvai and Pungor, 1987, 1989). Researchers eventually realized, however, that static tissue levels of neurotransmitters probably did not reflect accurately the dynamics of the neurotransmitters in the extracellular space. One recent sampling innovation, *in vivo* microdialysis perfusion, allows the direct measurement of neurotransmitter levels in brain extracellular fluid (Ungerstedt, 1984; Robinson and Justice, 1991; Rollema *et al.*, 1991; Louilot *et al.*, 1994). But such measurements put major constraints on the analytical instrument. Often sub-picogram levels of neurotransmitters have to be measured in the presence of many other endogenous metabolites and in small sample volumes (1-20μL). As samples are typically collected every 10-20 minutes, the analytical technique also has to be routine, with little maintenance and down time. Although some of the early "low volume" amperometric thin-layer detectors were successfully used for some microdialysis analyses, their inherent instability and poor limits of detection often made such routine analyses problematic. A superior approach uses the flow-through coulometric electrode. This electrode offers both routine stability and extreme sensitivity (e.g. <300fg/20μL sample [0.1nM] with a s/n ratio of 3:1 for the catecholamines). Recently, the sensitivity of the thin-layer amperometric electrode has been improved and, when forced to become coulometrically efficient by coupling to capillary columns, achieves a typical detection limit of <5fg/200nL sample with a s/n ratio of 3:1, making it ideal for the analysis of low volume samples such as microdialysis perfusates (Cullison *et al.*,1996). In just two decades there has been a millionfold improvement in sensitivity over the original amperometric detector (Kissinger *et al.*, 1973)! The differences between the amperometric and coulometric detector will be addressed more fully below.

As with UV-based spectrophotometric approaches, it soon became obvious that single channel ECD measurement could be problematic; that is, misidentification of analytes was too easy, possible coelutions were hard to detect, and a great deal of potential qualitative information was being lost. The spectrophotometric situation was improved somewhat by dual wavelength detectors, but it was not until the development of the photodiode array (PDA) that spectrophotometric techniques were revolutionized. With their potent resolution, PDAs overcame many of the disadvantages of single and dual wavelength approaches, but even today, they still lack sensitivity. Electrochemical detectors also underwent a similar development.

[1] HPLC-ECD is not exclusively a technique for neuroscientists. It has been used successfully for the analysis of hundreds of analytes, such as biochemical metabolites, pharmaceutical agents, natural products, food constituents, contaminants, and pollutants (Jane *et al.*, 1985; Musch *et al.*, 1985; Radzik and Lunte, 1989; Smyth, 1987). To review the major impact that HPLC-ECD has had on such analyses is far beyond the scope of this article.

Early approaches tried to improve the resolution of single channel ECDs. Theoretically, it is possible to sweep the applied potential to the working electrode and differentiate coeluting peaks based on their relative ease of oxidation (or reduction), but in practice this technique suffers from a number of inherent difficulties that have prevented it from being widely used.[2] Rapidly sweeping the applied potential to the working electrode causes major transient capacitive currents, a problem that renders the measurement of low analyte concentrations impossible. Researchers have developed several ways to offset the effects of the charging currents. These include using square wave voltammetry (Samuelsson *et al.*, 1980; Reardon *et al.*, 1984), fast potential scan on microelectrodes (White *et al.*, 1986), and coulostatic voltammetry (Delahay, 1962a,b; Reinmuth, 1962).

The use of square wave voltammetry is able to combine the charging current discrimination of differential pulse with the rapid potential scan ability of cyclic voltammetry. This technique purportedly is also to discriminate against some forms of background current and to reduce electrode fouling inherent in some other techniques. White *et al.* (1986) used fast scan cyclic voltammetry coupled with subtraction of a background scan (on the baseline in the absence of an eluting analyte) to eliminate the effects of the charging current. Others have used a coulometric voltammetric detector (Delayhay, 1962a,b; Barnes and Nieman, 1983) to overcome the capacitive currents. It has even been used in a dual electrode cell (serial coulostat/ potentiostat) to measure thiol/disulfide mixtures (Trubey and Nieman, 1986). The coulostatic technique has shown some promise for those using normal chromatography (Mikkelson *et al.*, 1985). Coulostatic voltammetry has been reviewed by van Leeuwen (1982).

Even though these methods have obtained discrimination against capacitive charging currents, they still lack the sensitivity inherent in fixed potential ECDs (Last, 1983). Noise arising from the potentiostat is a primary culprit since it must be able to effect potential changes very quickly. A rapid change in potential does not allow for analog filtering of the signal and results in significant noise. In general, the sensitivity of these types of detectors are ten times or more less sensitive than an equivalent fixed potential ECD. Also, the pulsed potential waveforms can lead to surface or double-layer reorganizations on some electrode surfaces that give sufficiently long-lived transients, which can interfere with faradaic relaxations and can lead to reduced performance (Last, 1983). The other major shortcoming of the swept potential methods is the lack of selectivity. As with the use of cyclic voltammetry, analyses of mixtures of compounds is generally impossible due to the overlap or near overlap of their reaction potentials. Thus, analytes that elute at similar retention times are in general impossible to resolve. This state of affairs is analogous to that which occurs with PDAs. And like PDAs, the use of complex mathematics such as factor analysis can be invoked to aid in the resolution of closely eluting and closely oxidizing analytes (Msimanga and Sturrock, 1990). However, the use of such involved mathematics has its own severe drawbacks (for example, Lindberg *et al*, 1986).

As with spectrophotometric approaches, dual electrodes offer many advantages over their single detector counterpart and, as discussed below, several have been tried and are in use today. Unfortunately, when it comes to electrode arrays, all voltammetric electrode designs are not created equal. There are two coulometric

---

[2] In addition, only recently has a commercial product become available, even though some form of this technique was first introduced three decades ago (Delahay, 1962a,b).

electrode array detectors currently available: the Coulometric Electrode Array System (CEAS) and the CoulArray detector. Both offer the qualitative data of PDA and the extreme sensitivity of ECD. The use of an array of up to 16 coulometric sensors in series, each working independently and at well-defined working potentials, enables compounds in a sample to be resolved chromatographically as well as electrochemically.

In this chapter we discuss the different types of ECDs available today, but before we can do this it is necessary to sketch a brief background of electrochemistry. We do not presume to give an in-depth review of electrochemistry, as it is simply far too broad a topic and there are already several excellent reviews (Kissinger, 1977, 1989; Rucki, 1980; Stulik and Pacakova, 1981; Horvai and Pungor 1987, 1989). The latter part of this chapter examines in detail the evolution of the single channel ECD into the coulometric array detector and its practical uses.

OVERVIEW OF VOLTAMMETRIC ELECTROCHEMISTRY

When certain compounds are subjected to a potential difference they become excited and undergo a molecular rearrangement with the loss (oxidation) or gain (reduction) of electrons. In amperometry, the potential is kept constant and the current produced from the electrochemical reaction is measured. This is called *potentiostatic amperometry*. As amperometry **is** the measurement of current in a general sense, all ECD detectors can be considered to be amperometric. Since the current is actually measured as a function of time, one possible term for this approach is *hydrodynamic chronoamperometry*. These definitions are fine when working with inefficient electrodes with an unknown proportion of the total analyte being measured. However, these definitions require further refinement when measuring signal from 100% of the analyte. When measuring peak areas, we are really measuring the total current as a function of time - that is, charge which is measured in coulombs. This specialized form of amperometry is called coulometry. Under conditions in which 100% of the analyte is being measured, it would be more appropriate to rename amperometry as coulometry, and amperometric detectors as coulometric detectors. The terms amperometric and coulometric mean different things to different people. To some coulometry means extreme sensitivity (improved s/n ratio); to others coulometry is one of selectivity or stability. Whether a coulometric transducer can be considered a highly efficient amperometric transducer or an amperometric transducer an inefficient coulometric one is pedantic (Kissinger, 1986). Coulometric electrodes often **are** found to be superior to amperometric ones in many ways and the practical advantages of coulometrically-efficient electrodes have now gained general acceptance in the marketplace even by previously unwavering critics (Kissinger, 1994). Under what circumstances an amperometric electrode can be forced to become coulometrically efficient is more fully discussed below.

Coulometric electrodes offer further benefits. Faraday's law (1) relates the response of the detector to the quantity of analyte undergoing electrolysis:

$$Q = n F N \qquad (1)$$

Q = charge transferred (coulombs); n = number of electrons transferred (equivalents/mole); F = Faraday's constant (coulombs/equivalent) ; N = moles of reactant (mole).

For a coulometrically-efficient electrode, the chromatographic peak area is measured in charge (current over time). Therefore (using Faraday's equation above), if the amount injected into the chromatograph is known, the peak area (charge) can be predicted. Alternatively, if the peak area is known, the exact amount of compound can be determined accurately. The amperometric thin-layer electrode would fail because the efficiency of electrolysis is unknown and is continuously decreasing.

Whether a molecule is electrochemically active or not depends critically upon its structure. Compounds tend to fall into three categories: those that readily react electrochemically, those that are more difficult to measure electrochemically, and those that are totally unreactive at usably applied potentials. Members of the first group tend either to have a hydroxyl or primary amine group capable of delocalizing charge through conjugation with the π electrons of an aromatic nucleus. Such mesomeric effects make electrochemical reactions more feasible by redistributing the charge over several canonical structures, thereby favoring the free energy of the reaction. Members of this reactive group typically include the phenols, catechols, phenylamines, and a variety of other aromatic structures as well as the aliphatic polyenes (such as carotenes and retinols).

Members of the second group include aliphatic compounds with hydroxyl, amine, or thiol groups. Their inability to delocalize charge renders them less able to undergo electrochemical detection than members of the first group, and their measurement usually requires a substantially higher potential to drive the reaction. For example, the oxidation of the aromatic compound phenylamine requires an applied potential of +600mV (vs. Pd) on a flow-through graphite working electrode, whereas the aliphatic compound diethylamine requires more energy with an applied potential of >+900mV (vs. Pd).

The third group contains hydrocarbons, alcohols, aldehydes, ketones and primary aliphatic amines.[3] For a review on the electrochemistry of different analyte classes, see Smyth (1987).

The ability to drive the reaction at higher potentials is not a problem for most potentiostats, but there are several important analytical repercussions. The use of a higher potential is usually problematic due to increased background current due to mobile phase oxidation (or reduction), leading to increased noise and a worsening of the s/n ratio; the greater chance of chromatographic interference; and shortened electrode-life. To compensate for these potential side effects, certain electrochemically inert compounds (or those requiring excessively high potentials) may be rendered electrochemically active at lower potentials by the use of novel working electrodes, by derivatization (e.g. amino acids (Dou *et al.*, 1990; Duda and Kissinger, 1993; Krull *et al.*, 1994) or by exposure to UV irradiation become transiently electrochemically active (Krull *et al.*, 1984) (See also Acworth and Waraska, this volume).

---

[3] Some of these compounds can be determined using pulsed amperometric detection at gold or platinum working electrodes. See Johnson and LaCourse (1990).

A major advantage of the electrochemical detector over other HPLC detectors is its extreme selectivity. The electrochemical detector is blind to many of the compounds in the sample, making chromatographic separation and analyte identification easier than with spectrophotometric detectors. However, even though ECD is extremely selective there still are many compounds in a biological sample that are electrochemically active and require separation before detection. One novel electrochemical approach in which the lack of specificity still remains a problem is *in vivo* voltammetry. Here a carbon fiber is implanted into a specific brain region, enriched in the analyte of interest, of a living animal. This fiber then can be subjected to a potential difference. Metabolites are oxidized and their levels determined *in situ*. Unfortunately, lacking a separation step, and thus suffering from poor selectivity, early experiments were hampered by interferences and compound misidentification. For example, the signal assumed to be from dopamine was in fact later found to be ascorbate and dopamine's metabolite dihydroxyphenylacetic acid (DOPAC), typically found in high levels within the extracellular space. *In vivo* voltammetry has been made more selective without the use of a chromatographic separation step by treating/coating the electrode to make it more specific for a given chemical species, by choosing the best measuring technique (for example differential pulse voltammetry), and by altering the extracellular level of the species of interest using pharmacology. For reviews on *in vivo* voltammetry see *Marsden et al.*, (1986), Myers and Knott (1986) and Justice (1987). Better selectivity may be obtained eventually by the use of biosensors (possibly utilizing immobilized receptors) but, ironically, as most would measure only one compound researchers may find them too restrictive.

If **all** the compounds typically found in a biological sample were EC-active, the ability to resolve them would be far beyond the capability even of the most state-of-the-art LC column. Such column limitations could be overcome to some extent by improving the resolving power of the detector.

Spectrophotometric-based HPLC techniques (UV absorption, fluorescence) measure a physical property of the molecule (e.g. absorption of energy by excitation of electrons or emission of energy as electrons return to the ground state). During the process the molecule remains chemically unaltered. Electrochemical detection, however, measures a compound by actually changing it chemically. As such, electrochemical measurement is a destructive method with relatively few compounds capable of returning (by reduction) to the original structure once oxidized. Compounds capable of returning to their original structure can give further selectivity to electrochemical detection. Figure 1 shows a variety of reversible and irreversible electrochemical reactions. Catechol, hydroquinone and resorcinol derivatives all show reversible redox cycling with the possibility of several electron transfers occurring per molecule. This ability not only helps to better identify a molecule in complex matrices but also leads to an amplification of the signal and an enhanced s/n ratio. These benefits will be discussed further below.

The correct choice of applied (redox) potential for the measurement of a particular analyte offering the best s/n ratio can be obtained either from the literature

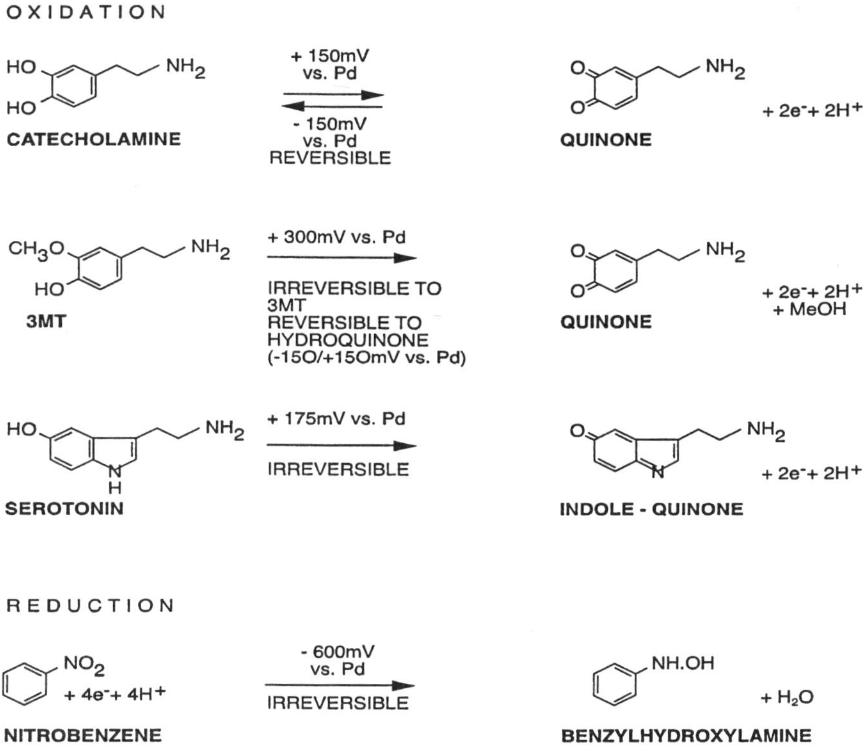

Figure 1. Changes in chemical structure when compounds are subjected to oxidation or reduction. Some compounds react reversibly and can be measured using redox cycling. Other compounds are irreversible and cannot be electrochemically changed to the original compound.

or by direct measurement using the electrochemical detector. The analyte can be subjected to either cyclic voltammetric (CV) analysis or hydrodynamic voltammetric (HDV) analysis. CV analysis measures the current produced as the applied potential to the working electrode is scanned. Although a wealth of information about a compound's oxidative (and reductive) potential and possible initial reaction mechanisms are obtained, few electrochemical detectors have the ability to do cyclic voltammetry. A more common approach is to construct an HDV curve. In this case a known concentration of analyte is measured on the HPLC-ECD system at a low potential, for example 0mV. The analyte is then measured at progressively higher potentials (for example 50mV increments) until the peak height (current) reaches a plateau. Conversely, if the compound is thought to undergo reduction, then its peak height is measured as the applied potential is incrementally decreased (for example from 0mV in -50mV increments). If the current is plotted as a function of potential, an HDV curve is produced. See Figure 2A. An HDV curve is analogous to the construction of an absorption spectrum (here the concentration of analyte is fixed and the absorbance is plotted as a function of wavelength). See Figure 2B. Although the HDV contains less information than the CV technique, it does allow easy measurement of both oxidation and reduction potentials, in which maximum signal and minimum noise can be obtained. It helps also to visualize how differences in electrochemical behavior can be the basis for enhanced resolution at multiple electrodes.

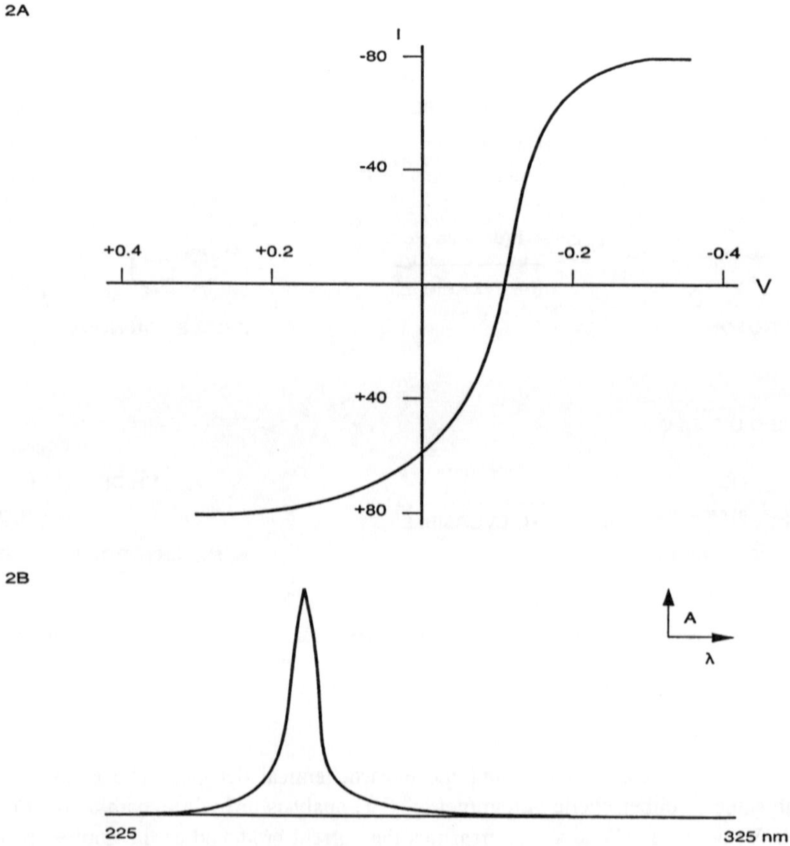

Figure 2. Typical hydrodynamic voltammogram (current-voltage curve) for an electrochemically reversible compound (upper figure; 2A). Spectrum for the same compound (lower figure; 2B).

NOISE IN THE HPLC-ECD SYSTEM

Noise in an HPLC-ECD system comes from many sources and can be either faradaic or non-faradaic in nature.[4] All mobile phases used in HPLC-ECD methods require that an electrolyte (such as salts of phosphate, citrate or acetate) be present to conduct charge. The electrolyte itself, however, can lead to noise.

---

[4] Non-faradaic noise does not obey Faraday's law. Examples include charging currents, electrical noise from the potentiostats, antenna effects, earth-loop effects, line noise, noise from connections, and thermal effects.

Faradaic noise obeys Faraday's law (see above) and includes electrolysis of the analyte of interest (signal) as well as electrolysis of unwanted reactants in the mobile phase (noise). To minimize faradaic noise, high-quality reagents must be used (typically low in divalent metal ions capable of undergoing changes in oxidation state, e.g. iron, chromium). All solutions used with the HPLC must be prepared with purified deionized water (typically > 18 Megohm-cm) which has undergone further polishing to remove trace organics (ESA, 1994). Organic solvents must be HPLC-ECD "compatible" and be devoid of organic peroxides and other electrochemically-active contaminants. A “guard cell,” an electrochemical cell capable of withstanding high pressure, can be used before the injector in order to “scrub” the mobile phase to reduce faradaic contaminants from it prior to sample introduction. In this case the “guard cell” must be of the coulometric design in order to remove the contaminants effectively.

The detection limit of an HPLC-ECD system is determined by the s/n ratio. For a flow-through graphite coulometric detector the faradaic signal will always be maximum. To further improve the s/n ratio, noise must be diminished by careful mobile phase production, keeping the system clean and free from leaks, using a low pulse pump, using non-metal tubing, placing the detector on an isolated power-supply, and operating at the optimal potential to maximize signal and minimize noise which, in turn, is obtained from the HDV.

COMPONENTS OF THE ELECTROCHEMICAL DETECTOR

The vast majority of commercially available electrochemical detectors use three electrodes: the working, the counter (auxiliary), and the reference. A fixed potential difference is applied between the working electrode and the reference electrode. This potential drives the electrochemical reaction at the working electrode's surface. The current produced from the electrochemical reaction at the working electrode is balanced by a current flowing in the opposite direction at the counter electrode. The reference electrode acts as a reference point for the redox couple. The current resulting from the electrochemical reaction is amplified and, when plotted as a function of time, appears as a peak on the recording device. Current measured at the working electrode surface results from not only the redox reaction of interest but also from unwanted redox reactions coming from the mobile phase (faradaic noise) and from the working electrode material (non-faradaic noise). Contribution from the latter two can be minimized by careful mobile phase production, HPLC system cleanliness, and the correct choice of working electrode material. See Noise in the HPLC System, above.

The working electrode

The most common working electrode materials are made of carbon - glassy carbon, carbon paste, pyrolytic carbon, and porous graphite. Metals such as platinum, gold, silver, mercury, and gold-amalgam are also common. The optimal working electrode choice is dependent upon many factors, including the usable potential range, involvement of the electrode in the redox reaction, and the kinetics of the electron transfer reaction (Rocklin, 1984). Other factors, such as the suitable choice of mobile phase, will also play a role. For example, a platinum-working electrode is a better choice for the measurement of hydrogen peroxide than carbon because the reaction at the platinum electrode is more efficient and is more favorable kinetically. A second example, carbon paste electrodes cannot be used with mobile phases containing high amounts of organic modifier for the electrode will dissolve unless a polymeric binder is used. One further example, mercury amalgams are usually used when detecting analytes that are difficult to reduce or for compounds containing thiol moieties.

In general, carbon-based working electrodes offer the widest range of potential (both negative and positive). The source of the carbon-based materials for working electrodes must be chosen carefully. Different sources of the carbon can result in different levels of performance. For instance, the fabrication conditions used in the production of glassy carbon are critical to performance factors such as noise and signal level. Also the porous graphite used in the flow-through coulometric cell must have certain demanding characteristics: the pore size must be small ($<1\mu m$) and uniform in order to obtain coulometric efficiency; the porosity (number of pores/unit volume) must be very high to keep back pressures low; finally, the electrochemical properties that lead to low background currents, low noise, and high efficiency must be carefully selected. Readers interested in learning more about working electrodes are referred to Rocklin (1984) and Johnson and LaCourse (1990).

The reference electrode

The potential of the working electrode is measured within the context of a known potential, which is in turn obtained from the reference electrode. The role of the reference electrode is to establish a stable half-cell potential. This electrode acts as a reference point along the potential axis by which the oxidizing or reducing power of the working electrode is judged. The potential axis is arbitrary and the reference electrode sets the zero point.

By far the most familiar reference electrode to electrochemists is the **standard hydrogen electrode (SHE),** from which many standard half-reaction redox potentials have been obtained and are to be found in many reference handbooks. The SHE is far too impractical for HPLC-ECD requiring hydrogen gas, so several other reference materials have been developed and are now in use.

The silver/silver chloride reference is a "wet" electrode of the metal/insoluble salt type and has an E^0 of +222mV vs. SHE at 25°C. It is usually placed downstream from the working electrode and makes "electrical" contact with the mobile phase via a microporous plug made from materials such as ceramic or Vicor®. Although more practical to use than the SHE, its relatively large size necessitates its placement far away from the working electrode. Such a configuration can lead to multiple junction potentials and iR drops, which ultimately lead to excessive noise, instability of the applied potential, and a nonlinear response. This reference electrode requires routine maintenance, is photosensitive, can be readily poisoned by sulfide and other anions, and may become blocked at the microporous plug.

Recently, Bohs *et al.* (1989) used a solid silver wire coated with silver chloride as a reference electrode. Although the reference can be placed close to the working electrode to overcome issues with junction potentials and iR drops, the electrode design necessitates the use of 10mM sodium chloride in the mobile phase. The reference is fairly unstable, capable of altering its reference potential by 60mV/month and shifting its potential (prolonged voids, system peaks resulting in unreliable quantitation) when exposed to solutions high in chloride ions - as with the analysis of microdialysis perfusates. The incorporation of 10mM sodium chloride into the mobile phase may diminish such problems but also may contribute to system noise and system corrosion.

A similar reference electrode to the wet silver/silver chloride electrode is the **mercury/mercurous chloride or calomel** electrode. The calomel electrode has an E^O of +245 to +336mV (dependent on the co-ion concentration) vs. SHE at 25°C. The construction of this electrode is more complex than the silver/silver chloride and suffers many of the same problems. In addition, it uses highly toxic salts.

Both the wet silver/silver chloride and calomel electrodes must be placed in a part of the HPLC system where the back pressure is relatively low - usually post column. Failure to do so will force mobile phase into the reference compartment through the porous plug, possibly rupturing the reference electrode. Electrodes designed to remove electroactive interferences from the mobile phase must be able to withstand high back pressures, as they are typically placed before the column. These electrodes cannot use either the wet silver/silver chloride or calomel references.

Another and perhaps more useful reference electrode is the **α-hydrogen/palladium** electrode (Matson, 1983). This reference is very small and can be placed in close proximity to the working and counter electrodes. This reference electrode is used in all commercially available flow-through porous graphite coulometric electrodes. It is maintenance free, stable, not easily poisoned and is capable of working at high pressures (≤6000 psi). For further reading, see Covington and Rebelo (1983) and Nevell and Walsh (1992).

A common practical problem occurs as a result of misunderstanding the importance of the reference electrode. Often a method developed on one particular HPLC-ECD system does not work well when transferred to another system using a different reference electrode. For example, catecholamines can be readily measured at 800mV against a silver/silver chloride reference but the same potential used with a palladium reference might cause the assay to fail from excessive noise and interferences, resulting in poor limits of detection. There is a difference of approximately 300mV between the silver/silver chloride and palladium reference electrodes, which means that if the working electrode was set to 800mV against a palladium reference then this would be equivalent to 1100mV vs. silver/silver chloride! Readers are strongly urged to check the reference type when trying published methods in their laboratory, produce an HDV, and then adjust the working potential of their system accordingly.

The counter electrode

This electrode offers a flow-path for electrons that are involved in the electrochemical oxidation (or reduction) occurring at the working electrode. The demands placed upon the counter electrode are far less stringent than for either the working or reference electrodes. All the counter electrode is required to do is supply (or remove) the equivalent number of electrons that pass through the working electrode. Counter electrodes are typically made from a noble metal such as platinum or palladium, and sometimes from stainless steel. The counter electrode is placed as close to the working electrode as practically possible. The shape, size, and location of the counter electrode can all affect signal, noise, and the number and magnitude of artifact peaks.

DIFFERENT DETECTOR CONFIGURATIONS

Although, theoretically, there are an infinite number of possible configurations for the relationship between the working, counter, and reference electrodes, only a few are found to be useful in practice. Electrochemical detectors tend to fall into one of three basic designs, in which mobile phase either flows past (thin layer), flows at (wall-jet), or flows through (frit) the working electrode. Electrochemical detectors can also be categorized as being either amperometric or coulometric (see above).

The thin-layer design

The first amperometric thin-layer transducer was developed in Adams' laboratory (Kissinger *et al.*, 1973) and is shown in cross section in Figure 3. A carbon paste electrode 0.1" in diameter was used with a 127μm Teflon® spacer. The active volume of the detector was 0.7μL (with a total of 8μL of dead volume from column to electrode). A wet silver/silver chloride reference electrode was placed downstream from the working electrode.

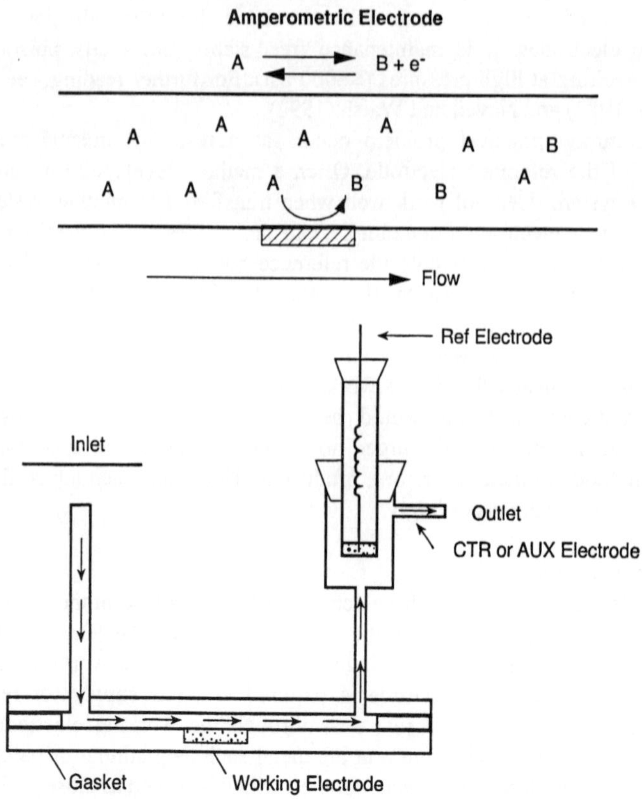

Figure 3. Amperometric thin-layer electrodes are electrochemically inefficient at normal HPLC flow rates. Only 5% of the total analyte is oxidized at the working electrode (shaded area). The remaining 95% passes through this detector unmeasured (upper figure). The lower figure shows a cross section through a typical thin-layer cell equipped with a wet silver/silver chloride reference electrode placed downstream from the working electrode.

The efficiency of thin-layer amperometric electrodes is typically below 5% at flow rates used with 3.0-4.6mm i.d. columns (Duda and Kissinger, 1993). This means that 95% of the number of molecules passing through the detector do so without being measured. This is obviously a very wasteful technique. Theoretically, the conversion efficiency of the thin-layer amperometric sensor can be improved by using a larger working electrode, decreasing the volume (decreasing the diffusion layer) of the detector, and decreasing the flow rate through the electrode (increasing an analyte's residency time). All of these factors increase the probability that a molecule reaches the working electrode's surface for electrochemical measurement. In practice, each of these factors have limitations. For example, Figure 4 shows that as the surface area of the thin-layer electrode is increased, the s/n ratio actually diminishes due to increased noise (Duda and Kissinger, 1993).

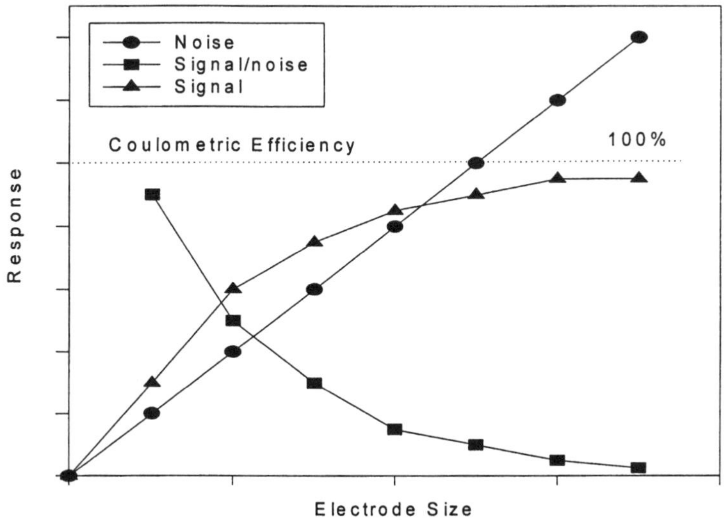

Figure 4. Abstract graph showing the interdependence between signal, noise and s/n ratio as a function of electrode size (area or length) for the thin-layer amperometric electrode. Notice that as the area of the thin-layer electrode is increased both signal and noise increase too. As the noise component increases more rapidly than the signal component, the s/n ratio actually decreases as the electrode area is increased. This same figure has wrongfully been applied to the flow-through coulometric electrode - see above for the explanation.

As the area of the thin-layer working electrode increases so do the conversion efficiencies of both the analyte and the mobile phase electrolytes as well as the noise. Therefore the s/n ratio does not increase as rapidly as would be predicted from the surface area alone because any gain in signal is out-weighed by increased noise. An additional problem with the thin-layer working electrode is that it can become contaminated easily by adsorption of analytes onto its surface. The smaller the electrode surface the better the s/n ratio but the greater the chance of contamination.

Although the detector can be taken apart for resurfacing and polishing, some amperometric detectors are notorious for requiring considerable time to restabilize, especially when the electrodes are used in applications requiring high sensitivity.

The volume of the detector can be decreased by using thinner gaskets. The noise is independent of the volume of the detector, whereas the signal increases as the volume decreases. For some thin-layer designs care must be taken that the gasket is not made too thin - the working electrode may short against the counter electrode. Also, thinner gaskets are more difficult to handle and are readily damaged.

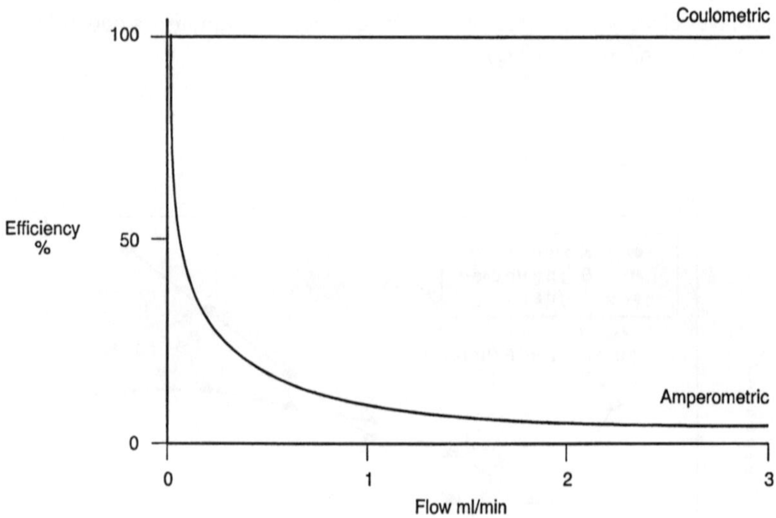

Figure 5. Graph showing electrode efficiency as a function of flow rate. Conversion efficiency decreases for the thin-layer amperometric electrode but remains at 100% for the coulometric electrode over the usable flow rates used with HPLC.

The conversion efficiency of the thin-layer amperometric detector is very sensitive to changes in flow rate - see Figure 5. Although small fluctuations in flow rate are often assumed to be relatively unimportant they can have marked effects on conversion efficiency. This can be very significant especially with microbore applications in which lower flow rates are used. Changes in an amperometric detector's efficiency due to flow rate effects and contamination of the working electrode's surface requires that standards be measured routinely throughout the analytical run. Furthermore, the use of column splitting to regulate the flow of mobile phase through a microbore column (Huang *et al.*, 1990) is rife with problems. Changes in an analyte's retention time on the microbore column due to aging of the flow regulating column or to temperature changes will lead to altered conversion efficiency and peak height (even for a standard of fixed concentration). Decreasing the flow through the detector will result in higher conversion efficiencies as the thin-layer amperometric detector becomes more coulometrically efficient. However, lowering the flow rate is not always chromatographically practical. As the flow rate decreases, retention times become prolonged and the peaks eluting from the column become broader. Using narrower columns permits the use of lower flow-rates (e.g. 3mm i.d. - 600μL/min; 1mm i.d. - 75μL/min; 0.3mm i.d. - 6μL/min) while keeping the linear velocity constant through the detector; thereby, good peak shape is maintained. This approach is practical only as long as the dead-volume from the column to the working electrode is minimized. The maximum s/n improvement comes from decreasing both the flow rate and detector volume. Readers interested in an in-depth discussion of electrode efficiencies are referred to Weber (1986) and Rocklin *et al.*, 1994.

Since Adams' original work, a variety of improved amperometric electrodes are now in use. Many different working electrodes have become available (carbon-based, gold, gold amalgam, lead alloy, silver, platinum, copper, nickel). Cell volume has been minimized. Detectors with solid state reference electrodes are now common (Bowers, 1991; Greaney *et al.*, 1993). Supra-coulometric thin-layer amperometric electrodes using 13μm thick gaskets and solid-state reference electrodes have been used with both microbore (Mishra *et al*, 1996) and capillary columns of 0.18mm internal diameter (Cullison *et al*, 1996). Cullison measured approximately 5fg/0.2μL of monoamines with a s/n ratio of 3:1 and with 1300% coulometric efficiency (Cullison *et al.*, 1996).

The wall-jet design

The wall-jet electrode is a modification of the thin-layer electrode. In this electrode, eluent from the column is directed perpendicularly to the plane of a thin-layer working electrode (Yamada and Matsuda, 1973). With this model, the theory states that by forcing the eluent stream onto the working electrode mass transfer of the analyte to this electrode's surface is increased. In this way signal amplitude is increased as the electrode becomes more coulometrically efficient. Practically, many of the so-called wall-jet electrodes currently available do not act like true wall-jet detectors but follow thin-layer characteristics (Gunasingham and Fleet, 1983; Elbicki *et al.*, 1984). Current "wall-jet" electrodes may offer only a slight improvement in sensitivity over the thin-layer design when used at typical HPLC flow rates.

The flow-through design

Several different coulometric electrodes have been developed with working electrodes made from tubular platinum or cadmium electrodes (Johnson and Larochelle, 1973; Taylor and Johnson, 1974; Davenport and Johnson, 1971), or from carbon, silver, or platinum gauzes (Takata and Muto, 1973; Takata and Fujita, 1975). Others have used flow-through detectors with working electrodes made from thin carbon plates (Fenn *et al.*, 1978), platinum tubes (Blaedel and Schieffer, 1974), columns packed with glassy carbon granules (Blaedel and Strohl, 1964), or granular amalgamated nickel (Roe, 1964). The first flow-through porous graphite frit type coulometric electrode, with >99% conversion efficiency, was developed by Matson (Andrews *et al.*, 1982; Matson *et al.*, 1982) and is shown in Figure 6. This practical flow-through design overcame the sensitivity issues and other problems of the larger surface area thin-layer designs (Blank, 1976; Lankelma and Poppe, 1976; Tjaden *et al.*, 1976; Kissinger, 1977; Schieffer, 1980; Roston and Kissinger, 1982).

The graphite flow-through working electrode has a very large surface area, but because it is a constrained cell design, the s/n ratio is vastly improved over a thin-layer working electrode of equivalent surface area. The flow-through graphite electrode is a three-dimensional electrode - no matter what direction a molecule

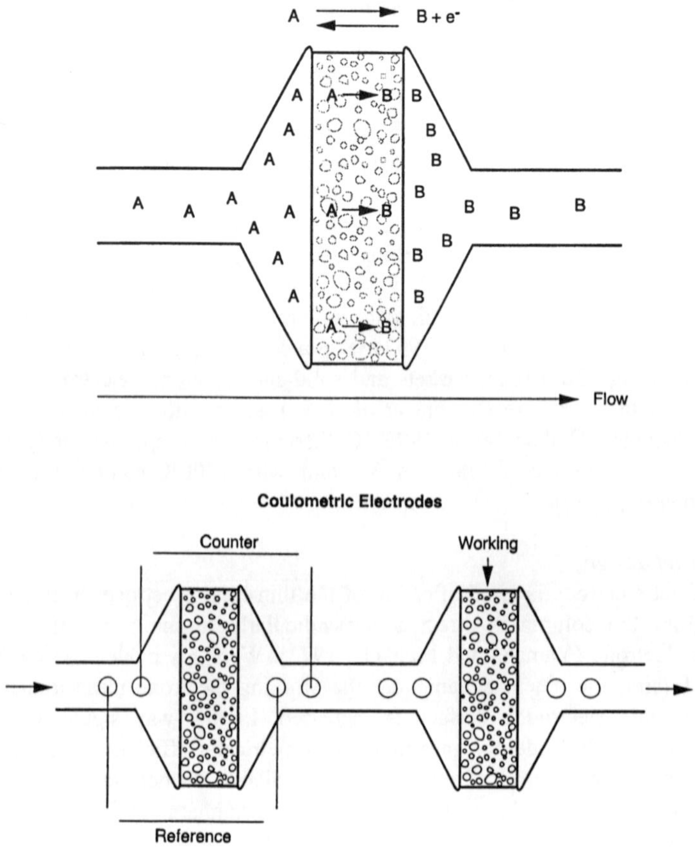

Figure 6. Due to their large surface area, flow-through coulometric electrodes are highly efficient with close to 100% of the analyte being measured (upper figure). The lower figure presents a cross section through a dual coulometric electrode and shows the placement of working counter and reference electrodes.

travels it has to encounter the working electrode surface. The thin-layer electrode is a two-dimensional electrode; analytes can encounter the working electrode only if they travel perpendicularly to the plane of the electrode and in a direction toward it. To obtain the same active surface area, a thin-layer electrode would have to be so large that any gain in signal would be outweighed by noise from the electrode (Duda and Kissinger, 1993). The porous graphite electrode has an extremely high electrode surface area to solution volume ratio, thereby improving the rate and efficiency at which an analyte undergoes electrochemical measurement. The s/n ratio actually improves with porous flow-through electrodes due to decreased noise. This is not so with the thin-layer design. There is also an HDV curve-sharpening effect from the large area and small volume of the cell. The majority of organic compounds roughly follows quasi-reversible kinetics, implying relatively broad HDV curves. Typically, in amperometric detectors, the cell constant (k), which mainly reflects the mass transport properties of the detector, is relatively small (0.5-1 sec^{-1}) (2). The HDV curve shape is dominated by the reaction kinetics at the electrode, giving the expected broad HDV shape.

$$k = AD/dV \qquad (2)$$

A = electrode area (cm^2); D = diffusion coefficient of analyte (cm^2/sec), V = cell volume (cm^3), d = Nernst or unstirred diffusion layer next to the electrode (cm).

In the flow-through coulometric cells, $k = 1 - 5 \times 10^4\ sec^{-1}$, the curve shape for many compounds becomes dominated by the cell kinetics, which effectively convert broad, quasi-reversible HDV curve shapes into sharper reversible curves (Figure 7).

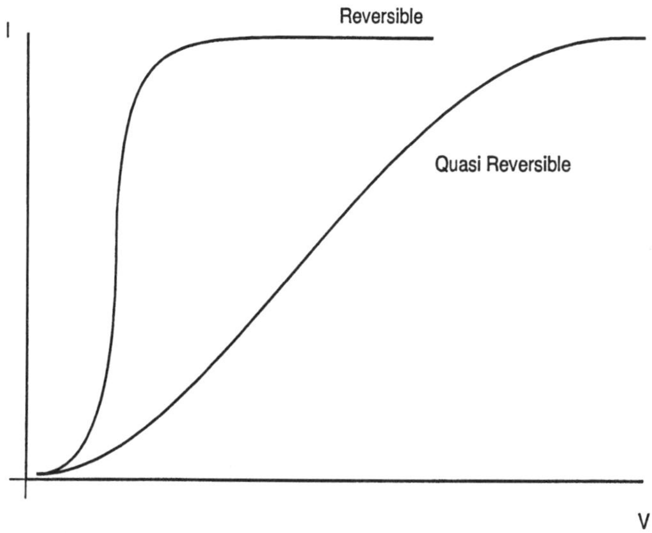

Figure 7. The broad HDV found for many quasi-reversible species measured using a thin-layer amperometric electrode are converted to the sharp HDV characteristic of reversible compounds measured using the coulometric flow-through electrode.

In addition to measuring 100% of the analyte passing through it, the signal from flow-through electrodes is flow-rate independent (over the practical flow-rate range used for HPLC) [see Figure 5 above], obviating the need to measure external standards frequently. In fact, because of this electrode's coulometric efficiency an external standard will give the same peak area from month to month and even from year to year (assuming that the standard does not decompose and the column resolution is maintained)! The flow-through electrode is little affected by adsorption and more than 95% of the electrode's surface has to be contaminated before a decrease in signal would be noticed. The solid-state α-palladium/hydrogen reference electrode and flow-through working electrode are both maintenance free and long-lived. The electrode rapidly stabilizes if the potential is changed as, for example, after electrochemical cleaning of the working electrode. Finally, because these electrodes are coulometrically efficient, they can be used with extreme efficacy in serial arrays - see below. Thin-layer amperometric and porous graphite flow-through coulometric transducers have been further compared recently (Acworth and Gamache, 1996; Acworth and Waraska, 1996).

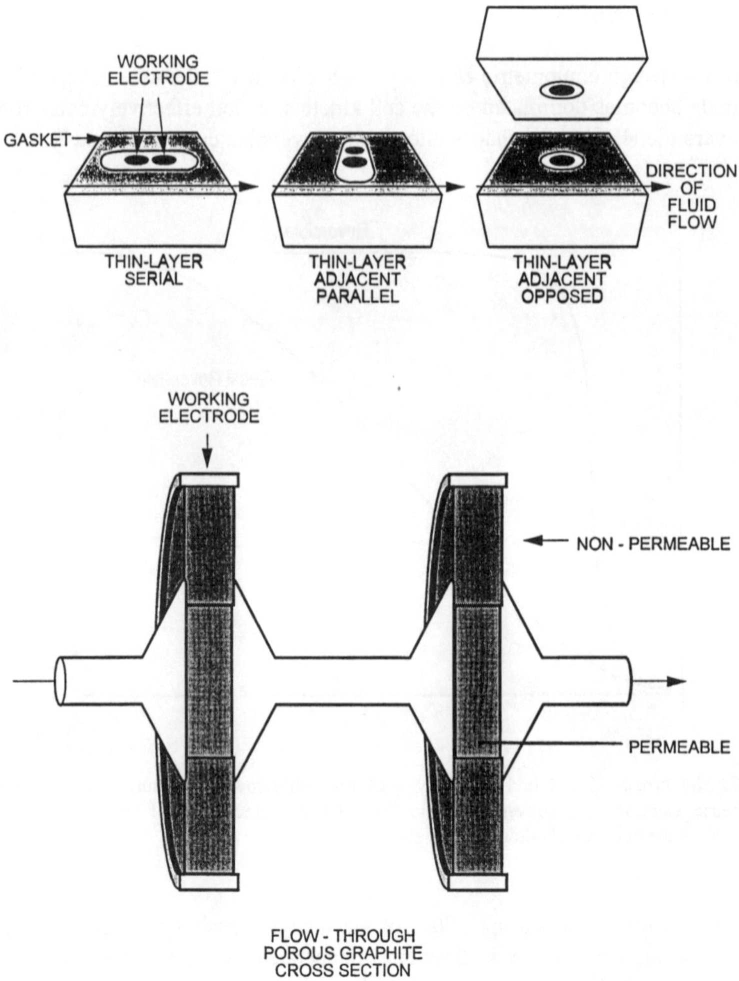

Figure 8. There are several dual electrode designs. Thin layer dual electrodes can be placed in series, in parallel or opposed Flow-through coulometric electrodes are available in series only. No attempts have been made to show the exact fluid path for each of the cell designs. The placement of reference and counter electrodes is not shown.

DUAL ELECTRODE DETECTORS

There are four different designs for dual electrode detectors: thin-layer in series, thin-layer parallel adjacent, thin-layer parallel opposed, and flow-through in series. See Figure 8.

The thin-layer serial detector
This involves the placement of two working electrodes in the flow channel and potentiostating each electrode independently (Blank, 1976; Roston and Kissinger, 1982; Lunte *et al.*, 1985). This configuration permits the products of the upstream electrode to be measured at the downstream electrode with the intention of better analytical selectivity. The benefits of this approach is critically dependent upon chromatographic resolution. As there is only 5% conversion at the upstream electrode, 95% of the unreacted species may overwhelm the measurement of the product at the downstream electrode when both electrodes are used either oxidatively or reductively. This means that for this approach to be effective the product from the first electrode must not co-elute with other species capable of undergoing electrolysis at the downstream electrode. (See Figures 9 and 10 and *"The flow-through detector in screen mode"* below for further details). The thin-layer serial detector may have practical applications when measuring analytes of differing oxidation potential that are chromatographically well resolved (interference will be minimized).

A second approach uses the redox mode. Here the first electrode is used to generate a species which can be detected at the second electrode set to the opposite polarity of the first. For example, Allison and Shoup used dual gold amalgam electrodes to measure thiols and disulfides (1983). Conversion of disulfides to thiols occurred at the first electrode set at -1100mV vs. Ag/AgCl while the resulting thiol was measured at the second electrode set at +150mV vs. Ag/AgCl. Another example would be the measurement of electrochemically freely reversible analytes co-eluting with irreversible ones. For example, the catecholamines could be oxidized at the upstream electrode and measured reductively at the downstream electrode. Possible co-eluants incapable of such redox cycling would therefore not be expected to interfere with the catecholamine measurement.

Roston and Kissinger attempted to use dual serial amperometric thin-layer electrodes to obtain HDV data on co-eluting species (1982). As the potential on each electrode has to be changed incrementally (with suitable periods for re-equilibration), this approach is obviously far from practical.

Coulometric/amperometric dual series electrodes have also been produced using the first (coulometric) electrode to effectively remove interferences and to lower background noise (Schieffer, 1980; Hepler and Purdy, 1980).

The thin-layer parallel adjacent detector
This detector also applies a potential to each electrode independently and is analogous to a dual-wavelength UV absorbance detector. The major advantage of this configuration is that two chromatograms are obtained simultaneously, thus saving time. Some idea of peak purity can be obtained from current ratios (Roston and Kissinger, 1982). This electrode design allows the measurement of readily oxidized

species (e.g. dopamine) in the presence of analytes that require higher potentials (e.g. tyrosine) but only if they are well resolved chromatographically.

When used in the differential mode (signal from the first electrode is electronically subtracted from that of the second electrode), improvements in noise and in signal have been observed (Brunt *et al.*, 1981). An additional advantage of this design is that oxidizable and reducible species can both be measured simultaneously

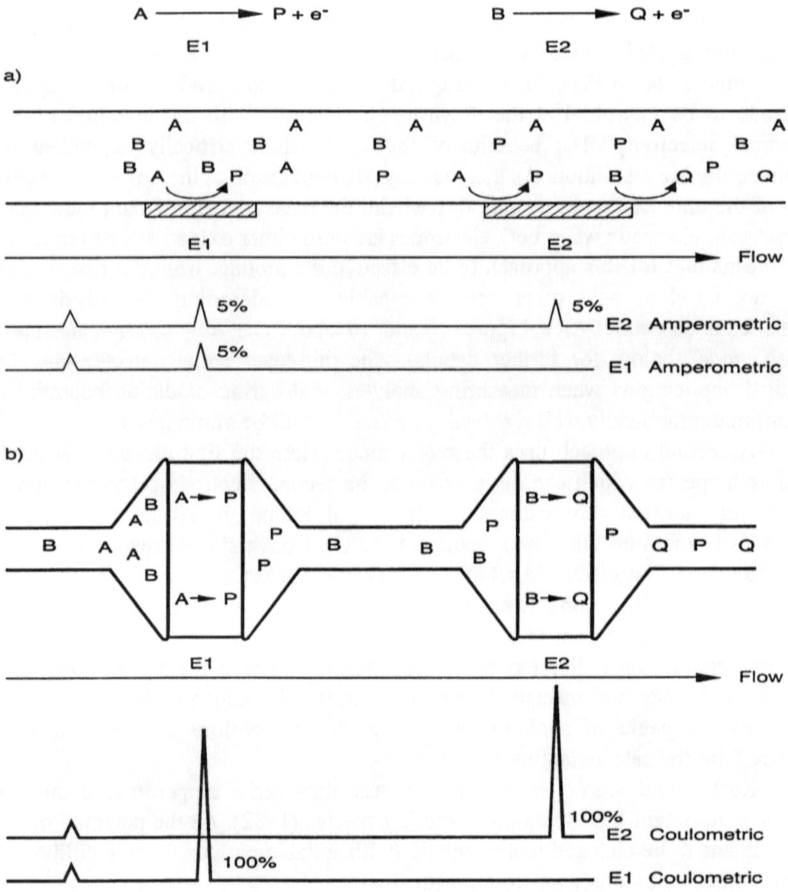

Figure 9. Resolving co-elutions. This figure illustrates differences between dual amperometric and dual coulometric electrodes for the measurement of compound A (oxidizing at 200mV) and compound B (oxidizing at 600mV). Figure 9a shows that with the inefficient amperometric electrodes, signal from compound A "spills over" onto the downstream electrode. If these compounds were poorly resolved chromatographically then measurement of compound B would be compromised. Figure 9b shows that with coulometric electrodes all of compound A is measured on the upstream electrode thereby effectively removing it from interfering at the downstream electrode. By such screening the coulometric approach overcomes poor chromatographic resolution. Notice that with the coulometric measurement the signal is twenty-fold that of the amperometric sensor.

in the same sample. A problem with this design is that the flow channel area has to be enlarged to accommodate two working electrodes; consequently, the peaks may become broader with poorer limits of detection.

The thin-layer parallel opposed detector

This detector was first popularized by Anderson and Reilley in the late 1960s (1965a,b; 1968; Reilley, 1968). With this detector, it is possible to cascade reversible redox reactions in order to amplify the detector current. The geometric requirements and mathematical model for such a "regenerative" detector can be found in articles by Weber and Purdy (1978; 1982). Such detectors have been reported to amplify the current by up to tenfold (McClintock and Purdy, 1981, 1983) and nineteenfold for

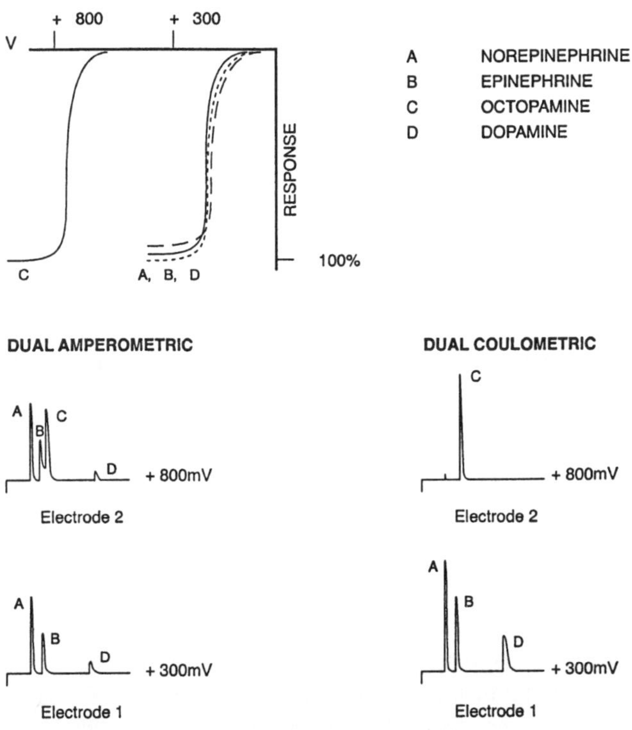

Figure 10. A practical example of screening. Differences in voltammetric behavior can be used to resolve different classes of compounds. Here, poor chromatographic resolution can be overcome by making use of electrochemical behavior. The HDVs show that catecholamines have an oxidation potential of 300mV whereas octopamine requires an oxidation potential of 800mV for measurement. Under the conditions of this experiment octopamine co-eluted with epinephrine. Although serial dual amperometric electrodes allowed measurement of the catecholamines (norepinephrine, epinephrine, and dopamine) on the first electrode, the signal from epinephrine corrupted the measurement of octopamine on the down-stream electrode (set at 800mV). This situation is markedly different with dual coulometric electrodes. Here, the first electrode measures the total signal for each of the catecholamines leaving the down-stream electrode free to measure signal from octopamine only. The chromatograms obtained from the coulometric electrodes shows marked improvement in signal over the amperometric ones.

flow rates of approximately 1μL/min (Goto *et al*., 1983). This detector can work only for electrochemically reversible compounds, and suffers from band-spreading as well as problems with maintaining a uniform current and potential distribution over the enlarged electrode surfaces.

Recently, improved limits of detection have been achieved by using a supracoulometric thin-layer cell coupled with either microbore (Mishra *et al*., 1996) or capillary chromatography (Cullison *et al*., 1996). Here, efficient redox cycling takes place between the working electrode and parallel counter electrode, producing approximate detection limits of <5fg on column.

The flow-through serial detector

This detector is by far the most versatile, offering true screen, redox, and differential measurement modes (see below). The porous graphite electrode wafers have paired reference and counter electrodes, ensuring that the current and potential distribution will be uniform and that the uncompensated cell resistance will be minimal. When a potential is first applied to the working electrode (or the potential is changed), charging currents are produced at the solution/electrode interface. Charging current flows and decays exponentially. The time to reach zero can be on the order of hours as with many thin-layer types with silver/silver chloride reference designs, meaning considerable analytical down-time will occur if this detector is turned off or the potential is changed. In the flow-through serial detector the paired reference electrodes are placed <2mm away from the working electrode and the time for charging currents to reach zero is only a few minutes. It is common to prolong the porous graphite electrode life by applying a potential difference of +1000mV to the working electrodes for 5 minutes to electrochemically clean the working electrode surface. High-sensitivity analyses can be restarted only minutes after the potential change.

The flow-through serial detector in screen mode. True screen mode can be accomplished only by coulometric electrodes. Easily oxidized (or reduced) compounds can be quantitated and removed at the upstream electrode, leaving the downstream electrode totally free to measure more difficultly oxidized (or reduced) analytes. Such selectivity could not be achieved if the electrodes were amperometrically inefficient. Although the easily oxidized species could be measured at the upstream electrode, 95% would remain to react at the downstream electrode. Figure 9 shows the electrochemical reactions for the readily oxidized compound A (+200mV) and for the less reactive compound B (+600mV) and shows chromatograms of A and B achieved using either a dual amperometric electrode detector (Figure 9A) or a dual coulometric electrode detector (Figure 9B). The dual coulometric electrode detector achieves true selectivity with compound A detected on electrode 1 only and compound B on detector 2 only. Although in this example compounds A and B are well resolved, problems due to poor chromatographic resolution would occur only with the dual amperometric detector. A practical example is shown in Figure 10 and shows how poor chromatographic resolution can only be overcome by dual coulometric electrodes. In this example the first coulometric electrode measures the readily oxidized catecholamines (norepinephrine, epinephrine, and dopamine), while the second coulometric electrode measures less readily oxidized tyramine, octopamine. This figure also shows what happens with amperometric dual coulometric electrode approach compensates for poor chromatographic resolution; the only recourse with the amperometric detector is to spend considerable time and effort testing several different mobile phases or columns.

The flow-through serial detector in redox mode. In this mode the upstream electrode generates a product which is subsequently detected at the downstream electrode. Although this approach can sometimes be achieved with dual thin-layer serial amperometric electrodes, the maximum signal allowable by Faraday's law will be achieved only with coulometrically efficient working electrodes. This mode increases the selectivity of the detector; only compounds that generate electroactive products stable enough to reach the second electrode will be measured (Andrews *et al.*, 1982). Figure 11 shows an application in which poor oxidative resolution of acetaminophen is overcome by reduction. The redox approach also has been used successfully to measure many metabolites (Andrews *et al.*, 1982) as well as dopamine in microdialysis perfusates (Robinson, 1990).

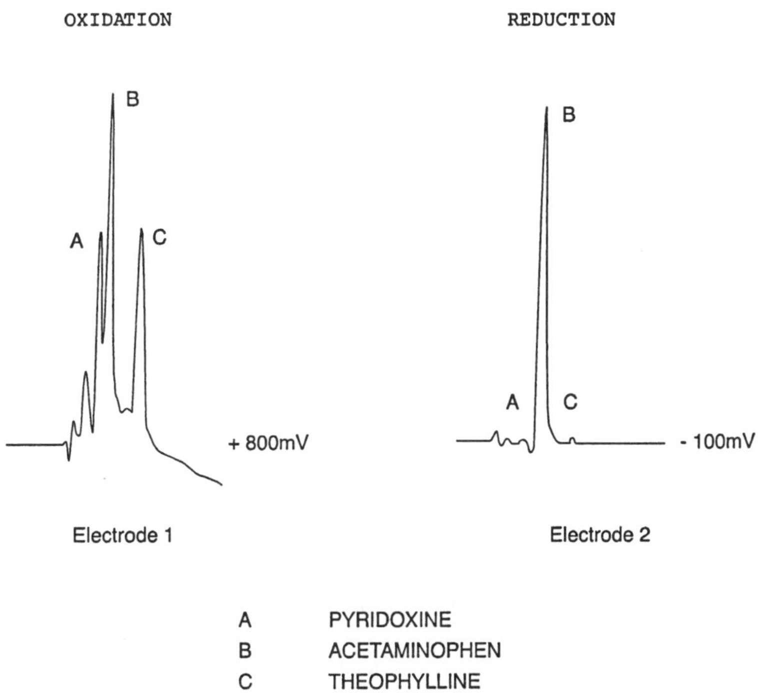

Figure 11. A practical example of the redox mode. Some compounds undergoing oxidation may be poorly resolved chromatographically. However, following initial oxidation chromatographic resolution may be improved using reductive measurement. This is the case when measuring acetaminophen in the presence of pyridoxine and theophylline. As only acetaminophen is reversible, the poor resolution found oxidatively can be improved by measuring it reductively.

The flow-through serial detector in differential mode. In the differential mode, current from each working electrode is monitored. The addition of these outputs can results in increased signal and reduced noise. Alternatively, subtraction also diminishes noise while permitting measurement of the difference in output between the two detectors (Andrews *et al.*, 1982; Lunte *et al.*, 1985).

MULTI-ELECTRODE ARRAYS

Multi-electrode arrays can be divided into those in which the arrays are connected together or "commoned" and those in which each electrode in the array is isolated. Both have been reviewed recently (Diamond, 1993). Commoned micro-arrays offer improved sensitivity due to diminished noise but may be fragile, difficult to use, and difficult to manufacture (Tallman and Peterson, 1990; Diamond, 1993; Niwa *et al.*, 1995). Isolated electrode arrays offer greater analytical flexibility.

There have been relatively few reports of isolated amperometric electrode arrays. All amperometric arrays, like the dual amperometric arrays described above, inherently suffer from poor selectivity (Figure 12). Fielden *et al.* (1994) designed and evaluated a variety of geometrically different arrays describing problems with anomalous behavior and difficulty in maintaining potentiostatic control across the arrays. Hoogvliet *et al.* (1991) developed a parallel sixteen carbon-paste working electrode array with common reference and auxiliary electrodes. Although they reported good sensitivity for monoamine measurement, their array lacked selectivity. Hoogvliet and co-workers also developed a novel four electrode parallel detector, using different working electrode materials (gold, gold-amalgam, carbon paste and modified carbon paste) which may show more promise in the future. Recently, a pseudo-microelectrode array of "80 individual electrodes" - capable of being set at incrementally higher potentials - has been developed with the ability to measure current at each electrode in real time (Aoki *et al.*, 1992). It has so far been used to measure ascorbic acid, uric acid, and catecholamines in plasma and urine. Turk and Blank (1989) used their neurobiological analyzer to measure a variety of monoamines and metabolites. Their analyzer consisted of four individual liquid chromatographs linked together by an autosampler; three of the chromatographs each used a four-glassy-carbon electrode cell, the fourth having a single platinum working electrode. Although their detector expanded the range of metabolites measured, their analyzer suffered from poor selectivity between co-eluting compounds and the considerable effort required to keep four individual chromatographs running synchronously.

Matson developed a serial array containing up to 16 coulometric flow-through electrodes, with each working electrode having its own paired reference and counter electrodes, and individual potentiostat (Matson *et al.*, 1984). This chromatographic system, originally called the neurochemical analyzer, had the ability to measure more than just neurochemicals; it was renamed the Coulochem® Electrode Array System (CEAS). Recently, an improved version, the CoulArray™ detector, has become available.

THE COULOCHEM ELECTRODE ARRAY SYSTEM

Overview

The CEAS has been in use by neuroscientists for more than ten years but still remains relatively unknown outside the field. A description of the instrument and its capabilities therefore is useful to help better understand the other chapters in this volume.

The use of two different wavelengths to measure absorbance ratio similarities between an external standard and an unknown in a sample has been used to access the purity of a peak eluting from the HPLC (Yost *et al.*, 1977). This in turn was expanded with the development of the photodiode array (PDA) in which 256 wavelengths can be monitored simultaneously - in real time - resulting in a third chromatographic dimension (Saitoh and Suzuki, 1979). Compounds thus can be characterized by wavelength vs. absorbance as a function of time (Figure 12). Although PDA has expanded the utility of the single channel UV detector, it still suffers from a poor ability to directly distinguish between two compounds eluting from the column at the same time. The absorbance spectrum of a compound is a result of the combined rotational, vibrational, and electronic transitions of the entire molecule and is thus a broad and relatively nonspecific relationship. The PDA detector barely enhances the real-time resolving power of the analytical column.

In the 1980s Matson *et al.* developed the voltammetric equivalent of the PDA in which a series of 16 coulometric sensors resolve compounds based on their redox potential, rather than on their wavelength (Matson *et al.*, 1984). Here co-eluting compounds were resolved by the coulometric array (as long as their half-wave potential differed by more than 60mV), consequently enhancing the resolving power of the analytical column (Figure 12). A compound's HDV is so characteristic that an array of coulometric electrodes is able to discriminate between different chemical isomers of the same compound even though their absorbance spectra may be equivalent. For example 2,3 dihydroxybenzoic acid oxidizes at +175mV (vs. Pd) whereas the 2,5 isomer requires an applied potential of +125mV (vs. Pd).

The CEAS is a fully integrated HPLC-ECD chromatograph. The software is responsible for control of the CEAS, including hardware (up to 16 working electrodes, pumps, high pressure gradient, switching-valve, autosampler, applied potential), data handling (peak identification, peak integration and data manipulation), and report generation. Each of the potenstiostats can apply a potential from -1000mV to +2000mV in increments of 1mV. There is full autoranging capability, permitting picogram and microgram quantities of analytes to be measured simultaneously - without the need for manual gain-range changes with the potential loss of data.

The coulometric array detectors are the only HPLC-ECDs that are capable of routine gradient operation. As the mobile phase changes during the gradient (e.g. in ion-strength, ion-pair, pH or percentage of organic modifier), the charging currents at the working electrode cause baseline drifts that can be so dramatic at higher sensitivity that analyte measurement is impossible. Such baseline changes can be overcome by autoranging and software manipulation. The CEAS software contains an advanced "self-correcting" algorithm that effectively removes baseline drift while keeping peak integrity intact (the chromatogram acts as its own blank) - see Figure 13.

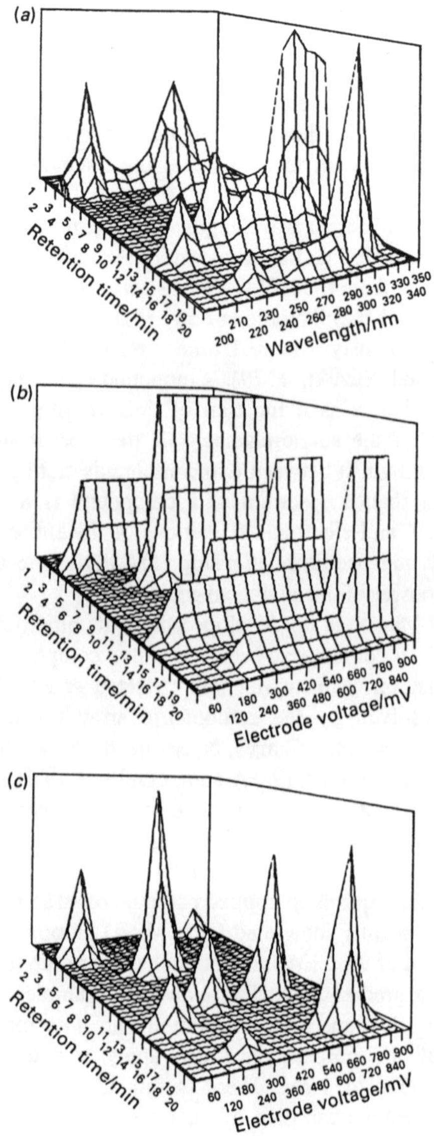

Figure 12. Three-dimensional representation of UV and EC array detector chromatograms. (a) A typical chromatogram from an UV PDA where peaks can be identified by both their retention time and absorbance characteristics at various wavelengths. Note that there is not true separation in the wavelength axis. (b) How the same hypothetical sample may look using an amperometric ECD. Note that once the oxidation potential for the compound is reached, all subsequent electrodes in the array give a similar peak thus obscuring resolution in the *z*-axis. (c) Finally, how the same sample would appear using a coulometric array ECD. Note that due to the 100% conversion of compounds at their oxidation potentials there is now full resolution in the electrode axis, revealing some compounds that were obscured in both (a) and (b). From Svendsen (1993). Reproduced with permission from The Analyst.

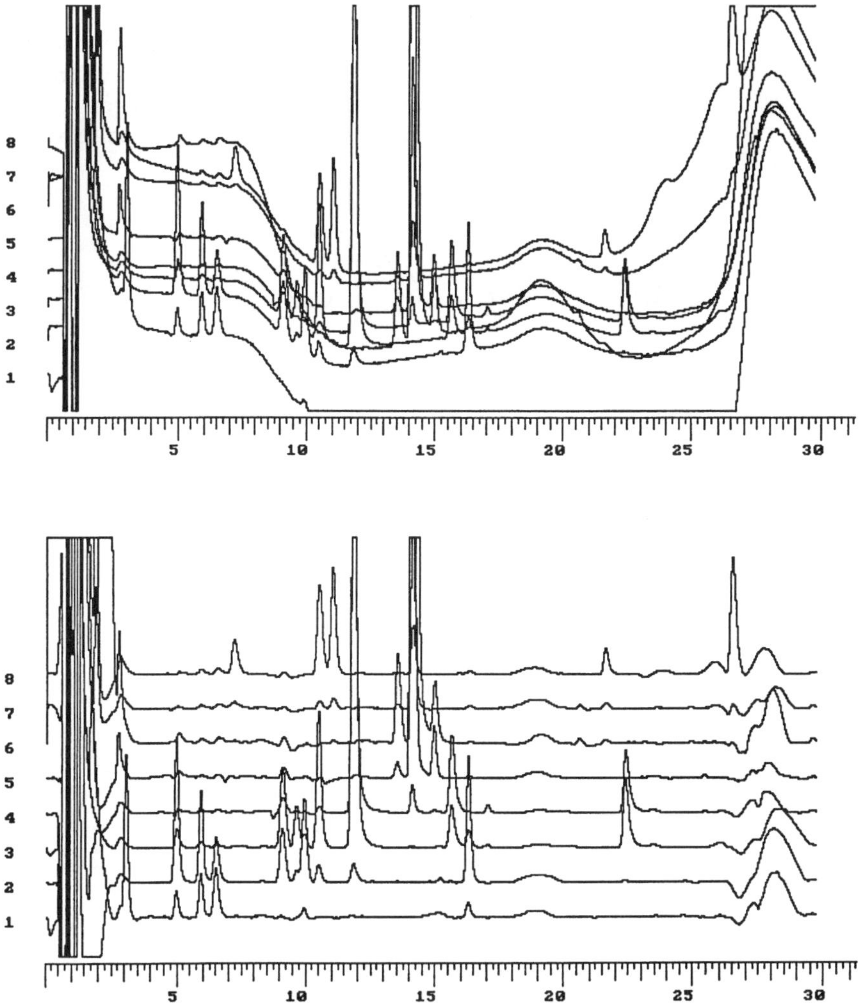

Figure 13. Electrochemical detection is not usually compatible with gradient HPLC as the detector baseline wanders due to changes in the conductance of the mobile phase flowing through the detector. The upper chromatogram shows the measurement of monoamines and metabolite standards during a methanol gradient (for better visualization only selected channels are shown). The lower chromatogram shows the same analysis except that the CEAS gradient-correction algorithm has been used. Even though some of the peaks appear to be truncated no data are lost due to autoranging.

Sensor Design

The coulometric array cell is built upon the dual serial flow-through graphite cell design described above - see Figure 14a. There are several characteristics inherent in the electrode design that make the porous graphite electrode attractive for arrays. First, they are coulometrically efficient; that is, a compound will present a maximum at one electrode and then drop to zero at subsequent sensors. Second, the band spreading, dead-volume and pressure characteristics of the cells are such that there is no significant degradation of the chromatographic separation at the last sensor in the array and no significant difference in registration time between the first and last electrode (Matson *et al.*, 1984). Third, HDVs, from flow-through coulometric electrodes, are sharper than those from amperometric electrodes for many compounds. Fourth, analytes can be measured at higher potentials with sensitivity, as noise will be diminished by the preceding sensors.

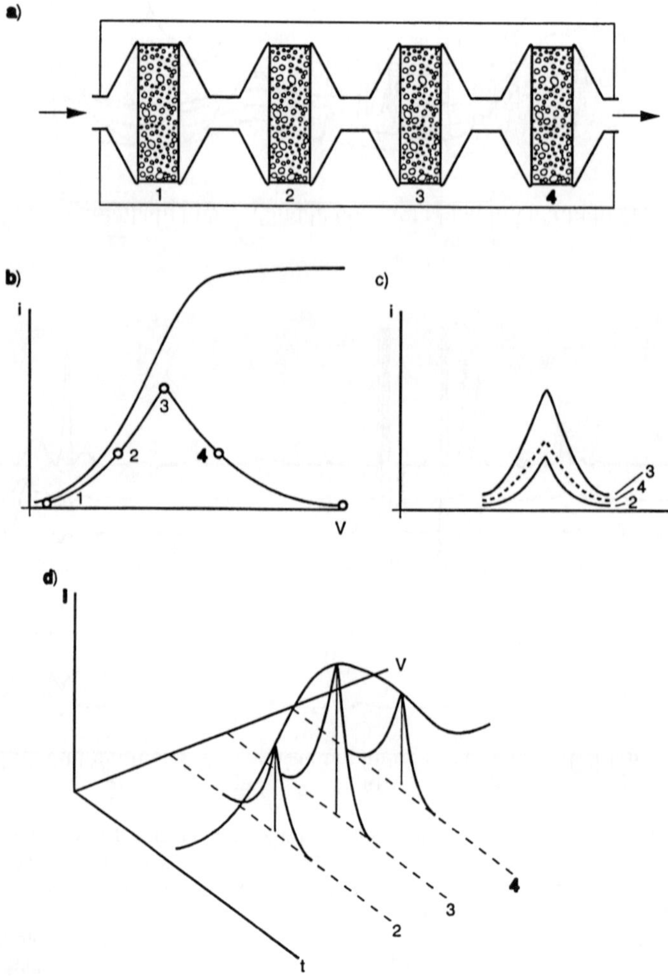

Figure 14. The coulometric array cell contains four serially placed flow-through graphite electrodes (a). (b) Signal from a single compound is measured on three consecutive electrodes (preceding - 2, dominant - 3 and following - 4) with applied potentials placed along the compound's potential axis. The response ratios of 2/3 and 4/3 (along with retention time) are used qualitatively to identify the compound. A plot of the cumulative detector response against applied voltage produces the compound's HDV. (c) Resolution of the compound across the array at its elution time. (d). Temporal and voltammetric resolution of the compound.

Array Resolution
With the coulometric array, compounds separated by chromatography are further resolved electrochemically. A common confusion with coulometric array analysis is the fact that not all of the signal is measured on a single electrode; that signal is "spread" over two or three consecutive electrodes. If maximum signal from a compound is the major objective, the applied oxidation potential to a single coulometric electrode should match the oxidation potential maximum of that compound (which in turn is obtained from its HDV). However, such a gain in sensitivity comes with a loss in electrochemical resolution. If maximum qualitative data is the goal, incrementally increasing potentials should be applied to a series of consecutive electrodes so that they span the analyte's potential axis. Signal at each of the consecutive electrodes can then be used qualitatively to identify that compound or to identify co-elutions. To better explain how the array functions see Figure 14. Figure 14b shows the current (cumulated sensor response) for a particular compound as a function of applied potential (incremented oxidative array across 4 adjacent electrodes) - its HDV. Figure 14b also shows each individual sensor's response. Typically, compounds are found to react at three adjacent sensors, termed preceding, dominant, and following. Figure 14c shows the chromatographic resolution of the compound (current as a function of time). Figure 14d shows three-dimensional resolution and the analyte's response is shown as a function of time and applied potential.

Perfect chromatographically co-eluting analytes can be resolved by the array as long as their half-wave potential is greater than 60mV. Thus, over a 1000mV range - assuming that each peak is baseline resolved, the resolution of a column can be increased potentially by as much as 16 times. However, such improvements are never so precise as some compounds, such as serotonin, present three oxidation waves across the array. The first wave is due to oxidation of the 5-hydroxyl group (dominant potential = +200mV); the second wave due to oxidation of the pyrrole ring (dominant potential = +600mV); and the third due to the amine (dominant potential = +800mV). Oxidation waves can be even more complex when compounds possess several moieties capable of undergoing electrochemical reaction (such as with the flavonoids and polyphenolics). Such characteristic behavior is of great benefit, however, and is used to better identify or "fingerprint" a compound. It is also extremely helpful when trying to measure an analyte in a complex matrix (Achilli *et al.*, 1993; Gamache *et al.*, 1993c).

The combination of retention time with voltammetric behavior has been used with some success to help identify certain aspects of the structure of an unknown analyte. For example, catechol analogs display a single oxidation at 100mV, phenols have a single oxidation at 600mV, hydroxyindoles show a biphasic wave at 200 and 600mV. Although not definitive, the CEAS can help researchers make informed guesses as to a molecule's possible structure.

Peak Identification
The most common approach to peak identification has the CEAS software initially identifying an analyte by its chromatographic retention time. Peak confirmation is then achieved by automatically comparing the HDV of the standard to the HDV of the peak in the sample. This comparison, called the "ratio accuracy," is an immediate numerical indicator of compound similarity, mis-identification, or co-elution

presence. The concept and importance of the ratio accuracy is handled in greater detail by Cullison and Gamache in this volume.

Different Types of Arrays

The CEAS has an array of eight, twelve, or sixteen electrodes. By far the most common array is the oxidation array (for example, the potentials of the sixteen electrode array are set from 0 to +900mV in 60mV increments). It is used to measure a wide variety of metabolites, including monoamines and metabolites, some amino acids and dipeptides, vitamins and cofactors, and many other electroactive species. Small incremental potential arrays are used to better characterize the voltammetric behavior of drugs and unknowns in a sample. Larger incremental potential arrays typically are used when trying to quantitate low levels of a compound - the majority of the signal will then be concentrated on the dominant electrode with little to no signal on the preceding and following electrodes. In this case the ratio accuracy will be zero and the array will act more like a series of advanced single-channel EC detectors with compounds identified solely on retention time. However, unlike an array of amperometrically-efficient electrodes, preceding electrodes in the coulometric array will diminish noise and remove interferences from subsequent electrodes.

To effectively remove irreversible compounds, the array can be set to a redox configuration, often with a large positive potential applied to the first electrode and a large negative potential applied to the second. The remaining electrodes in the array can then be set to alternate between positive and negative potentials (dependent upon the analyte), and the signal from low concentrations of reversible compounds can then be amplified (Matson *et al.*, 1984).

With applied potentials ranging from -1000 to +2000mV in 1mV increments and the ability to apply potentials to up to 16 electrodes individually means that the number of different arrays is virtually limitless but the optimal array choice will be dependent upon the objective of the analysis. The array can be developed to improve limits of detection, remove unwanted matrix effects, overcome coelution problems, and fully characterize the analyte electrochemically.

Software

Data are obtained from all electrodes simultaneously (typically 2 reads/second) and directly stored as the unalterable raw chromatogram on the hard drive. The software allows collection of up to 16 channels of data synchronously. Although the software never allows corruption of the original chromatogram, manipulations such as baseline subtraction, removal of gradient baseline artifacts, noise, or subtraction of one chromatogram from another (to examine differences) are permitted with the altered chromatogram being stored under a different file. The software can be used either manually or in a fully automated mode. Although it is possible to manually integrate every peak in the chromatogram (a urine sample analyzed using a 30-minute gradient method with 16 working electrodes typically generates hundreds of peaks), it is common practice to use the peak identification and integration algorithms within the software. Following integration, a peak is first identified based on retention time and then further characterized by its electrochemical behavior across the array. The software automatically scans a user-created library which can contain dozens of analytes replete with pertinent information (name, retention time, electrochemical behavior, concentration). An unknown peak in the sample is identified and given a name and concentration only if its elution time (retention time within 2%) and HDV (<20% shift) are equivalent to a compound contained within the library. The software generates a report containing name, concentration, retention time, retention time error, peak area (or height), and ratio accuracies. Integration, peak identification, and report generation can be done after completion of a series of samples or alternatively can be carried out automatically on-line after the completion of each chromatogram.

Data can be exported into a variety of formats - for spreadsheets, into Lotus123™ (Lotus Inc., Cambridge, Massachusetts); database software, such as Oracle™ (Oracle Corporation, Redwood Shores, California); or statistical packages such, as Einsight™ (Infometrics Inc., Seattle, Washington), which is used for pattern recognition.

Hardware

Components of the CEAS have been customized for HPLC-ECD. The pumps are controlled by the CEAS software via an RS232 connection. Each pump has two low-volume displacement pistons (10μL/stroke), giving a noise profile matched to the CEAS's noise filters, enabling the measurement of lower s/n ratios. Gradient elution is achieved using two pumps and a high pressure gradient mixer. Samples can be loaded onto the system using either a manual injector or an autosampler. This autosampler can inject from 1μL to 1000μL and can perform pre-column derivatization procedures. The column, pulse dampers, and electrodes are all housed in a thermal chamber (controlled to within 1°C) to ensure precise control of retention times and ratio accuracies. Most of the fluid path for the CEAS is made from PEEK (to effectively remove as much of the iron from the system as possible), which decreases the background noise and prevents auto-oxidation of labile molecules.

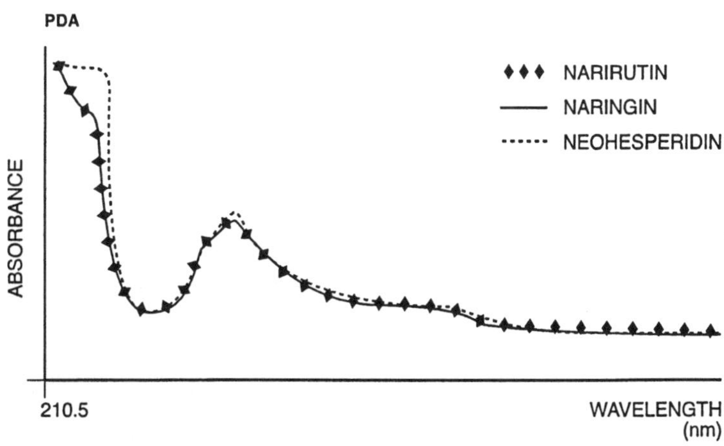

Figure 15. PDA traces for three flavonoid compounds. The spectrum for narirutin perfectly overlaps that of naringin. Reprinted from, Rouseff, R. L. (1988). Figure 3, page 60. In: *Adulteration of Fruit Juice Beverages*. Nagy, S., Attaway, J. A., and Rhodes, M. E. (eds.). With permission from Marcel Dekker Inc.

Comparison of the CEAS to PDA

As with PDA, the CEAS produces an abundance of analytical data. However, the quality and quantity of the data and the process of acquisition differ between the techniques.

As discussed above, the major difference between having potential or wavelength as the third chromatographic dimension is one of selectivity. Small changes in structure may have little effect on a molecule's UV spectrum but can have marked effects on its HDV. Figure 15 shows that the three structurally similar flavonoids - narirutin (naringenin-7-rutinoside), naringin (naringenin-7-neohesperidoside) and neohesperidin - all possess very similar spectra. However, naringin and hesperidin (hesperetin-7-rutinoside) possess very distinct HDVs and are easily resolved using the CEAS (Figure 16). The CEAS is at least three orders of magnitude more sensitive than the best PDA with low femtomole (pg on column) detection limits. The differences and similarities between the CEAS and PDA have previously been discussed (Acworth and Gamache, 1996).

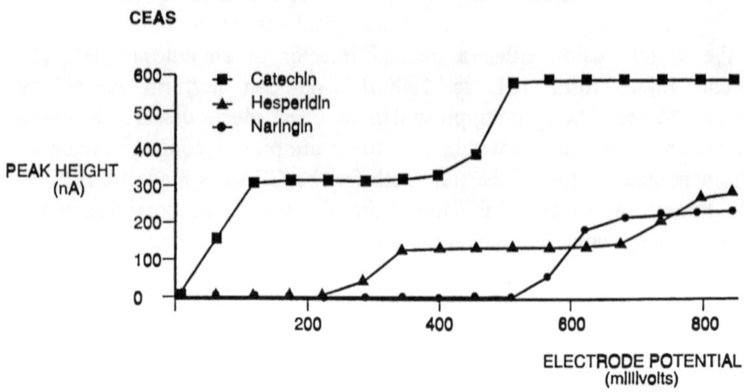

Figure 16. Hydrodynamic voltammograms for three flavonoid compounds. Although, as shown in Figure 15, it may be difficult to resolve them spectrophotometrically, their HDVs' are markedly different and can effectively be used for resolution.

Literature Review

Over the past decade the CEAS has been used to measure a variety of compounds. Although initially used by neuroscientists to measure analytes from the catecholaminergic and serotoninergic pathways, researchers from other disciplines have realized the full potential of multi-parameter data.

The brain contains and produces a wide variety of compounds, including neurotransmitters, neuromodulators, and neurohormones, many of which are electrochemically active. Study of this neurochemical world has been limited in the past by the resolution of the currently available analytical techniques. Typically, researchers could separate and detect only 1 to 5 compounds simultaneously. Furthermore, these compounds were identified solely by retention time, while ignoring the possibility of misidentification and co-elution. This situation was markedly changed in the late 1980s by the CEAS. Researchers could now

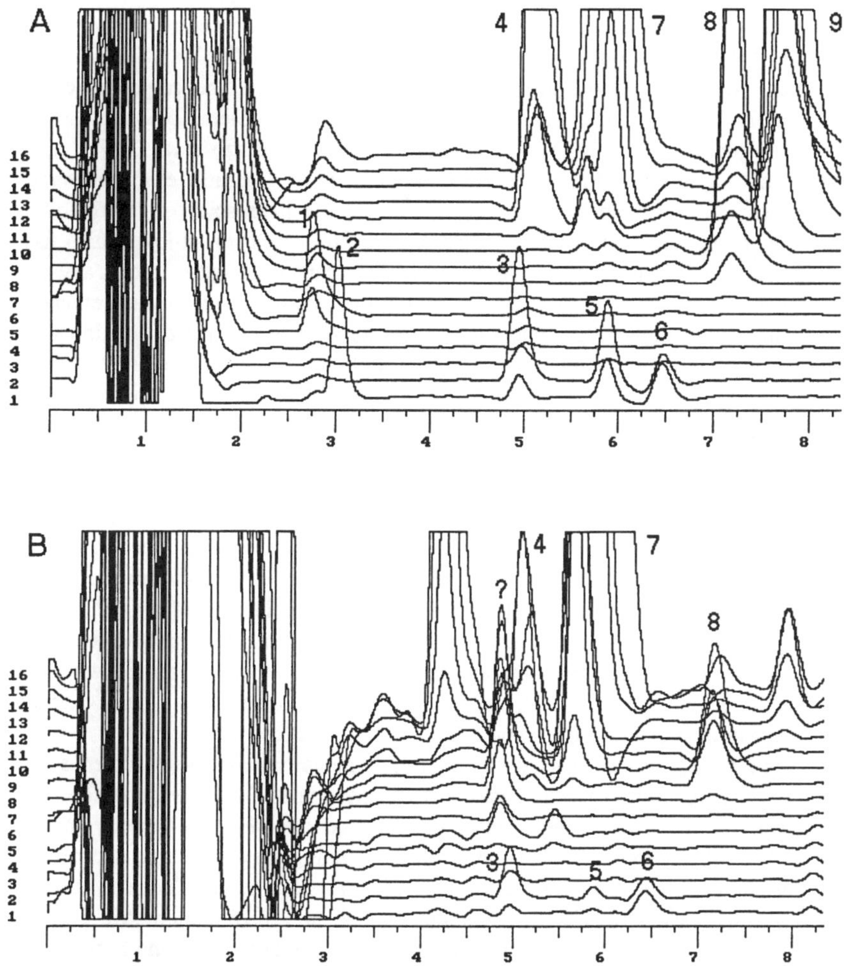

Figure 17. Coulometric array chromatogram of standard mixture (A) and rat posterior cortex homogenate (B). Analytes were resolved using a methanol gradient and sixteen channel array (0 to 900mV in 60mV increments). Analytes measured were: 1 - MHPG, 2 - Homogentisic acid, 3 - Norepinephrine (NE) , 4 - Guanosine (GN), 5 - Dihydroxyphenylacetic acid, 6 - L-Dihydroxyphenylalanine, 7 - Methionine, 8 - 4-Hydroxyphenyllactic acid. 8 - 4-hydroxybenzoic acid. Although NE and GN co-eluted they were well resolved voltammetrically (A). However, in the tissue homogenate an unknown peak (?) also co-eluted with NE and GN. Ratio accuracies for NE were 0.960 (preceding/dominant; channel 1:2) and 0.640 (following/dominant; channel 3:2). The low value for the second ratio accuracy was due to signal from ? corrupting the HDV for NE. The CEAS immediately informs the user of such co-elutions. A single electrode HPLC system would not be able to measure such co-elutions leading to the overestimation of NE tissue levels.

simultaneously measure a wide variety of compounds from several different neurotransmitter and other biochemical pathways, with sensitivity and accuracy.

Researchers could immediately see whether their peak of interest was real or whether a co-elution was present. For example, Figure 17 shows the capability of the CEAS to measure norepinephrine (NE) in posterior cortex homogenates. In this example another unknown metabolite co-elutes with NE. The use of conventional single-electrode techniques would lead to gross overestimation of the brain's NE content. However, due to a marked decrease in the ratio accuracy, the coulometric array immediately notifies the researcher of the coelution problem. NE can now be accurately quantitated using the upstream electrodes only. Rizzo *et al.* (1991), used a thirty-minute gradient CEAS method to accurately resolve and electrochemically characterize 33 compounds of general interest to the neuroscientist. With low pg limits of detection this method could be readily used to study brain tissue homogenates, CSF, and even microdialysis perfusates. Recently, more advanced techniques have been used which permit the simultaneous measurement of monoamine neurotransmitters and metabolites as well as derivatized amino acids, using column switching coulometric array technology (Gamache *et al.*, 1993b). This method not only produces qualitative data but also overcomes any variability in the data which may result when multiple analytical systems are used. It is also interesting to note that Gamache *et al.*, reported voltammetric resolution of serotonin and the dopamine metabolite, 3-methoxytyramine), even though these compounds co-eluted chromatographically. By using such coulometric array detection to analyze microdialysis perfusates it is now possible to not only study the interaction between different and distinct neurotransmitter pathways, but also to measure drug levels (on-line pharmacokinetics) and drug effects in the same sample (Acworth *et al.*, 1994b). This approach can be further refined by relating such drug-induced metabolic changes to the alteration in behavior of the awake animal. Further details on the use of the CEAS in the neurosciences can be found in this volume (Huettl and Gerhardt; Langlais; Maruyama and Naoi; Naoi; Sparks *et al.*; Takeuchi *et al.*).

The CEAS has been used in the environmental laboratory to screen a large number of compounds, such as herbicides and pesticides. With its great sensitivity and selectivity, Bernwieser and Sontag (1992) used the CEAS to directly measure levels of the plant growth inhibitor amitrole in drinking water. Without the need for trace enrichment steps it is possible to overcome errors due to sample preparation and to shorten analysis time. Using a gradient method and a 16-coulometric electrode array, Achilli *et al.* (1995) were able to directly measure 27 phenols and herbicides in ground water with limits of detection 200 times lower than permitted legal residue levels. These researchers even found trace levels of 6 phenylurea herbicides and phenolic derivatives in bottled mineral water. Recently the CEAS has been used in conjunction with UV photolysis to measure a wide variety of explosives and residues in contaminated ground water (Acworth and Waraska, this volume). The sensitivity of the CEAS makes trace enrichment unnecessary.

From quality control to screening for novel drug candidates, the CEAS enables routine low level measurement of drugs and metabolites even in complex biological matrices. For example, the CEAS was used to measure possible neurotoxins (tryptophan aminal and 3-phenylaminoalanine) in contaminated tryptophan samples (Ryan *et al.*, 1993). It is interesting to note that even in tryptophan samples deemed pure by HPLC-spectrophotometric techniques with the greater sensitivity of the CEAS, 350-400 contaminants were found at trace levels. Ryan reported that the analysis of both known and unknown compounds and the use of pattern recognition

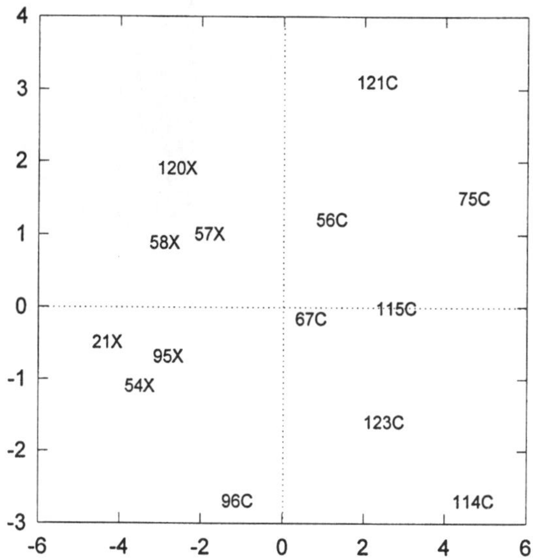

Figure 18. Principle component analysis of control (C) and case (X) tryptophan samples allegedly associated with Eosinophilia-Myalgia Syndrome in humans. Analytes were measured using a phosphate-methanol-acetonitrile gradient and 900mV coulometric array (Ryan *et al.*, 1993). All samples were kindly donated by Drs. G. Gleich and A. Mayeno (Mayo Clinic, Rochester, MN). All samples produced approximately 350-400 (levels of 92% of the peaks were <0.01% of the tryptophan level). Differences and similarities between the metabolite pattern (containing both known and unknown compounds) for each sample were compared using pattern recognition software (Einsight™). This figure clearly shows separate clusters for case and control samples. Furthermore, the patterns for the case samples were similar as shown by the tightness of the cluster.

software could immediately identify a sample as being from control or contaminated sample lots (Figure 18). Furthermore, Svendsen (1993) reported that plasma levels of a drug thought to be pure by radioimmunoassay were in fact shown by the CEAS to be a mixture of the parent drug and its principal metabolite. Results based on the radioimmunoassay data would have grossly overestimated the true plasma levels. The CEAS has been used to measure several different drug classes. Ryan *et al.* (1992) used a global method to simultaneously measure the plasma levels of a wide variety of benzodiazepines, tricyclic antidepressants, and fluoxetine (Figure 19). This method permitted low pg sensitivity, rapid screening of samples, and the ability to detect co-elutions using ratio accuracy data. Using a different gradient method, Achilli and Cellerino have used the CEAS to measure a number of illicit drugs in body fluids (this volume). A very exciting and rapidly developing research area is the study of "nutraceuticals" - naturally-occurring phytochemicals with purported physiological function. With hundreds of structurally-related chemicals typically found at low levels, few techniques can be used to generate good qualitative and quantitative data.

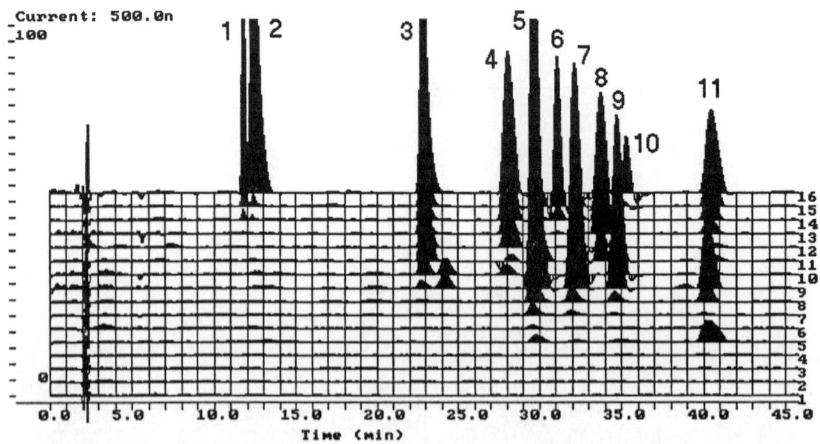

Figure 19. Simultaneous measurement of different drug classes using gradient chromatography and coulometric array detection (20-1000mV in 70mV increments) (Ryan *et al.*, 1992). 1 - Oxazepam, 2 - Chlordiazepoxide, 3 - Flurazepam, 4 - Doxepin, 5 - Desipramine, 6 - Nortriptyline, 7 - Imipramine, 8 - Amitriptyline, 9 - Trimipramine, 10 - Fluoxetine, 11 - Chlorpromazine.

Such capability is beyond the reach of even the best PDA detectors. For example, epigallocatechin and epicatechin are two flavonoids with possible anticancer activity found at relatively high levels in green tea. Lee *et al.* (1995) used the CEAS to directly measure levels of these compounds in tea, plasma, and even urine samples without major sample clean-up and concentration procedures. Recently, Gamache has developed a method using a coulometric electrode array capable of simultaneously measuring several phytoestrogens and estrogens in uterine and plasma samples at sensitivity, without the extensive clean-up procedures required for mass-spectrophotometric analysis (see Gamache *et al.* this volume). Certain phytoestrogens such as daidzein and daidzin are found in soy products and are reported to have anticarcinogenic properties. Studies are now underway to examine the abundance of these phytoestrogens in soy and their metabolism in rat and human. In this volume Gamache *et al.* discuss the role of the CEAS in measuring a wide variety of metabolites, such as the flavonoids and polyphenolics, which are thought to protect against free-radical attack during oxidative stress. Furthermore, Gamache has extended his phytoestrogen assay to include measurement of two mammalian lignans, eneterolactone and enterodiol and has measured passage of both phytoestrogens and mammalian lignans through out the blood-brain barrier (Gamache et al., 1996a,b).

Certain samples, such as plasma and urine, require extensive and time-consuming clean-up procedures prior to analysis in order to concentrate trace levels of analytes and remove interferences. The CEAS has become increasingly popular in clinical laboratories for its stability, sensitivity and selectivity, which translate into savings of both cost and time. Urinary metanephrine and monoamine acid metabolites are now two routine tests used in cancer screening with no known endogenous or drug interferences (Cheng *et al.*, 1992; Gamache *et al.*, 1993b). A method for the measurement of catecholamine levels in urine has now been completed.

Recently, the analytical repertoire of the coulometric array detector has been further expanded. The derivatizing agent, NBDCl, has been used to measure levels of imino acids, which are important to the understanding of collagen-related diseases. Simultaneous photolysis and ECD now allow routine measurement of compounds that are not normally electrogenic, such as many pesticides and explosives (see Acworth and Waraska, this volume).

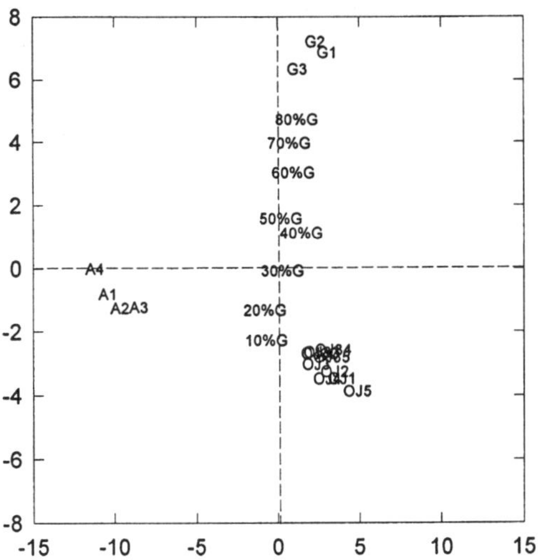

Figure 20. Levels of known and unknown metabolites were obtained for each sample using gradient HPLC and coulometric array detection. The inter-sample similarity and differences in metabolite patterns were obtained using pattern recognition software (Einsight™). Principle component analysis showed separate clusters for orange (OJ), apple (A) and grapefruit (G). Measurement of adulteration is clearly demonstrated by the separation of grapefruit-orange mixes from the OJ cluster. Addition of as little as 10% grapefruit juice to orange juice changes the metabolite pattern of the orange juice enough to permit discrimination (Gamache *et al.*, 1993c, 1995)..

From its initial use for the measurement of a handful of known compounds, the CEAS is now used routinely to examine the complex pattern of metabolites in many different samples. Such patterns have proven useful to verify adulteration in fruit juice (Gamache *et al.*, 1993c, 1995) (see Figure 20). Preliminary experiments by Matson suggest that the pattern of electrogenic homocyclic and heterocyclic compounds in crude oil may be indicative of the country of origin of the crude oil. This suggests that by using pattern recognition "unclaimed" oil slicks at sea may be able to be retraced to the perpetrators. Pattern recognition may also play a role in verifying the purity of bottled wine and detecting whether adulteration with grape juice or blending with cheaper wine has taken place. Pattern recognition has also been used to study the aging process in bottled beer, to verify contamination of over-the-counter tryptophan supplements (Ryan *et al.*, 1993, and above), and - probably what is the most exciting - to characterize and follow the progression of human neurodegenerative diseases. Over the past several years Matson and his colleagues have used the CEAS technology with advanced pattern recognition software to measure changes in the pattern of metabolites in samples obtained from individuals suffering from Parkinson's, Huntington's, and Alzheimer's diseases. The present neurodegenerative database contains samples from over 190 individuals and represents over 13,000 individual data points (Milbury *et al.*, 1994). It is the hope of the authors that eventually the predisposition of an individual to particular diseases will be reflected in the complex pattern of electrogenic biomolecules found in blood (or other physiological fluids) and diagnosed using the CEAS. Such early diagnosis will ensure early intervention and treatment. The practical uses of pattern recognition are presented in more detail in the chapter by Milbury (this volume).

Readers interested in a more detailed list of published references using the CEAS are referred to Table 1.

Table 1.
Published papers using Coulometric array technology

GENERAL REVIEWS	Matson *et al.*, 1984, 1987; Svendsen, 1993; Naoi *et al.*, 1993
METHODOLOGIES	Ikeda *et al.*, (1988); Takeda *et al.*, (1989); Swartz *et al.*, (1990); Volicer *et al.*, (1990); Rizzo *et al.*, (1991); Gamache *et al.*, (1993b); Welch *et al.*, (1993); Acworth *et al.*, (1994a,b); Turiak and Volicer, (1994)
NEUROSCIENCE	Volicer *et al.*, (1985); Langlais *et al.*, (1988); Ikeda *et al.*, (1989); Lucot *et al.*, (1989); Takaragi *et al.*, (1989); Wolff *et al.*, (1989); Beal *et al.*, (1990a,b); Bird *et al.*, (1990); Godefroy *et al.*, (1990); Matson *et al.*, (1990); Ogawa *et al.*, (1990); Acworth *et al.*, (1991); Langlais *et al.*, (1991); Mori *et al.*, (1991); Tohgi *et al.*, (1991a,b); Uemura *et al.*, (1991); Yamomoto, (1991a,b); Beal *et al.*, (1992); Jinnah *et al.*, (1992); Kabuto *et al.*, (1992); LeWitt *et al.*, (1992); Maruyama *et al.*, (1992a); Ogawa *et al*, (1992); Pelligra *et al.*, (1992); Satoh *et al.*, (1992); Siuciak *et al.*, (1992); Suzuki and Mori, (1992); Langlais *et al.*, (1993); Lavoie *et al.*, (1993); Mecocci *et al.*, (1993); Vriend *et al.*, (1993). Milbury *et al.*, (1994).
LESIONING AND TOXICOLOGY	Langlais *et al.*, (1990); Makino *et al.*, (1990); Beal *et al.*, (1991); Connor *et al*, (1991); Finn *et al.*, (1991); Mair *et al.*, (1991); Altar *et al.*, (1992); Chen *et al.*, (1992); Connor *et al.*, (1992a,b); Maruyama *et al.*, (1992b); Vecsei *et al.*, (1992); Beal *et al*, (1993); Hill *et al.*, (1993); Langlais *et al.*, (1993); Maruyama *et al.*, (1993a,b); Naoi *et al.*, (1993). Achilli *et al.*, (1995).

INVERTEBRATE NEUROSCIENCE	Shimizu and Takeda, (1991b); Shimizu *et al.*, (1991b); Takeda and Svendsen, (1991); Takeda *et al.*, (1991); Shimizu *et al.*, (1992); Sparks and Geng, (1992); Takeda, (1992a,b); Shimizu *et al.*, (1993).
FOOD	Bernwieser and Sontag, (1992); Achilli *et al.*, (1994); Bernweiser *et al.*, (1994); Ryan *et al.*, (1994).
CLINICAL	Cheng *et al.*, (1992); Gamache *et al.*, (1993).
PHARMACEUTICAL	Gamache *et al.*, (1992, 1993b); Ryan *et al.*, (1992); Acworth *et al.*, (1993); Ryan *et al.*, (1993); Gamache *et al.*, (1994); Achilli *et al.*, (1996)

CONCLUSION

The demands placed on any piece of analytical equipment is ultimately dependent upon the field of interest. For example, neuroscientists are continuously in need of instrumentation that offers ever improving limits of detection, often in combination with the desire for rapid analysis. Others, such as food technologists, may have to identify and quantitate a single compound in a complex mixture of similar species. Still others are now realizing that the key to their research no longer lies in the absolute level of a few compounds but in the complex pattern of metabolites found in their sample. The answer to many of these demands is to be found in HPLC with coulometric array detection. This technique offers the sensitivity and selectivity of HPLC-ECD with higher resolution and comparable stability to the PDA detector. The secret to all these capabilities is a direct consequence of the electrode design. Not all voltammetric electrodes used with HPLC are created equal, and electrode efficiency can have very important ramifications. The flow-through coulometric electrode is 100% efficient, making it very sensitive. Furthermore, the coulometric electrode is not as noisy as some thin-layer electrode proponents and manufacturers would have us believe. Consequently, the s/n ratio of the coulometric cell is far better than the thin-layer design. Additionally, this efficiency permits effective screening resulting in unique voltammetric resolution. Analytes can now be identified both by retention time and voltammetric behavior across an array. This second voltammetric axis is not attainable with an array of amperometric electrodes. More importantly, the current thin-layer amperometric electrode design is not suitable or practical for placement in an array.

The first iteration of the coulometric array technology was the CEAS. This is now being replaced by the CoulArray™ detector. With quieter potentiostats resulting in better s/n ratios, improved software, and the capability of using a wide variety of coulometric and amperometric electrodes, this detector is addressing many of the demands of the researcher. Further improvements in the limits of detection will also come from lower volume cells, permitting the use of narrower bore columns, and from microelectrode arrays, in which diminished noise will lead to superior s/n ratios.

More rapid analyses will be the result of improvements in column technology. With the availability of 1.5μm particle-packed columns (Micra Scientific, Inc., Northbrook, IL), routine analyses are now completed within approximately a quarter of the time for the equivalent assays using columns packed with 3μm particles. Although these smaller particles show promise for separations of hydrophobic compounds, the poor carbon-loading of the particles used in the commercially available columns to date means that good resolution of hydrophilic compounds, such as the catecholamines, is still not fully realized.

Improvements in the resolution of the coulometric array eventually will be achieved by increasing sampling rates and, possibly, using more electrodes in the array.

With better understanding of the implications and importance of the pattern of metabolites in a sample, the concerns of simultaneously measuring hundreds of compounds and handling such data will be much less daunting. Already such metabolic patterns are routinely used by the food industry to investigate product adulteration as well as product stability and aging, and to examine competitors' products. With the construction of large metabolite databases the authors hope that it will soon be routine practice to screen human body fluids - especially those of the newborn - against such databases as an indicator of disease status. Although much time and effort currently is being spent on mapping human DNA, perhaps it would be better to study the pattern of metabolites in a sample. After all, this is just a working expression of an organism's genome.

REFERENCES

Achilli, G., Cellerino, G. P., Gamache, P. H. and Melzi d'Eril, G. V. (1993). Identification and determination of phenolic constituents in natural beverages and plant extracts by means of a coulometric electrode array system. J. Chromatogr. **632**, 111-117.

Achilli, G., Cellerino, G. P. and Melzi D'Eril, G. (1994). Determination of amines in wines by high-performance liquid chromatography with electrochemical coulometric detection after precolumn derivatization. J. Chromatogr. **661**, 201-205.

Achilli, G., Cellerino, G. P., Melzi d'Eril, G. and Bird, S. (1995). Simultaneous determination of 27 phenols and herbicides in water by high-performance liquid chromatography with multi-electrode electrochemical detection. J. Chromatogr. **697**, 357-362.

Achilli, G., Cellerino, G. P., Melzi d'Eril, G.V. and Tagliaro, F. (1996). Determination of illicit drugs and related substances by high-performance liquid chromatography with an electrochemical coulometric array detector. J. Chromatogr. **729**, 273-277.

Acworth, I. N. and Gamache, P. H. (1996). The coulometric electrode array: for use in HPLC analysis. Part 1. Theory. American Lab., **5**, 33-37.

Acworth, I. N. and Waraska, J. (1996). Electrochemical detectors for HPLC. Electrode efficiency makes all the difference. American Lab., submitted.

Acworth, I. N., Kreutz, M., Gamache, P. and Lynch, M. (1991). Borna disease virus causes major alterations in central monoamine metabolism. The Pittsburgh Conference on Analytical Chemistry and Applied Spectroscopy, Chicago, Illinois. Paper No. 450P.

Acworth, I. N., Yu, J., Gariepy, K., Gamache, P. H., Ryan, E., Hull, K. and Maher, T. (1993). Monitoring drug levels and their effects on transmitter release *in vivo*: Utility of HPLC array-ECD analysis and microdialysis sampling procedures. The Pittsburgh Conference on Analytical Chemistry and Applied Spectroscopy, Atlanta, Georgia. Paper No. 121P.

Acworth, I. N., Kuo, M., Ryan, E. and Gamache, P. H. (1994a). Analysis of catechins, phenolic and polyphenolic compounds in tea using HPLC and coulometric array detection. The Pittsburgh Conference on Analytical Chemistry and Applied Spectroscopy, Chicago, Illinois. Paper No. 391aP.

Acworth, I. N., Yu, J., Ryan, E., Gamache, P., Hull, K. and Maher, T. (1994b). Simultaneous measurement of monoamine, amino acid, and drug levels, using high performance liquid chromatography and coulometric array detection. J. Liq. Chromatogr. **17**, 685-705.

Allison, L. A. and Shoup, R. E. (1983). Dual-electrode liquid chromatography detector for thiols and sulfides. Anal. Chem. **55**, 8-12.

Altar, C., Jakeman, L., Acworth, I. N., Soriano, R. and Dugich-Djordjevic, M. (1992). Regionally restricted loss and partial recovery of nigrostriatal dopamine input following intrastriatal 6-hydroxydopamine. Neurodegeneration **1**, 123-133.

Anderson, L. B. and Reilley, C. N. (1965a). Thin-layer electrochemistry: Steady-state methods of studying rate processes. J. Electroanal. Chem. **10**, 295-305.

Anderson, L. B. and Reilley, C. N. (1965b). Thin-layer electrochemistry: Use of twin working electrodes for the study of chemical kinetics. J. Electroanal. Chem. **10**, 538-552.

Andrews, R. W., Schubert, C., Morrison, J., Zink, E. W. and Matson, W. R. (1982). Dual electrode cells for LCEC. Am. Lab. **2**, 140-151.

Aoki, A., Matsue, T. and Uchida, I. (1992). Multichannel electrochemical detection with microelectrode array in flowing streams. Anal. Chem. **64**, 44-49.

Barnes, A. C. and Nieman, T. A. (1983). Coulostatic pulse amperometry for liquid chromatography/electrochemistry detection. Anal. Chem. **55**, 2309-2312.

Barth, H. G., Barber, W. E., Lochmuller, Ch. H., Majors, R. E. and Regnier, F. E. (1986). Column liquid chromatography. Anal. Chem. **58**, 211R-250R.

Beal, M. F., MacGarvey, U. and Swartz, K. J. (1990a). Galanin immunoreactivity is increased in the nucleus basalis of Meynert in Alzheimer's disease. Ann. Neurol. **28**, 157-161.

Beal, M. F., Matson, W. R., Swartz, K. J., Gamache, P. H. and Bird, E. D. (1990b). Kynurenine pathway measurements in Huntington's disease striatum: evidence for reduced formation of kynurenic acid. J. Neurochem. **55**, 1327-1339.

Beal, M. F., Swartz, K. J., Finn, S. F., Mazurek, M. F. and Kowall, N. W. (1991). Neurochemical characterization of excitotoxin lesions in the cerebral cortex. J. Neurosci. **11**, 147-158.

Beal, M. F., Matson, W. R., Storey, E., Milbury, P., Ryan, E. A, Ogawa, T. and Bird, E. D. (1992). Kynurenic acid concentrations are reduced in Huntington's disease in cerebral cortex. J. Neurolog. Sci. **108**, 80-87.

Beal, M. F., Brouillet, E., Jenkins, B., Ferrante, R., Kowall, R., Miller, J., Storey, E., Srivastava, R, Rose, B. and Hyman, B. (1993). Neurochemical and histological characterization of striatal excitotoxic lesions produced by the mitochondrial toxin 3-nitropropionic acid. J. Neurosci. **13**, 4181-4192.

Bernwieser, I. and Sontag, G. (1992). HPLC with electrode array detection: determination of amitrole in drinking and ground water. Fresenius J. Anal. Chem. **347**, 559-561.

Bernwieser, I., Patzold, R., Galensa, R. and Sontag, G. (1994). HPLC mit coulometrischer elektroden-array-detektion: bestimmung von 1-O-*trans*-p-cumaraoylglycerol in sorghumhirse, mais und bier. Z. Lebensm Unters Forsch **198**, 40-43.

Bird, E. D, Matson, W. R., Beal, M. F. and Ogawa, T. (1990). Patterns of tyrosine and tryptophan metabolites in controls and various degenerative disorders. In: Basic Clinical and Therapeutic Aspects of Alzheimer's and Parkinson's Disease, **1**. T. Nagatsu (Ed.). Plenum Press, New York, USA, pp. 529-535.

Blaedel, W. J. and Strohl, J. H. (1964). Continuous quantitative electrolysis. Anal. Chem. **36**, 1245-1251.

Blaedel, W. J. and Schieffer, G. W. (1974). A turbulent tubular electrode. Anal. Chem. **46**, 1564-1567.

Blank, C. L. (1976). Dual electrochemical detector for liquid chromatography. J. Chromatogr. **117**, 35-46.

Bohs, E., Linhares, M. C. and Kissinger, P. T. (1994). The unicell: an amperometric flow cell for on-line sensing at μL/min flow rates. Curr. Sep. **13**, 6-8.

Bowers, M. L. (1991). A new analytical cell for carbohydrate analysis with a maintenance-free reference electrode. J. Pharmac. Biomed. Anal. **9**, 1133-1137.

Brunt, K., Bruins, C. H. P. and Doorbos, D. A. (1981). Comparison between a differential amperometric detector in the reductive mode and a UV detector in high performance liquid chromatography with vitamin K3 as the test compound. Anal. Chim. Acta **125**, 85-91.

Chen, J. C., Schnepper, P. W., To, A. and Volicer, L. (1992). Neurochemical changes in the rat brain after intraventricular administration of tryptamine-4,5-dione. Neuropharmacol. **31**, 215-219.

Cheng, M. H., Lipsey, A. I., Lee, J. and Gamache, P. H. (1992). Automated analysis of urinary VMA, HVA and 5-HIAA by gradient HPLC using an array of eight coulometric electrochemical detectors. Lab. Robotics 4, 297-303.

Collins, S. D. (1993).Practical limits for solid-state reference electrodes. Sensors and Actuators B, **10**, 169-178.

Connor, D. J., Langlais, P. J. and Thal, L. J. (1991). Behavioral impairments after lesions of the nucleus basalis by ibotenic acid and quisqualic acid. Brain Res., **555**, 84-89.

Connor, D. J., Dietz, S., Langlais, P. J. and Thal, L. J. (1992a). Behavioral effects of concurrent lesions of the nucleus basalis magnocellularis and the dorsal noradrenergic bundle. Exp. Neurol. **116**, 69-75.

Connor, D. J., Thal, L. J., Langlais, P. J., Mandel, R. J. and Masliah, E. (1992b). Independent effects of age and nucleus basalis magnocellularis lesion: maze learning, cortical neurochemistry and morphometry. Behav. Neurosci. **106**, 776-788.

Covington, A. K. and Rebelo, M. J. F. (1983). Reference electrodes and liquid junction effects in ion-selective electrode potentiometry. Ion-Sel. Electrode Rev. **5**, 93-128.

Cullison, J., Bowers, M. and Acworth, I. N. (1996). Supra-coulometric electrochemical detection of catecholamines at sub-5 fg levels. The Pittsburgh Conference on Analytical Chemistry and Applied Spectroscopy, Chicago, Illinois. Paper No. 35P.

Davenport, R. J. and Johnson, D. C. (1974). Determination of nitrate and nitrite by forced-flow liquid chromatography with electrochemical detection. Anal. Chem. **46**, 1971.

Delahay, P. (1962a). A new electroanalytical method: coulostatic or charge step method. Anal. Chim. Acta **27**, 90-93.

Delahay, P. (1962b). Fundamentals of coulostatic analysis. Anal Chem. **34**, 1267-1271.

Diamond, D. (1993). Electroanalysis. **5**, 795-802.

Dou, L., Mazzeo, J. and Krull, I. S. (1990). Determination of amino acids, peptides and proteins by HPLC combined with electrochemical detection. BioChromatogr. **5**, 74-96.

Duda, C. T. and Kissinger, P. T. (1993). Determination of biogenic amines, their metabolites, and other neurochemicals by liquid chromatography/electrochemistry. In: Methods in Neurotransmitter and Neuropeptide Research, Part 1. **11**. Techniques in The Behavioral and Neural Sciences. Parvez, S. H., Naoi, M., Nagatsu, T., and Parvez, S. (Eds.), Elsevier Science Publishers BV Amsterdam, pp. 1-82.

Elbicki, J. M., Morgan, D. M. and Weber, S. G. (1984). Theoretical and practical limitations on the optimization of amperometric detectors. Anal. Chem. **56**, 978-985.

ESA (1994). Water polishing. Coulochem technical note #70-1668, ESA, Inc.

Fenn, R. J., Siggia, S. and Curran, D. J. (1978). Liquid chromatography detector based on single and twin electrode thin-layer electrochemistry: Application to the determination of catecholamines in blood plasma. Anal. Chem. **50**, 1067-1073.

Finn, S. T., Hyman, B. T., Storey, E., Miller, J. M. and Beal, M. F. (1991). Effects of aging on quinolinic acid lesions in rat striatum. Brain Res. **562**, 276-280.

Gamache, P., Ryan, E. and Acworth, I. N. (1992). HPLC with coulometric array detection for the study of disease and drug related changes in metabolism. The Pittsburgh Conference on Analytical Chemistry and Applied Spectroscopy, New Orleans, Louisiana. Paper No. 476P.

Gamache, P. H., Kingery, M. and Acworth, I. N. (1993a). Urinary metanephrine and normetanephrine determined without extraction by using liquid chromatography and coulometric array detection. Clin. Chem. **93**, 1825-1830.

Gamache, P. H., Maher, T. J., Setchell, K. D. R., Wu, T.-H. and Acworth, I. N. (1996). The transfer of non-steroidal dietary estrogens into the brain. 26th Annual Meeting of The Society for Neuroscience, November 16-21. Accepted.

Gamache, P. H., Maher, T. J., Setchell, K. D. R., Wu, T.-H. and Acworth, I. N. (1996). The passage of non-steroidal dietary estrogens through the blood-brain barrier. The 7th International Conference on *In Vivo* Methods. Santa Cruz de Tenerife, Spain. September 21-24.

Gamache, P., Ryan, E., Svendsen, C. N., Murayama, K. and Acworth, I. N., (1993b). Simultaneous measurement of monoamines, metabolites and amino acids in brain tissue and microdialysis perfusates. J. Chromatogr. **614**, 213-220.

Gamache, P., Ryan, E. and Acworth, I. N. (1993c). Analysis of phenolic and flavonoid compounds in juice beverages using HPLC with coulometric array detection. J. Chromatogr. **635**, 143-150.

Gamache, P. H, Ryan, E. and Acworth, I. N. (1994). Analysis of phenolics and flavonoids having potential biological activity: examination of human urine. The Pittsburgh Conference on Analytical Chemistry and Applied Spectroscopy, Chicago, Illinois. Paper No. 253P.

Gamache, P. H., Acworth, I. N., Lynch, M. L. and Matson, W. R. (1995). Coulometric array detection for HPLC in the analysis of juice products. In: Methods to Detect Adulteration of Fruit Juice Beverages. **1**. Nagy, S. and Wade, R. L. (Eds.). Agscience Inc., Auburndale, Florida USA, pp. 120-144.

Godefroy, F., Matson, W. R., Gamache, P. H. and Weil-Fugazza, J. (1990). Simultaneous measurements of tryptophan and its metabolites kynurenine and serotonin in the superficial layers of the spinal dorsal horn. A study in normal and arthritic rats. Brain Res. **526**, 169-172.

Goto, M., Zou, G. and Ishii, D. (1983). Determination of catechols in human serum by micro high performance liquid chromatography with micro precolumn and dual electrochemical detection. J. Chromatogr. **275**, 271-281.

Greaney, M. D, Marshall, D. L., Bailey, B. A. and Acworth, I. N. (1993). Improved method for the routine analysis of acetylcholine release *in vivo*: quantitation in the presence and absence of esterase inhibitor. J. Chromatogr. **622**, 125-135.

Gunasingham, H. and Fleet, B. (1983). Wall-jet electrode in continuous monitoring voltammetry. Anal. Chem. **55**, 1409-1414.

Hepler, B. R. and Purdy, W. C. (1980). The amperometric detection of thyroid hormones following reverse-phase high-performance liquid chromatography. Anal. Chim. Acta **113**, 269-276.

Hill, B. A., Kleiner, H. E., Ryan, E. A, Dekant, W., Dulik, D. M., Monks, T. J. and Lau, S. S. (1993). Identification of multi-S-substituted conjugates of hydroquinone by HPLC-coulometric electrode array analysis and mass spectroscopy. Chem. Res. Toxicol. **6**, 459-469.

Hoogvliet, J. C., Reijn, J. M. and van Bennekom, W. P. (1991). Multichannel amperometric detection system for liquid chromatography and flow injection analysis. Anal. Chem. **63**, 2418-2423.

Horvai, G. and Pungor, E. (1987). Electrochemical detection in liquid chromatography. Chromatogr. **3**, 15-21.

Horvai, G. and Pungor, E. (1989). Electrochemical detectors in HPLC and ion chromatography. Critical Revs. Anal. Chem. **21**, 1-28.

Huang, T., Shoup, R. E. and Kissinger, P. T. (1990). Flow splitting method for using Sepstick® microbore columns with conventional lc pumps - application to microdialysis. Curr. Sep. **10**, 16-19.

Ikeda, R., Hikoya, S., Hikiba, S., Ikeuchi, N., Tezuka, T., Sakaue, N., Shimizu, M. and Miura, S. (1988). Quantification of catecholamine, indolamine and their metabolites in cerebrospinal fluid using HPLC with coulometric electrochemical detection. Jap. J. Psychiat. Neurol. **42**, 675-676.

Ikeda, R., Svendsen, C. N., Shimizu, M. and Miura, S. (1989). Rat brain concentrations of catecholamines, indoleamines, their metabolites and other compounds after microwave irradiation. Jap. J. Psychiatr. Neurol. **43**, 731-732.

Jane, I., McKinnon, A. and Flanagan, R. J. (1985). HPLC analysis of basic drugs on silica columns using non-aqueous ionic eluents. II. Application of UV, fluorescence and electrochemical oxidation detection. J. Chromatogr. **323**, 191-225.

Jinnah, H. A, Langlais, P. J. and Friedmann, T. (1992). Functional analysis of brain dopamine systems in genetic mouse model of Lesch-Nyhan syndrome. J. Pharmacol. Expr. Thera. **268**, 596-607.

Johnson, D. C. and Larochelle, J. (1973). Forced-flow liquid chromatography with a coulometric detector. Talanta **20**, 959-971.

Johnson, D. C. and LaCourse, W. R. (1990). Liquid chromatography with pulsed electrochemical detection at gold and platinum electrodes. Anal. Chem. **62**, *589A-597A*.

Justice, Jr., J. B. (Ed.) (1987). Voltammetry in The Neurosciences: Principles, Methods and Applications. Humana Press, Clifton, New Jersey.

Kabuto, H., Yokoi, I., Suzuki, S., Takei, M. and Mori, A. (1992). Tryptophan and its metabolite levels in the brain of El mice, and the anticonvulsant effect of citalopram. Neurosci. **18**, 193-198.

Kemula, W. (1952). Rocz. Chem. **26**, 281-297.

Kissinger, P. T. (1977). Amperometric and coulometric detectors for high-performance liquid chromatography. Anal. Chem. **49**, 447A-456A.

Kissinger, P. T. (1986). "Coulometric" vs. "amperometric"? (Its the wrong question). Curr. Sep. **7**, 91-92.

Kissinger, P. T. (1989). Biomedical applications of liquid chromatography-electrochemistry. J. Chromatogr. **488**, 31-52.

Kissinger, P. T. (1994). The unijet. Curr. Sep. **13**, 31.

Kissinger, P. T., Refshauge, R., Dreiling, R. and Adams, R. N. (1973). An electrochemical detector for liquid chromatography with picogram sensitivity. Anal. Letts. **6**, 465-477.

Krull, IS., Ding, X-D, Selavka, C. and Nelson, R. (1984) Electrochemical detection in HPLC. L.C. **2**, 214-221.

Krull, IS., Deyl, Z. and Lingeman, H. (1994). General strategies and selection of derivatization reactions for liquid chromatography and capillary electrophoresis. J. Chromatogr. **659**, 1-17.

Langlais, P. J., Mair, R. G., Anderson, C. D. and McEntree, W. J. (1988). Long-lasting changes in regional brain amino acids and monoamines in recovered pyrithiamine treated rats. Neurochem. Res. **13**, 1199-1206.

Langlais, P. J. and Mair, R. G. (1990). Protective effects of the glutamate antagonist MK-801 on pyrithiamine-induced lesions and amino acid changes in rat brain. J. Neurosci. **10**, 1664-1674.

Langlais, P. J., Wardlow, M. L. and Yamamoto, H. (1991). Changes in CSF neurotransmitters in infantile spasms. Pediatric Neurol. 7, 440-445.

Langlais, P. J., Thal, L., Hansen, L., Galasko, D., Alford, M. and Masliah, E. (1993a). Neurotransmitters in basal ganglia and cortex of Alzheimer's disease with and without Lewy bodies. Neurol. **43**, 1927-1934.

Langlais, P. J., Connor, D. and Thal, L. (1993b). Comparison of the effects of a single and combined neurotoxic lesions of the nucleus basalis magnocellularis and dorsal noradrenergic bundle on learning and memory in the rat. Behav. Brain Res. **54**, 81-90.

Lankelma, J. and Poppe, H. (1976). Design and characterization of a coulometric detector with a glassy carbon electrode for high-performance liquid chromatography. J. Chromatogr. **125**, 375-388.

Last, T. A. (1983). Coulostatic voltammetric liquid chromatography detector. Anal. Chem. **55**, 1509-1512.

Lavoie, M. P., Palmatier, M., Gentile, F. T., Kaplan, F. A., Fiore, D. M., Hazlett, T. F., Bell, W. J. and Flanagan, T. R. (1993). Two PC12 pheochromocytoma lines sealed in hollow fiber-based capsules tonically release L-dopa *in vitro*. Cell Transplant. **2**, 163-173.

LeWitt, P. A., Galloway, M. P., Matson, W. R., Milbury, P., McDermott, M., Srivastava, D. K. and Oakes, D. (1992). Markers of dopamine metabolism in Parkinson's disease. Neurol. **42**, 2111-2117.

Lindberg, W., Ohman, J. and Svante, W. (1986). Multivariate resolution of overlapping peaks in liquid chromatography using diode array detection. Anal. Chem. **58**, 299-303.

Louilot, A., Durkin, T., Spampinato, U. and Cador, M. (Eds.) (1994). Monitoring Molecules in Neuroscience. Publi Typ a Gradignan.

Lucot, J., Crampton, G. H., Matson, W. R. and Gamache, P. H. (1989). Cerebrospinal fluid constituents of cat vary with susceptibility to motion sickness. Life Sci. **44**, 1239-1245.

Lunte, C. E., Kissinger, P. T., and Shoup, R. E. (1985). Difference mode detection with thin-layer dual-electrode liquid chromatography/ electrochemistry. Anal. Chem., **57**, 1541-1546.

Mair, R. G., Langlais, P. J., Knoth, R. L. and Rabchenuk, S. A. (1991). Impairment of olfactory, auditory, and spatial serial reversal learning in rats recovered from pyrithiamine-induced thiamine deficiency. Behav. Neurosci. **105**, 360-374.

Makino, Y., Ohta, S., Tasaki, Y., Tachikawa, O., Kashiwasake, M. and Hirobe, M. (1990). A novel and neurotoxic tetrahydroisoquinoline derivative *in vivo*: formation of 1,3-dimethyl-1,2,3,4-tetrahydroisoquinoline, a condensed product of amphetamines, in brain of rats under chronic ethanol treatment. J. Neurochem. **55**, 963-969.

Marsden, C. A., Brazell, M. P. and Maidment, N. T. (1984) An introduction to *in vivo* electrochemistry. In: Measurement of Neurotransmitter Release In Vivo. IBRO Handbook Series, Methods in The Neurosciences, Marsden C. A. (Ed.), John Wiley and Sons, UK, pp. 127-151.

Maruyama, W., Naoi, M., Takahashi, A., Watanabe, H., Konagaya, Y., Mokuno, K., Hasegawa, S. and Nakahara, D. (1992a). The mechanism of perturbation in monoamine metabolism by L-dopa therapy: *in vivo* and *in vitro* studies. J. Neural. Trans. **90**, 183-197.

Maruyama, W., Nakahara, D., Ota, M., Takahashi, T., Takahashi, A., Nagatsu, T. and Naoi, M. (1992b). N-methylation of dopamine-derived 6,7-dihydroxy-1,2,3,4-tetrahydroisoquinoline, (R)-salsolinol, in rat brains: *In vivo* microdialysis study. J. Neurochem. **59**, 395-400.

Maruyama, W., Nakahara, D., Dostert, P., Hashiguchi, H., Ohta, S., Hirobe, M., Takahashi, A., Nagatsu, T. and Naoi, M. (1993a). Selective release of serotonin by endogenous alkaloids, 1-methyl-6,7-dihydroxy-1,2,3,4-tetrahydroisoquinolines, (R)- and (S)-salsolinol, in the rat striatum: *in vivo* microdialysis study. Neurosci. Letts. **149**, 115-118.

Maruyama, W., Takahashi, T., Minami, M., Takahashi, A., Dostert, P., Nagatsu, T. and Naoi, M. (1993b). Cytotoxicity of dopamine-derived 6,7-dihydroxy-1,2,3,4-tetrahydroisoquinolines. Adv. Neurol. **60**, 224-230.

Matson, W. U.S. Patent 4,404,065 (1983).

Matson, W. R., Andrews, R. W., Ball, J., Skinner, D., Vitukevich, R. and Zink, E. W. (1982). A new electrochemical HPLC detector. The Pittsburgh Conference on Analytical Chemistry and Applied Spectroscopy. Atlantic City, New Jersey. Paper No. 565.

Matson, W. R., Langlais, P., Volicer, L., Gamache, P. H., Bird, E. and Mark, K. A. (1984). n-Electrode three-dimensional liquid chromatography with electrochemical detection of neurotransmitters. Clin. Chem. **30**, 1477-1488.

Matson, W. R., Gamache, P. H, Beal, M. F. and Bird, E. D. (1987). EC array sensor concepts and data. Life Sci. **41**, 905-908.

Matson, W. R., Bouckoms, A., Svendsen, C., Beal, M. F. and Bird, E. D. (1990). Generating and controlling multiparameter data bases for biochemical correlates of disorders. In: Basic Clinical and Therapeutic Aspects of Alzheimer's and Parkinson's Disease, **1**. T. Nagatsu (Ed.). Plenum Press, New York, USA, pp. 513-516.

McClintock, S. A. and Purdy, W. C. (1981). A bi-potentiostat for four electrode electrochemical detectors. Anal. Lett. **14** (B10), 791-798.

McClintock, S. A. and Purdy, W. C. (1983). A dual-working electrode electrochemical detector for liquid chromatography. Anal. Chim. Acta **148**, 127-133.

McClintock S. A., and Purdy, W. C. (1986). Developments in electrochemical detection for liquid chromatography. In: Electrochemical detection in medicine and chemistry. Progress in HPLC. **2**. H. Parvez, M. Bastart-Malsot, S. Parvez, T. Nagatsu and G. Carpentier (Eds.), VNU Science Press, Utrecht, The Netherlands, pp. 37-51.

Mecocci, P., MacGarvey, U., Kaufman, A., Koontz, D., Shoffner, J., Wallace, D. and Beal, M. F. (1993). Oxidative damage to mitochondrial DNA shows marked age-dependent increases in human brain. Ann. Neurol. **34**, 609-616.

Mikkelsen, S. R., McClintock, S. A. and Purdy, W. C. (1985). Single pulse coulostatic analysis. Canadian Chemical Conference, 1-5 June, Kingston, Ontario, Canada.

Milbury, P. E., Beal, M. F. and Matson, W. R. (1994). The potential of degenerative disorder diagnosis based on biochemical pattern or profile analysis. The fourth International Conference on Alzheimer's Disease and Related Disorders. Minneapolis, Minnesota.

Mishra, P. K., Burger, R. L., Cullison, J. K., Bowers, M. L., Acworth, I. N. and Jobe, P. C. (1995). Simultaneous measurement of sub-femtomole amounts of norepinephrine, dopamine and serotonin in microdialysis samples using HPLC-EC. 25th Annual Society for Neuroscience Meeting, November 11-16, San Diego, California. Paper No. 447.4.

Mori, A., Suzuki, S. and Kabuto, H. (1991). Genetically higher level of guanine and guanosine in the brain of epileptic El mice. Med. Sci. Res. **19**, 445-446.

Msimanga, H. Z. and Sturrock, P. E. (1990). Application of factor analysis on high-performance liquid chromatography data using a swept-potential electrochemical detection system. Anal Chem. **62**, 2134-2140.

Musch, G., De Smet. M. and Massart, D. L. (1985). Expert system for pharmaceutical analysis. I. Selection of the detection system in HPLC analysis: UV *versus* amperometric detection. J. Chromatogr. **348**, 97-110.

Myers, R. D., and Knott P. J. (Eds.) (1986). Neurochemical Analysis of The Conscious Brain: Voltammetry and Push-Pull Perfusion. Annals of The New York Academy of Sciences. **473**. New York Academy of Sciences, New York, New York.

Naoi, M., Dostert, P., Yoshida, M. and Nagatsu, T. (1993). N-methylated tetrahydroisoquinolines as dopaminergic neurotoxins. Adv. Neuro. **60**, 212-217.

Naoi, M., Maruyama, W., Acworth, I. N., Nakahara, D. and Parvez, H. (1993). Multi-electrode detection systems for determination of neurotransmitters. In: Methods in Neurotransmitter and Neuropeptide Research, Part 1. **11**. Techniques in The Behavioral and Neural Sciences. Parvez, S. H., Naoi, M., Nagatsu, T., and Parvez, S. (Eds.), Elsevier Science Publishers BV, Amsterdam, pp. 1-39.

Nevell, T. G. and Walsh, F. C. (1992). Reference electrodes. Trans. Inst. Met. Finish. **70**, 144-147.

Niwa, O., Tabei, H, Solomon, B. P., Xie, F. and Kissinger, P. T. (1995). Improved detection limit for catecholamines using liquid chromatography-electrochemistry with a carbon interdigitated array microelectrode. J. Chromatogr. **670**, 21-28.

Ogawa, T., Saso, B., Beal, M. F., Swartz, K., Matson, W. R. and Bird, E. D. (1990). Tryptophan metabolites in Parkinson's brain. In: Basic Clinical and Therapeutic Aspects of Alzheimer's and Parkinson's Disease, **1**. T. Nagatsu (Ed.). Plenum Press, New York, USA, pp. 329-332.

Ogawa, T., Matson, W. R., Beal, M. F., Myers, R. H, Bird, E. D., Milbury, P. and Saso, S. (1992). Kynurenine pathway abnormalities in Parkinson's disease. Neurol. **42**, 1702-1706.

Pelligra, R., Norton, R. D, Williamson, R., Leon, H. A. and Matson, W. R., (1992). Rett's syndrome: stimulation of endogenous biogenic amines. Neuropediatr. **23**, 131-137.

Radzik D. M. and Lunte S. M. (1989). Application of liquid chromatography/electrochemistry in pharmaceutical and biochemical analysis: a critical review. Crit. Rev. Anal. Chem. **20**, 317-352.

Reardon, P. A., O'Brien, G. E. and Sturrock, P. E. (1984). A swept-potential electrochemical detector for flow streams. Anal. Chim. Acta. **162**, 175-187.

Reilley, C. N. (1968). Electrochemistry using thin-layer cells. Rev. Pure Appl. Chem. **18**, 137-151.

Reinmuth, W. H. (1962). Theory of coulostatic impulse relaxation. Anal. Chem. **34**, 1272-1276.

Rizzo, V., Melzi D'Eril. G., Achilli, G. and Cellerino, G. P. (1991). Determination of neurochemicals in biological fluids by using an automated high-performance liquid chromatographic system with a coulometric array detector. J. Chromatogr. **536**, 229-236.

Robinson, T. E. (1990). Coulochem Applications: Microdialysates. ESA, Inc.

Robinson, T. E. and Justice Jr J. B. (eds.)(1991*)*. Microdialysis in The Neurosciences. Techniques in The Behavioral and Neural Sciences. 7. Elsevier Science Publishers BV, Amsterdam.

Rocklin, R. D. (1984). Working-electrode materials. L.C. **2**, 588-593.

Rocklin, R. D., Tullsen, T. R. and Marucco, M. G. (1994). Maximizing signal-to-noise ratio in direct current and pulsed amperometric detection. J. Chromatogr. **671**, 109-115.

Rollema, H., Westerink, B. and Drijfhout W. J. (eds.) (1991*)*. Monitoring Molecules in Neuroscience. Proceedings of The 5th International Conference on *In Vivo* Methods, Krips Repro.

Roston, D. A. and Kissinger, P. T. (1982). Series dual-electrode detector for liquid chromatography. Anal. Chem. **54**, 429-434.

Rucki, R. J., (1980). Electrochemical detectors for flowing liquid systems. Talanta. **27**, 147-156.

Ryan, E., Gamache, P. H. and Acworth, I. N. (1992). A single method for the measurement of different drug classes: the use of HPLC with coulometric array detection. The Pittsburgh Conference on Analytical Chemistry and Applied Spectroscopy, New Orleans, Louisiana. Paper No. 473P.

Ryan, E., Acworth, I. N., Gleich, G. J, Gamache, P. H. and Mayeno, A. N. (1993). Eosinophilia-myalgia syndrome: multi-electrode analysis of control and contaminated tryptophan samples. 23rd Annual Society for Neuroscience Meeting, November 7-12, Washington D.C. Paper No. 771.9.

Ryan, E., Dickmann, R., Acworth, I. N. and Gamache, P. H. (1994). Measurement of vitamins and endogenous anti-oxidants in food and other biological matrices. The Pittsburgh Conference on Analytical Chemistry and Applied Spectroscopy, Chicago, Illinois. Paper No. 254P.

Saitoh, R. and Suzuki, H. (1979). Multiwavelength detection for liquid chromatography with a repeat-scanning ultraviolet-visible spectrophotometer. Anal. Chem., **51**, 1683-1687.

Samuelsson, R., O'Dea, J. and Osteryoung, J. (1980). Rapid scan square wave voltammetric detector for high-performance liquid chromatography. Anal. Chem. **52**, 2215-2216.

Satoh, J., Gallyas, F., Endoh, M., Yamamura, T., Kunishita, T. and Tabira, T. (1992). Coexistance of cholinergic, catecholaminergic, serotoninergic, and glutamatergic neurotransmitter markers in mouse clonal hybrid neurons derived from the septal region. Neurosci. Res. **32**, 127-137.

Schieffer, G. W. (1980). Dual coulometric-amperometric cells for increasing the selectivity of electrochemical detection in high-performance liquid chromatography. Anal. Chem. **52**, 1994-1998.

Shimizu, T., Mihara, M. and Takeda, N. (1991a). High-performance liquid chromatography of biogenic amines in the corpus cardiacum of the American cockroach, *Periplaneta americana*. J. Chromatogr. **539**, 193-197.

Shimizu, T. and Takeda, N. (1991b). Biogenic amine levels in the haemolymph of the cabbage armyworm larvae (*Mamestra brassicae*) following injection of octopaminergic insecticides. Z. Naturforsch. **46C**, 127-132.

Shimizu, T., Yazawa, M. and Takeda, N. (1992). Aromatic amino acids in the leaves of the *Morus alba.* Sericologia **32**, 633-639.

Shimizu, T., Tatsuki, Y. and Takeda, N. (1993). Aromatic amino acids in the venom of the braconid parasitoid, *Apanteles kariyai.* Z. Naturforsch. **48C**., 108-109.

Siuciak, J. A., Gamache, P. H. and Dubocovich, M. L. (1992). Monoamines and their precursors and metabolites in the chicken brain, pineal and retina: regional distribution and day/night variations. J. Neurochem. **58**, 722-729.

Smyth, W. F. (1987). Voltammetric determination of molecules of biological, environmental, and pharmaceutical importance. Crit. Rev. Anal. Chem. **18**, 155-208.

Sparks, T. C. and Geng, C. (1992). Analysis of biogenic amines in the central nervous system of the tobacco hornworm by high-performance liquid chromatography with 16-sensor electrochemical detection. Anal. Biochem. **205**, 319-325.

Stulik, K. and Pacakova, V. (1981). Electrochemical detection techniques in high-performance liquid chromatography. J. Electroanal. Chem. **129**, 1-24.

Suzuki, S. and Mori, A. (1992). Regional distribution of tyrosine, tryptophan, and their metabolites in the brain of epileptic El mice. Neurochem. Res. **17**, 693-698.

Svendsen, C. N. (1993). Multi-electrode array detectors in high-performance liquid chromatography: A new dimension in electrochemical analysis. Analyst **118**, 123-129.

Swartz, K. J., Matson, W. R., MacGavery, U., Ryan, E. A and Beal, M. F. (1990). Measurement of kynurenic acid in mammalian brain extracts and cerebrospinal fluid by high-performance liquid chromatography with fluorometric and coulometric electrode array detection. Anal. Biochem. **185**, 363-376.

Takaragi, S., Horivichi, T., Miyahira, H., Hamano, I., Endo, A., Egawa, B., Yamamoto, H., Meguro, T. and Yamada, K. (1989). Cerebrospinal fluid neurotransmitters of the newborn with neonatal asphyxia. Acta Paediatr. Jpn. **31**, 745-747.

Takata, Y. and Muto, G. (1973). Flow coulometric detector for liquid chromatography. Anal. Chem. **45**, 1864-1868.

Takata, Y. and Fujita, K. (1975). Application of a coulometric detector to rapid ion-exchange chromatography. J. Chromatogr. **108**, 255-263.

Takeda, N. (1992a). Biogenic monoamine system detected simultaneously in the neural complex of the Ascidian, *Ciona intestinalis*. Comp. Biochem. Physiol. **103C**, 489-493.

Takeda, N. (1992b). Biogenic monoamine levels in the central nervous system of the sea hair, *Apalysia kurodai*. Comp. Biochem. Physiol. **103C**, 511-519.

Takeda, H., Shibuya, T. and Matsumiya, T. (1989). Pharmacobio-Dyn. **12**, s-3.

Takeda, N. and Svendsen, C. L. (1991). Monoamine concentrations in *Hydra magnipapillata*. Hydrobiologia **216/217**, 549-554.

Takeda, N., Takaoka, H., Shimizu, T., Yazawa, M. and Yagi, S. (1991). Biogenic amine levels in the central nervous system and haemolymph of the silkworm, *Bombyx mori*. Comp. Biochem. Physiol. **99C**, 588-593.

Tallman, D. E. and Peterson, S. L.(1990). Electroanalysis **2**, 499-506.

Taylor, L. R. and Johnson, D. C. (1974). Determination of antimony using force-flow liquid chromatography with a coulometric detector. Anal. Chem. **46**, 262-266.

Tjaden, U. R., Lankelma, J., Poppe, H. and Muusze, R. G. (1976). Anodic coulometric detection with a glassy carbon electrode in combination with reversed-phase high-performance liquid chromatography. Determination of blood levels of perphenazine and fluphenazine. J. Chromatogr. **125**, 275-286.

Tohgi, H., Abe, T., Takahashi, S., Nozaki, Y. and Kikuchi, T. (1991a). Concentrations of tyrosine, L-dopa, dopamine and 3-O-methyldopa in the cerebrospinal fluid of Parkinson's disease. Neurosci. Letts. **127**, 212-214.

Tohgi, H., Abe, T., Takahashi, S., Nozaki, Y. and Kikuchi, T. (1991b). The significance of 3-O-methyldopa concentrations in the cerebrospinal fluid in the pathogenesis of wearing-off phenomenon in Parkinson's disease. Neurosci. Letts. **132**, 19-22.

Trubey, R. K. and Nieman, T A. (1986). Rapid scanning coulostatic chromatography/electrochemistry in dual electrode cells. Anal. Chem. **58**, 2549-2554.

Turiak, G. and Volicer, L. (1994). Stability of *o*-phthalaldehyde-sulfite derivatives of amino acids and their methyl esters: electrochemical and chromatographic properties. J. Chromatogr. **668**, 323-329.

Turk, D. J. and Blank, C. L. (1989). Development of a neurobiological analyzer for the simultaneous determination of multiple components in tissue, CSF, blood and urine samples using liquid chromatography with multiple, amperometric electrochemical detection. 19th Annual Society for Neuroscience Meeting, Oct. 29th - Nov. 3rd, Phoenix, Arizona. Paper No. 491.14

Uemura, Y., Miller, J., Matson, W. R. and Beal, M. F. (1991). Neurochemical analysis of focal ischemia in rats. Stroke **22**, 1548-1553.

Ungerstedt, U. (1984). Measurement of neurotransmitter release by intracranial dialysis. In: Measurement of Neurotransmitter Release *In Vivo*. IBRO Handbook Series, Methods in The Neurosciences, Marsden, C. A. (Ed.). John Wiley and Sons, New York, USA, pp. 81-105.

van Leeuwan, H. P. (1982). Coulostatic pulse techniques. In: Electroanalytical Chemistry, A.J. Bard (Ed.). Marcel Dekker, New York, USA, pp. 171-188.

Varadi, M., Feher, Zs. and Pungor, E. (1974). Determination of purine bases by means of chromatovoltammetry. J. Chromatogr. **90**, 259-265.

Vecsei, L., Miller, J., MacGarvey, U., and Beal, M. F. (1992). Effects of kynurenic acid and probenecid on plasma and brain tissue concentrations of kynurenic acid. Neurodegeneration **1**, 17-26.

Volicer, L., Langlais, P. J., Matson, W. R., Mark, K. A. and Gamache, P. H. (1985). Serotoninergic system in dementia of the Alzheimer type. Arch. Neuro. **42**, 1158-1161.

Volicer, L., Matson, W. R., Schnepper, P., Gamache, P., Katz, D. I. and Wolfe, N. (1990). Electrochemical examination of major known components in cerebrospinal fluid by three dimensional liquid chromatography. Brain Res., **526**, 169-172.

Vriend, J., Alexiuk, N. A. M., Green-Johnson, J. and Ryan, E. (1993). Determination of amino acids and monoamine neurotransmitters in caudate nucleus of seizure-resistant and seizure-prone BALB/c mice. J. Neurochem. **60**, 1300-1307.

Weber, S. G. (1986). Detection based on electrical and electrochemical measurements. In: Detector for Liquid Chromatography.. Yeung, S. (Ed.). **7**. Chemical Analysis. A Series of Monographs on Analytical Chemistry and its Application. Eldring, R. J., and Wine-Fordner, J. D. (Series Eds.). John Wiley and Sons, New York, USA, pp. 229-289.

Weber, S. G. and Purdy, W. C. (1982). Electrochemical detection with a regenerative flow cell in liquid chromatography. Anal. Chem. **54**, 1757-1764.

Welch, R. W., Acworth, I. N. and Levine, M. (1993) Coulometric electrochemical detection of hydroxyproline using 7-chloro-4-nitrobenzo-2-oxa-1,3-diazole. Anal. Biochem. **210**, 199-205.

White, J. G., St. Claire, R. L. and Jorgenson, J. W. (1986). Scanning on-column voltammetric detector for open-tubular liquid chromatography. Anal. Chem. **58**, 293-298.

Wolff, J. A, Fisher, L. J., Xu, L., Jinnah, H. A., Langlais, P. J., Iuvone, P. M., O'Malley, K. L., Rosenberg, M. B., Shimohama, S., Friedmann, T. and Gage, F. H. (1989). Grafting fibroblasts genetically modified to produce L-dopa in a rat model of Parkinson disease. Proc. Natl. Acad. Sci. **86**, 9011-9014.

Yamada, J. and Matsuda, H. (1973). Limiting diffusion currents in hydrodynamic voltammetry. J. Electroanal. Chem. **44**, 189-198.

Yamamoto, H. (1991a). Changes in CSF neurotransmitters during the first year of life. Pediatr. Neurol. **7**, 406-410.

Yamamoto, H. (1991b). Studies on CSF tryptophan metabolism in infantile spasms. Pediatr. Neurol. **7**, 411-414.

Yost, R., Stoveken, J. and McClean, W. (1977). Positive peak identification in liquid chromatography using absorbance ratioing with a variable-wavelength spectrophotometric detector. J. Chromatogr. **134**, 73-82.

Progress in HPLC-HPCE, Vol. 6, pp. 51–74
I.N. Acworth *et al.* (Eds)

QUALITATIVE UTILITY OF COULOMETRIC ARRAY DETECTION: THEORETICAL AND APPLIED ASPECTS.

JOHN K. CULLISON and PAUL H. GAMACHE

ESA, Inc., 22 Alpha Road, Chelmsford, MA 01824-4171

INTRODUCTION

Some of the more important performance characteristics of any separation and detection scheme include: quantitative accuracy and sensitivity, the qualitative utility of the data, and the number of analytes measured per unit time. The versatility of HPLC allows optimization of many of these parameters through choice of detector, operating conditions, mobile and stationary phases. However, even with highly efficient columns and optimal separation conditions, chromatographic resolution and detector selectivity are often insufficient for many analyses and do not alone provide adequate qualitative information. Sample pre-purification (e.g. solid and liquid phase extraction), derivatization, and pre-concentration are often used to increase analytical sensitivity and selectivity. Detectors that provide a multi-dimensional spectrum or fingerprint of each analyte based on their physicochemical properties are very useful in providing qualitative data. Mass spectrometers, photodiode arrays (PDA) and electrochemical coulometric arrays are key examples of detectors which can be used on-line for these purposes.

Liquid chromatographic array detection is popular because of its high informational content and its ability to resolve co-eluting compounds in complex matrices. For example, array technology can be used in pharmaceutical drug purity and stability testing. Drug impurities and breakdown products can often have similar retention times to their parent compounds. Wavelength ratioing is often used with PDA detectors (White and Catterick, 1987). In wavelength ratioing one chooses an absorbance maximum of the analyte and another wavelength where the analyte also absorbs, but where many compounds also absorb as well (e.g. 280nm). The ratio of response at these two wavelengths will have a specific value for 100% purity, and will deviate from this value if other analytes are present.

Another type of chromatographic array detector with high popularity for specific applications is the coulometric electrode array (Matson *et al.*, 1984). This detector has a series of post-column flow-through porous graphite sensors. Each sensor is set to a different potential. The resulting chromatogram is a multi-dimensional plot of current versus time versus potential. Coulometric sensors are unique because each sensor performs a complete electrolysis of the analyte. Because of this property, an analyte that reacts at an upstream electrode can be resolved from a co-eluting analyte that reacts at a downstream electrode. The resolution of these co-eluting compounds will be in the potential domain of the multi-dimensional chromatogram. These detector arrays have the high sensitivity traditionally associated with electrochemical detectors (ECDs) but they can also provide information concerning peak purity.

Most commercially available (e.g. thin-layer amperometric) ECDs are limited to a single potential measurement which greatly diminishes the informational content of the data. It is not practical to make a thin-layer amperometric array. The highly inefficient thin-layer detectors would have unreacted analyte “spill over” to downstream electrodes. This makes resolution in the potential domain difficult. Moreover, the thin-layer electrodes are often impractical for multi-electrode applications since long stabilization times and frequent maintenance are required. The reference electrodes used in the coulometric array detectors are maintenance free. The more traditional "wet" Ag/AgCl reference electrodes commonly used with thin-layer detectors require significant maintenance and are not practical for use under pressure therefore further limiting their practical utility in array configurations (see Chapter 1, this volume, for further details).

The coulometric electrode array detector contains up to 16 highly stable and efficient sensors (or channels) in series, each having a separate applied potential. In one chromatographic run, analytes can be distinguished on the basis of their thermodynamic oxidation/reduction (redox) potentials as well as their retention time. Peak ratioing can also be used with coulometric array, allowing one to confirm or assess the degree of resolution of co-eluting analytes that have similar formal potentials (see below). On-line voltammograms generated with coulometric arrays can often aid in the assignment of structural identity to unknown peaks.

The purpose of this paper is to provide an overview of fundamental electrochemical concepts as applied to liquid chromatography and in particular coulometric arrays. The major focus will be on obtaining qualitative information from the chromatograms, the factors that influence the potential of a redox couple, and on how one applies the ratioing technique with coulometric arrays.

ELECTROCHEMICAL DETECTION IN LIQUID CHROMATOGRAPHY

Since the work of Wiktor Kemula in 1952 and Ralph Adams in 1973 on applying carbon based electrodes for oxidative amperometric detection, electrochemistry has become one of the more popular detection methods in liquid chromatography. ECDs are relatively inexpensive, highly selective and extremely sensitive. Picomolar detection limits are currently possible, with linear dynamic ranges of over 6 orders of magnitude. There have been a number of excellent electrochemical reviews in the past (Štulík and Pacáková, 1982; Horvai and Pungor, 1989; Weber, 1986). Because of the historical prevalence of amperometric cells, these reviews have concentrated on the inefficient amperometric ECDs. This paper will focus more heavily on coulometric array detectors.

Electrochemical thermodynamic theory

To understand the operation of coulometric arrays it is necessary to understand a few of the basic principals of electrochemistry. Consider the reaction:

$$\text{RED} \leftrightarrow \text{OX} + \text{ne}^- \qquad (1)$$

(where: RED is the reduced analyte species, OX is the oxidized species, and ne^- is the number of electrons taking part in the reaction). If the rate of electron transfer in both directions is rapid with respect to the time scale of the measurement, the extent to which the analyte is oxidized or reduced can be expressed by the Nernst equation:

$$E = E^{0} - RT/nF \ln[(a_{RED})/(a_{OX})] \qquad (2)$$

(Where: E = the equilibrium potential in volts; E^o = the standard potential for the redox reaction. [The standard potential is the potential that would be measured if the *activities* of all reactants and products were unity. This is the potential provided in most published tables]; R = gas constant [8.31 VC Kelvin^{-1} mol^{-1}]; F = Faraday's constant [9.65 x 10^4 Celsius mol^{-1}]; n = number of electrons transferred in the redox reaction; a_{RED} = activity of the reduced analyte; a_{OX} = activity of the oxidized analyte).

At 25°C the Nernst equation can be expressed as:

$$E = E^{0} - 0.0592/n \log[(a_{RED})/(a_{OX})] \qquad (3)$$

Most HPLC-ECDs use a 3 electrode configuration: a reference; working; and counter (auxiliary) electrode. The reference electrode has a defined potential with respect to the mobile phase solution. A potential difference is then typically maintained at a constant value between the working and reference electrodes. This potential difference is the driving force for oxidation or reduction of the analyte. The current which flows from the counter electrode is used to maintain this potential difference. Therefore, as analyte diffuses to the working electrode surface enough current will flow in *both* directions to adjust the activity of the reacting species *at the electrode surface* to the thermodynamic equilibrium represented in the Nernst equation. In Figure 1, a theoretical plot of current due to oxidation or reduction of analyte as a function of the applied working electrode potential shows an exponential curve. This curve is essentially a representation of the relationship described in the Nernst equation (3).

Different chemical structures will give rise to different voltammetric curves. Primarily these differences are a function of the relative ease of oxidation and reduction of analytes which results in differences in the position of each analyte along the potential axis as shown in Figure 2. As described below, the experimentally (i.e. from HPLC-ECD) derived relationship between applied potential and analyte response (current) is called an hydrodynamic voltammogram (HDV). These plots which are closely related to the fundamental Nernstian relationship are the highly valuable fingerprints which are used for qualitative analysis with coulometric arrays. These differences in the voltammetric curves are one of the reasons for the high selectivity obtained with HPLC-ECD.

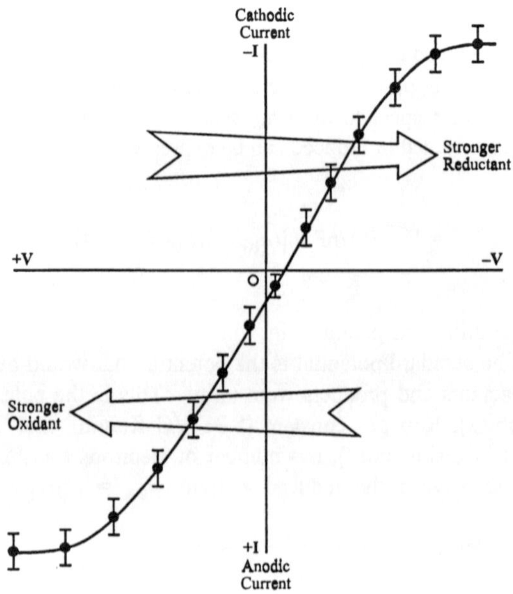

Figure 1. A theoretical plot of the current-voltage relationship (CV) or hydrodynamic voltammogram of a substance showing both oxidative and reductive behavior. By convention the lower left quadrant is the oxidation quadrant.

Analyte detection results from the changes in current due to the electrochemical reaction of the analyte at the surface of the working electrode. For a coulometric flow-through electrode, electrolysis takes place so that *all* of the analyte is adjusted to the Nernstian ratio expressed in the above equation. However, for an amperometric (thin-layer) electrode, only the analyte next to the surface (typically a very small percentage of the total) will be at the Nernstian equilibrium. The use of two or more post-column amperometric electrodes therefore provides limited selectivity and qualitative usefulness due to analyte "spill over" from one thin-layer electrode to the other.

The concept of activity was introduced in Equations 2 and 3 in order to stress that experimental variables that affect the analytes' activities will also influence their redox potentials (see below). It is, however, difficult to think in terms of activities and more common to think in terms of concentration because most of the time activity coefficients are unknown. For this reason it is often more convenient to think in terms of the *formal potential*, $E^{o'}$. This is the potential at which the *concentrations* of the reactants and products are such that the ratio [RED]/[OX] = 1. In other words,

$$E^{o'} = E^{o} - 0.0592/n \log[(\gamma_{RED})/(\gamma_{OX})] \qquad (4)$$

(where: γ = activity coefficient)

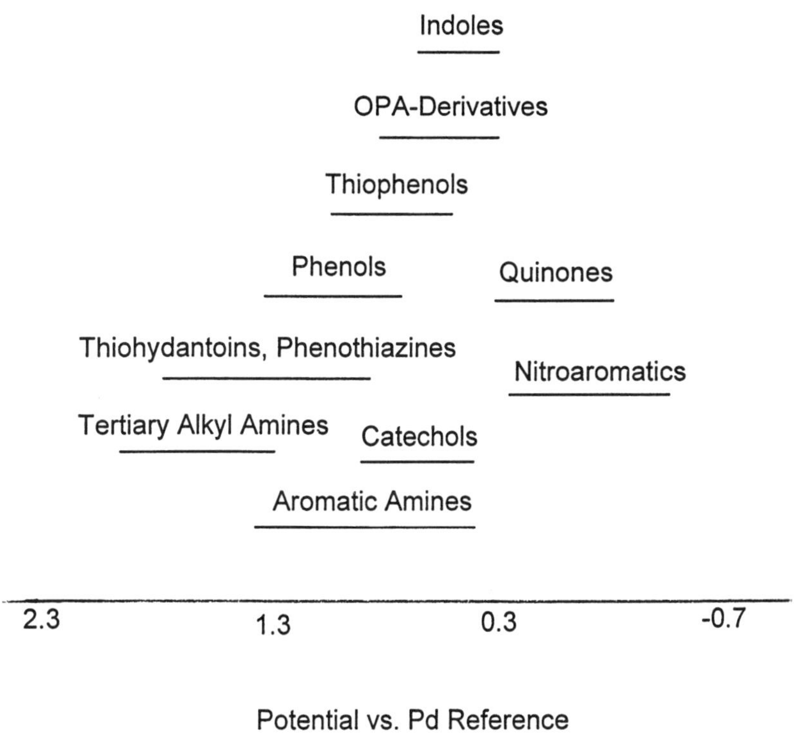

Figure 2. The relative ease of oxidation or reduction of analytes as a function of potential (presented here in Volts) is dependent upon the structure of the analyte being examined.

Using the formal potential, the Nernst equation can then be expressed in the popular form:

$$E = E^{o'} - 0.0592/n \log[RED]/[OX] \qquad (5)$$

Factors affecting an analyte's formal potential

The discussion above should have made it clear that anything that influences the activity of an analyte will in fact influence the apparent formal redox potential, $E^{o'}$. The formal potential is the redox potential under specified conditions such as pH, ionic strength, and concentration of complexing agents. In biochemistry the formal potential refers to the redox potential at pH = 7. One has to be careful when using formal potentials and understand under what conditions the stated formal potential applies because very often the standard conditions are very different from those used for HPLC analysis. For example, while standard conditions relate to a temperature of 25°C, chromatographic temperature varies from application to application. It should therefore be kept in mind that the apparent redox potential or apparent formal potential will also change with temperature.

Analytes that participate in acid base chemistry and have protons involved in their redox reactions will have apparent formal potentials that are also pH dependent. When the electrochemical reaction involves a proton, or when the reactants and products are acids or bases, the redox potential is pH dependent (Harris, 1982). Consider the reaction:

$$RED \leftrightarrow OX + 2e^{-} + 2H^{+} \quad (6)$$

In this case the Nernst equation can be written:

$$E = E^{o'} - 0.0592/n \log[RED]/[OX][H^{+}]^{2} \quad (7)$$

or for n = 2

$$E = E^{o'} - 0.0592\,pH - 0.0592/n \log[RED]/[OX] \quad (8)$$

From equation (8) it is evident that a shift in 1 pH unit will shift the redox potential by 59.2mV. In HPLC-ECD, the working electrode is typically set at a potential just beyond the current-voltage plateau for the analyte of interest (see Figures 1 and 3). Too low of a potential will result in low response while too high of a potential setting will result in increased background noise and decreased selectivity. Importantly, any shift in the analyte's current voltage curve due to experimental variables will result in altered response and signal to noise ratio. Changes in mobile phase pH which shift the redox behavior of an analyte can therefore significantly alter the response.

Other variables such as ionic strength of the mobile phase, temperature, organic modifiers, and possibly ion-pairing reagents may also slightly shift the apparent formal potential. These changes will be minor (a few millivolts at the most), but these effects may be significant when using HPLC-ECD methods. Practically speaking these variables will typically have more effect on the chromatography than on the redox potential of the analyte. For example, a change in 1 pH unit might shift the uncompensated redox potential by 59mV, but if the pH of the mobile phase is near the pK of the analyte, retention may be drastically altered. However, in order to develop a rugged HPLC-ECD method it is very important to closely consider the variables that affect both chromatographic and detector performance. Important variables are: a properly chosen buffer for the pH of the analysis, a well controlled temperature, and a mobile phase that is made reproducibly and using the highest grade reagents. One other variable that can be important is whether one degasses and whether one degasses on-line or off-line. As described below, oxygen in the mobile phase can shift the reference potential and thus shift the apparent redox potential of the analyte. If one degasses off-line one could observe changes in the response over time due to temporal changes in the oxygen level in the mobile phase. We recommend degassing on-line to keep the oxygen level as constant and reproducible as possible.

Experimental variables that affect the voltammetric behavior of analytes can therefore alter response at a given electrode. With the coulometric array these effects must also be considered in the context of qualitative analysis and these concepts will be discussed below. It is important to mention other considerations including the effects of variables on baseline currents and background noise. For example, shifts in the redox potential of trace contaminants in the mobile phase can cause changes in baseline currents and background noise. For further discussion of these effects see Chapter 1, in this volume.

The coulometric detector has a patented maintenance free palladium (Pd) quasi-reference electrode. It has a potential of approximately 0.25 to 0.35V vs. Ag/AgCl or about 0.45-0.55V vs. NHE (the normal hydrogen electrode). It is important to keep in mind that many of the ECD methods in literature use a Ag/AgCl reference electrode. Remember that the reference electrode provides a defined potential against which the working electrode is set. If one knows the redox potential versus Ag/AgCl from literature, then one can subtract about 300mV from this value to get "a ball park" redox potential versus Pd. This approximate detector potential setting should then be optimized for the particular HPLC conditions (see discussion of HDVs below). The ESA Pd reference electrode has the unique property of being pH sensitive. The reference electrode shifts about 60 mV/pH unit. The changes in the redox potential of an analyte as a result of a change in pH can therefore be compensated for by an accompanying shift in the reference potential. This novel reference electrode used in the coulometric array cells can thus minimize the effects of pH variation on HPLC-ECD. This approach overcomes significant shifts in analyte response due to pH shifts that are common with thin-layer detectors. The potential of the Pd wire has been ascribed to be a mixed potential effect Hoare, 1964: Vracar, *et al.*, 1987; Shao, *et al.*, 1994). The details are complex but the potential is sensitive to oxygen levels. Its sensitivity to oxygen in the mobile phase is important. One should try to keep the oxygen level in the mobile phase constant by degassing the mobile phase on-line.

Diffusion or kinetic control of electrochemical reactions

An electrochemical reaction can be either diffusion controlled or under kinetic limitations. If the electrochemical reaction is diffusion controlled, then the analyte will adjust to the Nernstian equilibrium as fast as it can diffuse to the electrode surface. The reaction is said to be under kinetic limitations if the rate of electron transfer is slower or comparable to the rate of diffusion. Reactions under kinetic limitations are said to be quasireversible or irreversible. The rate of electron transfer is exponentially related to the overpotential (Bard and Faulkner, 1980). The overpotential is how far past the formal potential one is. Therefore, an analyte under kinetic limitations can be driven at diffusion controlled rates by applying higher overpotentials. For additional information on kinetic limitations at working electrode surfaces see Rocklin (1989).

Another way to overcome kinetic limitations is to increase the length of time or the number of encounters an analyte has with an electrode surface. Porous graphite electrodes allow one to overcome kinetic limitations because of the large surface area to volume ratio of the electrode (Matson *et al.*, 1984). The large surface area to volume ratio limits the diffusional space of the analyte within the porous electrode thus increasing the number of encounters an analyte has with the electrode surface. These phenomena allow analyte reactions which are quasireversible to irreversible at amperometric electrodes to appear reversible at coulometric electrodes. In Rocklin's (Rocklin, 1989) discussion of working electrode materials he points out that increasing the surface area of the working electrode to achieve coulometric efficiency results in poorer signal to noise ratios due to a more rapidly increasing noise component. While increasing surface area of a thin-layer electrode does appear to show this effect, this is an oversimplification of the relationship between signal and noise obtained with a porous coulometric electrode. While there are many factors that determine the signal to noise ratio of a particular analysis, the coulometric design routinely provides better signal to noise performance than thin-layer detectors.

QUALITATIVE ANALYSIS OF ELECTROACTIVE ANALYTES

For many years analytical chemists have used electrochemistry for qualitative analysis. While there are many different ways one can use electrochemistry to glean information about a particular analyte solution or system, the discussion below will focus on two of the most popular approaches - cyclic voltammetry (CV) and hydrodynamic voltammetry (HDV). These approaches both examine the relationship between the potential of a cell and redox reactions of analytes. One can often identify a molecule simply by the potential at which that molecule is either oxidized or reduced. The potential of oxidation or reduction is related to the molecule's standard redox potential or formal potential. Every electroactive compound has a unique redox potential. The redox potential depends on the type of atoms and the structure of those atoms within a compound or more specifically on the energy of the highest occupied molecular orbital and the lowest unoccupied molecular orbital (Bard and Faulkner, 1980). Figure 2 illustrates the potential windows for the oxidation of various functional groups of a molecule. One can see that electrochemistry can provide selectivity over a wide range of potentials and that the presence of specific functional groups on a molecule can often be ascertained from the redox potential of an unknown.

The potential windows shown in Figure 2 are the approximate standard potentials of the functional groups. These molecules may be more or less kinetically hindered at different electrode surfaces. In other words, one may have to apply a larger over-potential at a carbon electrode than at a platinum electrode for certain applications. It is advised that the user perform an HDV experiment (see below) on the analytes of interest whenever a new method is being developed. Table 1 shows representative potentials of various functional groups from literature using coulometric array detection. These experimental data under a given set of chromatographic conditions can often be related back to the formal potentials as shown in Figure 1. The presence of electron withdrawing or donating groups and resonance stabilization will also shift the oxidation or reduction potential of analogous structures. Aromatic structures capable of charge delocalization often facilitate redox reactions. For example, the geometric isomers of dihydroxybenzene in Table 1 show large differences in oxidation potential between *meta* and both the *ortho* and *para* isomers. The latter forms are more capable of charge delocalization and are therefore more easily oxidized.

Table 1.
Oxidation and reduction potentials with coulometric array detection

Structural Class	**Functional Group**	**Typical Dominant Oxidation or Reduction Potential (mV vs Pd)**	**Examples**
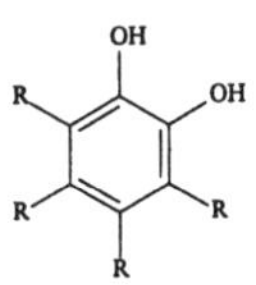	Catechol *o*-Dihydroxybenzene	+120	Catecholamines Catechins 2,3-DHBA Chlorogenic acid
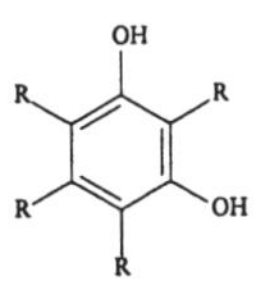	*m*-Dihydroxybenzene	+500	Anthocyanins Genistein
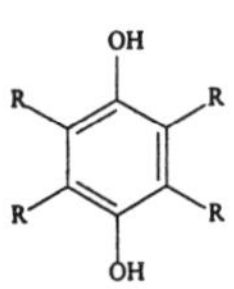	*p*-Dihydroxybenzene	+120	Hydroquinone 2,5- DHBA (dihydroxybenzoic acid)
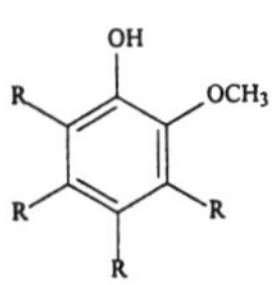	Methoxycatechol	+400	MHPG HVA Metanephrines Vanillic acid
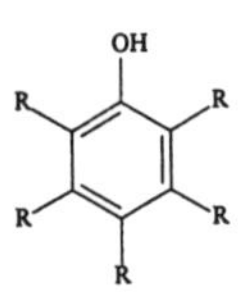	Phenol	+540	Tyrosine Salicylate Chlorophenols

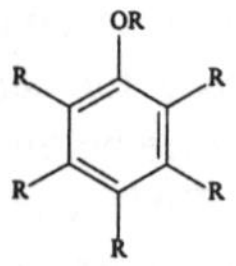	Phenyl alkyl ether	+700	Methoxylated flavones Propranolol
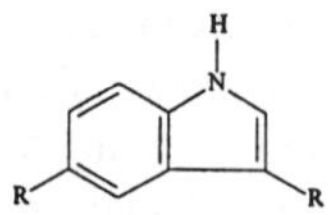	Indole	+540	Tryptophan Melatonin Pindolol
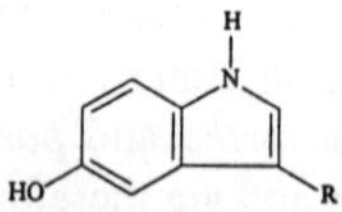	Hydroxyindole	+240 and +540	Serotonin 5HIAA
R—S—R	Thiol Thioether	+700	Glutathione Homocysteine Methionine
R—S—S—R	Disulfide	+900	Cystine Oxidized glutathione
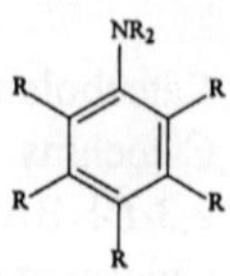	Aromatic amine	+600	*p*Aminobenzoic acid Kynurenine Phenylurea herbicides
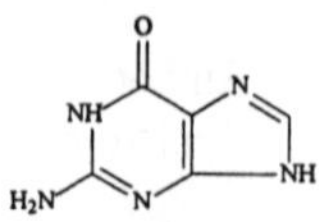	Purine	+540	Guanine Xanthine
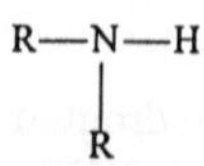	Secondary amine	+1000	Nortriptyline
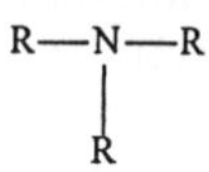	Tertiary amine	+540	Imipramine Desipramine
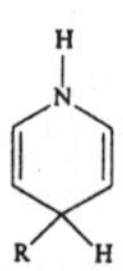	Secondary amine	+330	Benidipine
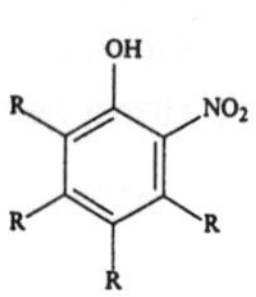	Nitrophenol	-500 and +600	Nitrotyrosine

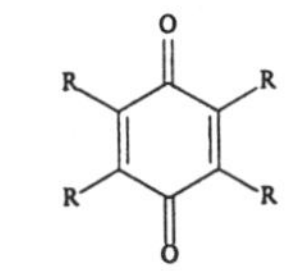	Quinone	-1000	Vitamin K1 Coenzyme Q10
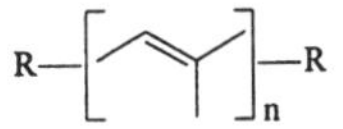	Polyenes	400 700	Carotenoids Retinoids

Cyclic Voltammetry

Traditionally, scanning methods such as cyclic voltammetry (CV) have been used to examine the redox characteristics of a particular compound (Sawyer and Roberts, 1974) CV experiments are normally performed off-line from a separation technique (e.g. HPLC) in a static electrochemical cell. The potential of the working electrode is scanned anodically and cathodically in a cyclic fashion and the current is plotted as a function of the potential. The advantages of CV include the ability to easily study a compound's redox characteristics in a variety of solvents (i.e. different pH, ionic and organic content etc.) as well as to examine redox behavior of a compound using different working electrode surfaces. CV experiments are particularly useful for studying the mechanisms of an electrochemical reaction. In addition to the qualitative utility of CV data, these experiments are commonly performed as a prelude to HPLC-ECD method development (i.e. for selection of working electrode material, mobile phase conditions, and approximate working potentials). The disadvantages of CV analyses are largely related to the use of a constantly changing potential and the buildup of reaction products (from the analyte, solution, and interfering solutes) near the electrode surface. These background effects of CV limit the sensitivity and selectivity of the technique (Sawyer and Roberts, 1974). For these reasons, CV experiments are typically performed using relatively high concentrations of analyte in pure solutions. A related technique using voltammetric scanning on-line with HPLC suffers from similar signal to noise limitations due to capacitive current changes when the electrode potential is scanned (Weber and Long, 1988).

Hydrodynamic Voltammetry

HPLC-EC detection is most commonly performed using a single channel cell at a fixed potential. A plot of the response (or current) for a particular analyte as a function of potential is commonly called an hydrodynamic voltammogram (HDV). An HDV experiment can be performed at much higher sensitivities than scanning methods since capacitive changes are minimized. A series of chromatograms are generated, each at a different fixed potential, over a range of potential settings while keeping the chromatographic conditions and mass of analyte injected constant. The response obtained for the analyte is then plotted as a function of the applied potential. An HDV plot is shown in Figure 1. Each point in this HDV represents one series of chromatographic runs where the potential of the detector was set to a specific value.

The shape and position of an HDV wave is a function of the electrochemical thermodynamic and kinetic behavior of the analyte under hydrodynamic conditions. The half wave potential ($E_{1/2}$) is commonly related to the thermodynamic formal potential of the analyte while the slope of the curve is primarily related to the kinetics of the electrochemical reaction. The potential where the current plateau is reached (E_{max}) is the value where the current is limited by analyte diffusion to the electrode surface. For electrochemical reactions with rapid kinetics (e.g. catechols, phenols) a very sharp HDV curve will be obtained, and the E_{max} will only be slightly past the $E_{1/2}$ potential. For analytes with slow electron transfer kinetics (e.g. macromolecules), E_{max} will be significantly different from $E_{1/2}$ and the HDV wave will appear more drawn out (Figure 3).

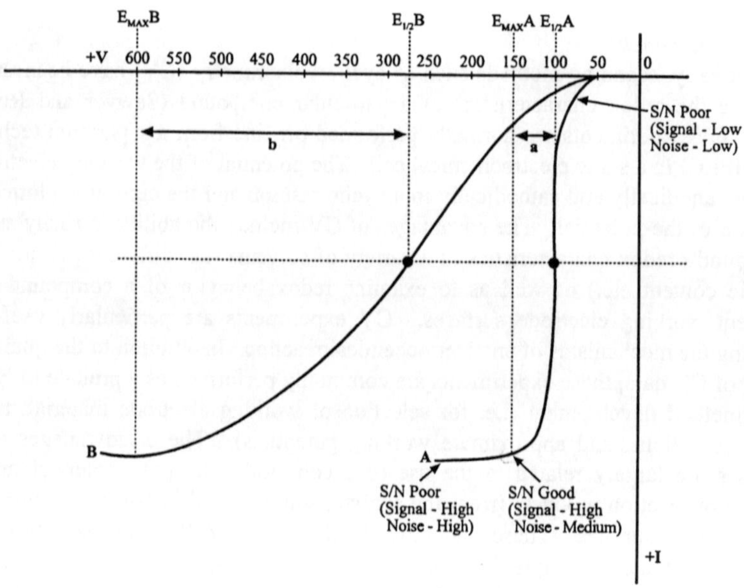

Figure 3. A lot of information is contained in the shape and position of the HDV. The HDV wave is a function of the electrochemical thermodynamic and kinetic behavior of an analyte under hydrodynamic conditions. The correct choice of applied potential will ensure a maximal signal to noise ratio.

For the sharper HDV shown in Figure 3, initial response (E_0) was obtained at +50mV reaching a plateau at +150mV (E_{max}). This information is very useful for method optimization. The potential used for routine chromatographic analysis is typically chosen after generation of an HDV for the analyte of interest. Before an HDV experiment is attempted the chromatographic conditions should be fully optimized. The chromatographic conditions should then remain unchanged since changes in mobile phase conditions or temperature may influence the voltammetric curve. To obtain maximum sensitivity, the working potential for this analyte should be about +30mV past E_{max} (to account for minor drift). However, in terms of selectivity, a lower potential may sometimes be optimal if interferences are expected. While some loss of signal may occur at lower potentials this may be offset by reduced background noise resulting in comparable or higher signal to noise ratios.

HDV data provide valuable information not only for detector optimization and troubleshooting, but also for investigation of interferences and many other qualitative applications. Many investigators have used HDV data to aid in: structural identification of unknowns, confirmation of peak identity, and to demonstrate peak purity. Although HDV data are extremely useful, the routine generation of these curves is extremely rare. The acquisition of each data point of an HDV involves an allowance of time for the compound to elute from the column as well as for stabilization of the detector cell at each subsequent potential setting. When using a single detector, HDV generation is a lengthy process highly dependent on analyte retention time, the number of data points obtained per applied potential, the potential window studied (e.g. Δ 700mV), the potential increment (typically +20 to +50mV) and the time necessary for detector stabilization (coulometric and wall-jet sensors stabilize more rapidly than amperometric sensors). These issues become more troublesome when more than one analyte is to be measured, when drift or changes in conditions are encountered (e.g. in gradient operation), and when the analyte stability is questionable. Nevertheless, HDV data can be extremely useful and is often necessary for most HPLC-EC analyses. The use of HDV data would undoubtedly be more widespread if it were not for the practical limitations of single channel HPLC-EC detectors.

On-line HDVs

Coulometric array detection is based on a previously developed dual electrode coulometric detector (Matson *et al.*, 1984, 1987). Coulometric array detectors have up to 16 coulometric electrodes in series set incrementally to a different potential across a potential window. Therefore, one obtains a chromatogram at each potential. A plot of these chromatograms versus channel number provides a multi-dimensional representation of current (response) versus time versus potential. Figure 4 shows such a chromatogram.

Since each of the electrodes can be set at a different applied potential, an HDV for all solutes can be generated in a single chromatographic run. This is accomplished by plotting the cumulative current as a function of channel (or potential) within a given time slice. This approach is therefore analogous to generation of an absorbance or mass spectrum of a given peak within a chromatogram.

From a qualitative standpoint, the benefit of coulometric array data is largely due to the ability to generate these hydrodynamic voltammograms "on-line" in combination with a separation step. With adequate chromatographic resolution, a single analyte within a complex mixture (e.g. a biological fluid) can be characterized voltammetrically without matrix effects or interferences from other electroactive compounds. Moreover, since each data point is typically obtained at a fixed cell potential, the high background (capacitive) currents, which occur when potentials are changed in CV experiments or with single channel HPLC-ECDs, are virtually eliminated. As described herein, HDV data can be routinely obtained on femtomole quantities of analyte within a biological matrix (e.g. urine, microdialysis perfusates, brain tissue extracts, plant extracts and natural products).

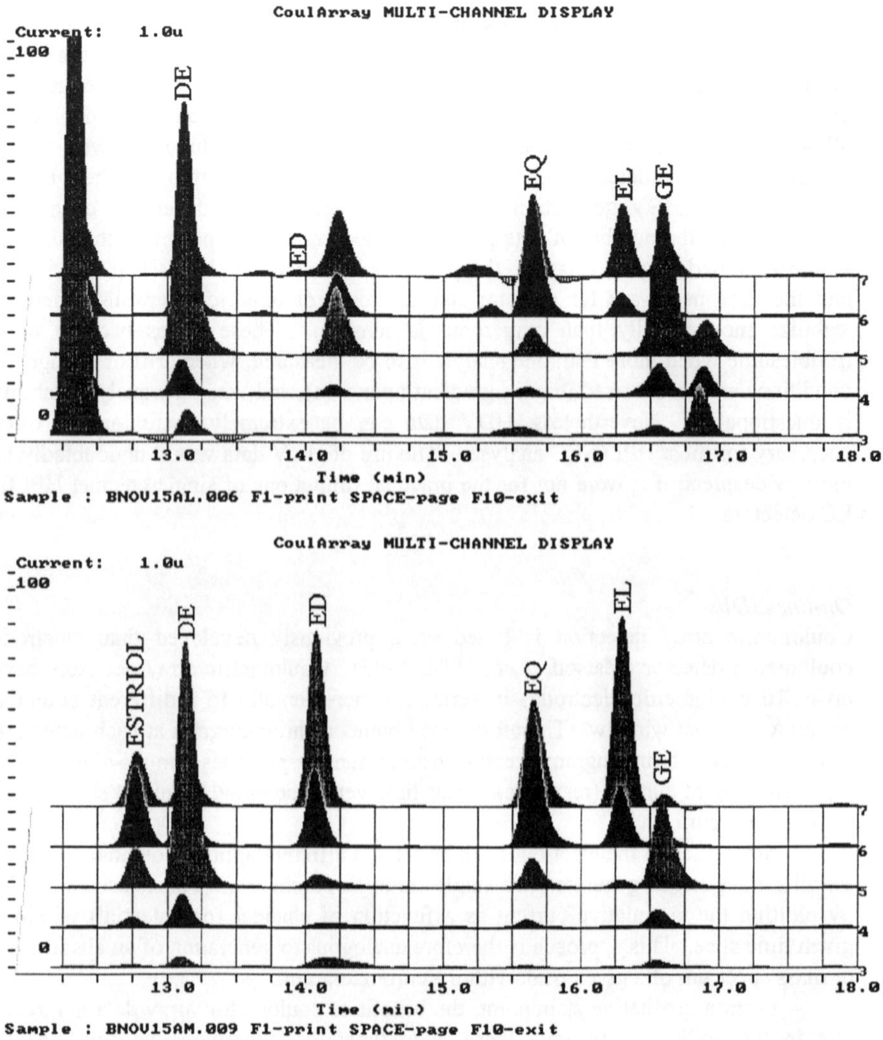

Figure 4. Segments of chromatograms obtained from an extracted standard (upper chromatogram) and urine (lower chromatogram) illustrating voltammetric resolution for each analyte across the electrode array.

The qualitative information generated with coulometric array detection is best utilized along with a clear understanding of some key practical considerations. These include: the effect of experimental variables on the reproducibility of a compound's voltammetric behavior and the differences in voltammetric behavior between analytes that require differentiation. Since this voltammetric data is typically obtained in conjunction with a chromatographic separation an analyte's retention time and/or capacity factor is complementary for compound identification. Also, the performance characteristics of the chromatographic system and the ruggedness of the chromatographic method (e.g. efficiency, resolution, reproducibility, artifacts, etc.) greatly impact the qualitative usefulness of the voltammetric data. Other considerations are related to secondary electrochemical reactions of analytes, their products and solvent. For example, dopamine (DA) when oxidized at pH above ca. 4.5 can rearrange to form other products, such as dopaquinone, which in turn can re-oxidize at downstream electrodes. For the great majority of analytes these reactions are not significant, however, and can often be eliminated by proper choice of conditions (e.g. by lowering the pH). At the same time, these effects are typically very reproducible and highly specific and can therefore provide useful qualitative and quantitative information.

DATA PROCESSING

Sixteen sensors are typically used for coulometric array detection. Although many different arrays are possible, the sensors are most often set in an oxidative array of +60mV increments with sensor 1 at 0mV and sensor 16 at +900mV. As a compound moves through the array it will be oxidized in a stepwise fashion along the current voltage curve until complete electrolytic conversion occurs at E_{max}. A co-eluting compound with a higher oxidation potential can then be detected selectively at the downstream sensors again in a stepwise fashion along its voltammetric curve. Complete electrolytic oxidation or reduction typically occurs over a range, E_0 to E_{max}, of 120 to 180mV. Therefore, with a 60mV difference between adjacent sensors, the response for a particular compound will occur across three adjacent sensors. The data analysis software defines this 3 sensor response as a peak cluster (Figure 5). The sensor having the highest response within a cluster is defined as the dominant channel. The adjacent upstream sensor is defined as the leading channel and the adjacent downstream as the following channel. The response ratios (height or area) between dominant/leading and dominant/following of a given cluster is therefore descriptive of a particular HDV. These ratios obtained from a pure standard can then be used to match a peak cluster of an unknown. Ratio accuracy is then calculated for the unknown as:

$$\frac{\text{Dominant response/Leading (or following response) from the unknown}}{\text{Dominant response/Leading (or following response) from the standard}}$$

where if the ratios are equivalent, then the ratio accuracy value = 1 (100%).

◄◄ Standard test Analyzing Sample UNKNOWN ►►

# File	1 test	2 A0727002.006	3 test	4 A0727002.006
Name	VMA	VMA	5-HIAA	5-HIAA
Conc	20.000	16.487	20.000	34.238
RT (rel)	3.592	3.617	10.892	10.983
Identify	<rt,height,ratio>	<rt,height,ratio>	<rt,height,ratio>	<rt,height,ratio>
Ch1	-	-	-	-
Ch2	-	-	-	-
Ch3	-	-	10.883: 658nA	10.975:1.07uA
Ch4	3.583: 138nA	3.608: 110nA	10.892:1.29uA	10.983:2.25uA
Ch5	3.583: 339nA	3.608: 275nA	10.892: 322nA	10.983: 571nA
Ch6	3.592: 493nA	3.617: 413nA	10.892:20.7nA	10.992:37.8nA
Ch7	3.600: 484nA	3.625: 397nA	10.917:4.05nA	10.992:11.5nA
Ch8	-	-	-	-

Print Report	Full/Brief	Default Reference	Default Identify	Cluster Window		Conc-by height	Force Unforce	Abort	Exit

3+ 3+ 0 + 49 = 55

Figure 5. Compounds in a sample are identified based on both retention time as well as voltammetric behavior across the array. Compounds are normally initially identified based on similarity in their retention time to that of an external standard. Peaks so chosen are then further characterized by their voltammetric behavior across the electrode array. In this example two analytes, VMA and 5-HIAA were identified in the external standard (Test) and in a human urine sample (A0727002.006) following simple sample preparation. As discussed in the text, the three sensor response (the shaded blocks) is a numerical representation of the hydrodynamic voltammetric behavior of an analyte. This on-line voltammetric behavior is an extremely powerful identifier as to the authenticity of an analyte

The chromatograms in Figure 4 show a comparison of the retention times and response profile obtained from a standard to that of a urine sample. An example of the ratio accuracy calculation for daidzein in this sample and the corresponding HDVs are shown in Figure 6. The comparison of HDVs show the plot of the individual peak heights (left axis) as well as the cumulative peak height (right axis) as a function of electrode potential. The latter format is more typical of HDVs presented in the literature. Calculation of ratio accuracy for daidzein is also shown where > 0.9 (>90%) was obtained for both dominant/leading and dominant/following ratios. If a co-eluting compound were present in the sample it would most likely be resolved voltammetrically across the array. This however depends on sample complexity, chromatographic resolution, and the potential increments used. If a lower oxidizing co-eluent were present which was close enough in potential to affect the leading channel, this would be reflected in a lower value for dominant/leading ratio accuracy. Likewise, an higher oxidizing co-eluent may effect the dominant/following ratio

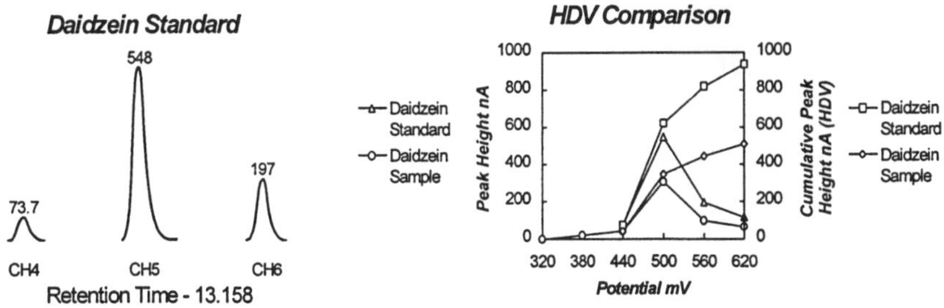

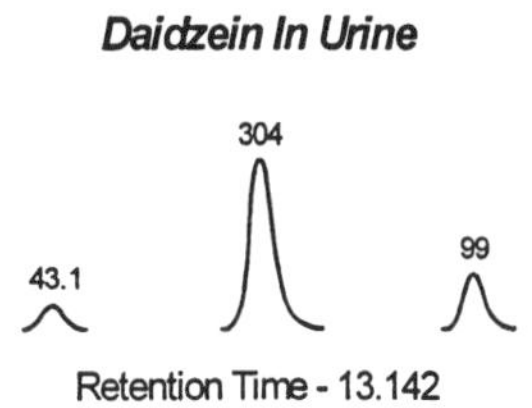

Ratio Accuracy Calculation

Channel	5/4	5/6
Standard	7.44	2.78
Sample	7.05	3.07
Accuracy	94.7	90.6

Figure 6. Obtaining qualitative information from the coulometric array. With a typical oxidative array compounds usually respond at three consecutive electrodes termed the preceding, dominant and following channels. The voltammetric response of an analyte across these three channels is a characteristic of that analyte and is independent of the analyte's concentration. For example the daidzein standard (top left) has a dominant response on channel 5 (+500mV) with less response on the preceding channel, 4 (+440mV), and following channel, 6 (+560mV). Although the coulometric electrodes are 100% efficient, the analyte still responds across three electrodes. The reason for this is that the electrodes are not set to the analyte's maximal oxidation potential but at voltages along its potential axis. This is illustrated in the upper right figure. Here a plot of peak height vs. applied potential shows the analyte's response across the array. The cumulative peak height vs. potential is the analyte's hydrodynamic voltammogram (HDV) and shows why the analyte is expected to respond over more than one channel. Qualitative information is obtained by comparing the standard analyte's HDV to that of the unknown in the sample (in this case daidzein in urine; bottom left quadrant). The system software splits the analyte's HDV into two segments. The analyte's response ratio on the preceding to dominant channels is compared to the unknown's response ratio for the same channels. As shown in the bottom right quadrant, division of the unknown's response ratio by that of the standard produces the first ratio accuracy. The second ratio accuracy is obtained by comparing the response ratio for the following and dominant channels. When expressed as a percentage this gives a numerical indicator as to the authenticity and purity of the unknown peak in the sample. When a peak is positively identified in the sample, both ratio accuracies will approach 100%.

accuracy. If either ratio accuracy value is acceptable, then it is most probable that the interference is not significant at the dominant channel. Quantitation would then be more accurate at the channels which have high ratio accuracy values. Acceptable values are typically greater than 0.7 (70%) (see below for discussion of voltammetric reproducibility). To improve the degree of confidence in peak purity when ratio

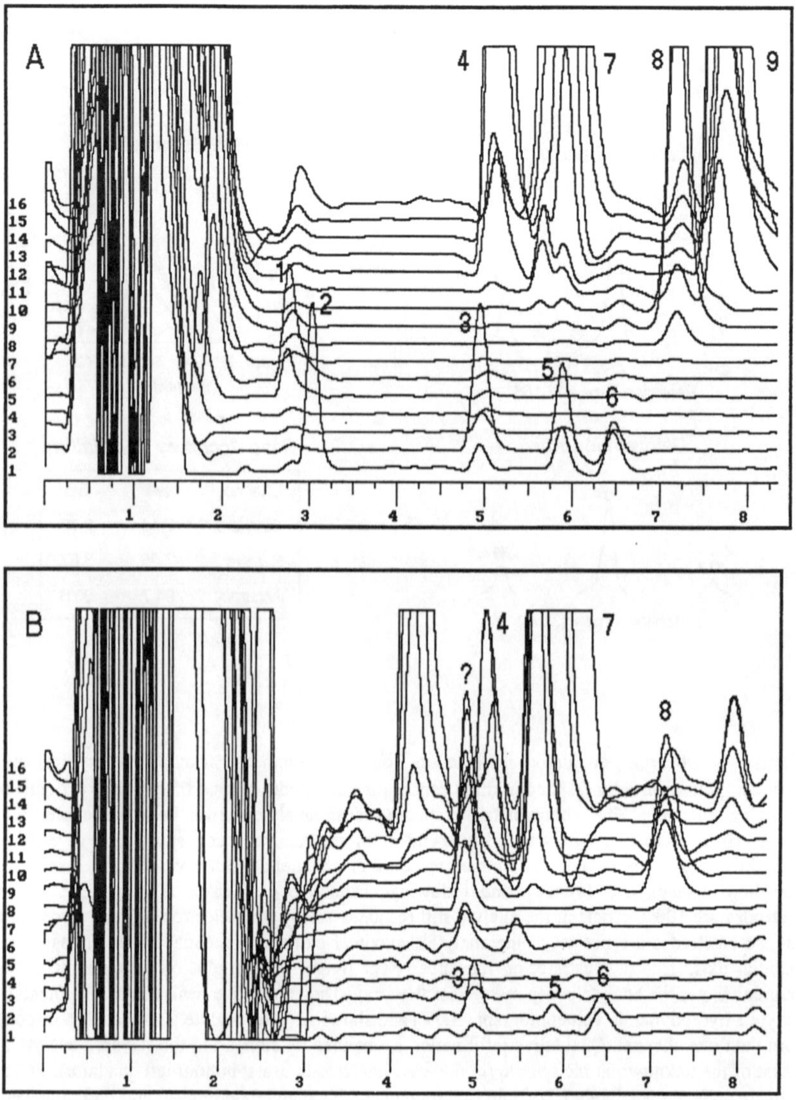

Figure 7. Gradient coulometric array chromatograms obtained for a standard (A) and rat posterior cortex homogenate (B). Analytes were resolved using an NBS (150 x 4.6mm, 5μm, C18) column using a binary gradient (phase A: 100mM sodium phosphate, 35μM sodium dodecyl sulfate, 50nM nitrilotriacetic acid, pH 3.35; phase B: 100mM sodium phosphate, 173μM sodium dodecyl sulfate, 50nM nitrilotriacetic acid, pH 3.45) and an oxidative array (0 to +900mV vs. Pd in +60mV increments). Both chromatograms are presented at the same sensitivity setting (200nA). In the first 8 minute segment the following peaks were identified: 1- 4-hydroxy-3-methoxyphenylglycol (MHPG), 2 - homogentisic acid (HGA), 3 - norepinephrine (NE), 4 - guanosine, 5 - 3,4-dihydroxyphenylacetic acid (DOPAC), 6 - 3,4-dihydroxy-L-phenylalanine (L-DOPA), 7 - methionine, 8 - 4-hydroxyphenyllactic acid (4-HPLA), 9 - 4-hydroxybenzoic acid (4-HBAC), ? - unknown. Even though there were several temporal co-elutions (e.g. NE and guanosine; DOPAC and methionine), the majority of analytes were resolved voltammetrically. However, the posterior cortical sample contained an unknown peak (?) eluting at approximately the same time as NE and guanosine but oxidizing between them. The first ratio accuracy for NE (response on channels 1/2) was 96% but the second ratio

accuracy (response on channel 3/2) was only 64% showing that response of the unknown analyte was corrupting the accurate measurement of NE at the higher potentials. Accurate determination of NE levels can readily be obtained from its response on the first two channels. With the coulometric array such co-elutions are readily apparent. However, with single electrode detectors, such co-elutions would go unnoticed leading to the measurement of falsely elevated NE levels.

accuracy values are questionable, smaller potential increments should be used to improve voltammetric resolution and/or alternative chromatographic conditions should be investigated to provide more separation. Therefore, ratio accuracy values are useful and immediate indicators of the presence of co-eluting solutes. In the standard shown in Figure 7A norepinephrine (NE) and guanosine co-elute in the standard but are clearly resolved voltammetrically ($\Delta E_{1/2} \cong 600$mV). The cortical tissue homogenate Figure 7B contains an additional co-eluting solute which oxidizes at intermediate potentials. Again, the differences between the voltammetric behavior of the unknown and analytes of interest are sufficient for resolution. With single channel detection, this co-eluent would likely go unnoticed and the measurement of NE would be highly susceptible to overestimation. This error can be particularly high if the electrode is set to also measure metabolites which typically require a higher potential. The ratio accuracy values for NE in this sample were above 0.9 (90%) for the first ratio accuracy but only approximately 0.6 (60%) for the second ratio accuracy indicating both the effectiveness of the electrode array in resolving this co-elution and the qualitative utility of the voltammetric data.

Voltammetric reproducibility

The qualitative utility of voltammetric data is highly dependent upon its reproducibility. As discussed above, many variables can affect the position and shape of a voltammetric curve. Control and or consideration of both transient effects (e.g. from sample matrix) and drift (temperature, electronic, etc.) is therefore very important. While the use of multiple sensors maintained at fixed potentials along an exponential current voltage curve is highly susceptible to small changes, in practice the ratio values for most analytes are very reproducible. Table 2 shows the dominant oxidation potentials for a number of compounds reported in literature. Although a variety of array settings and mobile phase conditions are represented, there is a high level of reproducibility demonstrated. This is particularly true when one considers that the dominant potential only is reported rather than the 3 channel cluster.

Since the coulometric electrode array presents voltammetric curves for every analyte in every sample, transient effects, drift, and sample matrix effects are also accounted for. Table 3 shows the within-run precision for one compound both in a biological matrix and in aqueous standards. Gamache and colleagues have previously reported ratio accuracy data for normetanephrine and metanephrine in unextracted urine (Gamache *et al.*, 1993b) and precision data for response ratios obtained for a variety of phenolics in juice beverages (Gamache *et al.*, 1995). These data together demonstrate that high voltammetric reproducibility can be routinely obtained even under gradient elution conditions, over a wide range of analyte quantities, and when injecting complex matrices. When conditions are adequately controlled (e.g. temperature and mobile phase(s) composition) ratio accuracy values of >0.7 (70%) for either dominant/leading or dominant/following typically indicate a high level of peak purity and a high level of confidence in quantitation at the dominant electrode. When higher levels of certainty are required, the analytical conditions (e.g. potential settings and increments, and chromatographic conditions) should be optimized for the highest level of precision in standards. Overspiking of samples and more frequent calibration can also be used to increase the level of confidence in the data.

Table 2.
Literature values for dominant oxidation potentials

Reference	1	2	3	4	5	6	7	8
Potential of Dominant Electrode (mV vs. Pd)								
Compound Name								
5HIAA	180	180	180	160	180		140	200
NE	60	60			60		70	100
DA	60	60			60		70	100
HVA	300	360		320	300		280	
TYR	540	600			600			600
TRP	540	600	600					600
5HTP	180	240	180		240			200
5HT	180	180	180		180		140	200
NM		300			300	360		350
MN		360				360		350
3OHKYN		120	180		180			
KYN		720	720		780			

1. Suiciak *et al*, (1992); 2. Rizzo *et al*, (1991); 3. Godefroy *et al*, (1990); 4. Cheng *et al*, (1992); 5. Naoi *et al*, (1993); 6. Gamache *et al*, (1993b); 7. Wolff *et al*, (1989); 8. Sparks *et al*, (1992).

The limits of linear response should also be considered when using voltammetric data. When very low levels of analyte are present, the leading and dominant channel response may fall below the limit of detection. In this case, response at the dominant sensor combined with the lack of detectable response on side channels provides added confidence in the measurement when compared to single channel detection. At elevated levels, typically >0.5µg, the upper limit of linearity may be exceeded and the response "spills over" to downstream electrodes. The voltammetric data would clearly indicate poor ratios and sample dilution or quantitation by sum across several electrodes is recommended. The ratio accuracy data obtained in microdialysis perfusates and brain tissue homogenates shows the ability to use voltammetric data from low pg to µg quantities of analyte (Gamache *et al*., 1993c).

Peak Identification

The use of voltammetric data to aid in the identification of unknowns has been widely reported. As with most qualitative tools, voltammetric data alone can rarely be used for definitive structural identification. In practice, voltammetric data provides structural information which is at least as specific as an UV/VIS absorbance spectrum. Very often, however, the chromatographic behavior of a solute combined with its voltammogram allows a high degree of structural determination. Using a global gradient method and a library of over 100 compounds, Matson and his colleagues have been able to quickly identify many unknown peaks present in cerebrospinal fluid, plasma, and other tissues (Matson *et al.*, 1987). They accomplished this by examining the relative retention characteristics of unknown peaks compared to those of known compounds together with the voltammograms derived from a 16 channel array. The electrochemical behavior is chiefly utilized to indicate the presence of certain functional groups where listed in order of ease of oxidation: catechol > hydroxyindole > methoxycatechol > indole > monophenol > thiol > methoxyphenol. Table 1 lists the approximate reaction potentials of these and other functional groups. Others have reported the utility of voltammetric data to provide structural classification of flavonoids (Gamache *et al.*, 1993a; Achilli *et al.*, 1993). Since the differences among this diverse class of phytochemicals is largely due to the aromatic hydroxyl and alkoxyl substitution patterns, the voltammograms provide highly specific information. Since more than one electroactive functional group is typically present on these molecules, highly characteristic multi-wave HDVs are generated. This provides a significant advantage over PDA where many of the absorbance spectra of analogous flavonoids are indistinguishable (see also Chapter 1, this volume). Coulometric array detection has the additional benefit over absorbance methods of providing qualitative data at very low levels of analyte. Coulometric arrays which combine incremental oxidative arrays and incremental reductive arrays can provide further specificity to structural analyses. Also, generation of these data under a variety of conditions of pH and organic modifier can provide additional useful information when trying to characterize an analyte.

Table 3.
Within-run voltammetric precision for normetanephrine

Matrix	0.1N HCl	Urine	Urine
Amount injected	10 ng	low amount	high amount
# of replicates	25	15	15
% Relative Standard Deviation			
Leading channel height	1.93	1.79	1.14
Dominant channel height	5.57	2.13	1.80
Following channel height	2.47	1.43	0.75
Dominant/leading ratio	4.82	3.60	1.56
Dominant/following ratio	2.89	2.55	1.95

When compared to mass spectrometry, voltammetric data is less specific. Mass spectrometry, particularly when combined with a chromatographic technique, often has limitations when attempting to perform structural identification. This is highly dependent on the instrumentation used, sample matrix, analyte properties, and amount of analyte available. These limitations may include: sensitivity, need for derivatization, expense, expertise, and extensive sample purification. There have been numerous examples in our laboratory where components present in a complex matrix could not be identified by classical techniques and were only identified using coulometric array. Others have also found coulometric array beneficial in compound identification or structural confirmation. Sparks and Geng (1992) describe the use of retention time, dominant channel potential and ratio accuracy for the identification of biogenic amines in insect tissue. Sontag and Bernweiser also use a similar approach for detection of phenolic esters in honey, 1-o-*p*-trans-coumeroyl glycerol as an adulterant in beer, phloretin as an adulterant of pear juice, amitrol in drinking water and heterocyclic amines in drinking water (see Sontag *et al.*, this volume). They also used oxidation-reduction behavior as a further means of gaining selectivity as well as peak confirmation. They found particular advantages with coulometric array when investigating trace levels of these substances in highly complex food matrices. Hill *et al* (1993) presented an excellent description of the use of ratio accuracy data for confirmation of the structural identity of hydroquinone thioethers in bile and urine. They used LC/MS to confirm the coulometric array data and found good qualitative agreement between the two techniques. Highly substituted conjugates did not, however, produce characteristic mass spectra. The authors postulated that these more polar analytes were not desorbed from the mobile phase matrix used with the mass spectrometer. The HPLC-coulometric array method however demonstrated that these compounds were indeed present. In many analyses coulometric array is therefore a complementary approach to compound identification. In many other analyses, depending on the analyte properties, the matrix and the amount of substance one is attempting to characterize, the coulometric array can be more informative than spectral data.

CONCLUSION

In conclusion, the coulometric array allows instantaneous voltammetric data for analytes on-line with HPLC. This data which is related to a compound's chemical structure can be used for qualitative purposes in a similar manner to photodiode array and mass spectrometry. A solid understanding of the structural properties which determine an analyte's redox characteristics and the experimental variables that effect them is very important in utilizing coulometric array data qualitatively. Some principle advantages of coulometric array are the ability to obtain qualitative data on very low analyte quantities (femtomoles) and in complex matrices (urine, natural products). Examples of where this has been successfully applied include structural identification of unknowns, confirmation of peak identity, demonstration of peak purity, as an aid to method development, and to study the electrochemical properties of a compound.

REFERENCES

Achilli, G., Cellerino, G. P., Gamache, P. H. and Melzi d'Eril, G. (1993). Identification and determination of phenolic constituents in natural beverages and plant extracts by means of a coulometric electrode array system. J. Chromatogr. **632**, 111-117.

Achilli, G., Cellerino, G. P., Melzi d'Eril, G. and Bird, S. (1995). Simultaneous determination of 27 phenols and herbicides in water by high-performance liquid chromatography with multi-electrode electrochemical detection. J. Chromatogr. **697**, 357-362.

Bard, A. W. and Faulkner, L.R. (1980). In: Electrochemical Methods: Fundamentals and Applications. John Wiley and Sons, New York.

Bernweiser, I., and Sontag, G. (1992). Determination of phenolic acids in vegetable oil using HPLC with multielectrode detection. Z. Lebensm Unters Forsch., **195** 559-561.

Cheng, M. H., Lipsey, A. I., Lee, J., and Gamache, P. H. (1992). Automated analysis of urinary VMA, HVA and 5HIAA by gradient HPLC using an array of eight coulometric electrochemical detectors. Lab. Robotics **4**, 297-303.

ESA (1993). Coulochem® EC Array System Application Note #70-1243.

Friedrich, O., and Sontag, G. (1989). Dual electrode electrochemical detector for HPLC. Determination of phenolic compounds in distilled alcoholic beverages. Fresenius Z. Anal. Chem. **334,** 59-63.

Gamache, P. H., Ryan, E., and Acworth, I. N. (1993a). Analysis of phenolics and flavonoid compounds in juice beverages using HPLC with coulometric detection. J. Chromatogr. **635,**143-150.

Gamache, P. H., Kingery, M. and Acworth, I. N. (1993b). Urinary metanephrine and normetanephrine determined without extraction by using liquid chromatography and coulometric array detection. Clin. Chem. **93,** 1825-1830.

Gamache, P. H., Ryan, E., Svendsen, C., Murayama, K., and Acworth, I. N. (1993c). Simultaneous measurement of monoamines, metabolites and amino acids in brain tissue and microdialysis samples. J. Chromatogr. **614,** 213-220.

Gamache, P. H., Acworth, I. N., Lynch, M. L. and Matson, W. R. (1995). Coulometric array detection for HPLC in the analysis of juice products. In: Methods to Detect Adulteration of Fruit Juice Beverages. 1. Nagy, S. and Wade, R. L. (Eds.). Agscience Inc., Auburndale, Florida, USA, pp. 120-144.

Godefroy, F., Matson, W. R., Gamache, P. H., and Weil-Fugazza, J. (1990). Simultaneous measurement of tryptophan and its metabolites kynurenine and serotonin in the superficial layers of the spinal dorsal horn. A study in normal and arthritic rats. Brain Res. **526,** 169-172.

Harris, D. C. (1982). Fundamentals of Electrochemistry. In: Quantitative Chemical Analysis. W. H. Freeman & Co., New York, USA, pp. 321-323.

Hill, B. A., Kleiner, H. E., Ryan, E. A., Dulik, D. M., Monks, T. J., and Lau, S. S. (1993). Identification of multi-S-substituted conjugates of hydroquinone by HPLC-coulometric array analysis and mass spectroscopy. Chem. Res. Toxicol. **6,** 459-469.

Hoare, J. (1964). The effects of metal dissolution on the rest potential in the palladium-oxygen-acid system. J. Electrochem. Soc. **111,** 610-615.

Horvai, G. and Pungor, E. (1989). Electrochemical detectors in HPLC and ion chromatography. Critical Rev. Anal. Chem. **21,** 1-28

Kemula, W. (1952). Rocz. Chem. **26,** 281-297.

Kissinger, P. T.; Refshauge, C.; Dreiling, R. and Adams, R. N. (1973). An electrochemical detector for liquid chromatography with picogram sensitivity. Anal. Lett. **6,** 465-477.

Lavoie, M. P., Palmatier, M., Gentile, F. T., Kaplan, F. A., Fiore, D. M., Hazlett, T. F., Bell, W. J., and Flanagan, T. R. (1993). Two PC12 pheochromocytoma lines sealed in a hollow fiber-based capsules tonically release L-dopa *in vitro*. Cell Transplantation **2,** 163-173.

Matson, W.R., Gamache, P.H., Beal, M.F., and Bird, E.D. (1987). EC array concepts and data. Life Sciences, **41,** 905-908.

Matson, W. R.; Langlais, P.; Volicer, L.; Gamache, P. H.; Bird, E. and Mark, K. A. (1984). n-Electrode three-dimensional liquid chromatography with electrochemical detection of neurotransmitters. (1984). Clin. Chem. **30,** 1477-1488.

Naoi, M., Maruyama, W., Acworth, I. N., Nakahara, D., and Parvez, H. (1993). Multi-electrode detection systems for determination of neurotransmitters. In: Methods in Neurotransmitter and Neuropeptide Research, Part 1. **11** Techniques in The Behavioral and Neural Sciences. Parvez, S. H., Naoi, M., Nagatsu, T. and Parvez, S. (Eds.), Elsevier Science Publishers BV, pp 1-39.

Rizzo, V., Melzi d'Eril, G., Achilli, G., and Cellerino, G. P. (1991). Determination of neurochemicals in biological fluids using an automated high-performance liquid chromatographic system with a coulometric array detector. J. Chromatogr. **536,** 229-236.

Rocklin, R. D. (1989). Working electrode materials. Proper selection for detectors used in liquid and ion chromatography. LC. **2,** 588-594.

Sawyer, D. T., Roberts, J. L. (1974). In: Experimental Electrochemistry for Chemists John Wiley-Interscience, New York, USA.

Shao, M., Xing, X., Liu, C.-C. (1994). pH measurements based on a palladium electrode. Electroanalysis **6,** 245-249.

Siuciak, J. A., Gamache, P. H., and Dubucovich, M. L. (1992). Monoamines and their precursors and metabolites in the chicken brain, pineal and retina: Regional distribution and day/night variations. J. Neurochem. **58,** 722-729.

Sontag, G., and Bernwieser, I. (1994). HPLC coupled with a coulometric electrode array detector. Determination of phloretin glucoside in juices for detection of adulteration. Deutsche Lebensmittel Rundschau, **90**, 72-74.

Sparks, T. C. and Geng, C. (1992). Analysis of biogenic amines in the central nervous system of the tobacco hornworm by high-performance liquid chromatography with 16-sensor electrochemical detection. Anal. Biochem. **205**, 319-325.

Štulík, K.; Pacáková, V. (1982). Electrochemical detection in high-performance liquid chromatography. Crit. Rev. Anal. Chem., **14**, 297-351.

Swartz, K. J., Matson, W. R., MacGarvey, U., Ryan, E. A. and Beal, M. F. (1990). Measurement of kynurenic acid in mammalian brain extracts and cerebrospinal fluid by high-performance liquid chromatography with fluorometric and coulometric electrode array detection. Anal. Biochem. **185**, 363-376.

Vracar, L. M., Sepa, D. B. and Damjanovic, A. (1987). Palladium electrode in oxygen saturated solutions: Rest potentials in solutions of different pH. J. Electrochem. Soc. **134**, 1695-1697.

Weber, S. G. (1986). Detection based on electrical and electrochemical measurements. In: Detectors for Liquid Chromatography. Yeung, E. S. (Ed.). 7. Chemical Analysis. A Series of Monographs on Analytical Chemistry and Its Application. Eldring, R. J. and Wine-Fordner, J. D. (Series Eds.). John Wiley and Sons,: New York, USA, pp. 229-289.

Weber, S. G.; Long, J. T. (1988). Detection limits and selectivity in electrochemical detectors. Anal. Chem. **60**, 903A-913A.

White, P.C.; Catterick, T. (1987). Evaluation of absorbance ratioing for solute identification in high-performance liquid chromatography using a diode array detector. J. Chromatogr. **402**, 135-147.

Wolff, J. A., Fisher, L. J, Xu, L., Jinnah, H. A., Langlais, P. J., Iuvone, P. M., O'Malley, K. L., Rosenberg, M. B., Shimohama, S., Freidman, T. and Gage, F. H. (1989). Grafting fibroblasts genetically modified to produce L-dopa in a rat model of Parkinson's disease. Proc. Natl. Acad. Sci. **86**, 9011-9014.

Progress in HPLC-HPCE, Vol. 6, pp. 75–98
I.N. Acworth *et al.* (Eds)

HPLC WITH ELECTRODE ARRAY DETECTION AND ITS APPLICATION IN ANALYTICAL FOOD CHEMISTRY

GERHARD SONTAG, IRMTRAUD BERNWIESER and CHRISTIAN KRACH

Institute for Analytical Chemistry, University of Vienna, Waehringerstrasse 38, A-1090 Vienna, Austria

INTRODUCTION

Within the last 25 years high performance liquid chromatography (HPLC) in combination with different detectors has become an integral part of modern food laboratories. Despite many improvements, there is still a demand for new sensitive and selective detection modes.

The development of the coulometric electrode array system (CEAS) with sixteen cells caused a significant change by adding a third dimension, the potential, to time and current. This allows the simultaneous use of 16 different potentials, which can give information about peak purity and peak identity. Data are not only available in the time and current (concentration) domain, but also in the potential region (redox or voltammetric behavior).

In the last few years coulometric electrode array detection coupled with HPLC has been mainly applied to neurological and clinical problems. Concerning food chemistry, only a few papers have been published (Achilli *et al.*, 1993, 1994; Gamache *et al.*, 1993). Our group has been working for more than ten years with electrochemical detectors in combination with HPLC and flow injection analysis (FIA). For a long time we used a variety of electrochemical approaches including single cell amperometric, polarographic and dual cell coulometric detection. We are now using the CEAS in order to detect important phenolic compounds in food (Bernwieser, 1992, 1994; Bernwieser and Sontag, 1992, 1993; Jörg, 1989, 1992; Jörg and Sontag, 1992, 1993), to prove adulteration of beverages (Bernwieser, 1992; Sontag and Bernwieser, 1994a, b) and to determine pesticides in drinking water (Bernwieser and Sontag, 1993). We are also interested in developing methods for determining whether food has been irradiated or not (Bernwieser *et al.*, 1995; Sontag and Bernwieser, 1994) and in detecting compounds which are responsible for mutagenic or carcinogenic effects of broiled and fried meat (Rauter, 1995).

All these problems require the determination of low analyte concentrations in complex matrices. Therefore after an effective separation a very sensitive and selective detector like the coulometric electrode array detector is important.

CHARACTERISTIC PHENOLIC COMPOUNDS IN HONEY AND VEGETABLE OIL

Honey and vegetable oils are well-known and popular foodstuffs. As long as consumers attach more and more importance to high quality and having a great selection of different varieties, reliable methods for evaluating the different kinds of honey and vegetable oils are required. Attention has been focused on measuring different ingredients or markers that can better characterise these foodstuffs. Phenolic acids (Cremer and Riedmann, 1964, 1965; Kozlowska *et al.*, 1983a, b; Steeg and Montag, 1987, 1988a, b) which occur in trace amounts, are good tools for this purpose. Phenolic esters (Graddon *et al.*, 1979; Speer and Montag, 1987; Steeg and Montag, 1987, 1988a, b) can also be used for identification. In previous investigations, GC and GC-MS methods were favoured, but HPLC in combination with UV-VIS or amperometric detection were also used. The problems with these methods are on the one hand, the time consuming and difficult sample preparation which may cause a lot of mistakes, and, on the other hand, the lack of selectivity of these detectors resulting in complex chromatograms.

A coulometric dual cell detector in combination with reversed phase (RP-) HPLC was successfully used for determination of phenolic acids in honey (Jörg, 1989, 1992; Jörg and Sontag, 1992, 1993). These substances allow a differentiation between blossom honey, honey-dew honey and chestnut honey, but it was not possible to differentiate between rapeseed and robinia honey, two kinds of blossom honey. The coulometric dual cell detector was also used for determination of phenolic acids in vegetable oil (Bernwieser, 1992, 1994) but the chromatographic separation was not as efficient as necessary and the determination of the characteristic substances was disturbed by coeluting compounds.

Determination of phenolic esters in honeys

The phenolic esters found in commercially available honeys were determined after extraction with ethylacetate (Jörg, 1989, 1992; Jörg and Sontag, 1992, 1993). A C-18-RP column was used to separate the compounds by HPLC. The electrodes of the CEAS were set at incrementally (+40mV) increasing potentials, covering a potential range between +300 and +900mV. The voltammetric information from the CEAS was used for peak identification. The phenolic esters were quantified (Table 1) and detection limits between 0.1 and 1μg/kg honey were found (Jörg and Sontag, 1992, 1993).

Examination of the results in Table 1 shows significant differences in metabolite levels found in the various sorts of honey. Such differences in levels allows, for example, differentiation between rape and robinia honey. Also the presence and the concentration of trans-methyl ferulate is characteristic for fir honey. A great advantage of this detection system, when compared to single cell detectors, is the possibility of determining very different concentrations of phenolic esters simultaneously and using only one measurement.

Table 1.
Content of phenolic esters (µg/kg) in honey. M4HB: Methyl 4-hydroxybenzoate, MVAN:Methyl vanillate, MSYR: Methyl syringate, MCOM: trans-p-Methyl coumarate, MFER: trans-Methyl ferulate.

Variety	M4HB	MVAN	MSYR	MCOM	MFER
Robinia	-	-	281.0	-	-
	-	3.5	93.4	-	-
Rape	177.0	31.3	3865	-	-
	126.2	36.7	4092	-	-
	184.8	237.4	4537	-	-
	167.3	35.2	5044	-	-
Chestnut	18.0	10.9	42.2	-	-
Clover	38.3	11.8	1592	-	-
Linden blossom	7.7	20.2	759.5	-	-
Dandelion honey	-	11.8	551.5	-	1.3
Sunflower	31.8	8.0	322.7	1.5	2.9
Orange blossom	194.2	18.8	-	-	7.9
Fir	-	14.7	539.5	-	1196

Determination of phenolic acids in vegetable oils
Phenolic acids are synthesized by plants, and their content and distribution pattern depend above all on the kind of plant. This distribution pattern should also be present in the oil extracted from the plant and can be used to distinguish between different kinds of vegetable oils.

After solid phase extraction (Diol cartridge) an aliquot of the vegetable oil extract was injected onto the CEAS. The compounds were separated on a C-18-RP column and determined by isocratic elution (methanol/phosphoric acid/sodium hydroxide) (Bernwieser, 1994; Bernwieser and Sontag, 1992). The potentials of the sixteen electrodes were set at potentials between 0 and +900mV, increasing in increments of +60mV. Phenolic acids were identified by comparing the retention times and the hydrodynamic voltammograms of standard and sample components. The effect of coeluted substances was diminished by detecting the compounds at different oxidation potentials, which made it possible to separate the substances in an electrochemical way. Table 2 shows the content of phenolic acids in several vegetable oils. The detection limits vary between 0.5 and 2.5µg/kg vegetable oil (Bernwieser and Sontag, 1992).

Table 2.
Content of phenolic acids (mg/kg) in different sorts of vegetable oil. 4HB: 4-Hydroxybenzoic acid, VAN: Vanillic acid, CAF: Caffeic acid, SYR: Syringic acid, PCOM: p-Coumaric acid, FER: Ferulic acid, SIN: Sinapic acid.

Variety	4HB	VAN	CAF	SYR	PCOM	FER	SIN
Sunflower	0.01	0.04	0.04	0.02	0.01	0.02	-
Palm	0.01	0.05	-	0.02	0.01	-	-
Coconut	0.26	0.67	0.09	0.46	0.02	-	-
Soybean	0.13	0.18	0.09	-	0.09	0.08	0.08
Rapeseed	0.08	0.19	-	0.54	0.08	0.68	15.54
Corn	0.83	1.12	0.09	0.40	1.63	5.79	-

The advantages of this method include the short and easy sample preparation, the possibility of separating and quantifying coeluting substances and the reliable identification of the compounds. Additionally, the described method allows a rapid differentiation between raw and refined vegetable oil.

DETECTION OF THE ADULTERATION OF BEVERAGES

The guarantee of the quality of beverages is an important assumption for the economical success of the beverage industry. Despite this fact, attempts are made throughout the world to produce beer or juices more cheaply. This leads to adulteration of various commercial products because cheaper raw materials may be employed. Therefore regulatory and industrial agencies are interested in methods to detect adulteration in order to preserve the quality and safety of beverages, e.g., beer or juices. One way of telling whether a product has been adulterated is to look for characteristic substances (indicators or markers), which make the detection of adulteration possible.

Determination of 1-O-trans-p-coumaroylglycerol
Normally beer is brewed with the addition of various cereals like barley or maize, but if a product is enlisted with the "Deutschem Reinheitsgebot" mark, only barley, hops and water are permitted during brewing. In maize, sorghum and beer brewed from

these components a characteristic compound, 1-O-trans-p-coumaroylglycerol, can be found and measured using HPLC with UV-detection (Fenz *et al.*, 1992; Fenz and Galensa, 1989, 1990). However the chromatograms produced using this approach are rather complex.

HPLC with the CEAS was used for the determination of 1-O-trans-p-coumaroylglycerol (Bernwieser *et al.*, 1992, 1994) because of its high selectivity and sensitivity. For some sorts of beer, even dilution with eluent (methanol/glacial acetic acid/bidistilled water) was necessary. For other samples, interfering compounds were removed by using C-18- and then N^{+}-cartridges (Pätzold *et al.*, 1992). A well-defined volume of the extract was injected onto a C-18 RP-column. For determination of 1-O-trans-p-coumaroylglycerol, the 16 coulometric electrodes were set at potentials between 0 and +900mV, increasing in +60mV steps. The eluted ester (Figure 1) was identified by matching its retention time to that of an external standard. Further confirmation of peak identity and purity came from the dominant oxidation potential and matching the voltammetric behavior of both the standard and sample component. In some samples a compound coeluted with the ester. However, it was no problem to identify 1-O-trans-p-coumaroylglycerol because this substance shows a very different voltammetric behavior. This method is so sensitive, that an addition of 1% maize or sorghum to the mash in the brewing process can be proven.

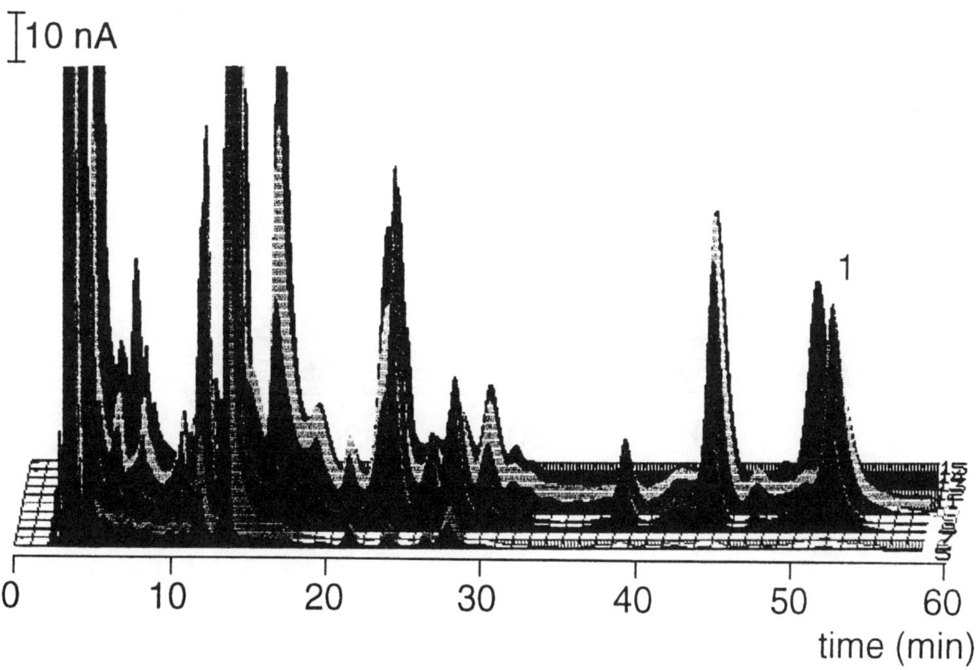

Figure 1. Electrode array chromatogram of an extract of maize beer. 1: 1-O-trans-p-coumaroylglycerol. Column: Bio Rad RoSiL C-18, 250 x 4.6mm, 5µm; eluent: methanol/glacial acetic acid/bidistilled water (150/20/830, v/v/v), adjusted to pH3.0 with sodium hydroxide (2M); flow rate: 1 mL/min; column temperature: 32°C; detector: CEAS; potential: 0 to +900mV in increments of +60mV; injection volume: 100µL.

Determination of phloretin glucoside

For control of the purity of fruit juices, good analytical methods are required. In the case of apple juice and products made of apples, characteristic substances such as phloretin glucoside and phloretin xyloglucoside have already been identified (Hollborn *et al.*, 1990; Wald and Galensa, 1989). These compounds are absent in pears and pear juices and therefore can be used to detect the adulteration of pear juice with apples.

Commercially available apple juices were investigated first in order to define an average content of phloretin glucoside. After a short sample preparation with a column filled with polyamide the extracts were injected into the HPLC system and analytes were separated on a C-18-RP column. Phloretin glucoside was identified by retention time, oxidation potential and its hydrodynamic voltammogram. Its content varied between 10.4 and 16.7mg/L (Sontag and Bernwieser, 1994a). The CEAS leads to very simple chromatograms (Figure 2).

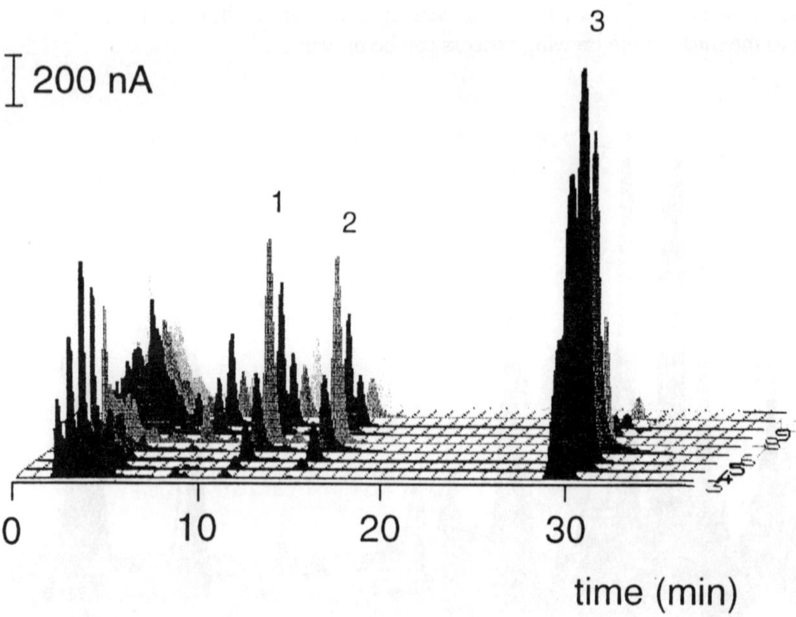

Figure 2. Electrode array chromatogram of an extract of a commercially available apple juice. 1: Phloretin xyloglucoside, 2: Phloretin glucoside, 3: Ethyl syringate (internal standard). Column: Merck LiChrospher 100 RP-18, 250 x 4mm, 5µm; eluent: methanol/glacial acetic acid/bidistilled water (350/20/630, v/v/v); flow rate: 1 mL/min; column temperature: 32°C; detector: CEAS; electrodes 1-3: + 400 mV, subsequent electrodes were in increments of +35 mV up to +855mV; injection volume: 100µL.

A pear juice sample produced in our laboratory was investigated in the same way but phloretin glucoside and phloretin xyloglucoside could not be found (Figure 3). However, in some commercially available pear juices and nectars these substances were present (Figure 4).

This indicates adulteration. The method is so sensitive that 1% of apple juice added to pear juice could be detected.

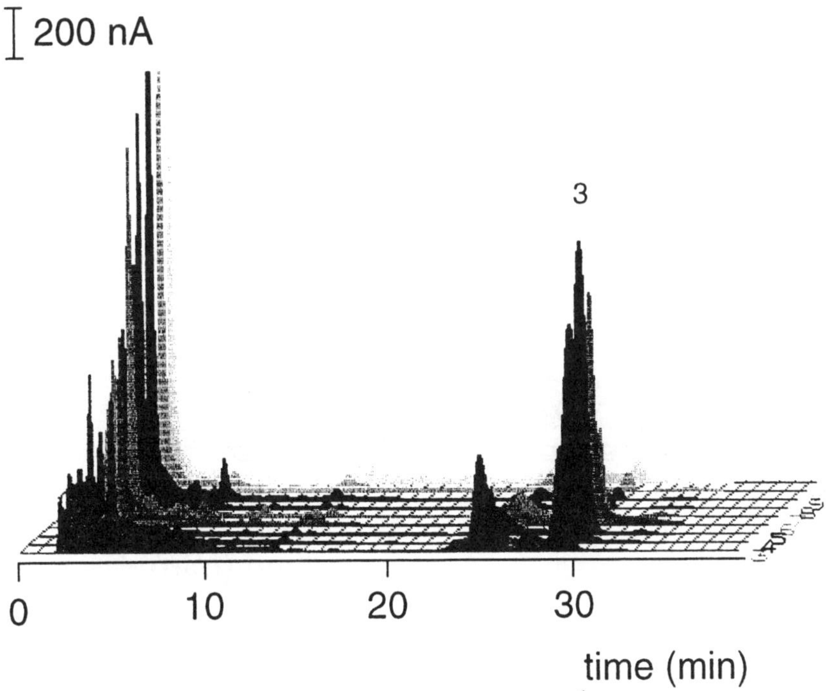

Figure 3. Electrode array chromatogram of an extract of self-produced pear juice. 3: Ethyl syringate (internal standard). Column: Merck LiChrospher 100 RP-18, 250 x 4mm, 5µm; eluent: methanol/glacial acetic acid/bidistilled water (350/20/630, v/v/v); flow rate: 1mL/min; column temperature: 32°C; detector: CEAS, potential: electrodes 1 to 3: +400mV, subsequent electrodes were in increments of +35mV up to +855mV; injection volume: 100µL.

DETERMINATION OF A HERBICIDE IN DRINKING WATER

The use of herbicides in agriculture may cause contamination of ground and drinking water. Therefore different limit values were defined in 1989 in the European Community Water Act (Water Act). The maximum individual pesticide concentration must be less than 0.1µg/L and the maximum admissible concentration of all pesticides should be less than 0.5µg/L. The plant growth inhibitor amitrole (3-amino-1,2,4-

triazole) is one of these compounds. It had already been determined by various analytical methods (Archer, 1984; Green and Feinstein, 1957; Lokke, 1980; Pachinger *et al.*, 1992; Pribyl *et al.*, 1978; Ternes and Rüssel-Sinn, 1987; Van der Poll *et al.*, 1988, 1990). However, none of these methods could be used without trace-enrichment steps.

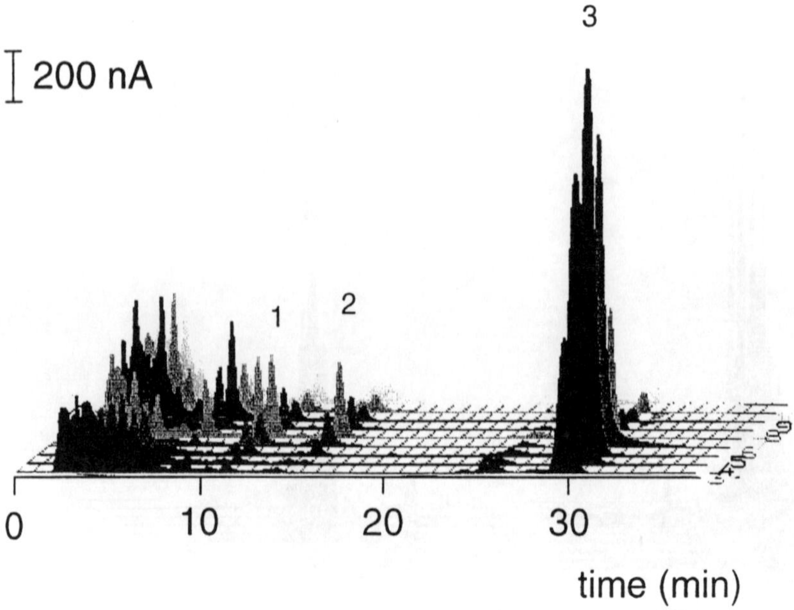

Figure 4. Electrode array chromatogram of an extract of a commercially available pear juice. 1: Phloretin xyloglucoside, 2: Phloretin glucoside, 3: Ethyl syringate (internal standard). Column: Merck LiChrospher 100 RP-18, 250 x 4mm, 5µm; eluent: methanol/glacial acetic acid/bidistilled water (350/20/630; v/v/v); flow rate: 1mL/min; column temperature: 32°C; detector: CEAS, electrodes 1-3 +400mV, subsequent electrodes were in increments of +35mV up to +855mV; injection volume: 100µL.

Determination of amitrole

Filtered samples of drinking and well water were injected into the HPLC system without any trace-enrichment. The compounds were separated on a C-18 RP-column and detected with the CEAS setting the potentials of the 1st to the 4th electrode at +550mV, and the potentials of the 5th to the 16th electrode increased in increments of +30mV from +580 to +910mV (Bernwieser and Sontag, 1993). In this way sample components oxidizable at potentials lower than +580mV could be eliminated. For quantitation of amitrole the standard addition method was performed. The detection limit of amitrole added to Viennese tap water was 0.05µg/L (Bernwieser, 1994; Bernwieser and Sontag, 1993).

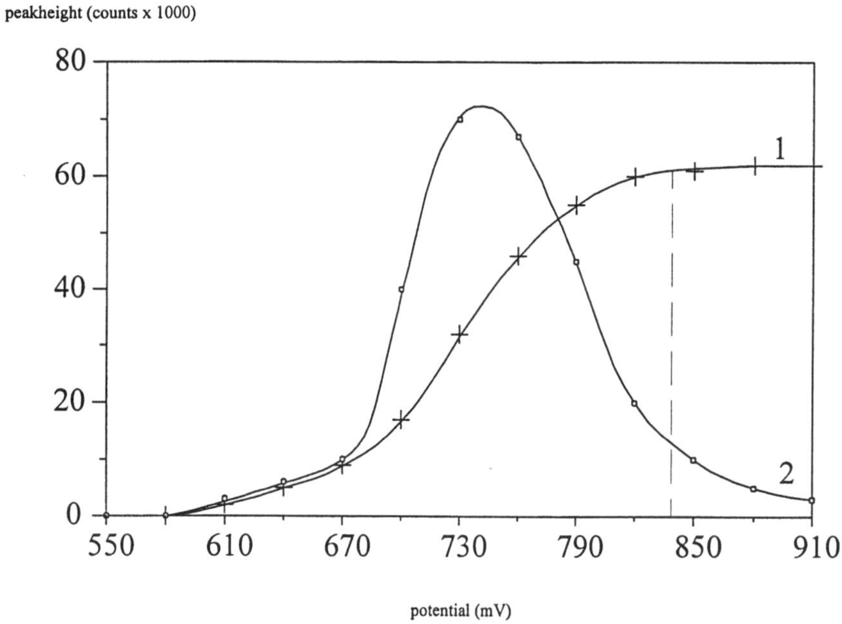

Figure 5. Hydrodynamic voltammograms of amitrole obtained with single cell detection (1) and with coulometric electrode array detection (2). Column: Bio-Rad RoSiL C-18, 250 x 4.6mm, 5µm; eluent: 550mL methanol, 1.4g sodium dihydrogenphosphate monohydrate and 1.2g sodium dodecylsulfate diluted with 450mL bidistilled water. The final pH 3.6 was achieved using ortho-phosphoric acid (83%); flow rate: 1mL/min; column temperature: 30°C; single cell detector: 0 to +1000mV in increments of +50mV; detector: CEAS, electrodes 1-4 +550mV, subsequent electrodes were in increments of +30mV to +910mV; injection volume: 100µL.

A great advantage with coulometric electrode array detection, in comparison to single cell detection, is that the optimal detection potential is approximately 100 mV lower than that for a single cell (Figure 5). This decreases the background current and leads to a better signal/noise ratio. A result of this highly sensitive detector is that it allows direct injection of the samples into the chromatographic system without any enrichment. In this way any sources of error due to sample preparation are diminished and the time of analysis is shortened.

DETERMINATION OF IRRADIATED PROTEIN RICH FOOD

Ionizing radiation has been proposed for the preservation of food, e.g., to inhibit sprouting of potatoes, to delay the ripening of food and vegetables, to control food born parasites and pathogenic organisms, and to extend the shelf life of food. Food irradiation is now recognized as a method for reducing postharvest food losses, ensuring hygienic quality and facilitating wider trade in foodstuffs.

Currently 37 countries have given a grant for collectively treating more than 40 different food products, treated using a well-defined dose range. The Expert Committee of the Wholesomeness of Irradiated Food states that "Irradiation of any food commodity up to an overall average dose of 10kGy introduces no toxicological hazard". Although opponents believe that irradiation of food is not safe, what is of greater concern is that vendors might irradiate contaminated food (e.g., with pathogenic micro-organisms) to disinfect it and then put it back on the shelf for resale.

Because of different national regulations, it is important to have a proper control of irradiated food in international trade. Therefore reliable analytical methods are desirable. First, it should be possible to state whether food has been irradiated or not. Second, correct labelling must be controlled, thereby giving the consumer confidence and the ability to choose between irradiated and non irradiated food.

Irradiation of meat leads to the production of H- and OH-radicals in the aqueous phase. These radicals react with organic molecules giving several reaction products. The kind of products depends on the absorbed dose and the conditions during irradiation. Irradiation of proteins initiates reactions like deamination, decarboxylation, cleavage of peptide bonds and so on. Phenylalanine is present in proteins and is very susceptible to irradiation. It reacts with OH-radicals to yield *o*-, *m*- and *p*-tyrosine. For a long time *o*- and *m*-tyrosine were thought to be absent in natural proteins and to be unique radiolytic products.

Since *o*-tyrosine can be more readily separated chromatographically from *p*-tyrosine than *m*-tyrosine, *o*-tyrosine was chosen as a potential indicator of irradiated food. Karam and Simic (1988a,b) determined *o*-tyrosine qualitatively and quantitatively after separation with capillary gas chromatography and selective ion monitoring (GC/MS). Meier *et al.* (1988, 1989, 1990) showed that *o*-tyrosine can be determined in chicken meat by HPLC with fluorescence detection without derivatization. They found, just as Karam and Simic, that the formation of *o*-tyrosine is proportional to the applied radiation-dose. Chuaqui Offermanns *et al.* (1991, 1993) reported that *o*-tyrosine would also meet the requirements of analytical detection methods for irradiation of food, despite the fact that in chicken meat it is a naturally occurring substance (Galceran *et al.*, 1993; Hart *et al.*, 1988).

Comparing the results of the published methods, one can confirm that they are inconsistent. This was our motivation to develop and test a new method based on HPLC with the CEAS, expecting that the selectivity of the detection system would lead to simpler chromatograms. The simultaneous use of 16 different cells (potentials) should give information about peak purity and peak identity to solve this difficult analytical problem. The use of *m*-tyrosine as a marker compound for irradiated food was also be investigated.

Determination of o- and m-tyrosine

Oxidative detection mode. A standard mixture of *o*-, *m*- and *p*-tyrosine was separated on a C-18-RP column using methanol/phosphoric acid as an eluent. First the substances were detected by use of oxidative potentials between +380 and +900mV (Bernwieser, 1992). The resulting 16 simultaneous chromatograms were stored, evaluated and nine channels plotted as a three dimensional picture of this analysis (Figure 6).

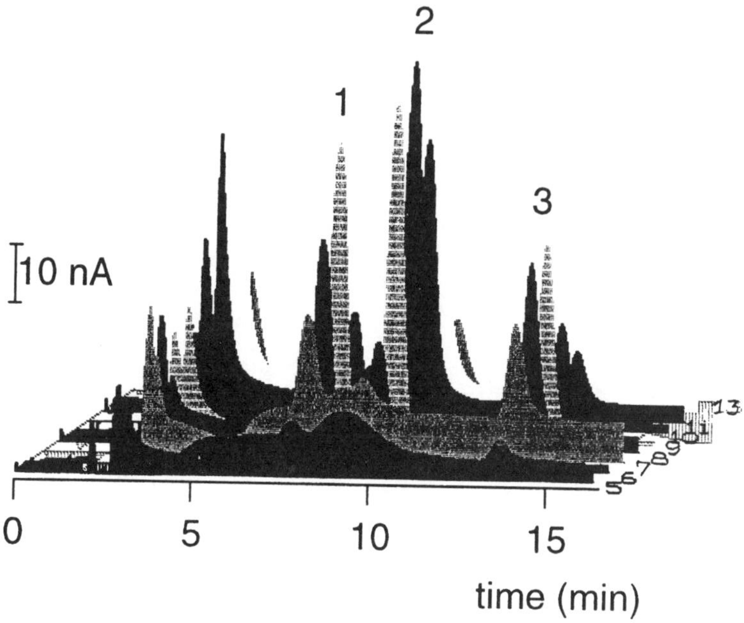

Figure 6. Electrode array chromatogram of a standard mixture of *p*-, *m*- and *o*-tyrosine. 1: *p*-tyrosine, 2: *m*-tyrosine, 3: *o*-tyrosine. Column: Merck LiChrospher 100 RP-18, 250 x 4mm, 5µm; eluent: methanol/phosphoric acid (0.0375M) /bidistilled water (50/910/40, v/v/v) adjusted to pH3.0 with sodium hydroxide (2M); flow rate: 1mL/min; column temperature: 32°C; detector: CEAS, potential: electrodes 1-3 +380mV, subsequent electrodes were in increments of +40mV to +900mV; injection volume: 100µL.

The substances of interest were characterized by retention times, dominant oxidation potentials and their hydrodynamic voltammograms (Figure 7).

Oxidative and reductive detection mode. It was also possible to oxidize *o*- and *m*-tyrosine completely by setting five working electrodes to +900mV. The resulting oxidation products were then reduced at the subsequent electrodes (+100 to -200mV) (Bernwieser, *et al.,* 1995). In this way hydrodynamic voltammograms for the reduction of the oxidation products were obtained. The hydrodynamic voltammograms of *o*- and *m*-tyrosine give useful data for confirmation of peak identity and peak purity. Figure 8 shows the hydrodynamic voltammogram of m-tyrosine for the oxidative and the reductive detection mode.

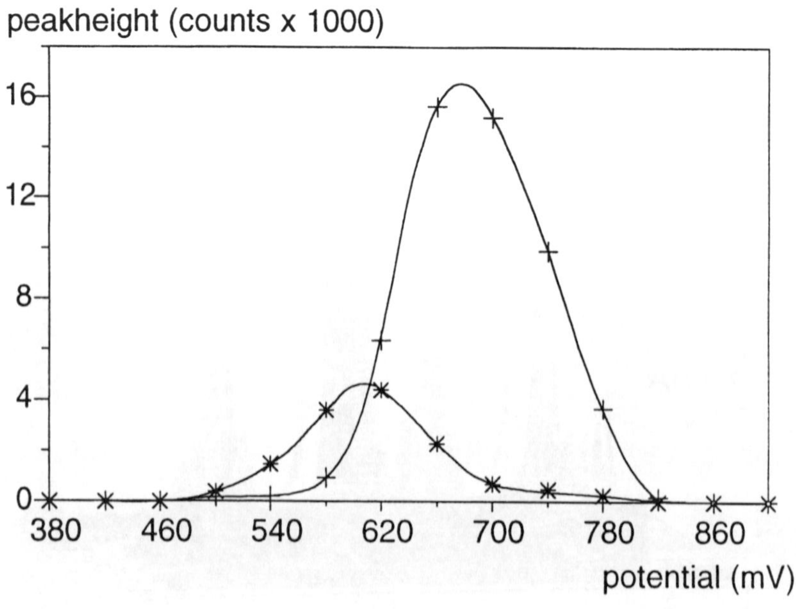

Figure 7. Hydrodynamic voltammograms of *o*- (*) and *m*- (+) tyrosine. Column: Merck LiChrospher 100 RP-18, 250 x 4mm, 5µm; eluent: methanol/phosphoric acid (0.0375M)/bidistilled water (50/910/40, v/v/v) adjusted to pH3.0 with sodium hydroxide (2M); flow rate: 1 mL/min; column temperature: 32°C; detector: CEAS, electrodes 1-3 +380mV, subsequent electrodes were in increments of +40mV from +420 to +900mV; injection volume: 100µL.·

Analysis of chicken meat

First non-irradiated and irradiated chicken meat was freeze dried and hydrolyzed with hydrochloric acid. The separation of *o*-, *m*- and *p*-tyrosine, not only from each other but also from interfering substances, was necessary. To ensure qualitative analysis, peak identity and peak purity were investigated by comparing the hydrodynamic voltammograms of standard compounds and sample components. With this method *o*-tyrosine was determined in irradiated and non-irradiated samples (Figure 9). Its concentration increased linearly with the applied radiation dose until 5kGy.

Up until now *m*-tyrosine could not be separated by GC- or HPLC methods from interfering substances in chicken meat. When chemical separation is incomplete the electrode array system offers the selectivity to complete this task. The sample components are completely oxidized by five electrodes at constant potential, then the oxidation products are selectively reduced by the subsequent electrodes. In this way it was possible to separate *m*-tyrosine from coeluting matrix compounds (Figure 10).

The concentration of *m*-tyrosine did not increase linearly with the applied radiation dose.

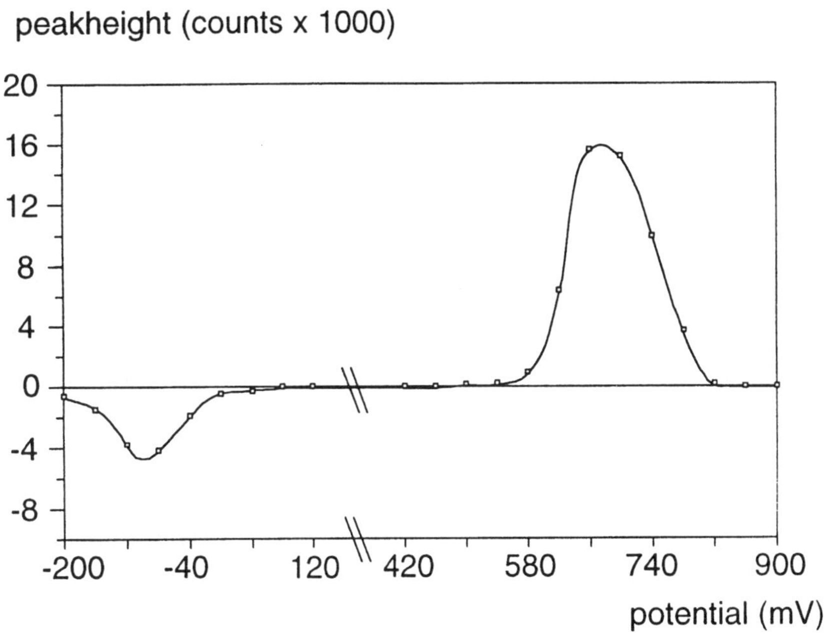

Figure 8. Hydrodynamic voltammogram of *m*-tyrosine in oxidative and reductive detection mode. Column: Merck Lichrospher RP-18; 250 x 4mm, 5µm; eluent: methanol/phosphoric acid (0.0375M)/bidistilled water (50/910/40, v/v/v), adjusted to pH3.0 with sodium hydroxide (2M); flow rate: 1mL/min; column temperature: 32°C; detector: CEAS, electrodes 1-5 +900mV, the subsequent electrodes were in increments of 30mV from +100 to -200mV; injection volume: 100µL.

Our investigations show that both *o*- and *m*-tyrosine isomers are present in unirradiated chicken meat. Therefore the background levels of these analytes must be determined exactly for samples of different origin. To correctly determine whether chicken meat has been irradiated or not, levels of *o*- and *m*-tyrosine formed in unirradiated chicken meat during sample preparation and storage should be investigated.

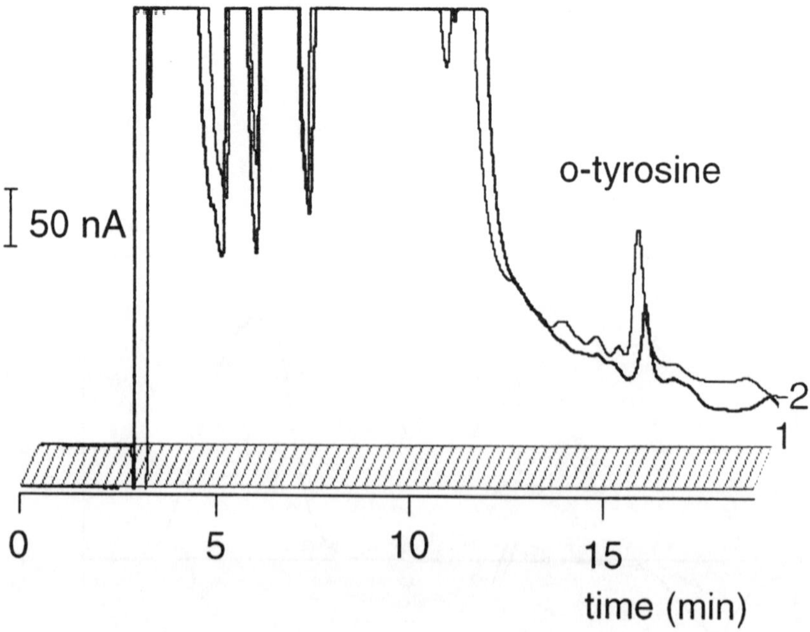

Figure 9. Electrode array chromatogram (channel 9) of an acid hydrolysate of chicken meat. 1: non-irradiated meat, 2: irradiated with a dose of 6 kGy. Column: Merck LiChrospher 100 RP-18, 250 x 4mm, 5µm; eluent: methanol/phosphoric acid (0.0375M)/bidistilled water (50/910/40, v/v/v); adjusted to pH3.5 with sodium hydroxide (2M); flow rate: 1mL/min; column temperature: 32°C; detector CEAS, electrode 1-3 +380mV, the subsequent electrodes were in increments of +40mV to +900mV; injection volume: 100µL.

The radiation dose response curve for the formation of *o*-tyrosine is linear up to 5 kGy. At higher radiation doses it deviates from linearity. This is caused by the oxygen concentration during irradiation (Bernwieser *et al.*, 1995) which influences the decomposition of phenylalanine and the formation of *o*-tyrosine. At higher oxygen concentrations more phenylalanine is transformed into *o*-, *m*- and *p*-tyrosine. Conversely more of the *o*- and *m*-tyrosine isomers are decomposed and other compounds are formed (Bernwieser, 1994).

The high selectivity of the CEAS makes it possible to determine *m*-tyrosine, which was not practical with other methods, because it would not be chromatographically separated from coeluting impurities.

The acid hydrolysis during sample preparation is a critical point in this analytical method. In the presence of oxygen during the hydrolysis step, tyrosine isomers and other compounds can be oxidized leading to complex chromatograms. Therefore an additional sample preparation had to be developed.

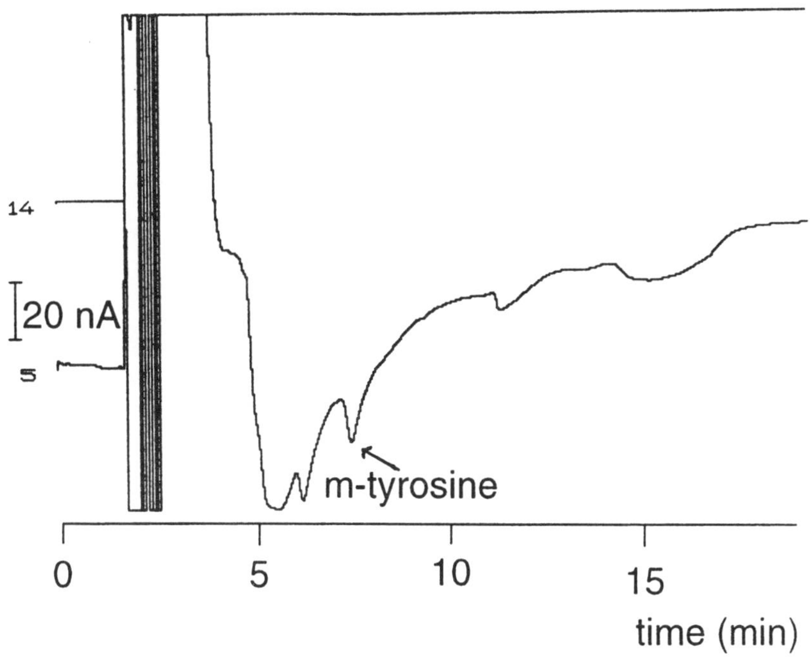

Figure 10. Electrode array chromatogram (channel: 5 and 14) of a hydrolysate of irradiated chicken meat; irradiation dose: 5 kGy; column: Merck LiChrospher 100 RP-18, 250 x 4mm, 5µm; eluent: methanol/phosphoric acid (0.0375M)/bidistilled water (50/910/40, v/v/v); adjusted to pH3.0 with sodium hydroxide (2M); flow rate: 1mL/min; column temperature: 32°C; detector: CEAS, electrodes 1-5 +900mV, electrodes 6-16 +100 to -200 mV in increments of 30mV; injection volume: 100µL.

Analysis of shrimp tissue. It is known (Vazquez-Ortiz *et al.*, 1995) that free amino acids are abundant in shrimp. Therefore it should be possible to determine the content of *o*- and *m*-tyrosine formed from free phenylalanine during radiolysis. This means that acid hydrolysis of the proteins is not necessary. The tyrosine isomers have only to be extracted from the shrimp with trichloroacetic acid. A chromatogram of a 20µL volume of the extract is shown in Figure 11.

A linear relationship between the content of *o*- and *m*-tyrosine and the radiation dose was found up to 6 kGy (Figure 12). In contrast to the acid hydrolysis method described above the background levels of *o*- and *m*-tyrosine could be readily determined and at very low levels (*o*-tyrosine: 0.1ppm, *m*-tyrosine: 0.01ppm).

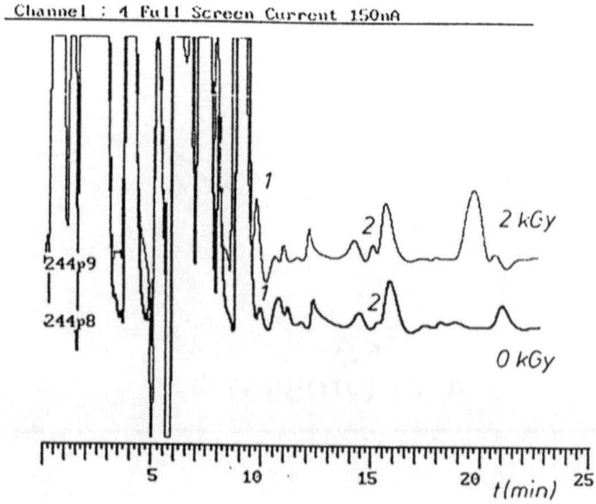

Figure 11. Electrode array chromatogram of a shrimp extract. 1: *m*-tyrosine, 2: *o*-tyrosine, Column: Inertsil RP-18, 250 x 2.1mm, 5µm; eluent: 94% 0.05M trichloroacetic acid adjusted to pH3.1 with 2M sodium hydroxide, 6% methanol; flow rate: 0.3mL/min; detector: CEAS, electrodes 1-8 +300 to +860mV in increments of +80mV; injection volume: 20µL.

HETEROCYCLIC AROMATIC AMINES IN FOOD PRODUCTS

Several heterocyclic aromatic amines [MeIQx (2-Amino-3,8-dimethylimidazo-(4,5-f)quinoxaline, MeIQ (2-Amino-3,4-dimethylimidazo(4,5-f)quinoline, IQ (2-Amino-3-methylimidazo(4,5-f))quinoline, PhIP (2-Amino-1-methyl-6-phenylimid-azo(4,5-b)-pyridine and Trp-P-1 (3-Amino-1,4-dimethyl-5H-pyrido(4,3-b)indole] can be generated by the heating of protein-rich food, e.g., after frying or broiling of meat. The extent of their formation depends on the processing temperature and the intensity of the brown colour of the food. These compounds may also be present in Maillard and Amadori-type processed flavours (Schwarzenbach and Gubler, 1992). These browning type food additives are used in meat, cacao, bread and many other culinary products.

Heterocyclic aromatic amines are known to be mutagenic and potentially carcinogenic (Sugimura, 1985; Sugimura and Sato, 1983). IQ, MeIQ and MeIQx have been reported to be among the numerous compounds responsible for the high mutagenicity of broiled or fried fish and meat (Edmonds *et al.*, 1986). The carcinogenicity of these substances has not yet been tested thoroughly, but there is convincing evidence that some members induce cancer in humans (Galceran *et al.*, 1993; Garner *et al.*, 1984). Despite the low concentrations which have been found, one has to remember that humans are continually exposed to these heterocyclic aromatic amines in heat processed food. Therefore the development of analytical methods for determination of these compounds is important in order to estimate the consumption and risk to which the consumer is exposed over a long period of time.

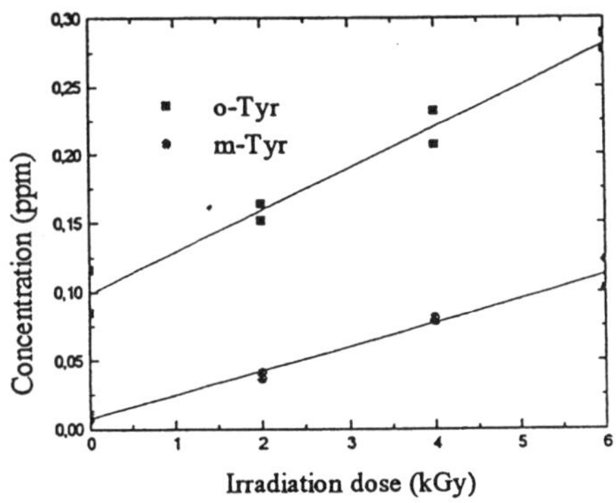

Figure 12. Radiation dose response curve of *o*- and *m*-tyrosine in shrimp irradiated up to 6kGy.

Analytical methods presently used for determination of heterocyclic aromatic amines include GC/MS (Grivas and Nyhammar, 1982; Gross, 1992; Murray *et al.*, 1988, 1989), HPLC/MS (Edmonds *et al.*, 1986; Milton *et al.*, 1987), HPLC with UV (Grivas and Nyhammar, 1982; Gross, 1990; Gross and Grüter, 1992) and fluorescence (Gross and Grüter, 1992; Knize *et al.*, 1992) detection. HPLC with electrochemical detection is a very promising method (Billedeau *et al.*, 1991; Galceran *et al.*, 1993; Grivas and Nyhammar, 1982: Takahashi *et al.*, 1985), since high selectivity and sensitivity are achieved.

Billedeau *et al.* (1991) showed that using dual-electrode detection, the selectivity and sensitivity of the determination can be further increased and the detection limits are lower than those of UV-detection. However, to get qualitative information, time-consuming measurements at different potentials are necessary. Although HPLC methods can be used for the analysis of heterocyclic amines in foods, due to the complexity of the food samples, there is a need for improvements in the selectivity of the analytical method.

Using HPLC with the CEAS the selectivity was markedly increased. With this detector using electrodes set at different potentials, excellent qualitative information can be obtained. Consequently, it was easier to prevent false peak identification of the amines and possible to quantitatively determine the identified heterocyclic aromatic amines.

Determination of some heterocyclic aromatic amines

Oxidative detection mode. A standard mixture of MeIQx, IQ, MeIQ was separated on a C-18-RP column by isocratic elution with acetonitrile/methanol/phosphoric acid and detected at potentials between +300 and +900mV (Figure 13). The amines show different voltammetric behaviour (Figure 14), which can be used for their identification and purity control.

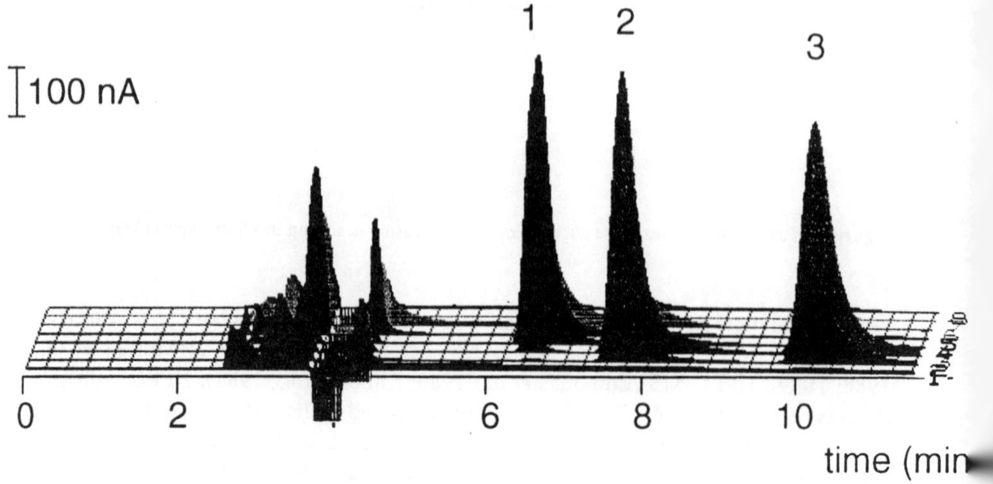

Figure 13. Electrode array chromatogram of a standard mixture of heterocyclic aromatic amines. 1: MeIQx, 2: IQ, 3: MeIQ. Column: Merck, LiChrosorb RP select B, 250 x 4mm, 5µm; eluent: acetonitrile/methanol/glacial acetic acid/bidistilled water (200/100/20/680, v/v/v/v); adjusted to pH5.1 with ammonium hydroxide (25%); flow rate: 0.8mL/min; column temperature: 25°C; detector: CEAS, electrodes 1-8 +400mV to +715mV in increments of +45mV; injection volume: 100µL.

Oxidative-reductive detection mode. While investigating the voltammetric behaviour of the heterocyclic amines (Rauter, 1995), some of them yielded oxidation products, which were reducible. In this approach the compounds are first oxidized in an eight electrode array (+400mV to +715mV). The oxidized products are then completely reduced in the subsequent electrodes (+105 to -165 mV). In this way it is possible to get information about the identity of a compound in real samples. Furthermore, the peak height ratio between anodic and cathodic peaks can contribute to confidence in peak purity.

Analysis of a commercial food product. When developing a method for the determination of heterocyclic aromatic amines, the main difficulties are that the content of these compounds is in the low ng/g range. Additionally the matrix is very complex. Therefore an efficient sample preparation (Gross, 1990; Gross and Grüter, 1992; Knize *et al.*, 1992) is necessary to permit the detection of several heterocyclic amines in one chromatographic experiment and to eliminate interfering substances. After a time-consuming sample preparation, the substances were separated on a C-18 RP column and detected in the oxidative/reductive detection mode (Figure 15).

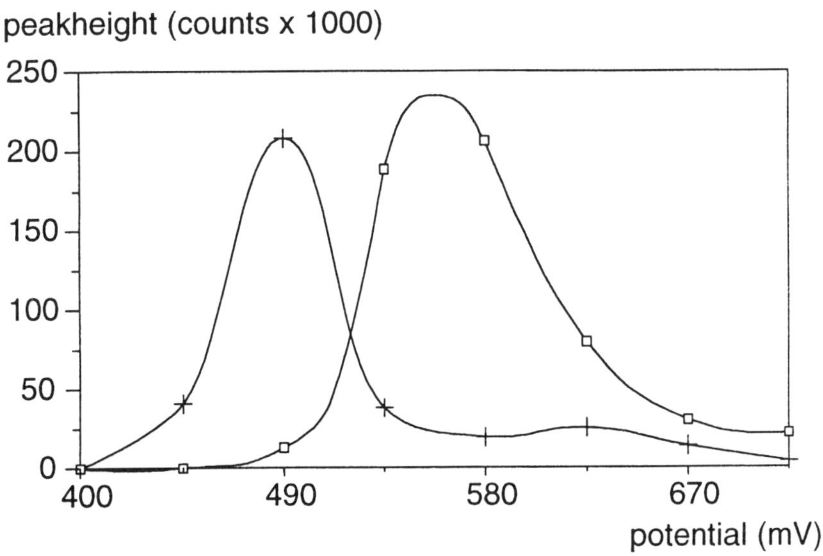

Figure 14. Hydrodynamic voltammograms of MeIQx ( ) and IQ (+). Column: Merck, LiChrosorb RP select B, 250 x 4mm, 5µm; eluent: acetonitrile/methanol/glacial acetic acid/bidistilled water (200/100/20/680; v/v/v/v); adjusted to pH5.1 with ammonium hydroxide (25%); flow rate: 0.8mL/min; column temperature: 25°C; detector: CEAS, electrodes 1-8 +400 to +715mV in increments of +45mV; injection volume: 100µL.

Usually peak confirmation is a critical problem when working with low levels of heterocyclic amines since coelution can readily occur. Using only retention time matching, it seems that IQ and MeIQ (Figure 15, second peak) are present in a commercial soup cube. With voltammetric matching the identification of MeIQ has to be rejected. This is also confirmed by matching the peak height ratios of anodic and cathodic peaks for the standard compounds and sample components.

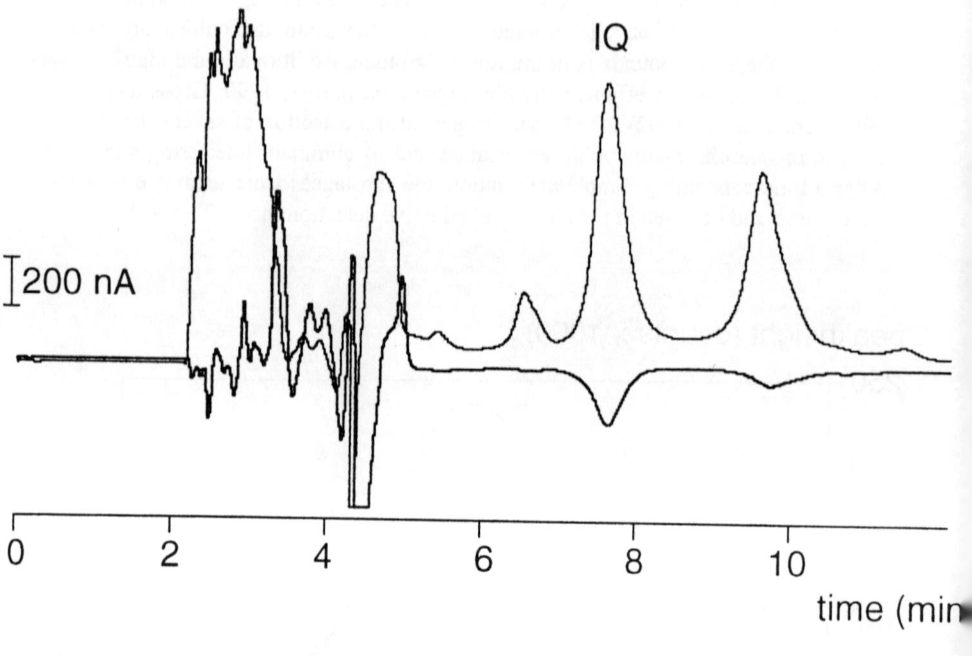

Figure 15. Electrode array chromatogram (oxidative/reductive detection mode, channel 3 and 12) of an extract of a soup cube. Column: Merck, LiChrosorb, RP select B, 250 x 4mm, 5μm; eluent: acetonitrile/methanol/glacial acetic acid/bidistilled water (200/100/20/680; v/v/v/v); adjusted to pH5.1 with ammonium hydroxide (25%); flow rate: 0.8mL/min.; column temperature: 25°C; detector: CEAS, electrodes 1-8 +400 to +715mV, electrodes 9-16 +150 to -165 mV in increments of 45mV; injection volume: 100μL.

In summary, the main advantages of the HPLC-CEAS for detection of heterocyclic aromatic amines include its high selectivity and its information on peak purity and peak identity. Currently we are working on quantification of these compounds in heat processed food.

CONCLUSION

Working with HPLC and coulometric electrode array detection the obvious advantages for the food chemist are:

When developing a new chromatographic method, often there is no information about which potential is needed for optimal detection of the compounds. After one injection setting the 16 cells at different potentials it is immediately possible to get this information.

The CEAS can monitor the sample components at more than one potential. This is useful when the analytes are different in their voltammetric behaviour and results in simultaneous chromatograms at several potentials.

The on-line generation of hydrodynamic voltammograms provide useful data for confirmation of peak identity and peak purity.

For a substance which forms a reducible oxidation product, the peak height ratios of oxidative and reductive detection modes can also give information on peak purity.

In the coulometric electrode array detector the substances are oxidized with highest sensitivity at their half wave potentials. This means that compounds can be detected at a lower potential in comparison to amperometric and coulometric single cell detection thereby producing a better signal-to-noise ratio.

Detection of coeluting substances can be suppressed if the compounds show different voltammetric behaviour. In this way an accurate quantitation can be made.

Acknowledgement
The authors would like to thank the "Fonds zur Förderung der wissenschaftlichen Forschung" (P-8308-CH) and the "Jubiläumsfonds der Österreichischen Nationalbank" (Nr. 5191 and 5764) for supporting this work.

REFERENCES

Achilli, G., Cellerino, G. P., Gamache, P. H. and Melzi d' Eril, G. V. (1993). Identification and determination of phenolic constituents in natural beverages and plant extracts by means of a coulometric electrode array system. J. Chromatog. **632**, 111-117.

Achilli, G., Cellerino, G. P. and Melzi d' Eril, G. V. (1994). Determination of amines in wines by high performance liquid chromatography with electrochemical coulometric detection after precolumn derivatisation. J. Chromatogr. **661**, 201-205.

Archer, A. W. (1984). Determination of 3-amino-1,2,4-triazole (amitrole) in urine by ion-pair high-performance liquid chromatography. J. Chromatogr. **303**, 267-271.

Bernwieser, I. (1992). Phenolsäuren in pflanzenölen. Thesis, University of Vienna.

Bernwieser, I. (1994). Coulometrische elektroden array detektion in kombination mit der HPLC. Thesis, University of Vienna.

Bernwieser, I., Pätzold, R., Galensa, R. and Sontag, G. (1994). HPLC mit coulometrischer elektroden array detektion. bestimmung von 1-O-trans-p-cumarylglycerol in Sorghumhirse, Mais und Bier. Z. Lebensm. Unters. Forsch. **198**, 40-43.

Bernwieser, I. and Sontag, G. (1992). Determination of phenolic acids in vegetable oil using HPLC with multielectrode detection. Z. Lebensm. Unters. Forsch. **195**, 559-561.

Bernwieser, I. and Sontag, G. (1993). HPLC with electrode array detection. Determination of amitrole in drinking and ground water. Fresenius J. Anal. Chem. **347**, 499-501.

Bernwieser, I., Sontag, G., Solar, S., Quint, R. M., Krajnik, P. and Getoff, N. (1995). Nachweis der bestrahlung von hühnerfleisch durch bestimmung von o- und m-tyrosin mit HPLC und elektroden array detektor. Ernährung/Nutrition **19**, 159-162.

Billedeau, St. M., Bryant, M. S. and Holder, C. L. (1991). Analysis of heterocyclic amines using reversed phase high performance liquid chromatography with electrochemical detection. LC-GC Int. **4**, 38-41.

Chuaqui-Offermanns, N. and Mc Dougall, T. M. (1991). Background levels and radiation dose yield of o-tyrosine in chicken meat. J. Food Protection **54**, 935-938.

Chuaqui-Offermanns, N., McDougall, T. M. and Guerrero, A. M. (1993). Validation of o-tyrosine as a marker for detection and dosimetry of irradiated chicken meat. J. Food Protection **56**, 47-50.

Cremer, E. and Riedmann, M. (1964). Identifizierung von gaschromatographisch getrennten aromastoffen in honigen. Z. Naturforsch. **19 B**, 76-77.

Cremer, E. and Riedmann, M. (1965). Gaschromatographische untersuchungen zur frage des honigaromas. Monatsh. Chem. **96**, 364-368.

Cremer, E. and Riedmann, M. (1965). Gaschromatographische untersuchungen des aromas von honigen. Z. Anal. Chem. **212**, 31-37.

Edmonds, C. G., Sethi, S. K., Yamaizumi, Z., Kasai, H., Nishimura, S. and Mc Closkey, J. A. (1986). Analysis of mutagens from cooked foods by directly combined liquid chromatography-mass spectrometry. Environm. Health Perspect. **67**, 35-40.

Fenz, R., Ernst, L. and Galensa, R. (1992). Phenolcarbonsäuren und ihre glycerinester in maisgrits. Z. Lebensm. Unters. Forsch. **194**, 252-258.

Fenz, R. and Galensa, R. (1989). Identification of 1-O-trans-p-coumaroylglycerol as an indicator of maize in beer. Z. Lebensm. Unters. Forsch. **188**, 314-316.

Fenz, R. and Galensa, R. (1990). HPLC-nachweis von mais als malzersatzstoff in bier. Dtsch. Lebensm. Rdsch. **86**, 107-111.

Galceran, M. T., Pais, P. and Puignou, L. (1993). High performance liquid chromatographic determination of ten heterocyclic aromatic amines with electrochemical detection. J. Chromatogr. **655A**, 101-110.

Gamache, P. H., Ryan, E. and Acworth, I. N. (1993). Analysis of phenolic and flavonoid compounds in juice beverages using HPLC with coulometric array detection. J. Chromatogr. **635**, 143-150.

Garner, R. C., Martin, C. N. and Clayson, D. B. (1984). In: Chemical Carcinogens, **1**, Searle, C.E. (Ed.), American Chemical Society, Washington, DC, 2nd ed. pp. 175-275.

Graddon, A. D., Morrison, J. D. and Smith, J. F. (1979). Volatile constituents of some unifloral Australian honeys. J. Agric. Food Chem. **27**, 832-837.

Green, F. O. and Feinstein, R. N. (1957). Quantitative estimation of 3-amino-1,2,4-triazole. Anal. Chem. **29**, 1658-1660.

Grivas, S. and Nyhammar, T. (1982). Analysis of mutagenic imidazo[4,5-f] quinolines and quinoxalines (IQ compounds). Mutat. Res. **142**, 5-8.

Gross, G. A. (1990). Simple methods for quantifying mutagenic heterocyclic aromatic amines in food products. Carcinogenesis **11**, 1597-1603.

Gross, G. A. and Grüter, A. (1992). Quantitation of mutagenic/carcinogenic heterocyclic amines in food products. J. Chromatogr. **592**, 271-278.

Hart, J., White, J. A. and Reid, W. J. (1988). Technical note: Occurrence of o-tyrosine in non irradiated foods. Intern. J. Food Sci. Technol. **23**, 643-647.

Hollborn, E., Wald, B., Galensa, R. and Herrmann, K. (1990). Nachweis von apfelbestandteilen in lebensmitteln mittels HPLC nach festphasenextraktion. Dtsch. Lebensm. Rdsch. **86**, 1-3.

Jörg, E. (1989). Phenolsäuren im honig, Thesis, University of Vienna.

Jörg, E. (1992). Phenolsäuren und phenolsäureester in verschiedenen honigsorten. bestimmung durch HPLC mit coulometrischer multielektrodendetektion. Thesis, University of Vienna.

Jörg, E. and Sontag, G. (1992). Determination of phenolic acids in honey by HPLC using coulometric dual electrode detection. Dtsch Lebensm. Rdsch. **88**, 179-183.

Jörg, E. and Sontag, G. (1993). Multichannel coulometric detection coupled with liquid chromatography for determination of phenolic esters in honey. J. Chromatogr. **635** 137-142.
Karam, L. R. and Simic, M. G. (1988a). Detecting irradiated foods: use of hydroxyl radical biomarkers. Anal. Chem. **60**, 1117A-1119A.
Karam, L. R. and Simic, M. G. (1988b). Ortho-tyrosine as a marker in post irradiation dosimetry (PID) of chicken. In: Health, Impact, Identification, and Dosimetry of Irradiated Foods. Report of a WHO Working Group, Bericht des Instituts für Strahlenhygiene des Bundesgesundheitsamtes ISH-125, 297-304, Neuherberg, FRG.
Knize, M. G., Felton, J. S. and Gross, G. A. (1992). Chromatographic methods for the analysis of heterocyclic amine food mutagens/carcinogens. J. Chromatogr. **624**, 253-265.
Kozlowska, H. Zadernowski, R. and Sosulski, F. (1983a). Phenolic acids in oil seed flours. Nahrung **27**, 449-453.
Kozlowska, H., Rotkiewicz, D., Zadernowski, R. and Sosulski, F. (1983b). Phenolic acids in rapeseed and mustard. J. Am. Oil Chem. Soc. **60**, 1119-1123.
Lokke, H. (1980). Determination of amitrole by ion-pair high-performance liquid chromatography. J. Chromatogr. **200**, 234-237.
Meier, W., Bürgin, R. and Fröhlich, D. (1988). Analysis of o-tyrosine as a method for the identification of irradiated chicken meat. Beta-gamma **1**, 34-35.
Meier, W., Bürgin, R. and Fröhlich, D. (1989). Nachweis von bestrahltem frischfleisch mittels o-tyrosin. Mitt. Geb. Lebensm. Hyg. **80**, 22-29.
Meier, W., Bürgin, R. and Fröhlich, D. (1990). Analysis of o-tyrosine as a method for identification of irradiated chicken and the comparison with other methods. Radiat. Phys. Chem. **35**, 332-336.
Milton, H., Bur, H. and Turesky, R. (1987). Thermospray liquid chromatographic analysis of mutagenic substances present in tryptophan pyrolyzates. J. Chromatogr. **394**, 201 -208.
Murray, S., Gooderham, N. J., Boobis, A. R. and Davis, D. S. (1988). Measurement of MeIQx and DiMeIQx in fried beef by capillary column gas chromatography electron negative ion chemical ionization mass spectrometry. Carcinogenesis **9**, 321-325.
Murray, S., Gooderham, N. J., Boobis, A. R. and Davis, D. S. (1989). Detection and measurement of MeIQx in human urine after ingestion of cooked meat meal. Carcinogenesis **10**, 763-765.
Pachinger, A., Eisner, E., Begutter, H. and Klus, H. (1992). A simple method for the determination of amitrole in drinking and ground water. Fresenius J. Anal. Chem. **342**, 413-415.
Pätzold, R., Fenz, R., Krause, M. and Galensa, R. (1992). Schnellmethode zum nachweis von mais als malzersatzstoff in bier. Dtsch. Lebensm. Rundsch. **88**, 390-391.
Pribyl, L. J., Herzel, F. and Schmidt, G. (1978). Beitrag zur rückstandsanalytik des aminotriazols. Fresenius Z. Anal. Chem. **289**, 81-85.
Rauter, S. (1995). Detection of heterocyclic aromatic amines in food. Thesis, University of Vienna.
Schwarzenbach, R. and Gubler, D. (1992). Detection of heterocyclic aromatic amines in food flavours. J. Chromatogr. **624**, 491-495.
Sontag, G. and Bernwieser, I. (1994a). HPLC coupled with a coulometric electrode array detector. Determination of phloretin glucoside in juices for detection of adulteration. Dtsch. Lebensm. Rdsch. **90**, 72-74.
Sontag, G. and Bernwieser, I. (1994b). HPLC mit coulometrischer elektroden array detektion. Einsatz in der lebensmittelanalytik. Lebensmittelchemie **48**, 127-129.
Speer, K. and Montag, A. (1987). Abbauprodukte des phenylalanins als aromakomponenten des honigs", Dtsch. Lebensm. Rdsch. **83**, 103-107.
Steeg, E. and Montag, A. (1987). Nachweis aromatischer carbonsäuren im honig. Z. Lebensm. Unters. Forsch. **184**, 17-19.
Steeg, E. and Montag, A. (1988a). Minorbestandteile des honigs mit aromarelevanz II. Dtsch. Lebensm. Rdsch. **84**, 147-150.
Steeg, E. and Montag, A. (1988b). Quantitative bestimmung aromatischer carbonsäuren in honig. Z. Lebensm. Unters. Forsch. **187**, 115-120.
Sugimura, T. (1985). Carcinogenicity of mutagenic heterocyclic amines formed during the cooking process. Mutat. Res. **150**, 33-41.
Sugimura, T. and Sato, S. (1983). Mutagens-carcinogens in foods. Cancer Res. **43**, 2415s-2421s.
Takahashi, M., Wakabayashi, K., Nagao, M., Yamamoto, M., Masui, T., Goto, T., Kinae, N., Tomita, I. and Sugimura, T. (1985). Quantification of 2-amino-3-methylimidazo[4,5-f]quinoline and 2-amino-3,8-dimethylimidazo[4,5-f] quinoxaline in beef extracts by liquid chromatography with electrochemical detection. Carcinogenesis **6**, 1195-1199.

Ternes, W. and Rüssel-Sinn, H.A. (1987). Determination of 3-amino-1,2,4-triazole (amitrole) in tissues with electrochemical detection by ion-pair high-performance liquid chromatography. Fresenius Z. Anal. Chem. **326**, 757-759.

Van der Poll, J.M., Vink, M. and Quirijns, J.K. (1988). Capillary gas chromatographic determination of amitrole in water with alkali flame ionization detection. Chromatographia **25**, 511-514.

Van der Poll, J.M., Vink, M. and Quirijns, J.K., (1990). Determination of amitrole in plant tissue and sandy soils by capillary gas chromatography with alkali flame ionization detection. Chromatographia **30**, 155-158.

Vázquez-Ortiz, F.A., Caire, G., Higuera-Ciapara, I. and Hernández, G. (1995). High performance liquid chromatographic determination of free amino acids in shrimps. J. Liquid Chromatog. **18**, 2059-2068.

Wald, B. and Galensa, R. (1989). Nachweis von fruchtsaftmanipulationen bei apfel- und birnensaft. Z. Lebensm. Unters. Forsch. **188**, 107-114.

Water Act (1989). Water supply (water quality) regulation. Department of Environment, London, UK.

Progress in HPLC-HPCE, Vol. 6, pp. 99–126
I.N. Acworth *et al.* (Eds)

THE MEASUREMENT OF MARKERS OF OXIDATIVE DAMAGE, ANTI-OXIDANTS AND RELATED COMPOUNDS USING HPLC AND COULOMETRIC ARRAY ANALYSIS

PAUL H. GAMACHE[1], DOUGLAS R. McCABE[1], HASAN PARVEZ[2], SIMONE PARVEZ[2] and IAN N. ACWORTH[1]

[1]*ESA Inc. 22 Alpha Road, Chelmsford, MA 01824-4171, and* [2]*Neuroendocrinolodie and Neuropharmacologie, du Developpement, Universite de Reims-UFR Sciences, Reims Cedex, France University of Reims, France*

INTRODUCTION

The old adage "moderation in all things" is perhaps best demonstrated by the toxicity of pure oxygen. Although it is obvious that oxygen is essential for life (at least for aerobic organisms), breathing pure oxygen at atmospheric pressure for longer than 48 hours will lead to respiratory distress and death (Huber and Drath, 1981). This toxicity is not due entirely to oxygen *per se* but more to the production of a variety of reactive oxygen species (ROS). Several ROS exist. Diatomic molecular oxygen (O_2) readily reacts to form partially reduced species, which are generally short-lived and highly reactive and include the superoxide anion ($O_2^{\bullet-}$), the hydroperoxyl radical ($HO_2^{\bullet}$) as well as the most reactive species, the hydroxyl free radical ($HO^{\bullet}$). ROS are produced as part of normal metabolism (e.g. superoxide is formed when xanthine is converted to uric acid by xanthine oxidase) and are also extremely useful (e.g. superoxide and hydrogen peroxide are the major "microbicides" produced by circulating phagocytic leukocytes) (Johnston *et al.*, 1975; Babior, 1978; Colton and Gilbert, 1987; Green *et al.*, 1987). Furthermore, the effectiveness of certain anticancer drugs is also dependent on their ability to produce free radicals (Kehrer, 1993).

The body is well-equipped with a variety of mechanisms that serve to control the natural level of ROS and keep them within well-defined limits. These include enzymes (superoxide dismutase, catalase and glutathione peroxidase), antioxidants (vitamins A, C, E, K, coenzyme Q10, uric acid, glutathione and other thiols, polyphenolics and flavonoids) and metal chelation (stopping free radical formation by preventing changes in the metal's oxidation state). These mechanisms act either by preventing excessive ROS formation or by rendering them inactive when overproduced. Under certain conditions, however, such as genetically predisposed individuals or those exposed to irradiation or environmental factors, ROS levels can be raised well beyond the capacity of the protective mechanisms to control them. This situation is called oxidative stress. With oxidative stress, excessive ROS levels attack and chemically alter a variety of macromolecules, including proteins, lipids, and DNA. Such damage can lead to a variety of diseases - reperfusion injury, tissue damage associated with alcoholism, atherosclerosis, cancer, cardiovascular diseases, inflammation and rheumatoid arthritis (Kehrer, 1993).

The ability of the body to prevent ROS attack or, once formed, to recognize and remove the products of ROS damage is a fascinating and very active area of research, with implications in aging, cancer, and a variety of neurodegenerative disorders such as Parkinson's and Alzheimer's diseases. Furthermore, as many of the anti-oxidants are found in the food substances, a person's diet is believed to have a profound affect on the body's ability to control ROS levels. Diets high in anti-oxidants and phytochemicals are now being investigated for their role in regulating diseases associated with excessive ROS levels. Readers interested in learning more about ROS, their metabolism, control, and diseases associated with their overproduction, are referred to Halliwell and Gutteridge (1989). Further information on the use of HPLC-ECD in ROS biochemistry can be found in Acworth and Bailey (1995).

This chapter examines the use of HPLC with coulometric array technology in the study of ROS damage and control. Four different methods are presented: 1) DNA damage assessed by DNA adduct formation; 2) simultaneous measurement of many naturally-occurring antioxidant phytochemicals - the polyphenolics and flavonoids; 3) the measurement of a novel group of phytochemicals, the phytoestrogens, which have been reported to both promote and to inhibit cancer, and to also have anti-oxidant capability; and 4) a method capable of measuring several different fat soluble anti-oxidant vitamins concurrently.

DNA ADDUCT MEASUREMENT

DNA is the genetic code of life. Its faithful translation into mRNA and transcription of the resulting mRNA into proteins is essential for normal growth and reproduction. It is imperative that the integrity and fidelity of the DNA molecule be maintained. Any modification to the bases found in the DNA molecule may result in a mutant protein with modified structure which, in turn, could lead to disease. To this end the body is equipped with a variety of mechanisms designed to identify, correct and repair such DNA mutations. Many examples exist: DNA polymerase I removes mismatched residues; thymine dimers produced by exposure to UV irradiation are removed by an exinuclease; thymine in DNA, which slowly deaminates to uracil, is removed in an enzymatic sequence consisting of uracil-DNA-glycosidase, an AP endonuclease, DNA polymerase I, and finally, DNA ligase. Moreover, it is very difficult to gauge the mutation frequency because many of the above products occur naturally.

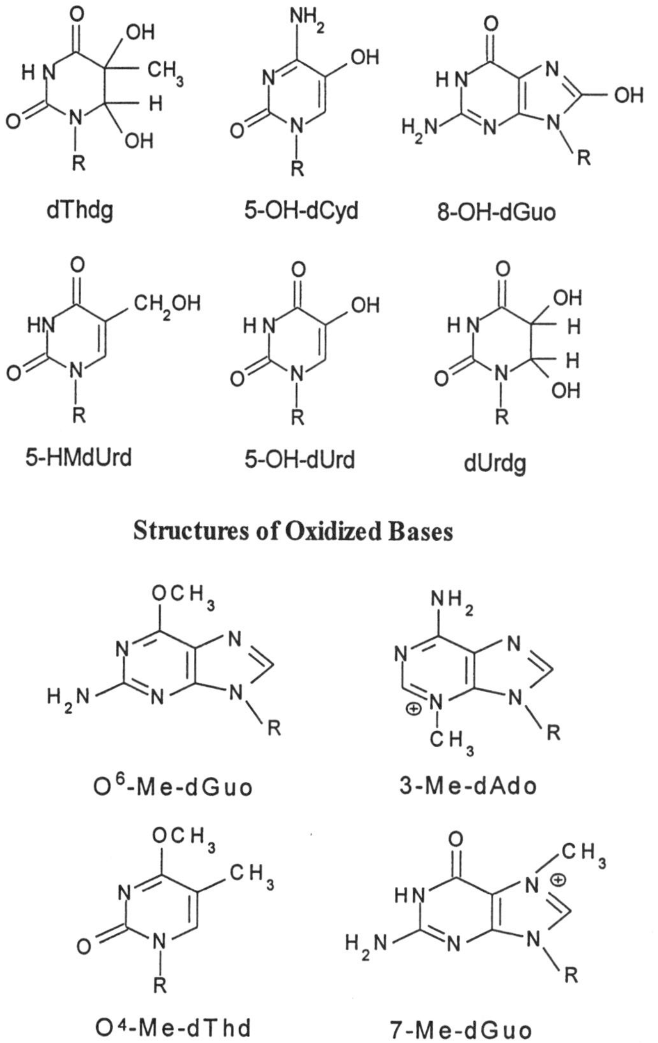

Figure 1. Structures of oxidized and alkylated DNA adducts. dThdg - 5,6-dihdroxy-5,6-dihydrothymidine; 5-OH-dCyd - 5-hydroxy-2'-deoxycytidine; 8-OH-dGuo - 8-hydroxy-2'-deoxyguanosine; 5-HMdUrd - 5-(hydroxymethyl)-2'-deoxyuridine; 5-OH-dUrd - 5-hydroxy-2'-deoxyuridine; dUrdg - 5,6-dihydroxy-5,6-dihydrouridine; O^6-Me-dGuo - O^6-methyl-2'-deoxyguanosine; 3-Me-dAdo - 3-methyl-2'-deoxyadenosine; O^4-Me-dThd - O^4-methyl-thymidine; 7-Me-dGuo - 7-methyl-2'-deoxyguanosine.

Whereas the action of ROS on DNA results in a variety of novel oxidation products, exposure of DNA to electrophiles (including alkylating agents) produces many different covalently modified bases (Marnett and Burcham, 1993) - see Figure 1. Production of DNA adducts have been reported to block or slow the rate of DNA replication which can prevent cell division or produce aberrant chromosomes (Moore and Strauss, 1979); misdirect incorporation of bases during replication, resulting in mutations (Echols and Goodman, 1991; Shibutani *et al.*, 1991; Strauss, 1991); and undergo hydrolysis, eventually leading to strand scission. Measurement of such adducts in tissue is being used currently as an indicator of DNA damage and exposure to mutagens.

One of the major products of oxygen radical attack is 8-hydroxy-2'-deoxyguanosine (8OH2DG) and is reported to be one of the most mutagenic lesions induced in DNA by ROS (Shibutani *et al.*, 1991; Ohtsuka and Nishimura, 1987; Nghiem *et al.*, 1989). Estimates of the daily production of oxidized adducts range from 10^4 to 10^6 per cell (Ames and Gold, 1991) with eight 8OH2DG adducts being formed for every 10^6 normal nucleotides in rat liver nuclear DNA, while 8OH2DG levels were sixteenfold higher in mitochondrial DNA (*Richter et al.*, 1988).

DNA adducts resulting from alkylating agents are the best-studied adducts to date and are formed by the action of exogenous agents (nitrosoureas and nitrosamines) and possibly by endogenous agents (betaine, choline and S-adenosylmethionine) on DNA. These adducts show markedly different mutagenicity, with 7-methyl-2'-deoxyguanosine being the weakest and 3-methyl-2'-deoxyadenosine and O6-methyl-2'-deoxyguanosine being the most potent (see Marnett and Burcham (1993) and references therein).

Separation and detection of DNA adducts has proven problematic due to their low concentration and their occurrence in complex biological matrices. A variety of approaches have been used to measure tissue DNA adduct levels, including immunochemical methods (van Delft *et al.*, 1993), GC/MS (Dizdaroglu and Gajewski, E., 1990), GC/MS/MS (Cushnir et al., 1993), ^{32}P post-labeling (Randerath *et al.*, 1981; Gupta, *et al.*, 1982) and HPLC with UV (Cathcart *et al.*, 1984).[1] Many DNA adducts are electrochemically active and can be measured using HPLC-ECD. This technique is selective, sensitive, has good dynamic range, requires limited sample preparation, and does not involve the use of radioisotopes or the purchase of expensive equipment, thereby overcoming many of the issues associated with the other analytical methodologies. For a review on HPLC-ECD see Acworth, Matson and Bowers (this volume).

Many different HPLC-ECD methods have appeared in the literature but most measure relatively few adducts. This is probably a result of issues with chromatographic and voltammetric resolution. For example, van Delft *et al.* (1993) used a wall-jet amperometric ECD to measure the formation of N7-guanine in salmon sperm DNA *in vitro*. With an applied potential of 1350mV it is unclear how long this assay would be of practical use before the working electrode would fail. Xie and Duda attempted to measure several guanine adducts and compared the use of a single electrode thin-layer detector and a wall-jet cell (1994). They elegantly illustrated the problem of using a single electrode to simultaneously measure a wide range of compounds with disparate voltammetric behaviors. In order to measure methyl-adducts, which are more difficult to oxidize, a high potential has to be applied to the thin-layer working electrode. This compensation produces more noise, resulting in a

[1] See also Cummings *et al.*, 1993 for a review on the use of HPLC for investigation of covalent modifications to nucleic acid.

decreased signal to noise (s/n) ratio. Furthermore, there is a greater probability that chromatographic interferences may occur.

Others have developed methods using dual coulometric electrodes to measure DNA adducts. These electrodes offer superior sensitivity and voltammetric selectivity and have been used to measure 8OH2DG in a variety of tissues (including urine) and disease states (Shigenaga *et al.*, 1990; Takeuchi and Morimoto, 1993; Fischer-Nielsen *et al.*, 1994; Shigenaga *et al.*, 1994; Adachi *et al.*, 1995). Ames and his group have also used a coulometric approach to measure a variety of novel DNA adducts, including 5-hydroxy-2'-deoxycytidine (Wagner *et al.*, 1992) and 1,N^2-propano adducts of 2'-deoxyguanosine (Douki and Ames, 1994). We have developed a gradient HPLC coulometric array method with low picogram sensitivity capable of simultaneously measuring eleven DNA adducts and parent bases.

The HPLC system

All samples were analyzed using a CoulArray™ system based on a CoulArray™ detector (model 5600). The system consisted of two model 580 dual piston pumps connected to an M800 dynamic gradient mixer and dual pulse dampers. Samples (10µL) were injected onto the system using a model 540 autosampler. Analytes were resolved at 1mL/min on a Meta-250 (250 x 4.6mm ID, 5µm, C18) column with integrated guard column. Analytes were detected electrochemically using an array of twelve graphite flow-through coulometric electrodes (with applied potentials set to 100, 150, 220, 290, 380, 510, 570, 630, 690, 750, 810, 870mV vs. palladium reference). The column and electrodes were housed in a thermal chamber maintained at 31°C. All equipment was from ESA, Inc. The fluid path was plumbed with inert PEEK tubing to help lower noise (background currents). CoulArray™ software was used to control the method as well as to collect and process data.

Mobile phases

Mobile phase A consisted of 100 mM sodium acetate (EM Science, New Jersey) pH 5.2 with phosphoric acid (85% Fisher HPLC Grade, New Jersey), containing 50 µL/L of the microbicide reagent, MB (ESA, Inc.). Mobile phase B consisted of 100 mM sodium acetate pH 5.2 with phosphoric acid, 25% methanol (v/v) (EM Science, New Jersey).

Standards

Compounds were purchased, dissolved and stored as reported in Table 1.

Results and discussion

Using this method a variety of DNA adducts and parent bases can be measured in under 25 minutes - see Figure 2. This approach utilizes the inherent differences between compounds' voltammetric behavior to add a further degree of resolution to the chromatography. The use of coulometric array detection enables the sensitive measurement of the readily oxidized compounds at the same time as the non-readily oxidized species. This method has an estimated limit of detection for 8-hydroxy-guanine, 5-hydroxy-cytidine and 8OH2DG of <5pg on column (s/n 3:1) and <50pg for all the other compounds. Furthermore this method does not suffer from a common problem seen with other methods. With inadequate resolution, high levels of guanine

Table 1.
DNA adduct standards source and storage information

Chemical	Chemical Source	Standard Preparation Solvent
guanine	Sigma Chemical Company St. Louis, MO	guanine adducts mobile phase*
8-hydroxy-guanine	Lancaster Synthesis, Inc. Windham, NH	guanine adducts mobile phase*
5-hydroxy-2'-deoxycytidine	Gift from R.A. Floyd OMRF Oklahoma City, OK	water
3-methyl-guanine	Chemsyn Laboratories Lenexa, KS	3 mL water 0.1 mL 10 N NaOH
7-hydroxy-ethyl-guanine	Gift from Xi Huang NYU Medical Center Tuxedo, NY	guanine adducts mobile phase*, pH 3.0
adenine	Sigma Chemical Company St. Louis, MO	guanine adducts mobile phase*
7-methyl-guanine	Chemsyn Laboratories Lenexa, KS	3 mL water 0.1 mL 10 N NaOH
guanosine	Sigma Chemical Company St. Louis, MO	0.5% HCl
2'-deoxyguanosine	Sigma Chemical Company St. Louis, MO	guanine adducts mobile phase*
8-hydroxy-2'-deoxyguanosine	Wako Chemicals Richmond, VA	0.1 M H_3PO_4
O6-methyl-guanine	Chemsyn Laboratories Lenexa, KS	0.1 M H_3PO_4

*Standards were dissolved in 100 mM sodium acetate, pH 5.2 with phosphoric acid, 10% MeOH (v/v). All standards were prepared at stock concentrations of 1.0 mg/mL and were stored at 4°C.

and adenine can overwhelm the measurement of the low levels of adducts (for example adenine usually interferes with the measurement of 7-methyl-guanine). With our method of both chromatographic and voltammetric resolution all compounds can be routinely measured without interference. This is very important as an adduct's level needs to be normalized by relating its concentration to those of the parent base. For example, Mecocci *et al.* (1993) reported that there is an age-dependent increase in levels of 8OH2DG in human brain tissue. These authors used a gradient Coulochem® electrode array system (CEAS, model 5500, ESA, Inc.) method to show that the mitochondrial ratio of 8OH2DG to deoxyguanosine was tenfold higher than in nuclear DNA and concluded that such DNA damage may contribute to the age-dependent increases in incidence of neurodegenerative diseases.

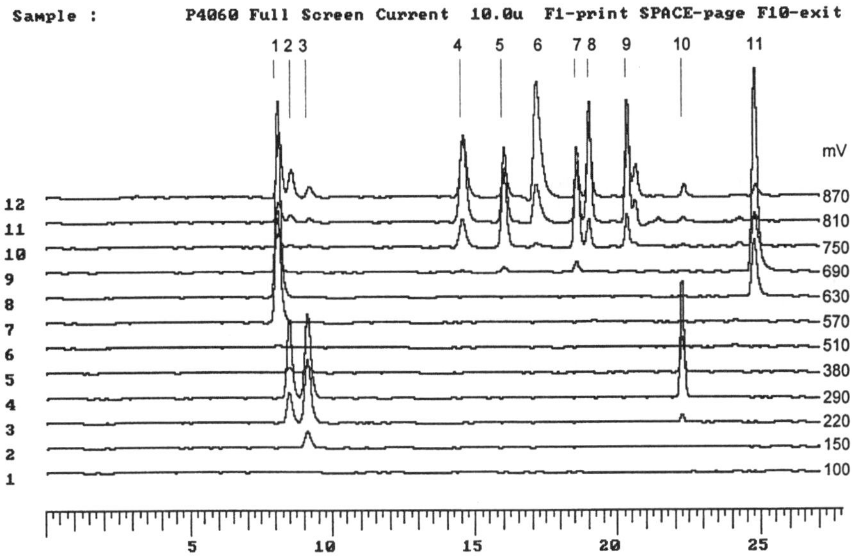

Figure 2. Analysis of DNA adducts using a gradient HPLC coulometric array method as outlined above. 1- guanine; 2- 8-hydroxy-guanine; 3- 5-hydroxy-2'-deoxycytidine; 4- 3-methyl-guanine; 5- 7-hydroxy-ethyl-guanine; 6- adenine; 7- 7-methyl-guanine; 8- guanosine; 9- 2'-deoxyguanosine; 10- 8-hydroxy-2'- deoxyguanosine; 11- O6-methyl-guanine.

POLYPHENOLIC AND FLAVONOID MEASUREMENT

The biochemistry of plants is a fascinating and complex subject. Many plant metabolites are bioactive and may have effects in animals. This is obvious for certain alkaloids such as ricin, curare, strychnine, atropine and aconitine can kill a person at low levels. Yet for many phytochemicals, bioactivity is far from clear. There is now considerable interest in quantifying phytochemicals especially since so many plant extracts are now being consumed by the general public as health aids and dietary supplements (Consumer Reports, 1995).

Although scientists have tried, for the sake of clarity, to classify plant metabolites into groups of common chemical structures, such cataloging remains artificial and can be confusing because certain compounds fall into more than one category. For example, quercetin is a polyphenolic yet it is usually regarded as a flavonoid.

Recent research has focused on two major groups of plant metabolites - the polyphenolics and flavonoids. Polyphenols (which used to be called vegetable tannins) are an abundant and complex group of secondary metabolites found in higher plants. They are water soluble and have molecular weights from 500 to 3000D. The polyphenol group is made up of three families of which the proanthocyanidins predominate. The other two families are derivatives of gallic acid and hexahydroxydiphenic acid.

The other single most important group of phenolics in food are the flavonoids, which mainly consist of catechins, proanthocyanins, anthocyanidins and flavones, flavonols and their glycosides. Catechins are widely distributed in plants and are especially abundant in tea leaves (making up approximately 30% of the dry leaf weight). Catechins present antioxidant and anticancer properties. The antioxidant capability of green tea is derived from epigallocatechin (EGC) (Lunder, 1992). Epicatechin gallate (ECG) and epigallocatechin gallate (EGCG) quench free radical chain reactions that damage membrane lipids; they also inhibit mutagenicity and DNA damaging activity (Osawa, 1992; Osawa *et al.*, 1992). Proanthocyanidins consist of chains of flavan-3-ol units and can complex strongly with proteins and carbohydrates (Butler and Rogler, 1992; Hasalam *et al.*, 1992). Anthocyanins are responsible for plant pigmentation, including red, orange, mauve, and blue shades. Flavones, flavanols and their glycosides are distributed widely in plants and it has been estimated that humans consume 1g of these compounds daily. The most common and most bioactive member is quercetin. Quercetin and other polyphenols, such as ellagic acid and chlorogenic acid, may play a major role in carcinogenesis by both reducing the availability of carcinogens and interfering with their biotransformation in the liver (Huan and Ferraro, 1992; Starvic *et al.*, 1992).

Polyphenolics and flavonoids, therefore, play several major roles. They are responsible for the sensory and nutritional quality of fresh and processed foods, and their stability. For example, as anti-oxidants they prevent the formation of oxidized lipids (malodors and tastes). One major property common to all polyphenols and flavonoids is their ability to scavenge free radicals, thereby preventing free radical damage. It is this ability that may account for their bioactivity and reported prevention of cancer and tumorogenesis.

A variety of approaches have been used to measure polyphenolics and flavonoids, including GC-MS and HPLC, the latter with a variety of different detection systems (visible, UV and PDA) (see Ho *et al*, 1992 and Nagy and Wade, 1995 and references therein). The use of these techniques for the quantification and characterization of polyphenols and flavonoids, however, often can be difficult due to sample complexity or trace substance levels. Although HPLC-ECD offers superior sensitivity over other techniques, single electrode approaches suffer from poor voltammetric resolution. The CEAS, with its array of coulometrically efficient electrodes, is ideally suited for the measurement of phytochemicals. The coulometric array offers the stability of HPLC-UV detectors with better resolution than HPLC-PDA (see Acworth, Matson and Bowers, this volume; Sontag and Bernwieser, this volume). The difference between one flavonoid and another sometimes is simply the shift of a hydroxyl group from one position to another. This would not be expected to affect the compound's UV absorbance spectrum, but it would markedly change its electrochemical voltammetric behavior. Coulometric array detection, therefore, can achieve greater resolution of analogous flavonoids and polyphenolics than PDA.

Several papers have reported the use of coulometric arrays to measure polyphenolics and flavonoids. Gamache *et al.*, (1993) developed a gradient method with an array of 16 coulometric electrodes capable of analyzing 27 compounds in 30 minutes in a 20μL sample. After simple sample filtration, metabolite patterns generated on the coulometric array were used in the study of fruit juice adulteration. This study was later extended further by the ability to classify orange juice based on varietal and also on geographical region (Gamache *et al.*, 1995). Achilli *et al.* (1993)

developed a similar gradient method capable of resolving 34 analytes in 45 minutes. This method studied metabolite profiles in wine, beer, lemon juice, and in plant extracts such as soy and tobacco. *Uchida et al.* (1992) used a coulometric array to examine the antioxidant levels during fermentation, storage, and packaging, and discussed the role of phytochemicals in beer stability.

Recently, Acworth *et al.* (1994) developed a gradient coulometric array method capable of measuring epicatechin (EC), ECG, EGC and EGCG within 30 minutes and used it to examine urinary levels of these compounds following green tea consumption. Lee *et al.* (1995) adapted this method to measure levels of these compounds in plasma and urine of human subjects consuming green tea. The limits of detection of these compounds were 0.5 to 1.5ng/ml of plasma or urine samples. Each of the four compounds showed markedly different metabolism with EC and ECG not being detected in plasma, while EGCG and ECG were not detected in urine.

ANALYSIS OF PHYTOESTROGEN LEVELS

Two groups of diphenolic compounds with purported biological activity have been found in human urine and other biological fluids - the lignans and the isoflavonic phytoestrogens (Adlercreutz *et al.*, 1986 and references therein). Human diet, especially those high in grain and other fiber-rich foods, contain plant lignans. The common plant lignan, matairesinol, is converted to the major mammalian lignan, eneterolactone, by intestinal bacteria. Enterolactone is further metabolized to another bioactive compound, enterodiol.

Researchers first became aware of the isoflavonic phytoestrogens following a massive outbreak of infertility in Australian sheep. The cause was linked to a phytochemical, formononetin, found in the clover feed (Price and Fenwick, 1985). This infertility was not due to formononetin *per se*, but to one of its metabolites produced by the action of ruminal bacteria. Although several phytoestrogens are produced including daidzein, equol and *O*-desmethyl- angolensin, it was equol that was found to be responsible for the permanent infertility of the ewes. Another phytoestrogen, genistein, has recently been found in cow's urine in addition to the other phytoestrogens mentioned above. Humans are also exposed to phytoestrogens (as evidenced by urinary levels) which are thought to be derived from either cow's milk (Bannwart, *et al.*, 1985) or by the action of gut bacteria on ingested soya protein (Setchell *et al.*, 1984). The physiological effects of the phytoestrogens on women, especially on their menstrual cycle, is still far from clear (Adlercreutz *et al.*, 1986, 1987).

The phytoestrogens are thought to play a role in prevention of some forms of hormone-dependent cancers including those of the breast, endometrium and ovary. Although phytoestrogens have clearly reduced the incidence of cancer in experimental animals their ability to reduce cancer in humans still remains anecdotal (see Steinmetz and Potter, 1991, and references therein). One possible mechanism of action involves their weak estrogenic activity. Messina and Messina (1991) suggest that phytoestrogens may bind to estrogen receptors thereby blocking the action of the more potent estrogens. Additionally, phytoestrogens may stimulate liver production of sex-hormone- binding globulin resulting in a decrease in biologically active free estrogen (Adlercreutz, 1990). Genistein interferes with the transmission of signals from cellular growth factor receptors by inhibiting tyrosine kinases and other enzymes (Adlercreutz, 1990; Messina and Messina, 1991). Genistein may also inhibit some cytochrome P450 enzymes (Messina and Barnes, 1991). One of the responsibilities of cytochrome P450 is the detoxification of xenobiotics. Unfortunately, a bi-product of this reaction is the production of hydrogen peroxide which can damage DNA and possibly lead to cancer (see above). Phytoestrogens may play a dual role in ROS control - inhibiting their production and reacting with them if produced (free radical quenching).

A variety of techniques have been used to measure lignans and phytoestrogens and include GC (Fotsis *et al.*, 1982), GC-MS (Adlercreutz *et al.*, 1983; Setchell *et al.*, 1983), HPLC-MS (Setchell *et al.*, 1987) and HPLC-UV (Wang *et al*, 1990; Supko and Philips, 1995). For general information on extraction and identification see Harris and Haggerty (1993). Only one paper could be found that measured these compounds using HPLC-ECD. Setchell, *et al.* (1987) compared isocratic HPLC-UV and isocratic HPLC-ECD methods and concluded that electrochemical measurement was 300-1000 fold more sensitive than the UV approach. Mammalian estrogens (estrone, estradiol, estriol and estetrol) also have been measured successfully using HPLC-ECD (Shimada *et al.*, 1979, 1981; Hayashi *et al.*, 1985).

We have developed a sensitive, gradient coulometric array method capable of measuring a variety of phytoestrogens (coumestrol, daidzein, daidzin, equol, and quercetin), estrogens (estrone, estradiol and estriol), and the synthetic estrogen analog - diethylstilbestrol - simultaneously.

The HPLC system

All samples were analyzed using a CoulArray™ detector (model 5600) based system. The system consisted of two model 580 dual piston pumps connected to an M800 dynamic gradient mixer and dual pulse dampers. Samples (50μL) were injected onto the system using a model 540 autosampler maintained at 20°C. Analytes were resolved at 0.8mL/min on an MD150 (150 x 3mm ID, 3 μm C18) column using the gradient presented in Table 2. Analytes were detected electrochemically using an array of eight graphite flow-through coulometric electrodes (with applied potentials set to 260, 320, 380, 440, 500, 560, 620 and -300 mV vs. palladium reference). The column and electrodes were housed in a thermal chamber maintained at 37°C. All equipment was from ESA, Inc. The fluid path was plumbed with inert PEEK tubing to help lower noise (background currents). CoulArray™ software was used to control the method as well as to collect and process data.

Table 2.
Gradient profile for the phytoestrogen method

Line	Time (min)	Activity	Comments
1	0.0	Flow	20%B 0.8 mL/min
2	0.1	Autozero	On
3	0.5	Autosampler	Inject
4	1.0	File	Start
5	25	Flow	100%B 0.8 mL/min
6	25	Flow	20%B 0.8 mL/min
7	30	Flow	20%B 0.8mlLmin
8	30	File	Stop
9	30	Method	Nite[1]

[1]Isocratic 50%B at 0.1 ml/min

Mobile phases
Mobile phase A consisted of 50mM sodium acetate (EM Science, New Jersey) pH 4.8 with acetic acid (J. T. Baker Chemical Company, New Jersey): methanol (EM Science) 80:20 (v/v). Mobile phase B consisted of 50mM sodium acetate pH 4.8: methanol: acetonitrile (EM Science), 40:40:20 (v/v/v).

Standards
Compounds were purchased, dissolved and stored as reported in Table 3.

Sample preparation
Female Sprague-Dawley rats were injected with either vehicle (dimethylsulfoxide, DMSO), 10μg diethylstilbestrol, 100μg coumestrol, or 1mg equol. One hour following drug administration, a plasma sample was obtained, the animals were then sacrificed and uterine tissue collected. The uterine tissue from several rats was pooled and homogenized in cold buffer (10mM Tris, 1.5nM EDTA, pH7.4), 30mg/mL (w/v) buffer. Homogenates were stored at -70°C until analysis.

Determination of free metabolite levels
A 0.2mL volume of tissue homogenate, plasma or standard mixture (1μg/mL of each analyte) was vortex-mixed for 10 minutes with 1.2mL of ethanol. Following centrifugation (12,000rpm, 4°C, 10 minutes), 1.0mL of supernatant was evaporated to dryness by vacuum centrifugation. Prior to analysis, the extracts were reconstituted as described below.

Table 3.
Phytoestrogen and estrogen standard, source and storage information

Name	Source/cat #	Solvent	Stock conc. (mg/mL)
coumestrol	Indofine / 02-0535	Ethanol / 1%DMSO	0.2
daidzein	Indofine / D-101	Ethanol	0.1
daidzin	Indofine / 02-1096	Ethanol	0.1
diethylstilbestrol	Sigma / D4628	Ethanol/ 0.2%DMSO	1.0
equol	Indofine / 02 -1268	Ethanol/ 0.2%DMSO	0.1
estradiol	Sigma / E-8875	Ethanol	0.1
estriol	Sigma / E-1253	Ethanol	1.0
estrone	Sigma / E-9750	Ethanol	1.0
genistein	Sigma / G6649	Ethanol	0.1
quercetin	Sigma / Q-0125	Methanol	1.0

Standards are stored at -20 °C in polypropylene tubes. Before use, the stock standards are sonicated in warm water for 10 minutes. Dilutions are made in mobile phase B - 50mM sodium acetate pH4.8:methanol:acetonitrile 40:20:20 (v/v/v). DMSO - dimethylsulfoxide.

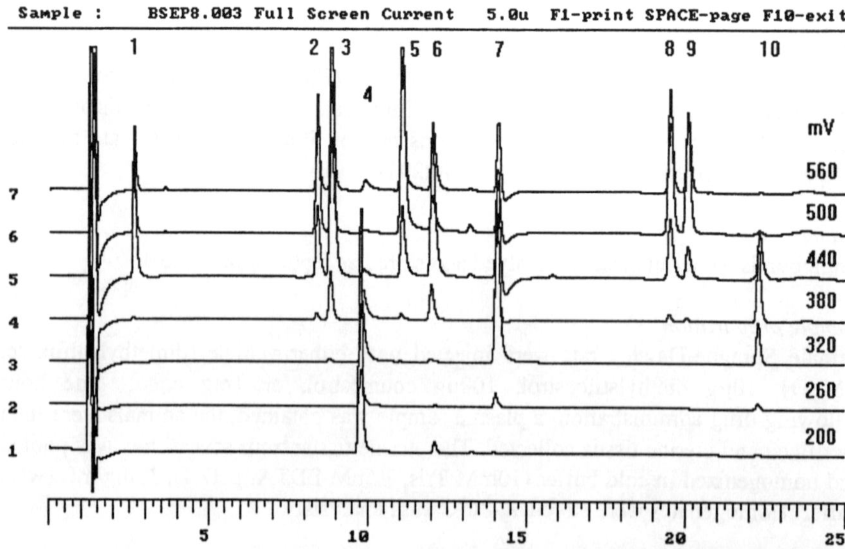

Figure 3. Analysis of phytoestrogens, estrogens and the synthetic estrogen analog - diethylstilbestrol using gradient HPLC with coulometric array detection. Method conditions are given above. 1- daidzin; 2- estriol; 3- daidzein; 4- quercetin; 5- equol; 6- genistein; 7- coumestrol; 8- estradiol; 9- estrone; 10- diethylstilbestrol.

Determination of total (conjugated and free) analyte levels

A 0.2mL volume of tissue homogenate, plasma, or standard mixture (1µg/mL of each analyte) was mixed with 50µL ß-glucuronidase (from *Helix pomatia* from Sigma; 100,000 units/mL ß-glucuronidase activity and 5000 units/mL sulfatase activity) and 0.2mL buffer (0.1M sodium acetate pH 5.0 containing 0.1% ascorbic acid (w/v) and 0.01% (w/v) EDTA). This mixture was incubated at 37°C for two hours before the addition of 1.2mL of ethanol to stop the reaction. Following vortex-mixing, the samples were centrifuged and evaporated to dryness as described above.

Evaporated extracts were dissolved in 0.4mL of sodium acetate buffer (0.1M sodium acetate pH 5.0 containing 0.1% (w/v) ascorbic acid and 0.01% (w/v) EDTA).

Results and discussion

Compounds were resolved in under 25 minutes and were identified both by their chromatographic retention times and voltammetric behavior across the array - see Figure 3. Maximal oxidation potentials (in mV) for each compound were: coumestrol = +500; daidzin = +620; diethylstilbestrol = +440; equol = +560; estradiol = +560; estriol = +560; estrone = +560; genistein = +500 and quercetin = +260.

Basal plasma free levels of genistein, quercetin, and equol were approximately 1.8, 11.2 and 1.7 ng/mL respectively. Following hydrolysis these levels increased markedly to 293, 183 and 22 ng/mL for genistein, quercetin and equol, respectively. Subcutaneous administration of diethylstilbestrol elevated both free and total diethylstilbestrol levels in plasma and uterine tissue. Similarly, administration of equol elevated its free and total levels in plasma and tissue. In animals receiving

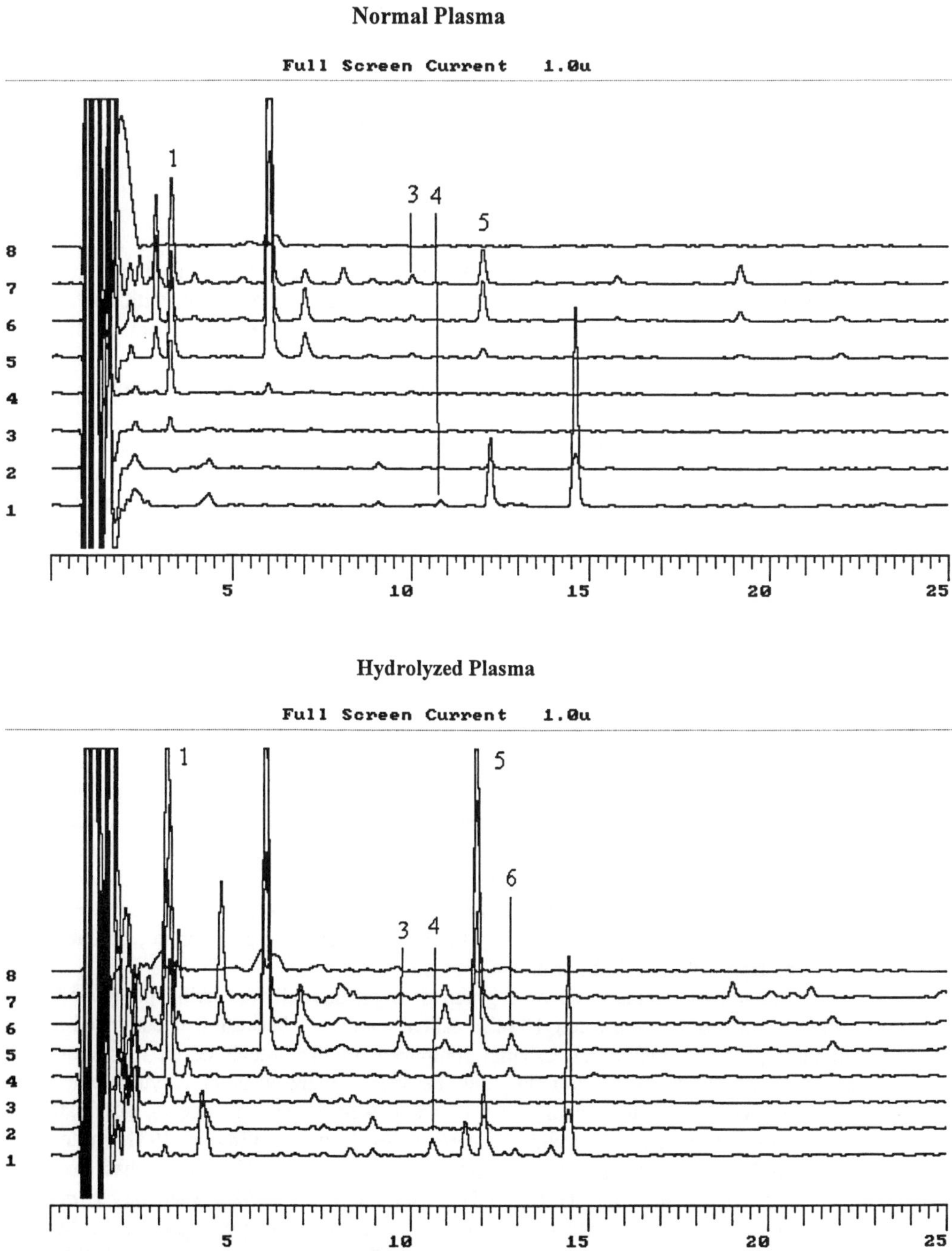

Figure 4. Analysis of phytoestrogens in plasma using a gradient HPLC and coulometric array method. The upper chromatogram shows the free levels of endogenous phytoestrogens and the lower chromatogram shows a marked increase in their levels following hydrolysis. A slightly different gradient profile was used than for Figure 3.

coumestrol, however, only conjugated plasma levels were elevated. See Figure 4. Readers are directed to Gamache (1995) for further details.

This global method will enable the study of both mammalian estrogens and phytoestrogens concurrently. This method is, at present, being modified to include resolution of the mammalian lignans, enterolactone and enterodiol.

DETERMINATION OF FAT SOLUBLE VITAMIN LEVELS

Fat soluble vitamins are a diverse group of essential cofactors divided into four families - A, D, E and K. Of these vitamins A, E, possibly K, along with another lipophilic compound, coenzyme Q10, all play a role in oxidative metabolism and free radical control.

Vitamin A (retinol) is formed in the intestine from its precursor ß-carotene. ß-carotene and the carotenoids as a whole are abundant in green plants, carrots, and other vegetables. Following ingestion, vitamin A is stored in the liver as fatty acid esters (e.g. palmitate). Although vitamin A's major role is in photoreception, it and other carotenoids (ß-carotene, lutein and lycopene) are all effective antioxidants and free radical scavengers (Frei and Ames, 1991; Rice-Evans and Diplock (1993) and references therein). There are several papers in literature using HPLC-ECD for the measurement of the carotenoids. Bryan *et al.* (1991) used normal phase HPLC with amperometry to measure a variety of retinoids, including vitamin A palmitate, and reported a lower detection limit of 1ng on column. MacCrehan and Schonberger (1987a) compared HPLC-ECD and HPLC-UV techniques and measured retinol and its isomers in human serum. They reported a lower detection limit of <200pg on column. Others have measured vitamin A1 using amperometry and have examined its plasma levels in control patients and those with liver cirrhosis (Wring *et al.*, 1988; Wring and Hart, 1989). MacCrehan and Schonberger (1987b) developed a gradient method capable of simultaneously measuring retinol, α-tocopherol, and ß-carotene. Although the method produced data comparable to those obtained from HPLC-UV/Vis its full potential could not be realized due to poor voltammetric resolution (the single thin-layer electrode had to be driven at 900mV vs. Ag/AgCl reference in order to detect retinol) and an inability to use more sensitive gain ranges, probably resulting from baseline drift during the gradient.

Vitamin E is made up of a group of eight naturally-occurring compounds called tocopherols. Coenzyme Q10 (also called ubiquinone because it is contained in all biological systems) is an essential component of the mitochondrial respiratory chain in which it acts as a redox link between flavoproteins and cytochromes. Together, vitamin E and coenzyme Q10 are powerful reducing agents, acting to prevent formation of lipid peroxides in membranes (Burton and Ingold, 1981; Burton *et al*, 1983; Halliwell, 1992; Hall and Traystman, 1992). Vitamin E has been measured either alone using HPLC-ECD (Castle and Cooke, 1985; Vatassary *et al.*, 1993) or in conjunction with coenzyme Q10. Simultaneous approaches use either amperometric detection (Lang and Packer, 1987) or coulometric redox cycling (Murphy and Kehrer, 1987; Edlund, 1988).

The global method described below permits the rapid separation and detection of retinol, retinyl acetate (internal standard), lutein, γ-tocopherol, α-tocopherol, retinyl palmitate, lycopene, α-carotene, β-carotene, coenzyme Q10, and vitamin K1, simultaneously. The use of gradient elution with voltammetric resolution, achieved using an array of eight coulometrically efficient electrodes, permits measurement of these compounds in extracts of human serum. An oxidative array is used for the measurement of retinoids, tocopherols, and carotenoids followed by reduction and re-oxidation for determination of vitamin K and coenzyme Q10. Control sera from the National Institute of Standards and Technology (NIST) were used to qualify the calibration. Linearity, limit of detection, within run precision, and comparison with control data are presented below.

The HPLC system

All samples were analyzed using a CoulArray™ system (model 5600). The system consisted of two model 580 dual piston pumps connected to an M800 dynamic gradient mixer and dual pulse dampers. Samples (10μL) were injected onto the system using a model 540 autosampler. Analytes were resolved on an MD-150 (150 x 3.0mm ID, 3μm, C18) column using the gradient presented in Table 4. Analytes were detected electrochemically using an array of eight graphite flow-through coulometric electrodes with applied potentials set to 200, 400, 500, 700, 800, -1000, 200, 500 vs. Pd. reference. All equipment was from ESA, Inc.

Mobile phases

Mobile phase A consisted of methanol:water:1.0M ammonium acetate pH4.4, 90:8:2 (v/v/v). Mobile phase B consisted of methanol:1-propanol:1.0M ammonium acetate pH4.4, 78:20:2, (v/v/v). All compounds were from either Fisher Scientific (New Jersey) or EM Science (New Jersey).

Table 4.
Gradient profile for the fat soluble vitamin method

Line	Time	Activity	Comments
1	0.0	Flow	0%B 0.8 mL/min
2	4.0	Flow	0%B 0.8 mL/min
3	4.2	Autozero	On
4	4.4	Autosampler	Inject
5	5.0	File	Start
6	15	Flow	80%B 0.8 mL/min
7	25	Flow	100%B 0.8 mL/min
8	32	Flow	100%B 0.8 mL/min
9	32	File	Stop
10	32	Method	Nite[1]

[1]Isocratic 50%B at 0.1 ml/min

Standards

Compound source, preparation, and alternate name are given in Table 5. Standards were stored at -20 °C in polypropylene tubes. Before use, the stock standards were sonicated in warm water for one hour. Dilutions were made in pure ethanol (Pharmco) containing 10 mg/L butylated hydroxyanisole (BHA) (Sigma, Missouri). Individual solutions of 10 μg/mL were thawed daily by sonication for 30 minutes in warm water. The exact concentrations for the standards that were more difficult to dissolve were directly obtained spectrophotometrically (Bacharach Coleman Model 35 spectrophotometer) - see Table 6. Compounds were first dissolved in ethanol lacking BHA before being assayed.

Table 5.
Fat soluble vitamin standards source and storage information

Name	Source/cat #	Alternate name	Solvent	Approx stock conc. (mg/mL)
retinol (all trans)	Sigma R7632	vitamin A	ethanol	1.0
retinyl acetate (all trans)	Sigma R4632		ethanol	1.0
lutein	Sigma X6250	xanthophyll, α - carotene-3, 3'diol	ethanol	0.1
γ-tocopherol	Sigma T1782		methanol	1.0
α-tocopherol	Sigma T3251	vitamin E	methanol	1.0
retinyl palmitate (all trans)	Sigma R3375		ethanol	5.0
lycopene	Sigma L9879	ψ, ψ-carotene	chloroform/ methanol -50/50 (v/v)	0.1
α - carotene	Sigma C0251			0.1
β - carotene	Sigma C0126		ethanol	0.1
coenzyme Q_{10}	Sigma C9629	ubiquinone -50	ethanol	0.1
vitamin K_1	Calbiochem 6811	phylloquinone	ethanol	1.0

Sample preparation:
NIST serum controls (Gaithersburg, Maryland) were reconstituted as directed in 1.0mL of water. A 0.2mL volume of serum (NIST or patient sample) or standard mixture (0.5 or 1.0μg/mL each) was mixed in an eppendorf centrifuge tube with 0.2 mL ethanol containing 10mg/L BHA and 10μL of a 10μg/mL retinyl acetate (internal standard) solution. After vortex-mixing for 1 minute, a 1.0mL volume of hexane was added, followed by an additional 10 minutes of mixing. After centrifugation (4,000rpm, 10 minutes), about 0.8mL of the supernatant was withdrawn into a second eppendorf tube and was evaporated to dryness by vacuum centrifugation (about 35 minutes). The residue was then dissolved in 0.2mL ethanol/BHA diluent by vortex-mixing for five minutes.

Results and discussion
All compounds were well resolved in under 30 minutes (Figure 5). A comparison of levels for a 10μg/mL standard, obtained from the spectrophotometric and coulometric techniques, showed fair agreement (apart from α-carotene, see below). A 1.0μg/mL mixture was also extracted and analyzed using the NIST high level as a calibrator. Results from the spectrophotometric assay, coulometric assay, and values extrapolated from the NIST reference material are presented in Table 7. The coulometric values were obtained by chromatographic peak area integration from a single coulometric electrode set at an oxidation potential above the current-voltage curve of all analytes. The concentration values were then calculated using Faraday's law (see Acworth,

Matson and Bowers, this volume). Based on the spectrophotometric data in Table 7, the coulometric assays suggests that 3 electrons take part in the electrochemical oxidation of lutein, α-carotene, and ß-carotene, and ten electrons in the oxidation of lycopene. Since the values obtained spectrophotometrically represent the total absorbance of the individual standard solutions without chromatographic separation, there may be some overestimation of the pure analyte. The value for α-carotene calculated against the NIST control shows poor agreement and required further study.

Table 6.
Spectrophotometric assay of 10 μg/mL solutions

Name	Wavelength (nm)	Absorbance	Absorbtivity dL/g-cm	Assayed concentration (μg/mL)
retinol (all trans)	325		1850	
retinyl acetate (all trans)				
lutein	445	3.71	2765	13.4
γ-tocopherol	298		91.4	
α-tocopherol	292		75.8	
retinyl palmitate (all trans)	325		975	
lycopene	472	0.253	3450	0.73
α - carotene	444	0.475	2800	1.70
β - carotene	452	0.253	3450	0.73
coenzyme Q10				
vitamin K1				

One possible explanation might be due to differences between the resolution of the HPLC-ECD and HPLC absorbance methods.

Table 7.
Comparison between spectrophotometric and coulometric techniques

Name	Spectro-photometric concentration (μg/mL)	Coulometric concentration	1 μg/mL std against NIST control x 10
retinol (all trans)			3.61
retinyl acetate (all trans)			
lutein	13.4	31.4/n	21.47
γ-tocopherol			9.03
α-tocopherol			7.42
retinyl palmitate (all trans)			7.35
lycopene	0.73	7.26/n	0.89
α - carotene	1.70	4.68/n	16.62
β - carotene	2.35	7.06/n	2.64
coenzyme Q_{10}			
vitamin K1			

Where n is the number of electrons taking part in the electrochemical reaction.

The intra-assay precision (n=18) was performed using extracted, pooled serum and was analyzed against the NIST high control sample. The RSDs generally varied between 0.6% to 2.4% (for retinol and lycopene, respectively). Coenzyme Q10 and α-carotene were slightly higher with RSDs of 6.3% and 7.8%, respectively. Linearity was studied using dilutions of a mixture of commercial standards. Dilutions were made in ethanol/BHA and were analyzed in duplicate. The concentration values plotted were obtained using the NIST high control as a reference. Linearities (R^2) for all metabolites ranged from 0.969 for ß-carotene to 0.9999 for retinol. From these

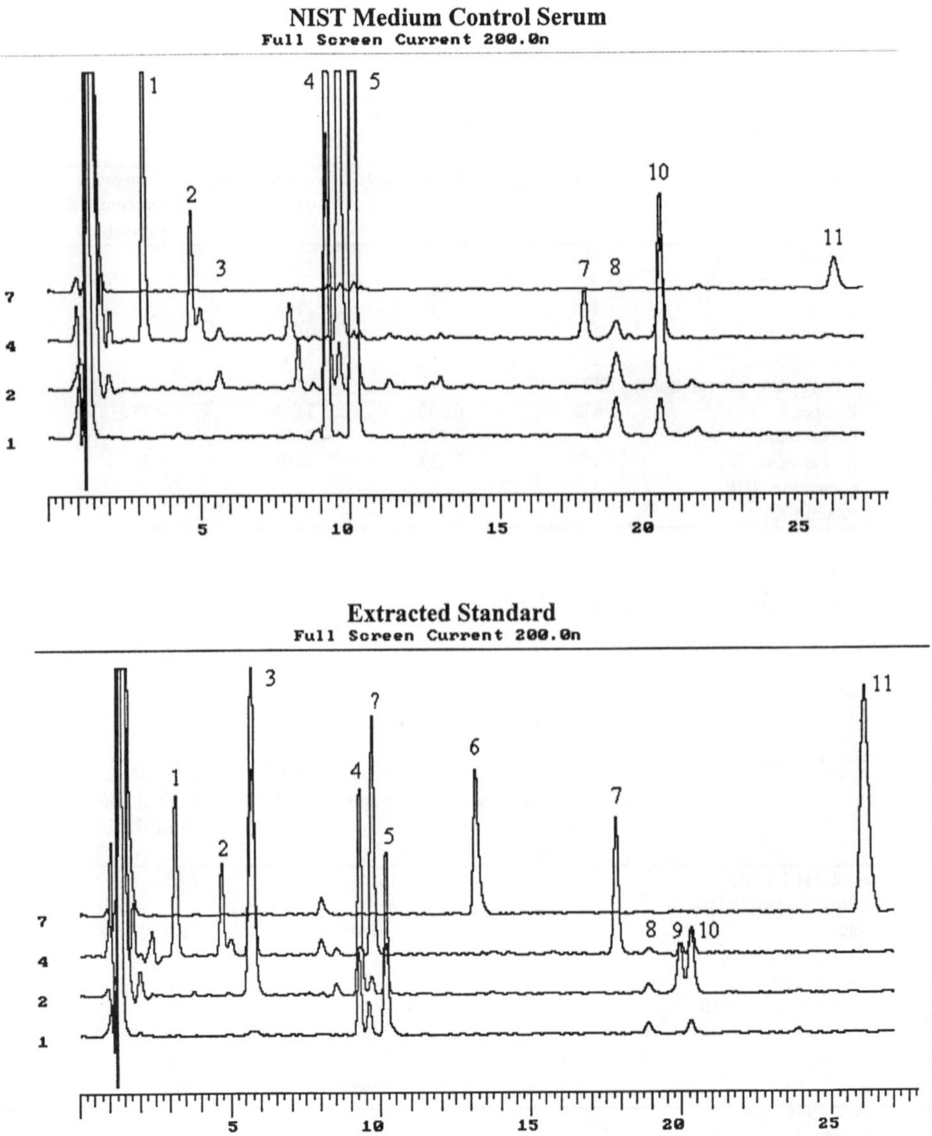

Figure 5. Analysis of fat soluble vitamins using gradient HPLC with coulometric array detection. The method conditions are given above. Only four channels are shown for clarity. Channel 6 is set at -1000mV to reduce compounds prior to their oxidation and measurement on channel 7 (+200mV). This redox approach allows the measurement of vitamin K1 and coenzyme Q10. The upper chromatogram presents the analysis of NIST medium control serum following hexane extraction, evaporation and reconstitution as

outlined above. This chromatogram is presented at a gain of 200nA. The lower chromatogram is for extracted standards (1µg/mL) and is presented at a gain of 200nA. 1- retinol; 2- retinyl acetate; 3- lutein; 4- γ-tocopherol; 5- α-tocopherol; 6- vitamin K1; 7- retinyl palmitate; 8- lycopene; 9- α-carotene; 10- β-Carotene; 11- coenzyme Q10.

dilutions the limit of detection was estimated to be <10pg (s/n = 3:1) on column for most of the metabolites. The limits of detection for lutein, retinol palmitate and lycopene was <19pg on column, while that of α-carotene was <75pg. These LODs are far better than those with the above mentioned approaches. The principal reason for improved s/n ratios is the marked reduction in noise when using electrochemical techniques and highly organic mobile phases.

Table 8.
Concentration, expected values and ratio accuracies for NIST medium and low calibrators

NIST- Medium

Name	Concentration ng/mL	Expected Value	Ratio Acc. 1	Ratio Acc. 2
retinol (all trans)	498	514	.957	.957.
retinyl acetate (all trans)	101 - IS		.913	
lutein	74.0	62.9	.776	
γ-tocopherol	2230	2300	.967	
α-tocopherol	10,500	10100	.722	
retinyl palmitate (all trans)	193	178	.780	
lycopene	263	300	.987	
α - carotene	33.7	30	0	
β - carotene	576	569	.954	.956
coenzyme Q10	95.7	NA	.900	
vitamin K	<0.98	NA		

NIST- Low

Name	Concentration ng/mL	Expected Value	Ratio Acc. 1	Ratio Acc. 2
retinol (all trans)	292	296	.963	.921
retinyl acetate (all trans)	96.85 - IS		.898	
lutein	71.5	62	.788	
γ-tocopherol	1660	1700	.942	
α-tocopherol	7490	7070	.572	
retinyl palmitate (all trans)	82.4	99	.965	
lycopene	163	200	.968	
α - carotene	24.5	20	0	
β - carotene	222	224	.941	0
coenzyme Q10	59.8	NA	.837	
vitamin K	LT 0.98	NA		

NA - not available.

Using the NIST high control with internal standard (retinyl acetate) correction, levels for NIST-medium and NIST-low were calculated. All compounds showed good matches to their expected values and all showed good ratio accuracies - see Table 8. The same procedure was then used to evaluate extracted human plasma samples. Two examples are shown, obtained from subjects on either a high dosage vitamin regimen (multivitamin pills with additional vitamin A, B and E pills) or a daily multivitamin supplement (single multivitamin pill). See Table 9 and Figure 6.

Table 9.
Plasma profiles of fat soluble vitamins, their concentrations, and ratio accuracies in a high-dosed and a normal-dosed subject

Human Serum Sample 1 - High Dose Vitamin Regimen

Name	Concentration ng/mL	Ratio Acc. 1	Ratio Acc. 2
retinol (all trans)	723	.992	1.00
retinyl acetate (all trans)	113.4 - IS	.890	
lutein	209	.737	
γ-tocopherol	322	.918	
α-tocopherol	26,000	.558	
retinyl palmitate (all trans)	382	.953	
lycopene	650	.947	
α - carotene	592	0	
β - carotene	3480	.830	.972
coenzyme Q10	94.5	.980	
vitamin K	<0.98		

Human Serum Sample 2 - Daily Multivitamin Supplement

Name	Concentration ng/mL	Ratio Acc. 1	Ratio Acc. 2
retinol (all trans)	505	.993	.947
retinyl acetate (all trans)	86.1 - IS	.913	
lutein	125	.797	
γ-tocopherol	926	.962	
α-tocopherol	10,800	.604	
retinyl palmitate (all trans)	8.40	.552	
lycopene	577	.985	
α - carotene	94.5	0	
β - carotene	122	.940	.972
coenzyme Q10	22	.964	
vitamin K	<0.98		

This method provides a comprehensive screening approach for fat soluble vitamins and antioxidants. It is the only method capable of measuring the retinoids, carotenoids, tocopherols, vitamin K, and coenzyme Q10 simultaneously. Some improvements may include improving the resolution of α and β carotene, lowering the concentration LOD by analyzing a higher volume than 10μL and/or reconstituting the sample with a lower volume, and addressing contaminants produced from the exposure of the plastic eppendorf tubes to hexane. Overall, this method is robust, easy to use, sensitive, has good linear response and gives accurate values for biological samples.

Human Serum Sample #1 - High Dose Vitamin Regimen

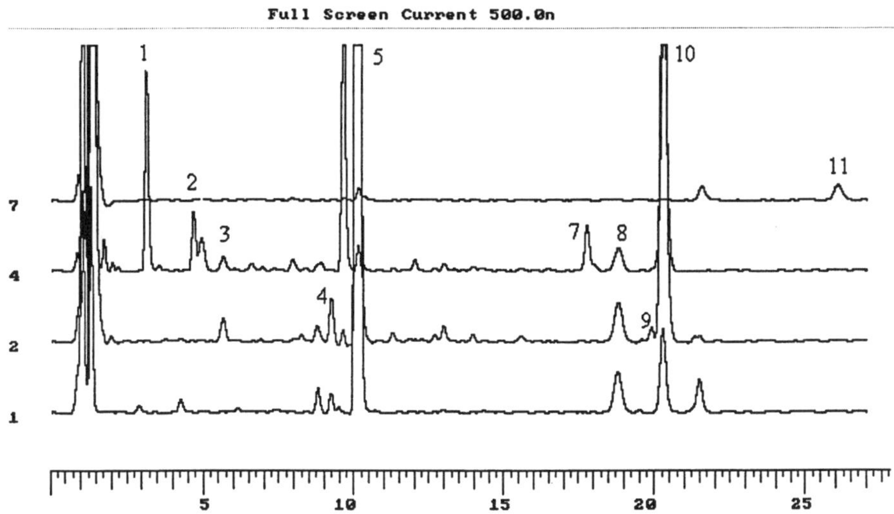

Human Serum Sample 2 - Daily Multivitamin

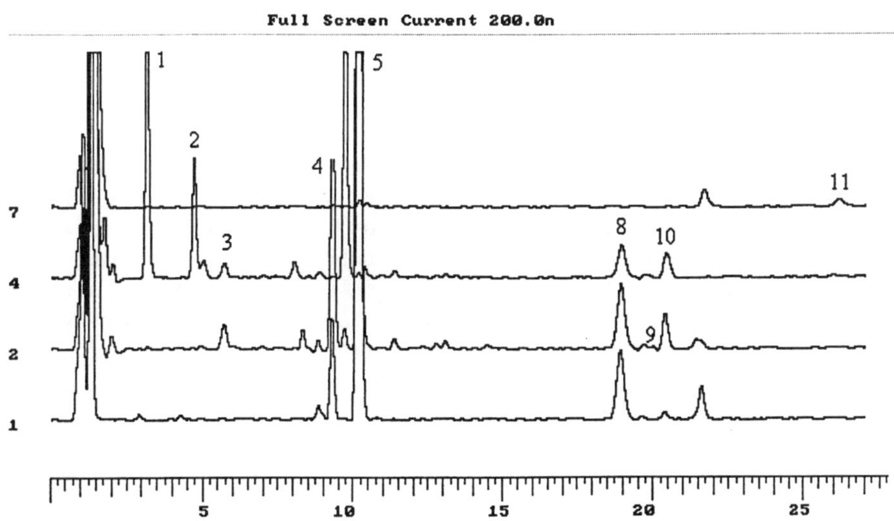

Figure 6. Analysis of fat soluble vitamins in extracted human plasma. The upper chromatogram shows the profile of analytes from a subject on a high dose vitamin regimen. The lower chromatogram is from a subject taking a single multivitamin pill per day. Notice that the upper chromatogram is obtained at the 500nA range whereas the lower is obtained at the 200nA range. As expected levels of most fat soluble vitamins were higher in the high dose subject - see Table 9.

CONCLUSION

Gradient HPLC with coulometric array detection enables the rapid and sensitive measurement of a wide variety of compounds involved either in the control of endogenous ROS levels or produced as a result of their action. On-line voltammetric analysis immediately highlights correct compound identification. This is especially important when trying to discriminate between compounds of similar structure such as with the plethora of plant compounds. Furthermore, the assignment of peak identity only when strict chromatographic and voltammetric criteria are met enables analyte measurement in complex matrices, often with limited sample preparation.

HPLC with coulometric array detection has also been used to measure a variety of other antioxidants and related compounds such as uric acid, xanthine, hypoxanthine, and oxidized and reduced glutathiones (Rizzo *et al.*, 1991; Svendsen, 1993; Milbury *et al.*, 1994). Recently, coulometric array detection has been used to investigate endogenous levels of the highly reactive and damaging hydroxyl free radical. Here, hydroxyl radicals react stoichiometrically with exogenous salicylic acid, forming 2,3- and 2,5-dihydroxybenzoic acids (Beal *et al.*, 1995). Measurement of DHBA/salicylate ratios is a direct indicator of hydroxyl free radical levels. Using the voltammetric resolution of the coulometric array it may also enable the use of other markers of hydroxyl free radical attack without the need for administration of salicylate, which may have pharmacological activity. The formation of the *m*- and *o*-isomers of tyrosine from endogenous phenylalanine and the production of 2,4-dihydroxyphenylalanine (2,4-L-DOPA) from endogenous tyrosine may be better indicators of hydroxyl radical damage *in vivo*.

Free radical attack of proteins yields a variety of oxidation products such as protein carbonyls, the formation of 2-oxohistidine from histidine and methionine sulfoxide from methionine, and the production of disulfides from thiols. One very interesting oxidation product is dityrosine, which has recently been discussed in detail by Giulivi and Davies (1994). Using a gradient HPLC coulometric electrode array method, Beal has recently started to examine the formation of dityrosine *in vivo* (1995).

Free radical attack of lipids produces lipid peroxides. Lipid peroxides disrupt the fluidity of membranes which, in turn, can affect membrane-based processes such as receptor and ion channel function. Although most lipids are electrochemically inert there are some exceptions. Mulchandani and Rudolph (1995) used ferrocene-linked voltammetry to measure linoleic acid and linolenic acid hydroperoxides. Lacking a separation step and having detection limits of 20nM, this approach may not be practical for the study of ROS effects on lipid metabolism *in vivo*. Leukotriene B_4 (Hermann *et al.*, 1988), phospholipid hydroperoxides (Yamada *et al.*, 1987), and cholesterol hydroperoxides (Korytowski *et al.*, 1993) can be measured directly using HPLC-ECD. In each of these cases, however, the chromatographic separation of the analytes was poor. Korytowski *et al.* (1995) measured cholesterol peroxides, fatty acid peroxides and two synthetic phosphatidylcholine peroxides using HPLC-ECD utilizing a mercury drop electrode. Although they reported limits of detection between 0.3 to 30 pmole, their chromatography was complicated by the use of four different isocratic conditions and the necessary use of the expensive and toxic mercury drop electrode. Other approaches have used indirect HPLC-ECD methods to measure lipids. For example, prostaglandins can be measured electrochemically as their 2,4-

dimethoxyanilide derivatives (Knospe *et al.*, 1988). Recently, Goldring *et al.* (1993) measured 4-hydroxynonenal, the highly reactive breakdown product of lipid peroxidation, as its 2,4-dinitrophenylhydrazine derivative. The resulting phenylhydrazone was measured on the second electrode of a dual coulometric electrode cell (the first electrode was used to screen interferences) with limits of detection <1pmole. Adaptation of this method to a coulometric array system will enable better selectivity and lower detection limits to be achieved. Furthermore, the use of gradient chromatography will make it possible to measure a wider variety of ROS-produced lipid fragments.

Nitric oxide is another naturally-occurring free radical which acts as a vasodilator and is purported to be a retrograde neurotransmitter with potential importance in memory (Snyder and Bredt, 1992; Kostka, 1995). Due to its short half-life (6 seconds), endogenous levels have been extremely difficult to measure. Nitric oxide readily reacts with the superoxide free radical to form the extremely reactive peroxynitrite molecule. Peroxynitrite (or the product of its reaction with carbonate, the peroxynitrite carbanion) is thought to be responsible for nitration of proteins (reacting particularly with aromatic amino acids) and DNA (e.g. the formation of 8-nitroguanine and 8-nitroguanosine). Consequently, the measurement of free 3-nitrotyrosine levels is being used as indicator of cellular nitric oxide levels. Both tyrosine and 3-nitrotyrosine (3NT) are measured using gradient HPLC with coulometric array detection. The peak for 3NT occurs at 24 minutes, has an oxidation potential maximum of 840mV, and is abolished by pretreatment with the reducing agent sodium dithionite (Beal *et al.*, 1995; Schulz *et al.*, 1995). These authors also discuss the importance of measuring the 3NT/tyrosine ratio to normalize for varying brain concentrations of tyrosine.

As the field of oxidative metabolism research progresses, investigators will need to continue to develop methods in order to analyze compounds related to the cause or prevention of oxidative stress related damage. Additionally, these newly discovered compounds will probably have to be measured at lower and lower limits of detection. HPLC with coulometric array detection allows measurement of oxidative stress related compounds at the necessary limits of detection in present and future research. ESA's CoulArray™ allows gradient HPLC-ECD to be employed, resulting in more analytes being separated and detected in one analytical run. The CoulArray's™ autoranging ability allows the simultaneous measurement of two analytes (e.g. an adduct and a parent base) which are detected on one channel and occur at concentrations differing by orders of magnitude. In conclusion, coulometric array detection, with its inherent low detection limits, coupled with the powerful separation capabilities of gradient HPLC, gives the researcher a powerful tool in the investigation of oxidative stress related compounds.

REFERENCES

Achilli, G., Cellerino, G. P., Gamache, P. H., and Melzi d'Eril., G. V. (1993). Identification and determination of phenolic constituents in natural beverages and plant extracts by means of a coulometric electrode array system. J. Chromatogr. **632**, 111-117.

Acworth, I. N., Kuo, M., Ryan, E., and Gamache, P. H. (1994). Analysis of catechins, phenolic and polyphenolic compounds in tea using HPLC and coulometric array detection. The Pittsburgh Conference on Analytical Chemistry and Applied Spectroscopy, Chicago, Illinois. Paper No. 391aP.

Acworth, I. N. and Bailey, B. A. (1995). ESA's Handbook on Oxidative Metabolism. Available from ESA, Inc., Chelmsford, MA, USA.

Adachi, S., Zeisig, M., and Moller, L. (1995). Improvements in the analytical method for 8-hydroxydeoxyguanosine in nuclear DNA. Carcinogen. **16**, 253-258.

Adlercreutz, H. (1990). Western diet and western diseases: some hormonal and biochemical mechanisms and associations. Scand. J. Clin. Lab. Invest. **201**, 3-23.

Adlercreutz, H., Fotsis, T., Bannwart, C., Wahala, K., Makela, T., Brunow, G., and Hase, T. (1986). Determination of urinary lignans and phytoestrogen metabolites, potential antiestrogens and anticarcinogens, in urine of women on various habitual diets. J. Steroid Biochem. **25**, 791-797.

Adlercreutz, H., Hockerstedt, K., Bannwart, C., Bloigu, S., Hamalainen, E., Fotsis, T., and Ollus, A. (1987). Effect of dietary components, including lignans and phytoestrogens, on enterohepatic circulation and liver metabolism of estrogens and on sex hormone binding globulin (SHBG). J. Steroid Biochem. **27**, 1135-1144.

Adlercreutz, H., Fotsis, T., Lampe, J., Wahala, K., Makela, T., Brunow, G., and Hase, T. (1993). Quantitative determination of lignans and isoflavonoids in plasma of omnivorous and vegetarian women by isotope dilution gas chromatography-mass spectrometry. Scand. J. Clin. Invest. **53**, 5-18.

Ames, B. N. and Gold, L. S. (1991). Endogenous mutagens and the causes of aging and cancer. Mutat. Res. Fundam. Mol. Mech. Mutagen. **250**, 3-16.

Babior, B. M. (1978a). Oxygen-dependent microbial killing by phagocytes. Part1. New Engl. J. Med. **298**, 645-668.

Babior, B. M. (1978b). Oxygen-dependent microbial killing by phagocytes. 2. New Engl. J. Med. **298**, 721-725.

Bannwart, C., Adlercreutz, H., Fotsis, T., Wahala, K., Hase, T., and Brunow, G. (1985). Identification of isoflavonic phytoestrogens and of lignans in human urine and in cow's milk by GC/MS. In: Advances in Mass Spectrometry-1985. Proceedings of the 10th International Mass Spectrometry Conference. Todd, J., (Ed.). John Wiley, Chichester, Sussex, pp. 95-106.

Beal, M. F. (1995). Personal communications.

Beal, M. F., Ferrante, R. J., Henshaw, R., Matthews, R. T., Chan, P. H., Kowall, N. W., Epstein, C. J., and Schulz, J. B. (1995). 3-Nitropropionic acid neurotoxicity is attenuated in copper/zinc superoxide dismutase transgenic mice. J. Neurochem. **65**, 919-922.

Bryan, P. D., Honigberg, I. L., and Meltzer, N. M. (1991). Electrochemical detection of retinoids using normal phase HPLC. J. Liq. Chromatogr. **14**, 2287-2295.

Burton, G. W., Cheesman, K. H., Doba, T., Ingold, K. U., and Slater, T. F. (1983). Vitamin E as an antioxidant *in vitro* and *in vivo*. In: Biology of Vitamin E. Porter, R and Whelan, J. (Eds.). Ciba. Found. Symp. Pitman. London, pp. 4-18.

Burton, G. W., and Ingold, K. U. (1981). Auto-oxidation of biological molecules. The antioxidant activity of vitamin E and related chain-breaking phenolic antioxidants *in vitro*. J. Am. Chem. Soc. **103**, 6472-6477.

Butler, L. G. and Rogler, J. C. (1992). Biochemical mechanisms of the antinutritional effects of tannins. In: Phenolic Compounds in Food and Their Effects on Health. ACS Symposium Series, Huang, M-T., Ho, C-T and Lee C. Y. (Eds.). Developed from a symposium sponsored by the Division of Agricultural and Food Chemistry of the American Chemical Society at the Fourth Chemical Congress of North America. ACS Press, Washington D.C., pp. 298-304.

Castle, M. C. and Cooke, W. J. (1985). Measurement of vitamin E in serum and plasma by high-performance liquid chromatography with electrochemical detection. Ther. Drug. Monit. 7, 364-368.

Cathcart, R., Schwiers, E., Saul, R. L., and Ames, B. N. (1984). Thymine glycol and thymidine glycol in human and rat urine: A possible assay for oxidative DNA damage. Proc. Natl. Acad. Sci. U. S. A., **81**, 5633-5637.

Colton, C. A., and Gilbert, D. L. (1987). Production of superoxide anions by a CNS macrophage, the microglia. FEBS Lett. **223**, 284-288.

Consumer Reports (1995). Herbal roulette. Gurin, J. (Ed.). Consumers Union of U. S., Yonkers, N. Y. pp. 698-705.

Cummings, J., French, R. C., and Smyth, J. F (1993). Application of high-performance liquid chromatography for the recognition of covalent nucleic acid modification with anticancer drugs. J. Chromatogr. **618**, 251-276.

Cushnir, J. R., Naylor, S., Lamb, J. H, and Farmer, P. B. (1993). Tandem mass spectrometric approaches for the analysis of alkylguanines in human urine. Org. Mass Spectrom. **28**, 552-558.

Dizdaroglu, M. and Gajewski, E. (1990). Selected-ion mass spectrometry: Assays of oxidative DNA damage. Methods Enzymol., **186**, 530-544.

Douki, T., and Ames, B. N. (1994). An HPLC-EC assay for 1,N^2-prapano adducts of 2'-deoxyguanosine with 4-hydroxynonenal and other α,β-unsaturated aldehydes. Chem. Res. Toxicol. **7**, 511-518.

Echols, H., and Goodman, M. F. (1991). Fidelity mechanisms in DNA replication. Ann. Rev. Biochem. **60**, 477-511.

Edlund, P. O. (1988). Determination of coenzyme Q_{10}, α-tocopherol and cholesterol in biological samples by coupled -column liquid chromatography with coulometric and ultraviolet detection. J. Chromatogr. **425**, 87-97.

Fischer-Nielsen, A., Jeding, I. B., and Loft, S. (1994). Irradiation-induced formation of 8-hydroxy-2'-deoxyguanosine and its prevention by scavengers. Carcinogen. **15**, 1609-1612.

Fotsis, T., Heikkinen, R., Adlercreutz, H., Axelson, M., and Setchell, K. D. R. (1982). Capillary gas chromatographic method for the analysis of lignans in human urine. Clin. Chim. Acta **121**, 361-371.

Frei, B., and Ames, B. N. (1991). Small molecule antioxidant defenses in human extracellular fluids. In: The Molecular Biology of Free Radical Systems. Scandalios, J. (Ed.). Cold Spring Harbor Laboratory Press, Cold Spring Harbor, New York.

Gamache, P. H. (1995). Analysis of phytoestrogens and polyphenolics in plant and animal tissue using HPLC with coulometric array detection. Proceedings for The Society of Experimental Biology and Medicine. The Third International Conference on Phytoestrogens, Little Rock, Arkansas. December 3-6.Gamache, P., Ryan, E., and Acworth, I. N. (1993). Analysis of phenolic and flavonoid compounds in juice beverages using high-performance liquid chromatography with coulometric array detection. J. Chromatogr. **635**, 143-150.

Gamache, P., Acworth, I. N., Lynch, M., and Matson, W. R. (1995). Coulometric array detection for HPLC in the analysis of juice products. In: Methods to Detect Adulteration of Fruit Juice Beverages. Nagy, S. and Wade, R. L. (Eds.). Agscience Inc., Auburndale, Florida, pp. 120-144.

Giulivi, C., and Davies, K. J. A. (1994). Dityrosine: A marker for oxidatively modified proteins and selective proteolysis. Methods Enzymol. **233**, 363-371.

Goldring, C., Casini, A., Maellero, E., Bello, B. D., and Comporti, M. (1993). Determination of 4-hydroxynonenal by high-performance liquid chromatography with electrochemical detection. Lipids **28**, 141-145.

Green, M. J., Hill, H. O. A., and Tew, D. G. (1987). The rate of oxygen consumption and superoxide anion formation by stimulated human neutrophils. FEBS Lett. **216**, 31-34.

Gupta, R. C. Reddy, M. V., and Randerath, K. (1982). ^{32}P-postlabeling analysis of non-radioactive aromatic carcinogen DNA adducts. Carcinogen. **3**, 1081-1092.

Hall, E. D., and Traystman, R. J. (1992). Secondary tissue damage after CNS injury. An Upjohn Publication. FRESA BioMedical Laboratories, Inc., Redmond, WA, pp. 1-28.

Halliwell, B. (1992). Lipid peroxidation, free-radical reactions, and human disease. An Upjohn Publication. FRESA BioMedical Laboratories, Inc., Redmond, WA, pp. 1-28.

Halliwell, B. and Gutteridge, J. M. C. (1989). Free Radicals in Biology and Medicine. Claridon Press, Oxford.

Harris, K., and Haggerty, W. J. (1993). Assays for potentially anticarcinogenic phytochemicals from flaxseed. Cereal Foods World, **38**, 147-151.

Haslam, E., Lilley, T. H., Warminski, E., Liao, H., Cai, Y., Martin, R., Gaffney, S. H., Goulding, P. N., Luck, G. (1992). Polyphenol complexation. In: Phenolic Compounds in Food and Their Effects on Health. ACS Symposium Series, Huang, M-T., Ho, C-T and Lee C. Y. (Eds.). Developed from a symposium sponsored by the Division of Agricultural and Food Chemistry of the American Chemical Society at the Fourth Chemical Congress of North America. ACS Press, Washington D.C., pp. 8-50.

Hayashi, N., Hayata, K., and Sekiba, K. (1985). Rapid and simultaneous measurement of estrone, estradiol, estriol and estetrol in serum by high performance liquid chromatography with electrochemical detection. Acta Med. Okayama, **39**, 143-153.

Herrmann, T., Steinhilber, D., and Roth, H. (1987). Determination of leukotriene B_4 by high-performance liquid chromatography with electrochemical detection. J. Chromatogr. **416**, 170-175.

Ho, C-T., Lee, C. Y, and Huang, M-T (eds.). (1992). Phenolic Compounds in Food and Their Effects on Health. **506** and **507**. ACS Symposium Series. Developed from a symposium sponsored by the Division of Agricultural and Food Chemistry of the American Chemical Society at the Fourth Chemical Congress of North America. ACS Press, Washington D.C.

Huan, M-T., and Ferraro, T. (1992). Phenolic compounds in food and cancer prevention. In: Phenolic Compounds in Food and Their Effects on Health. ACS Symposium Series, Huang, M-T., Ho, C-T and Lee C. Y. (Eds.). Developed from a symposium sponsored by the Division of Agricultural and Food Chemistry of the American Chemical Society at the Fourth Chemical Congress of North America. ACS Press, Washington D.C., pp. 8-34.

Huber, G. L., and Drath, D. B. (1981). In: Oxygen and Living Processes, an Interdisciplinary Approach. Gilbert, D. L. (Ed.). Springer-Verlag, New York.

Johnston, R. B. Jr., Keele, B. B. Jr., Misra, H. P., Lehmeyer, J. E., Webb, L. S., Baehner, R. L. and RaJagopalan, R. (1975). The role of superoxide anion generation in phagocyte bactericidal activity. Studies with normal and chronic granulomatous disease leukocytes. J. Clin. Invest. **55**, 1357-1372.

Kehrer, J. P. (1993). Free radicals as mediators of tissue injury and disease. Crit. Rev. Toxicol. **23**, 21-48.

Knopse, J., Steinhilber, D., Herrmann, T., and Roth, H. (1988). Picomole determination of 2,4-dimethoxyanalides of prostaglandins by high-performance liquid chromatography with electrochemical detection. J. Chromatogr. **442**, 444-450.

Korytowski, W., Bachowski, G. J., and Girotti, A. W. (1993). Analysis of cholesterol and phospholipid hydroperoxides by high-performance liquid chromatography with mercury drop electrochemical detection. Anal. Biochem. **213**, 111-119.

Korytowski, W., Geiger, P. G., and Girotti, A. W. (1995). High performance liquid chromatography with mercury cathode electrochemical detection: application to lipid hydroperoxide analysis. J. Chromatogr. **670**, 189-197.

Kostka, P., (1995). Free radicals (nitric oxide). Anal. Chem., **67**, 411R-416R.

Lang, J. K., and Packer, L. (1987). Quantitative determination of vitamin E and oxidized and reduced coenzyme Q by high-performance liquid chromatography. J. Chromatogr. **385**, 109-117.

Lee, M-J., Wang, Z-Y, Li, H., Chen, L., Sun, Y., Gobbo, S., Balentine, B. A., and Yang, C. S. (1995). Analysis of plasma and urinary tea polyphenols in human subjects. Cancer Epidemiol. Biomark. Prevent. **4**, 393-399.

Lunder, T. L. (1992). Catechins of green tea. In: Phenolic Compounds in Food and Their Effects on Health. ACS Symposium Series, Huang, M-T., Ho, C-T and Lee C. Y. (Eds.). Developed from a symposium sponsored by the Division of Agricultural and Food Chemistry of the American Chemical Society at the Fourth Chemical Congress of North America. ACS Press, Washington D.C., pp. 114-120.

MacCrehan, W. A., and Schonberger, E. (1987a). Reversed-phase high-performance liquid chromatographic separation and electrochemical detection of retinol and its isomers. J. Chromatogr. **417**, 65-78.

MacCrehan, W. A., and Schonberger, E. (1987b). Determination of retinol, α-tocopherol, and ß-carotene in serum by liquid chromatography with absorbance and electrochemical detection. Clin. Chem. **33**, 1585-1592.

Marnett, L. J., and Burcham, P. C. (1993). Endogenous DNA adducts: Potential and Paradox. Chem. Res. Toxicol. **6**, 771-785.

Mecocci, P., MacGarvey, U., Kaufman, A., Koontz, D., Shoffner, J., Wallace, D., and Beal, M. F. (1993). Oxidative damage to mitochondrial DNA shows marked age-dependent increases in human brains. Ann. Neurol. **34**, 609-616.

Messina, M., and Barnes, S. (1991). The role of soy products in reducing risk of cancer. J. Natl. Cancer Inst. **83**, 541-546.

Messina, M., and Messina, V. (1991). Increasing use of soyfoods and their potential role in cancer prevention. J. Am. Dietetic Assoc. **91**, 836-840.

Milbury, P. E., Beal, M. F., and Matson, W. R. (1994). The potential of degenerative disorder diagnosis based on biochemical pattern or profile analysis. The Fourth International Conference on Alzheimer's Disease and Related Disorders. Minneapolis, Minnesota.

Moore, P., and Strauss, B. S. (1979). Sites of inhibition of in vitro DNA synthesis in carcinogen and UV-treated ΦX174DNA. Nature **278**, 664-666.

Mulchandani, A., and Rudolph, D. C. (1995). Amperometric determination of lipid hydroperoxides. Anal. Biochem. **225**, 277-282.

Murphy, M. E., and Kehrer, J. P. (1987). Simultaneous measurement of tocopherols and tocopheryl quinones in tissue fractions using high-performance liquid chromatography with redox cycling electrochemical detection. J. Chromatogr. **421**, 71-82.

Nagy, S. and Wade, R. L. (Eds.). (1995). Methods to Detect Adulteration of Fruit Juice Beverages. Agscience Inc., Auburndale, Florida.

Ngheim, Y., Cabrera, M., Cupples, C. G., and Miller, J. H. (1989). The *mut*Ygene: A mutator locus in *E. coli* that generates G.C ⇒ T.A transversions. Proc. Natl. Acad. Sci. U. S. A. **85**, 2709-2713.

Ohtsuka, E., and Nishimura, S. (1987). Misreading of DNA templates containing 8-hydroxydeoxyguanosine at the modified base and at adjacent residues. Nature **327**, 77-99.

Osawa, T. (1992). Phenolic antioxidants in dietary plants as antimutagens. In: Phenolic Compounds in Food and Their Effects on Health. ACS Symposium Series, Huang, M-T., Ho, C-T and Lee C. Y. (Eds.). Developed from a symposium sponsored by the Division of Agricultural and Food Chemistry of the American Chemical Society at the Fourth Chemical Congress of North America. ACS Press, pp. 135-149.

Osawa, T., Ramarathnam, N., Kawakishi, S., and Namiki, M. (1992). Antioxidant defense systems generated by phenolic plant constituents. In: Phenolic Compounds in Food and Their Effects on Health. ACS Symposium Series, Huang, M-T., Ho, C-T and Lee C. Y. (Eds.). Developed from a symposium sponsored by the Division of Agricultural and Food Chemistry of the American Chemical Society at the Fourth Chemical Congress of North America. ACS Press, Washington D.C., pp. 122-143.

Price, K. R, and Fenwick, G. R. (1985). Naturally occurring oestrogens in foods - a review. Food Addit. Contam. **2**, 73-106.

Randerath, K., Reddy, M. V., and Gupta, R. C. (1981). ^{32}P-labeling for DNA damage. Proc. Natl. Acad. Sci. U. S. A. **78**, 6126-6129.

Rice-Evans, C. A., and Diplock, A. T. (1993). Current status of antioxidant therapy. Free Rad. Biol. Med. **15**, 77-96.

Richter, C., Park, J. -W., and Ames, B. N. (1988). Normal oxidative damage to mitochondrial and nuclear DNA is extensive. Proc. Natl. Acad. Sci. U. S. A. **85**, 6465-6467.

Rizzo, V., Melzi D'Eril, G., Achilli, G., and Cellerino, G. P. (1991). Determination of neurochemicals in biological fluids by using an automated high-performance liquid chromatographic system with a coulometric array detector. J. Chromatogr. **536**, 229-236.

Schulz, J. B., Matthews, R. T., Muqit, M. M. K., Browne, S. E., and Beal, M. F. (1995). Inhibition of neuronal nitric oxide synthase by 7-nitroindazole protects against MPTP-induced neurotoxicity in mice. J. Neurochem. **64**, 936-939.

Setchell, K. D. R., Lawson, A. M., McLaughlin, L. M., Patel, S., Kirk, D. N., and Axelson, M. (1983). Measurement of enterolactone and enterodiol, the first mammalian lignans, using stable isotope dilution and gas chromatography mass spectrometry. Biomed. Mass. Spectrum. **10**, 227-235.

Setchell, K. D. R., Borriello, S. P., Hume, P., and Axelson, M. (1984). Nonsteroidal estrogens of dietary origin: possible roles in hormone-dependent disease. Am. J. Clin. Nutr. **40**, 569-578.

Setchell, K. D. R., Welsh, M. B., and Lim, C. K. (1987). High-performance liquid chromatographic analysis of phytoestrogens in soy protein preparations with ultraviolet, electrochemical and thermospray mass spectrometric detection. J. Chromatogr. **386**, 315-323.

Shibutani, S., Takeshita, M., and Grollman, A. P. (1991). Insertion of specific bases during DNA synthesis past the oxidation-damaged base 8-oxodG. Nature **349**, 431-434.

Shigenaga, M. K., Park, J-W, Cundy, K. C., Gimeno, C. J. and Ames, B. N. (1990). *In vivo* oxidative DNA damage: measurement of 8-hydroxy-2'-deoxyguanosine in DNA and urine by high-performance liquid chromatography with electrochemical detection. Methods Enzymol. **186**, 521-530.

Shigenaga, M. K., Aboujaoude, E. N, Chen, Q., and Ames, B. N. (1994). Assays of oxidative DNA damage biomarkers 8-oxo-2'-deoxyguanosine and 8-oxoguanine in nuclear DNA and biological fluids by high-performance liquid chromatography with electrochemical detection. Methods Enzymol. **234**, 16-33.

Shimada, K., Tanaka, T. and Nambara, T. (1979). Separation of catechol estrogens by high-performance liquid chromatography with electrochemical detection. J. Chromatogr. **178**, 350-354.

Shimada, K., Tanaka, T. and Nambara, T. (1981). Determination of isomeric catechol estrogens in pregnancy urine by high-performance liquid chromatography with electrochemical detection. J. Chromatogr. **223**, 33-39.

Snyder, S. and Bredt, D. (1992). Biological roles of nitric oxide. Sci. Am. **266**. 68-77.

Strauss, B. S. (1991). The 'A rule' of mutagenic specificity: A consequence of DNA polymerase bypass of non-instructional lesions? BioEssays **13**, 79-84.

Stavric, B., Matula, T. I., Klassen, R., Downie, R. H., and Wood, R. J. (1992). Effect of flavonoids on mutagenicity and bioavailability of xenobiotics in foods. In: Phenolic Compounds in Food and Their Effects on Health. ACS Symposium Series, Huang, M-T., Ho, C-T and Lee C. Y. (Eds.). Developed from a symposium sponsored by the Division of Agricultural and Food Chemistry of the American Chemical Society at the Fourth Chemical Congress of North America. ACS Press, Washington D.C., pp. 239-249.

Steinmetz, K. A., and Potter, J. D. (1991). Vegetables, fruit, and cancer. *Cancer Causes and Control* **2**, 427-440.

Supko, J. G., and Philips, L. R. (1995). High-performance liquid chromatography assay for genistein in biological fluids. J. Chromatogr. **666**, 157-167.

Svendsen, C. N., (1993). Multi-electrode array detectors in high performance liquid chromatography: A new dimension in electrochemical analysis. Analyst **118**, 123-129.

Takeuchi, T., and Morimoto, K. (1993). Increased formation of 8-hydroxy- deoxyguanosine, an oxidative DNA damage, in lymphoblasts from Fanconi's anemia patients due to possible catalase deficiency. Carcinogen. **14**, 1115-1120.

Uchida, M., Kataoka, Y., and Ono, M. (1992). Determination of reducing substances in beer and wort by HPLC with 16 electrode electrochemical detection and its application to the practical brewing. The 2nd, BCOA meeting, St Louis, Missouri.

van Delft, J. H. M., van Winden, M. J. M., van den Ende, A. M. C. and Baan, R. A. (1993). Determining N7-alkylguanine adducts by immunochemical methods and HPLC with electrochemical detection: Applications in animal studies and in monitoring human exposure to alkylating agents. Environ. Health Perspect. **99**, 25-32.

Vatassery, G. T., Smith, W. E., and Quach, H. T. (1993). A liquid chromatographic method for the simultaneous determination of α-tocopherol and tocopherolquinone in human red blood cells and other biological samples where tocopherol is easily oxidized during sample treatments. Anal. Biochem. **214**, 426-430.

Wagner, J. R., Hu, C-C., and Ames, B. N. (1992). Endogenous oxidative damage of deoxycytidine in DNA. Proc. Natl. Acad. Sci. U. S. A. **89**, 3380-3384.

Wang, G., Kuan, S., Francis, O. J., Ware, G. M., and Carmen, A. S. (1990). A simplified HPLC method for the determination of phytoestrogens in soybean and its processed products. J. Agric. Food Chem. **38**, 185-190.

Wring, S. A., Hart, J. P., and Knight, D. W. (1988). Voltammetric behavior of all-trans-retinol (vitamin A1) at a glassy carbon electrode and its determination in human serum using high-performance liquid chromatography with electrochemical detection. Analyst **113**, 1785-1789.

Wring, S. A., and Hart, J. P. (1989). Electrochemical determination of all-trans-retinol, and correlation with retinol binding protein in liver cirrhosis. Med. Lab. Sci. **46**, 367-369.

Yamada, K., Terao, J., and Matsushita, S. (1987). Electrochemical detection of phospholipid hydroperoxides in reverse-phase high performance liquid chromatography. Lipids **22**, 125-128.

Progress in HPLC-HPCE, Vol. 6, pp. 127–144
I.N. Acworth *et al.* (Eds)

CEAS GENERATION OF LARGE MULTI-PARAMETER METABOLIC DATABASES FOR DETERMINING CATEGORICAL PROCESS INVOLVEMENT OF BIOLOGICAL MOLECULES

PAUL E. MILBURY

Long Range Research, ESA. Inc., 22 Alpha Road, Chelmsford, MA 01824-4171

INTRODUCTION

Diversity is the primary characteristic of life which permits evolution and it is the diversity in lifeforms which appears to us as complexity or perhaps even as chaos. Yet it is amazing, with all the interwoven complexity which exists in ecosystems and within the biochemistry of individual organisms within these ecosystems, that humans have evolved to think predominantly in a linear mode. For the vast majority of people, mere mention of "an n dimensional network of biochemical pathways" elicits a blank expression. The effect of the passage of time in our lives and, for the greater part of our evolution, the effects of action and linear time related consequence has lead us to think predominantly in linear terms. Throughout most of our existence, linear thinking has served us well. Over the past several centuries, however, a change has been taking place as, out of necessity, we abandon old concepts and recognize the multidimensional interconnections of our universe.

In the fields of medicine and biochemistry, enormous strides have been made in solving mysteries of diseases which have plagued mankind. As we look at that progress in retrospect, however, we see that most of that progress has been achieved in cases of diseases or disorders involving few etiologic factors. The diseases and disorders which continue to elude cures seem to be multietiologic in origin and multifaceted in symptoms. Biomedical researchers are finding they can no longer avoid confronting the multidimensional nature of many disorders. We are now studying diseases where the number of variables involved are beyond the capacity for three dimensional depiction and hence beyond our visual reality. In the realm of hypervariate data, modes of thinking must change from perceptual to cognitive, while we realize that evolution has limited us to the capability of considering only subsets of the data. We realize that relationships exist with other variables or data-sets beyond three dimensions. In this light, metabolism might properly be envisioned as a network of complex interconnected biochemical reactions capable of producing effects which are measurable far from a primary sight of reaction. This network approach can be used to great advantage.

In the Long Range Research Department at ESA, Inc., we have been studying the utility of constructing large metabolic databases for categorical determination in degenerative disorders and in highlighting those compounds which show significant process involvement in determining disease categories. We have constructed metabolic databases from plasma, urine, and cerebrospinal fluid utilizing the Coulometric Electrode Array System's (CEAS) ability to quickly and efficiently separate and analyze oxidizable compounds chromatographically. Although sample preparation may vary slightly, the basic approach to analysis is still evolving but is fundamentally the same for sample sets across a wide variety of disorder studies.

The problem of immobility osteopenia

Immobility osteopenia represents one of many chronic disorders where the utility of constructing large metabolic databases can be demonstrated. When men and women are convalescent, or as we undertake space travel, bone is inevitably lost. The reasons are not clear but exercise or altered skeletal loading alone is insufficient to prevent bone loss, or osteopenia. Bone is dynamic and continually undergoes remodeling. A picture has emerged from research which defines bone homeostasis as the result of a complex balance between the actions of osteoclasts, which degrade bone, and osteoblasts, which synthesize bone. In some endocrine conditions, such as post menopausal osteoporosis or immobility osteopenia, the osteoblast become less efficient at rebuilding bone. As long as normal hormonal balance exists, sufficient exercise is continued, and proper nutrition is provided to sustain protein, calcium, and vitamin D metabolism, bone mass and strength are retained. Inevitably increased age or immobility leads to bone loss. A wide variety of theories have been proffered suggesting that in some manner altered regulation of parathyroid hormone or expression of a faulty src proto-oncogene affects proper osteoclast function. After several decades of research, the mechanisms of bone loss still remain unclear. The emphasis in the field lately has turned to identify markers of bone synthesis, degradation, or remodeling. There are many recent overviews of the status of the search for markers of bone metabolism (Kleerekoper, 1994; Riis, 1991; Delmas, 1991; Tamburino and Fiore, 1990 ; McGowan, 1993; Ebling *et. al.*, 1992).

Measurement of bone biological markers and markers of protein degradation can offer important information, supplementing imaging technology and aiding researchers and clinicians who study and manage musculo-skeletal degenerative processes. Numerous markers are emerging as indicators of bone and muscle quality, however many are assayed only by individual immunoassays. We propose here that some of the current markers of bone and muscle metabolism can be assayed simultaneously utilizing the CEAS. Furthermore, the effort also affords an opportunity to produce databases of both identified and unidentified compounds reflecting biochemical patterns indicative of the metabolic condition. The resulting databases may be more useful in assessing overall metabolic status than are measures of individual analytes. As an added benefit, such databases may highlight previously unsuspected pathways, new potential markers, or new potential endogenous pharmaceuticals.

The basic concepts of CEAS technology (e.g., separating co-eluting compounds by both retention time and detector potentials and the verification of chemical identity by detector channel response ratios) have been extensively described (Beal *et al.*, 1990; Matson, *et al.*, 1987; Matson, *et al.*, 1990; LeWitt, *et al.*, 1992), as have the general protocols and controls necessary for creating reliable databases. The use of databases to describe markers and evaluate biochemical mechanisms has also been studied (Beal, *et al.*, 1990; LeWitt, *et al.*, 1992). For the proposed work, several significant variations and hardware/software modifications have been developed to address specific problems of analytical matrices, the fragility of analytes of interest, and variations within and among the subjects studied.

Of bears and men

A fundamental axiom of basic research holds that study of an exception to a rule of nature leads to better understanding of the rule. The wintering bear, unlike bedridden humans or any other immobilized or hibernating animal, remarkably avoids the debilitating effects of immobility osteopenia (Haller and Zimny, 1977; Lundberg, *et al.*, 1976; Azzi, *et al.*, 1979; Nelson, *et al.*, 1983; Nelson, *et al.*, 1984; Floyd, *et al.*, 1990). The bear's metabolism dramatically changes during the seasons in order to facilitate accumulation of sufficient stored fat to assure survival during a winter period of fasting and dormancy which can even extend to six months in some regions of the world. The bear's stored fat is utilized as a source of energy and metabolic water, and to provide resources to maintain essential functions. In contrast, humans can neither fast for six months nor remain relatively immobile without risking starvation and suffering bone loss. It is reasonable to conclude that studying seasonal metabolic changes in bears may provide clues to understanding the mechanism by which they avoid bone loss.

MATERIALS AND METHODS

During a six month study supported by the National Institutes of Health, over 450 bear plasma samples and 140 human bedrest study samples were analyzed. The bear samples were provided by The Virginia Polytechnic Institute and State University Department of Fisheries and Wildlife, the State of Maine Department of Inland Fisheries and Wildlife, and Washington State University, Department of Natural Resource Sciences. The samples represent trajectories of individual bears over all four seasons for both black and grizzly bears. The National Aeronautics and Space Administration provided human plasma samples for analysis from ongoing and prior NASA bedrest studies. The samples represented initial and bedrest times out to 30 days, with a limited number of post experiment recovery samples.

During the initial study, over 900 two-hour chromatographic runs were completed in under six months utilizing a single CEAS unit and a single operator. Raw chromatographic data was collected from the operating instrument via a local area network while data reduction and analysis were conducted on a networked 486 computer. The resulting database reflects the relative levels of over 500 oxidizable compounds in NASA human bedrest study and seasonal bear plasma samples.

Chemicals, standards and sample preparation
Chromatograms were analyzed for authentic standards using the compounds listed in Table 1. Currently these include compounds describing tyrosine and tryptophan peptides and metabolic pathways, purines, vitamins, flavones, phenolics, flavonoids, drugs (e.g., benzodiazepines, tricyclics, antihistamines, acetaminophen, salicylate) and markers of bone and muscle metabolism (e.g., carnosine, osteocalcin, and methylhistidine). Pyridinium cross-link products of collagen were graciously provided by Dr. David Eyre, University of Washington, or obtained from Metra Biosystems, Inc., Mountain View, CA Human Gla Protein fragments were obtained from Bachem California, Torrance, CA and from Sigma, St. Louis, MO. All other standards were obtained from Sigma, St. Louis, MO and stored at -80°C until needed. Stock solutions were made by dilution of 100mg in 100mL of normal saline containing 1% (w/v) phosphoric acid. Subsequent standard dilutions were made into normal saline containing 10^{-3}mg/L ascorbic acid as an anti-oxidant.

Approximately 3mL of blood was collected from bears or humans into EDTA-containing vacutainer tubes (7.2mg) and chilled on ice prior to centrifugation in a clinical centrifuge at 2000g, and then at 5000g to collect platelets and platelet-free plasma. The plasma was frozen and shipped on dry ice to ESA, Inc. where it was stored at -80°C until analysis. An acetonitrile(An)/acetic acid(HAc) precipitation and extraction procedure was utilized for plasmas in which 250µL of plasma at 0°C was precipitated with 1mL of An/0.4% HAc at -20°C. A 1mL volume of centrifuged supernatant was removed and evaporated to dryness under vacuum in a Centrivap Concentrator (Labconco), and reconstituted in 200µL of mobile phase A. This protocol conserves reactive species such as ascorbate, 6-hydroxydopamine, and homogentisic acid at 1ng/mL concentrations. A 100µL volume of reconstituted extract was then placed in each of two autosampler vials, one of which was immediately analyzed and the other frozen at -80°C for future confirmation analysis. Prior to a 50µL injection onto the system, samples were maintained at 4°C by circulating coolant.

Chromatographic methods and mobile phases
Chromatographic solvents were obtained as follows: isopropyl alcohol from Chempure, C.M.S., Houston, TX; methanol and Ac from EM Science, Gibbstown, NJ; lithium hydroxide from Sigma, St. Louis, MO; and glacial acetic acid and pentane sulfonic acid from J.T. Baker, Phillipsburg, NJ.

Plasma compounds in virtually all preparative schemes for high-pressure liquid chromatography (HPLC) on C18 columns have a deleterious long-term effect on the stability of retention times and response potentials. The situation is particularly difficult when simplified preparative protocols are used in an effort to conserve the original status of reactive analytes. For this reason, a novel mobile phase pair was developed: mobile phase A (11g/L of pentane sulfonic acid (PSA) at pH3.00 with acetic acid) and mobile phase B (0.1M lithium acetate at pH3.00 with HAc in 80/10/10 methanol/An/isopropanol (MeOH/An/IPA)). The PSA (mobile phase A) demonstrates an improved ability to solubilize and remove protein and peptide fragments from both HPLC (C18) columns and coulometric detectors while the high organic modifier (mobile phase B) effectively removes residual lipids and polysaccharides. Sulfonic acids are, however, inherently contaminated, necessitating the development of a cleaning protocol in which the PSA/HAc concentrated buffer

(4L of 400g/L PSA) was electrolyzed over pyrolytic graphite at a potential of 1000mV vs. a palladium reference [Pd(H)].

The chromatographic method involved a 120 minute complex gradient from 0% B to 100% B, with flow rate adjusted to compensate for aziotropic viscosity effects. The gradients used are depicted in Figure 1. Gradient operation was provided by two HPLC pumps (ESA, Inc.).

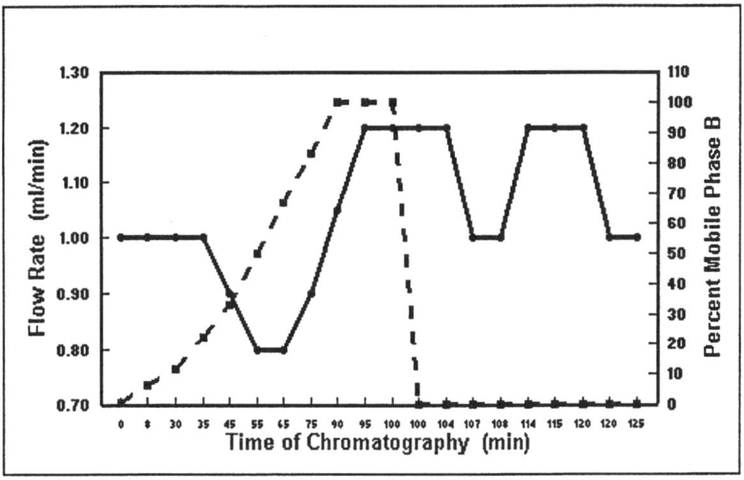

Figure 1. Mobile phase gradient (dashed line = percent mobile phase B) and flow rate profile (solid line = flow rate in mL/min).

Despite the meticulous use of precleaning protocols, double distilled deionized water, and selected organic modifiers, spurious peaks still occur late in the gradient of blank runs. This problem has been addressed by the development of a device based on electrochemically activated porous carbon with similar sorption characteristics to C18. This "peak suppresser/gradient mixer" (PS/GM) was placed in the flow stream before the HPLC injector. The device incorporated a 2cm length of 1cm diameter C18 pre column which was integrated with a 2.5cm of porous carbon with flow modifying bore holes which served to trap and then spread mobile phase contaminants. When these are released to the mixing section, during the gradient run, they are mixed to a peak width of approximately 140 seconds at half height. This effectively reduces a mobile phase derived contaminant signal to a wave that is later eliminated during data reduction by the CEAS analytical software. Analytes occurring in this region of the chromatography can then be measured on top of this baseline. The mixed gradient was converted from the PS/GM to a PEEK lined pulse damper (ESA, Inc.) prior to flowing through the autosampler injector (ESA, Inc.) and onto the C18 columns. Sample extracts were separated on dual PTFE lined HR80 columns (ESA, Inc.) containing 3μm ODS particles and each measuring 80mm x 4.6mm I.D.

Analyte detection was accomplished with a 16-channel coulometric electrode array from -100 to +990mV to detect both reducible and oxidizable compounds.

PS/GM, pulse damper, columns and detectors were contained within a temperature controlled enclosure maintained at 35°C. System functions were controlled by CEAS software installed on a 386 microcomputer.

Data reduction and analysis

Over the course of the study chromatographic retention times, monitored by pure standards and identified sample compounds, did not vary by more than 1%. The absolute qualitative channel ratio responses did not vary by more than 20% and were controlled by inclusion of authentic standards to within 5%. Confirmation of sample chromatographic peak identities was made by spiking with the relevant authentic standard. Final confirmation was made by comparison of the matching ratio (*R*) of the standard and the sample peaks. *R* represents the ratio between the dominant oxidation channel and juxtaposition subdominant channels. A given compound is oxidized at a specific potential and therefore any compound can be described by a retention time and a potential. In practice, compounds are oxidized on a dominant detector set near its oxidation potential and exhibit a smaller response on the prior and following detector. The ratio exhibited between the dominant and adjacent detector response is characteristic of a given compound and changes in the ratio, when a standard is close in concentration to a sample compound, indicates a coeluting contaminant is present (Beal, *et al.*, 1990)

Typically ca. 600 compounds can be resolved in plasma at a 20 nanoampere gain range. Data from each detector analog signal was converted and combined with other detector data to construct a time-potential map which was then compared with the time/potential map for authentic standards. Analytical values were calculated for sample peaks based on matches under restrictions for retention time, detector ratios and, to a lesser degree, peak heights according to defined priorities. Final results were calculated as ng per mL of sample based on standard responses. Sensitivity limits were defined as the amount of analyte that could be detected while maintaining a SD of ±0.5 times the amount injected and were typically 0.1-100pg on column, depending on the analyte.

During this study, an effort was made to collect the information represented by the unidentified compounds detectable in the patterns of the chromatography. The approach is relatively straightforward. The CEAS software, in the course of normal data reduction, constructs digital time/potential maps of the 16 channel chromatographic data. With sufficient space in the computer memory, a peak table can be constructed which represents all of the potentially detectable compounds in the plasma sample. Under the conditions of this study, 603 compounds were potentially detectable. In practice, however, there are limitations to the reliability of automated data analysis due to chromatographic stability and the amount of verification time that can be devoted to particular projects. Under the fundamental logic described earlier, a decision was made to detect those compounds which changed significantly (by chance or not) in the NASA human bedrest experiment HR113, and then to determine how these analytes changed during the winter torpor period in the bear. This was further restricted to those analytes which could be reliably analyzed by automated analysis.

TABLE 1.
Known compounds assayed

2HPAC	2-hydroxyphenylacetic acid	URIC	Uric Acid
30MD	3-0-Methyldopa	VMA	Vanillylmandelic acid
3MT	3-Methoxytyramine	XAN	Xanthine
4HBAC	4-Hydroxybenzoic acid	Xanth	Xanthosine
4HPAC	4-Hydroxyphenylacetic acid	3H-MAN	Hydroxymandelic acid
4HPLA	4-Hydroxyphenyllactic acid	3H4HMAN	3,4-Dihydroxymandelic acid
5HIAA	5-Hydroxyindoleacetic acid	3H4M-PHENEH	3-Hydroxy-4-methoxy- phenethyl amine
5HT	5-Serotonin	3HPAC	M-Hydroxyphenylacetic acid
5HTOL	5-Hydroxytryptophol	4H3M-MAN	4-Hydroxy-3-methoxy-mandelic acid
5HTP	5-Hydroxytryptophan	5MCYS	5-Methyl-L-cysteine
AM	Acetaminophen	5MTPOL	5-Methoxy-tryptophol
ASC	Ascorbic Acid	5MTRP	5-Methoxy-DL-tryptophan
CYS	Cysteine	5MTRPTAM	5-Methoxytryptamine
DA	Dopamine	6HDa	6-Hydroxydopamine
DOPAC	Dihydroxyphenylacetic acid	6HMEL	6-Hydroxymelatonin
EPI	Epinephrine	7MG	7-Methylguanine
G	Guanine	AN	Anthranillic acid
GR	Guanosine	FERUL	Ferulic acid
GSH	Glutathione (reduced)	HVERAT	Homoveratic acid
GSSG	Glutathione (oxidized)	HVOL	Homovanillyl alcohol
HGA	Homogentisic acid	IAA	Indole-3-acetic acid
HVA	Homovanillic acid	ILACT	DL-Indole-3-lactic acid
HX	Hypoxanthine	IPPA	Indole-3-proprionic acid
KYN	Kynurenine	ISATIN	Isatin
LD	L-DOPA	ISOPRO	Isoproterenol
MEL	Melatonin	MDOPA	α-Methyl-L-DOPA
MET	Methionine	METHOXAMINE	Methoxamine
MHPG	Methoxy-hydroxyphenyl glycol	NMeDA	N-Methyl-dopamine
MN	Metanephrine	PYRIDOXAL	Pyridoxal
NA5HT	Acetylserotonin (N)	T3	3,3,5-Triiodo-DL-thyronine
NE	Norepinephrine	VA	Vanillic acid
NM5HT	Methylserotonin (N)	α-Me-HIST	α-Methyl-histidine
NMN	Normetanephrine	1-Me-HIST	1-Methyl-histidine
TPOL	Tryptophol	3-Me-HIST	3-Methyl-histidine
TRP	Tryptophan	7-Me-XAN	7-Methyl-xanthine
TRYP	Tryptamine	NaHIST	N-Acetyl-histidine
TYR	Tyrosine	ANS	Anserine
TYRA	Tyramine	CARN	Carnosine
		HCARN	Homocarnosine

A compound table was generated from a pool of all samples in the HR113 study with concentrations set to 100. Subsequent sample analysis therefore generated reported values as a percent of pool values. This table was used to analyze, with initially heavy manual oversight, all other pools and a few samples within the study. The CEAS analytical software has a built in "learning" capacity which is inherently part of the "standards" definition function of the analysis. As the operator oversees a few analyses, decisions can be made about parameters such as referencing retention times to other compounds or what degree of variation from the channel ratio's which will be tolerated. Conflicts and ambiguity in analysis are monitored and resolved during this test phase of the analysis. Eventually, the pool standard table will "learn" how to reliably find a majority of the potential analytes in the samples. In the case of HR113, the pools "standards" were capable of reliably finding in all samples 506 of the potentially 603 detectable analytes. The table was trimmed to include only 506 compound peaks and the analysis was then run automatically using each of the pools within the study as a standard and the reported values were captured in files suitable for downloading into a database.

Once the data report files were generated, they were picked up with minor software programming for inclusion into orthogonal databases such as Oracle© or imported into Lotus 123© spreadsheets via macros which would import the text and parse results into proper locations within spreadsheets. For small studies this is a practical approach which is less expensive and more user friendly than construction of orthogonal databases. Over much larger studies, orthogonal databases may offer more useful storage and access functions. A software utility program has now been created to facilitate importation of CEAS data into popular databases.

The collected data was inspected for discrepancies which were easily found by checking for high standard deviations for compound values within samples analyzed by multiple pools. Decisions were made regarding the correct analytical value from predominant data values and reliability factors derived from prior analytical experience with the compound in question. The original chromatographic data was also inspected for indications of retention time shifts or anomalies in baseline construction during data reduction. Once the database was "vetted" for any analytical errors, statistical analysis proceeded in an effort to identify those compounds which showed significant differences between the categories in question. In this case those categories represent ambulatory and bedrest conditions.

Statistical approaches to the data

Many categorization efforts involve comparisons between a trait of interest and each of a number of markers for evidence of linkage. Such multiple tests or comparisons greatly increase the Type I error rate compared to the single-test situation. The most direct way to obtain multiple-test error rates is to employ a Bonferroni correction, where the single-test alpha is divided by the number of independent (non redundant) comparisons (k) to yield a multiple-test alpha that protects against even one fortuitous association with any of the markers. Most commercially available statistical packages in a moderate price range do not have the capacity to input Type I error corrections into databases of greater than 100 variables. As a result most multi-variant studies must rely on multiple independent protocols to establish linkage or use results for hypothesis generating purposes only.

This study assumed that peripheral serum samples from bears during winter torpor must contain metabolites expressing a fundamentally different pattern than that of humans restricted to bedrest. The study was designed to test the concept that the ambulatory and bedrest metabolic patterns, representing hundreds of compounds, could be collected and compared, highlighting known and/or unidentified compounds. The limitations of the Bonferroni solution dictates that if univariate nonparametric analyses are to be performed on the entire database of known and unidentified compounds, given ca. 500 variables, a corrected p value of 8.3×10^{-5} would be necessary to establish significance.

Multivariate data analysis is currently an emerging field in and of itself. Many subtle traps can lead to both over and under estimates of significance. These are the subject of intense study and discussion. Until many of these "traps" are defined, the best analysis will continue to involve multiple independent studies designed to confirm findings. Nevertheless, all statisticians consulted during this study concurred that it is probable that most of the shifts in the metabolite levels observed are indeed due to inactivity. While the Bonferroni issues related to Type I errors dictate caution, the concept that significant findings can be invalidated by the measurement and consideration of additional variables is not philosophically satisfying. All mathematical models are limited in a finite number of variables and a finite arithmetic precision. Application of new approaches are evolving from dynamic modeling of chaos, randomness and determinism, which hold promise of resolve problems associated with the analysis of all large multivariant databases.

Alternate statistical approaches to multivariate data analysis.
The rationale behind multivariate data analysis is that univariate methods, while well developed and proven for a wide array of applications, can too easily overlook features in complex data sets. Multivariate approaches are often dictated by the nature of the data. When the collection of measurements are too numerous or too intercorrelated, the data set must be processed as a unit and thus, matrix methods of data analysis are more appropriate.

Two proven analytical methods for exploratory data analysis of large multivariate databases included hierarchical cluster analysis and principal component. These methods allow observation of trends and/or clustering in data, as well as detection of relationships among the variables measured (Cingolani, 1985). The algorithms for these methods are designed to reduce large and complex data sets into subsets of measures allowing insight into any correlation that exist among the samples and the variables in the data set. Steps required and often overlooked in such analysis involve definition of the problem and organization of the data prior to performance of data analysis. CEAS protocols allow quick construction of databases from the chromatographic data analysis results which carry identification tags. With simple sorts of the keying data tags, the database constructed is readily available for statistical analysis. Many statistical software companies offer commercial programs for conducting hierarchical cluster and principal component analysis. Infometrics offers software such as Einsight© and BMDP offers modular software which includes cluster analysis.

Most multivariate procedures do not permit cases with incomplete data, especially when standardization methods are used. In cluster analysis, computation of cluster centers, variances, and Euclidean distances use all values which are present and, when data is missing, the mean and standard deviation for one variable in a cluster may be computed using a different number of values than those used for another variable. To avoid this problem in cases where a value could not be determined by manual verification of original chromatographic analysis, the missing value was added as the mean category value determined by the data present. Thus, the contribution of a missing value to the square of the distance is the average of the contributions of the non-missing values. Because of the initial selection criterion for determining the variables for chromatographic analysis, the existence of missing values was not a significant issue arising in the statistical analysis. In all computations an initial weight of one is assume for each case. However, methods were sought which would weight variables in the database on their degree of variance for use in the cluster center computations.

For these types of analysis the measured variables are best related such that the range and magnitude of values across these variables is similar. Variables measured in units across a wide range can have profound results in clustering algorithms. When one variable in a data set dominates, the variance in this variable alone may skew the processing results such that any interpretation will focus primarily on its influence, in spite of any diagnostic significance in other variables. In some cases where this dominance may be the characterizing phenomenon in the data set, no preprocessing may be necessary. However, in most analyses, some form of preprocessing is necessary to avoid these affects.

Centering and scaling operations including mean-centering (subtract the average from the values of each variable), variance scaling (divide by the standard deviation of each variable), range scaling (subtract the minimum, divide by the maximum-minimum), and autoscaling (the sequential application of mean-centering and Variance scaling). In all analyses for this study, an autoscaling approach was utilized (Manly, 1986)

Hierarchical cluster analysis

Multivariate data can be represented on an n-dimensional plot, where n is the number of variables measured on samples. Each sample then is a single point in this n-space, and pattern recognition methodologies are designed to highlight relationships among these points. In hierarchical cluster analysis, Euclidean distances between the samples (or variables) in a data set are calculated. Samples whose distances are relatively small are likely similar, at least with respect to the measurements taken. Dissimilar samples result in relatively larger distances. A dendogram can be constructed reflecting the set of inter-sample Euclidean distances, by linking samples and clusters of samples, as a function of the distances. Iterative linking of the individual samples, by linking the most similar samples, and recalculation of distances for the remaining samples can result in a dendogram, whose branch lengths are representative of the similarity of samples to each other. If real differences between categories exist between the variance weighted variables, then a pattern of clustered groupings will emerge.

Based on prior assumption of categories, the analysis results can be judged for validity and then inferences can be made regarding the contributions of each analyte or variable to the categorization. Those compounds contributing most heavily to the categorization can be assumed to have potential process involvement in bedrest induced changes in metabolism. Hartigan (1975) has reviewed the clustering algorithms utilized in this analysis.

Figure 2 illustrates, by dendogram, the categorization achieved utilizing the database constructed for the NASA HR113 Human Bedrest Study. Similarity of samples increases as pairs are matched toward the left. It is apparent that the control samples are separated from the 16 day bedrest and 1 day recovery samples. The first day recovery metabolic samples are not separated from the 16 day bedrest samples, despite NASA's evidence that cardiac function and baroreflex measures in this experiment had recovered. It must be noted that the recovery samples indicated here contain both the exercised and non-exercised subjects. It was anticipated that metabolic measures might lag cardiac functions in recovery and it is unfortunate that later recovery samples were not available from HR113 to confirm this hypothesis. Subsequent investigation of the trends for compounds indicated that recovery is underway for many of the compounds showing change during the 16 day bedrest period. Once having determined categorical separation is possible by hierarchical cluster analysis, it is then possible to determine by careful inspection of the principal component analysis data the compounds driving the separation.

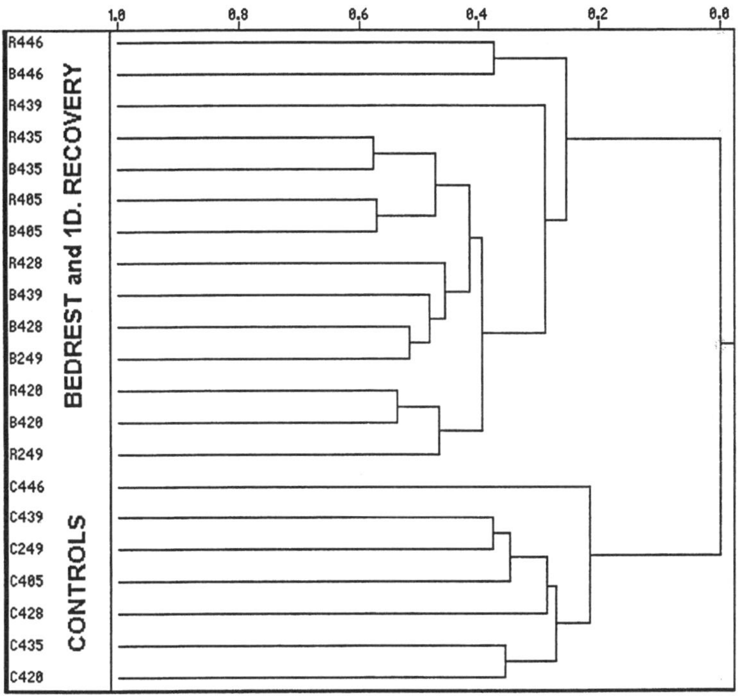

Figure 2. Dendogram of NASA HR113 hierarchical cluster analysis.

Principal component analysis

Principal component analysis methods reorganizes multivariate data by constructing a series of data axes that are linear combinations of the variables in the data set. These newly created axes have the property of being orthogonal to one another. In addition, the axes that are created are prioritized by the amount of variance in the data for which they can account. As a result, the first principal component describes more of the variance in the data set than does the second principal component. Principal component analysis is fundamentally the same as factor or eigenvector analysis. The relationships between samples remain unchanged during this processing; however, by ordering eigenvectors by the variance which they are capable of explaining, the most important differences between samples can be represented in a low dimensionality graphic output. The minimal dimensions necessary to represent any significant information in the data are the principal components. All other factors usually reflect noise in the data (Malinowski and Howery, 1980; Manley, 1986). The algorithm used in the analysis of this data is the non-iterative partial least squares method.

Principal component analysis results in the calculation values for eigenvalues, scores, and loadings. Eigenvalues reflect the magnitude in variance contained in the corresponding eigenvector, which ultimately are a reflection of the variance derived from the data. Evaluation of the eigenvalues is often sufficient to determine the optimal number of factors to retain the results in order to adequately describe the data. Scores and loadings were utilized to determine the relative importance of each variable in the data to the factor analysis. This is a less traditional use of these values but was effective in identifying those compounds which were most heavily involved in describing the differences between controls and bedrest samples.

Neural network

Once a database of multivariate metabolic data has been constructed and some insight into the nature of the relationships in the data are known, neural network analysis can be brought into use for categorization of future samples, perhaps in a diagnostic function. The database is used to "train" the neural-network algorithms and the patterns of future samples can then be recognized by the algorithms. While it is instructive and valuable to proceed with identification of unknown compounds in the database which contribute to categorizations, the identities of these compounds are not required to make use of them as markers or potential markers of a medical condition or potential outcome.

RESULTS OF THE DATABASE ANALYSIS

Precision of individual components in the database, monitored by pools run as samples throughout the study, varied between ±3 and ±25% coefficient of variation. In no case was the precision of the assay a contributing factor in the analysis of significance. Over the six month course of the study, chromatographic retention times did not vary more than 1%. The absolute qualitative ratio response did not vary by more than 20% and was controlled by inclusion of authentic standards to within 5% (Milbury, 1992).

Results of NASA's HR113 analysis demonstrated that 506 of a possible 603 oxidizable compounds could be reliably and automatically assayed with the CEAS technology. Cluster analysis showed clear categorical separation between bedrest and ambulatory samples (Figure 3).

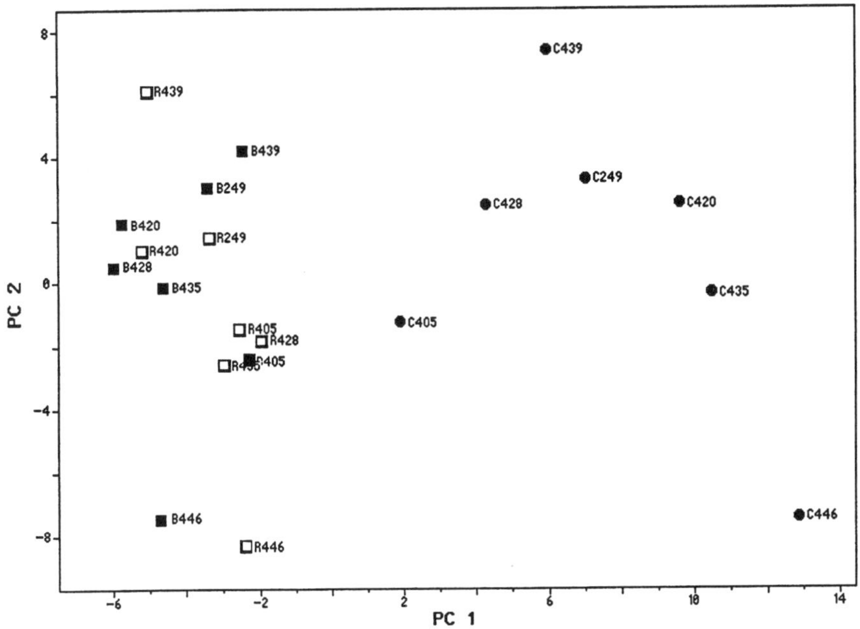

Figure 3. Principle component analysis showing components 1 vs. 2 from NASA HR113 metabolic database. Control (ambulatory pre-bedrest) samples are depicted by filled circles and 16 day bedrest samples are shown by filled squares. One day recovery (samples depicted by open squares) was insufficient to show a return to pre-bedrest metabolism, however, all means where changes were significant showed a trend toward pre-bedrest values.

From these analyses we were able to determine that the data effectively describes two distinct clusters from the HR113 data set which also included the 1 day recovery samples. As anticipated from prior work at ESA, Inc., studying patterns of neurometabolites in degenerative neurological disorders, the control samples exhibit a greater degree of variability than do the bedrest samples as indicated by the tightness of the groupings (representative of eigenvectors). From comparison of the loadings for the corresponding principal components, it is possible to determine the compounds contributing most heavily to the eigenvectors (Figure 4). Compounds located at the extremes of the chart have greater contribution in the calculation of the eigenvectors. These compounds include serotonin, ascorbate, xanthine, hypoxanthine, compounds #366, #350, #181 and some 40 other compounds listed in Table 2. All totaled, forty-five identified and unidentified compounds contributed to categorization and showed significant differences by analysis of variance between human ambulatory and bedrest categories.

Cross species comparisons of human and bear metabolic patterns

In the analysis of the bear plasma samples, particular attention was paid to the compounds that changed significantly in the NASA bedrest studies. Significant differences were observed in 10-30 known compounds and 11-38 unidentified compounds, depending on study and seasonal category. A final database was created containing 40 known and 120 unidentified peaks that could be automatically and reliably assayed in all bear samples from all seasons and all human samples. These constituted the primary data set for an interspecies comparison. It should be noted that a fundamental decision was made at this point to, once again, narrow the compounds under investigation to those compounds which were easily found in both species and to concentrate on those which showed potential human involvement. The CEAS analytical software can also be utilized effectively to search for unique markers or compounds which are absent from one or the other species and which may or may not change with category. Several such unique and seasonally variable compounds were noted in the bear. These compounds, while not the direct focus of this study, are of interest as they may represent new endogenous pharmaceuticals (Milbury *et al.*, 1995).

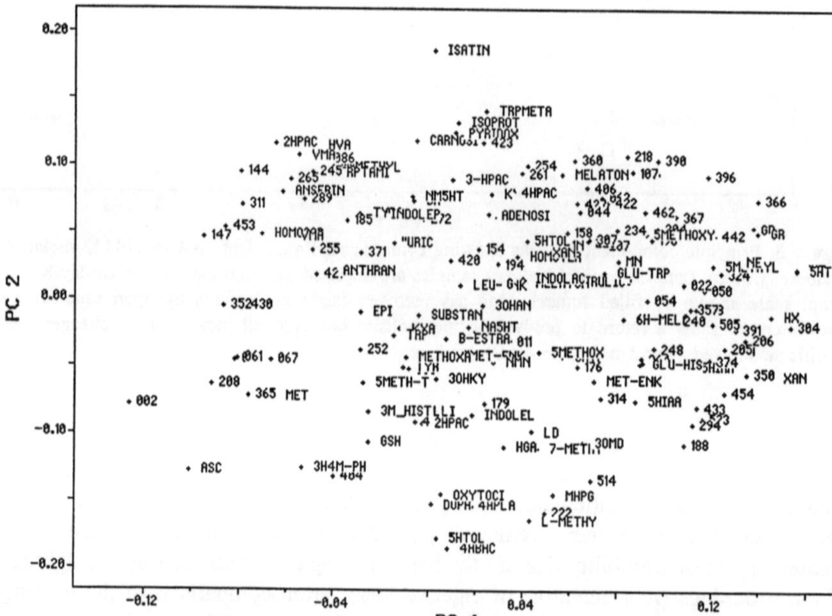

Figure 4. Analyte loadings for principle components 1 and 2 from NASA HR113 analysis.

Six identified compounds changed significantly, both in the bear and the human. Interestingly, these six compounds changed in the opposite direction in the bears from that observed in the human bedrest samples. Three known compounds, guanosine, serotonin, and 5-hydroxy indole acetic acid, appeared to change in the same manner, decreasing in both human subjects and the bear, as each went into immobility. Ascorbate and homocarnosine increased in the human bedrest subjects but remained unchanged in the bear. Xanthine, hypoxanthine, 5-methylcysteine and 6-hydroxymelatonin decrease in the human during bedrest but remain unchanged in the bear.

Compounds not changing significantly in the human subjects but increasing in the wintering bear include: tyrosine, homovanillic acid, 1-methyl histidine, 3-methyl histidine, 3-OH kynurenine, and the tripeptide, glu-his-pro. While remaining unchanged in the resting human, 3-O-methyl-DOPA showed a rise in the fall bear samples, decreasing into the winter months and then increasing as spring approached. Those compounds decreasing in the bear but remaining unchanged in humans include: anserine, 2-hydroxy-phenylacetic acid, and 3-hydroxy-phenylacetic acid.

Thirty-three unidentified compounds demonstrated possible involvement in bedrest induced metabolic pattern changes by Mann-Whitney U test with a $p<0.05$. Twelve achieved p values between 0.00044 and 0.01. Fifteen showed differential changes between bedrest human and the denning bear samples. As mentioned, by statistical theory, it is probable that some are by chance alone, however subsequent independent bedrest studies are currently underway to verify compound involvement. Each unidentified compound is labeled with a reference designation and is known by its retention time, relative migration to marker compounds, dominant oxidation potential and its oxidation channel ratios. Four such compounds appear to change diametrically in the human and bear. Another nine compounds, changing significantly in the human bedrest subject, are handled differentially by the wintering bear. Twenty unidentified compounds with p values of <0.05 in the HR113 study show a high degree of variability in the bear. Fresh bear samples and further manual verification of the automated bear analysis will be necessary to determine the related significance of these compounds.

Quality control issues related to oxidizable metabolic patterns

At the initiation of this project, samples from two previously conducted NASA bedrest protocols were received from the Ames Research Facility. HR62 plasma samples had been frozen at -70°C since 1987. HR71 was conducted in 1989. These samples were 5 to 6 years old at the time of analysis and a concern was present regarding their history and usefulness. During the short initial study, it was not possible to ascertain a full understanding of the effects of sample age on the patterns of metabolites. An opportunity was taken during the study to collect both bear and human plasma samples for storage and periodic analysis to ascertain the effect of age and storage conditions on CEAS generated metabolic patterns. Oxidative changes occurring in frozen materials is not an entirely new concern.

At the time of analysis, little could be done to resolve issues related to sample age except to make the analysis and inspect the results. The data strongly suggested degradation had occurred as reflected by the absence of many of the most easily oxidized compounds, such as homogentisic acid. The degree of variability was disconcerting. When databases of samples from old NASA bedrest experiments were compared to databases of fresh samples from neurological degenerative disorder studies, the difference in the quality of the results was clear. Similar observations were made in the libraries of previously collected bear plasma. Data derived from sample groupings dating back to 1984 from the Maine Department of Inland Fisheries and Wildlife exhibited higher variability than more recent samples proffered by the Virginia Polytechnic Institute. Within the Virginia samples, the 1990 data exhibited more variance than did the recent 1993 samples. The same observation was made regarding the Washington State University samples. As a result, data from older samples was used only for confirmation of compounds where stability could be demonstrated in time degradation studies over several days at room temperature. The issue of sample age is not unexpected, but it does argue that samples for metabolic database construction should be collected and analyzed within six to twelve months unless temperature control, preferably below -80°C (under nitrogen), is maintained. It is advisable under all circumstances to conduct the chromatography as soon as practicable.

Sample collection issues were also raised relative to the variability which might result from collecting bear blood in the field under difficult and highly variable conditions. To investigate these issues, a small laptop Eppendorf centrifuge was fitted with a small DC\AC power converter deriving its power from a 12V DC battery. This "field" centrifuge was compact and light enough to be brought to the den sights in Maine permitting standardized blood centrifugation and sample preparation within 15 minutes of drawing. In a controlled experiment, blood was also carried back under several protocols of temperature and delay to a remote laboratory for routine preparation, storage and delivery to ESA, Inc. for analysis. Samples at each sight were prepared as plasma for storage and as an Ac preparation ready for evaporation in the protocol described under the methods section above.

Little difference was seen in the patterns between the immediate preparation and after a two hour delay, when samples remain cool. Several delayed samples showed a small but insignificant elevation in 5HT indicating possible hemolysis or platelet lysis. Subsequent repeated analysis show no detectable changes in patterns over a six month period for samples stored at -80°C. Aliquots of this experiment will be analyzed over the next several years to assertion storage age affects.

Evaluation of other markers of bone turnover

Evaluation of the database, in conjunction with collaborators, raised the possibility that several of the compounds which demonstrated differential change in the bedrest and bear studies could be potential markers of bone metabolism. Investigators in the field have proposed a number of compounds which are currently considered to be the best potential markers of bone metabolism. Recently, methods were presented for high performance liquid chromatographic separation and coulometric electrochemical detection of many of these bone turnover markers, collagen crosslink and protein breakdown markers in plasma and urine. Simultaneous assay is now possible for hydroxyproline, hydroxylysine, lysyl- and hydroxylyslypyridinoline (PYR and PDP), osteocalcin (BGP) 45-49 and 7-19 fragments, as well as measures of muscle breakdown including homocarnosine, carnosine, acetyl histidine and the methylated histidines, anserine, xanthine and hypoxanthine (Milbury and Matson, 1994). As studies continue to confirm the bedrest and wintering bear data presented here, these markers of bone turnover will be followed as authentic standards and checked against several of the unidentified compounds singled out in the initial database studies.

CONCLUSION

The data observed and reported here without an *a priori* hypothesis support several current theories regarding torpor in euthermic species, starvation biology, and maintenance of anabolic metabolism. Consideration of entire pathways, especially of the nitrogen cycle compounds, should yield more insight into the regulation of protein breakdown and turnover. The database is being enlarged to clarify and confirm these findings prior to a chemical isolation effort of several of the unidentified compounds included in the database. This effort holds promise of leading to new pharmaceutical and/or dietary strategies for amelioration of the debilitating effects of extended bedrest and immobility. Immobility osteopenia is used here as only one example of a disorder where large metabolic database construction can aid in advancing knowledge. Similar studies reported within this collection and elsewhere show the utility of such approaches investigating degenerative neurological disorders and in developing a degenerative disorder diagnosis based on metabolic database frequency distribution (LeWitt, *et al.*, 1992; Ogawa, *et al.*, 1992; Beal, *et al.*, 1992; Milbury, *et al.*, 1994). Another study utilizing CEAS generated metabolic databases investigates neurometabolite alterations in Rett syndrome (Pelligra, *et al*, 1992). There are currently projects underway to construct databases for chronic fatigue syndrome and midbrain injury in children, paying particular attention to oxidative stress markers. These are all metabolically complex conditions where samples can be collected from easily definable categories of illness and where control samples are available.

Metabolic profiles and pattern recognition has enormous potential in providing a quantitative measurement of organism physiological age (Robinson and Robinson, 1994). Recently support has been growing for long-term studies of aging which will utilize eventual outcome to direct retrospective studies of extensive biomedical databases in a search for possible indicator markers of outcome. CEAS generated metabolic databases might not only provide potential outcome markers but also additional valuable information about the natural metabolic changes occurring with increasing age or during the progression of chronic illnesses.

REFERENCES

Azzi, F., Mannix, J. E., Howard, D. and Nelson, R. A. (1979). Effect of winter sleep on pituitary-thyroid axis in American black bear. Am. J. Physiology **237**, E227-E230.

Beal, M. F., Matson, W. R., Story, E., Milbury, P., Ryan, E. A., Ogawa, T. and Bird, E.D. (1992). Kynurenic acid concentrations are reduced in Huntington's disease cerebral cortex. J. Neurol. Sci. **108**, 80-7.

Beal, M. F., Matson, W. R., Swartz, K. J., Gamache, P. H. and Bird, E. D. (1990). Kynurenine pathway measurements in Huntington's disease striatum: evidence for reduced formation of kynurenic acid. J. Neurochem. **55**, 1327-1339.

Cingolani, M. L., Re, L. and Rossini, L. (1985). The usefulness, in pharmaceutical classification, of complementary pattern-recognition techniques and structure modeling as afforded by the iterative collation of multiple-trial data in data banks. Pharmacol. Res. Commun. 17, 1-22.

Delmas, P. D. (1991). Biochemical markers of bone turnover: methodology and clinical use in osteoporosis. Am. J. Med. **91**, 59S-63S.

Ebling, P. R., Peterson, J. M. and Riggs, B. L. (1992). Utility of type I procollagen propeptide assays for assessing abnormalities in metabolic bone diseases. J. Bone. Miner. Res. **7**, 1243-1250.

Floyd, T., Nelson, R. A. and Wynne, G. F. (1990). Calcium and bone metabolic homeostasis in active and denning black bears (*Ursus americanus*), Clin Orthop, **255**, 301-309.

Haller, A.C. and Zimny, M.L. (1977). Effects of hibernation on interradicular alveolar bone. J. Dent. Res. **56**, 1552-1557.

Hanson, D. A., Weis, M. A., Bollen, A. M., Maslan, S. L., Singer, F. R. and Eyre, D. R. (1992). A specific immunoassay for monitoring human bone resorption: Quantitation of type I collagen cross-linked N-telopeptides in urine. J. Bone. Miner. Res. **7**, 1251-1258.

Hartigan, J. A. (1975). Clustering Algorithms, John Wiley & Sons, New York.

LeWitt, P. A., Galloway, M. P., Matson, W., Milbury, P., McDermott, M. and Srivastava, D. K. (1992). Markers of dopamine metabolism in Parkinson's disease. The Parkinson Study Group. Neurology **42**, 2111-2117.

Lundberg, D. A., Nelson, R. A., Wahner, H. W. and Jones, J. D. (1976). Protein metabolism in the black bear before and during hibernation. Mayo Clin. Proc. 51, 716-722.

Malinowski, E. R. and Howery, D. G. (Eds.) (1980). Factor Analysis in Chemistry, John Wiley & Sons, New York.

Manly, B. F. (Ed.) (1986). Multivariate Statistical Methods. Chapman and Hall, London.

Matson, W. R., Gamache, P. G., Beal, M. F. and Bird. E. D. (1987). EC array sensor concepts and data. Life Sci. 41, 905-908.

McGowan, J. A. (1993). Osteoporosis: Assessment of bone loss and remodeling. Aging Clin. Exp. Res. **5**, 81-93.

Milbury, P. E. and Matson, W. R. (1994). Coulometric electrochemical detection of markers of bone turnover and products of collagen crosslink and muscle protein breakdown. Thirty-fourth Annual Meeting of The American Society for Cell Biology. Dec 10-14, San Francisco, California.

Milbury, P. E., Beal, M. F. and Matson, W. R. (1994). The potential of degenerative disorder diagnosis based on biochemical pattern or profile analysis. Fourth International Conference on Alzheimer's Disease and Related Disorders, July 29 to Aug. 3, Minneapolis, Minnesota.

Milbury, P. E, Vaughan, M. R., Farley, S., Matula, G. J. Jr., Convertino, V. A. and Matson, W. R. (1995). A Comparative Bear Model for Immobility Osteopenia. Tenth International Conference on Bear Research and Management, July, Fairbanks, AK.

Milbury P. E. (1992). A Comparative Bear Model of Immobility Osteopenia. NIA/NIH SBIR Phase I Final Report and Phase II Application (#R44 AG13317-02).

Nelson, R. A., Beck, T. D. I. and Steiger, D. L. (1984). Ratio of serum urea to serum creatinine in wild black bears. Science **226**, 841-842.

Nelson, R.A., Steiger, D.L. and Beck, T.D.I. (1983). Neuroendocrine and metabolic interactions in hibernating black bear. Acta Zool. Fenn. **174**, 137-141.

Ogawa, T., Matson, W. R., Beal, M. F., Myers, R. H., Bird, E. D., Milbury, P. and Saso, S. (1992). Kynurenine pathway abnormalities in Parkinson's disease. Neurology **42**, 1702-1706.

Pelligra, R., Norton, R. D., Wilkinson, R., Leon, H. A. and Matson, W. R. (1992). Rett syndrome: stimulation of endogenous biogenic amines. Neuropediatrics **23**, 131-137.

Riis, B. J. (1991). Biochemical markers of bone turnover in diagnosis and assessment of therapy. Am. J. Med. **91**, 64S-68S.

Risteli, J., Niemi, S., Elomaa, I. and Risteli, L. (1989). Bone resorption assay based on a peptide liberated during type 1 collagen degradation. J. Bone Miner Res. **4**, 47-55.

Robinson, A. B. and Robinson, L. R. (1994). Quantitative Measurement of Human Physiological Age by Profiling of Body Fluids and Pattern Recognition. In: Practical Handbook of Human Biologic Age Determination. Balin, A.K. (Ed.). CRC Press, Inc., Boca Raton, FL. pp. 265-284.

Tamburino, G. and Fiore, C. E. (1990). Plasmatic markers of bone metabolism. Ann. Ital. Med. Int. **5**, 296-302.

Progress in HPLC-HPCE, Vol. 6, pp. 145–168
I.N. Acworth *et al.* (Eds)

THE APPLICATION OF n-EC-HPLC TO THE ANALYSIS OF INSECT BIOGENIC AMINES

THOMAS C. SPARKS, CHAOXIAN GENG, ROBERT P. GAJEWSKI, JOHN R. SKOMP, NOELLE A. GRANGER[1] and HOWARD W. FESCEMYER[2]

DowElanco, Discovery Research, 9330 Zionsville Road, Indianapolis, IN. [1]*Department of Anatomy, University of North Carolina, Chapel Hill, NC,* [2]*Department of Entomology, Clemson University, Clemson, SC*

INTRODUCTION

The biogenic amines of insects function in a variety of roles including those of neurotransmitter, neuromodulator and neurohormone (Orchard, 1982). Over the past two decades a variety of methods have been used for the identification of insect biogenic amines including radioenzymology, spectrofluorometry, gas-chromatography negative ion chemical mass spectrometry (GC-NICIMS), and HPLC with electrochemical detection (HPLC-EC) (Evans *et al.*, 1985; MacFarlane *et al.*, 1990). Based on these methods, a variety of biogenic amines have been identified in insects including octopamine, tyramine, dopamine and 5-hydroxytryptamine (5-HT) and, to a lesser extent, norepinephrine and epinephrine (Brown and Nester, 1985; Evans, 1985; Evans *et al.*, 1985).

Insects present a variety of challenges for any of these technologies in that tissue samples can be very small, in some cases only a few milligrams, and the quantities of biogenic amines are correspondingly low, often in the range of low picomoles per whole tissue. Complicating these considerations may be the need to quantify several biogenic amines, precursors and/or metabolites simultaneously in the same sample, the requirement of easy sample preparation, and the desirability of providing simple, accurate compound identification. While any one of the above mentioned technologies can fulfill some of these requirements/goals, until recently, no one methodology could provide a reasonable expectation of meeting all of these needs. GC-NICIMS and dual channel HPLC-EC provide many of these needs, but each has its limitation. GC-NICIMS requires an elaborate extraction and derivatization procedure (MacFarlane *et al.,* 1990), as well as ready access to a CG-MS. HPLC-based methods involving electrochemical detection predominate due to the sensitivity of this method and the ability to monitor several monoamines simultaneously (Evans *et al.*, 1985). However, identification of the monoamines remains principally co-chromatography-based, relying on retention time relative to an internal standard.

A recent improvement upon this one-dimensional approach to biogenic amine characterization is the development of an array-based EC-detector system incorporating up to 16-channels coupled to an HPLC system (n-EC-HPLC) (Matson *et al.*, 1984, 1987). Such an approach allows a more precise identification due to the availability of two additional parameters beyond retention time: peak height ratio and maximum oxidation potential (Beal *et al.*, 1990; Matson *et al.*, 1987). This technology has only recently been applied to the study of biogenic amines in insects, and it is the goal of this discussion to provide an overview of insect-based studies and impact of n-EC-HPLC technology in discerning biogenic amine profiles, synthesis and metabolism.

INSECT BIOGENIC AMINES

Over the past 30 years a body of information on the nature, biosynthesis and metabolism of insect biogenic amines has accumulated from the use of an assortment of techniques including spectrofluorometry, radioenzymology and HPLC-ECD (Evans *et al.*, 1985). Based on these studies, octopamine is thought to be present in a variety of insect species and in neural, as well as some non-neural, tissues, such as salivary glands, Malphigian tubules, fat body and abdominal wall muscle (Evans, 1985). Likewise, dopamine (Brown and Nester, 1985; Hopkins *et al.*, 1984; Kruegar *et al.*, 1990) and 5-HT (Brown and Nester, 1985) have been reported to be present in numerous insect species. In addition to octopamine, dopamine and 5-HT, many studies have reported the presence of the catecholamine norepinephrine (Brown and Nester, 1985). Epinephrine has been only rarely reported (Brown and Nester 1985), and its actual presence in insects is uncertain (Evans, 1985). In mammalian systems norepinephrine and epinephrine are the predominate phenethylamines, while octopamine is a minor component (Evans, 1985). In contrast, octopamine predominates in insects and invertebrates, while norepinephrine generally constitutes a minor component (Evans, 1985; Vaughan, 1988), and is often entirely absent.

Biosynthetic and metabolic pathways for these biogenic amines have been proposed (Brown and Nester, 1985; Downer and Hiripi, 1994; Evans, 1985; Vaughan, 1988). The biosynthesis of octopamine appears to follow tyrosine - tyramine - octopamine (Downer and Hiripi, 1994; Vaughan, 1988) (Figure 1). In general, the biosynthetic pathways for dopamine and 5-HT in insects are thought to be similar to mammalian pathways; 5-HT is probably derived from tryptophan via 5-hydroxytryptophan (Brown and Nester, 1985; Downer, 1990; Vaughan, 1988) (Figure 2), while for dopamine the pathway seems to be tyrosine - DOPA - dopamine (Brown and Nester, 1985; Downer, 1990) (Figure 3). An alternative pathway involving tyramine as the intermediate has also been suggested (Vaughan, 1988). Although DOPA is readily detected in the hemolymph, where it is known to be involved in tanning and cuticle formation (Hopkins and Kramer, 1992), it is often not detected in samples of neural tissues. In part, this may be the result of highly active aromatic amino acid decarboxylase (i.e., DOPA-decarboxylase) enzymes, which prevent the accumulation of DOPA in insect tissues. In insects, the aromatic amino acid decarboxylases may be more specific than their mammalian counterparts in that the insect aromatic amino acid decarboxylase may catalyze the conversion of DOPA to dopamine and 5-hydroxytryptophan to 5-HT, while a distinct enzyme (tyrosine decarboxylase) catalyzes the conversion of tyrosine to tyramine (Downer and Hiripi,

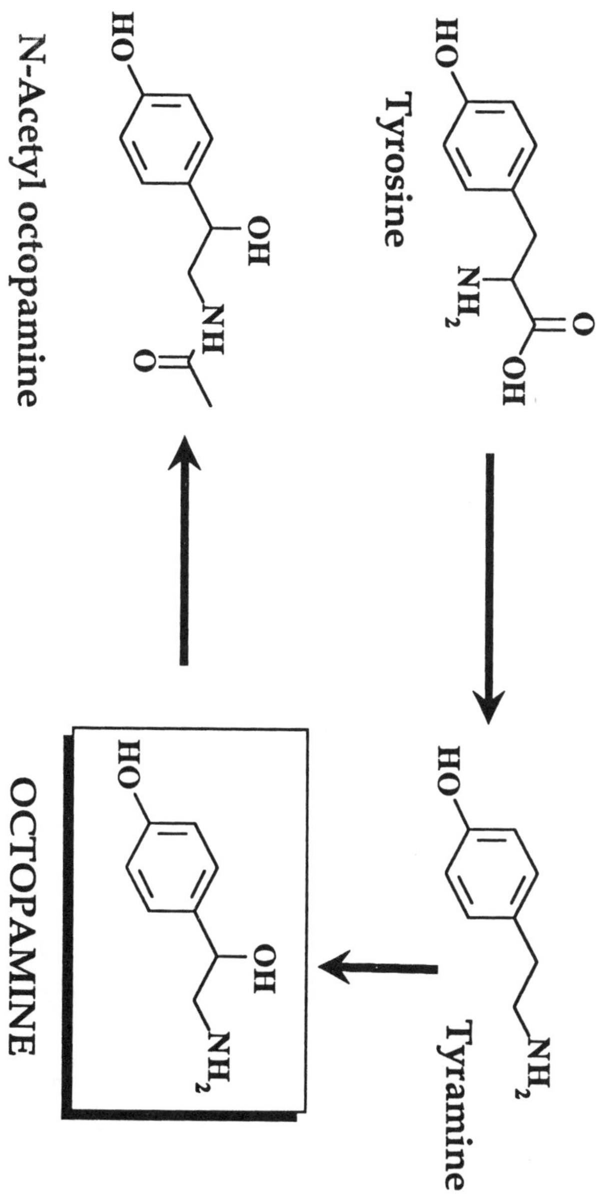

Figure 1. Structures of octopamine, its biosynthetic pathway and its *N*-acetyl metabolite.

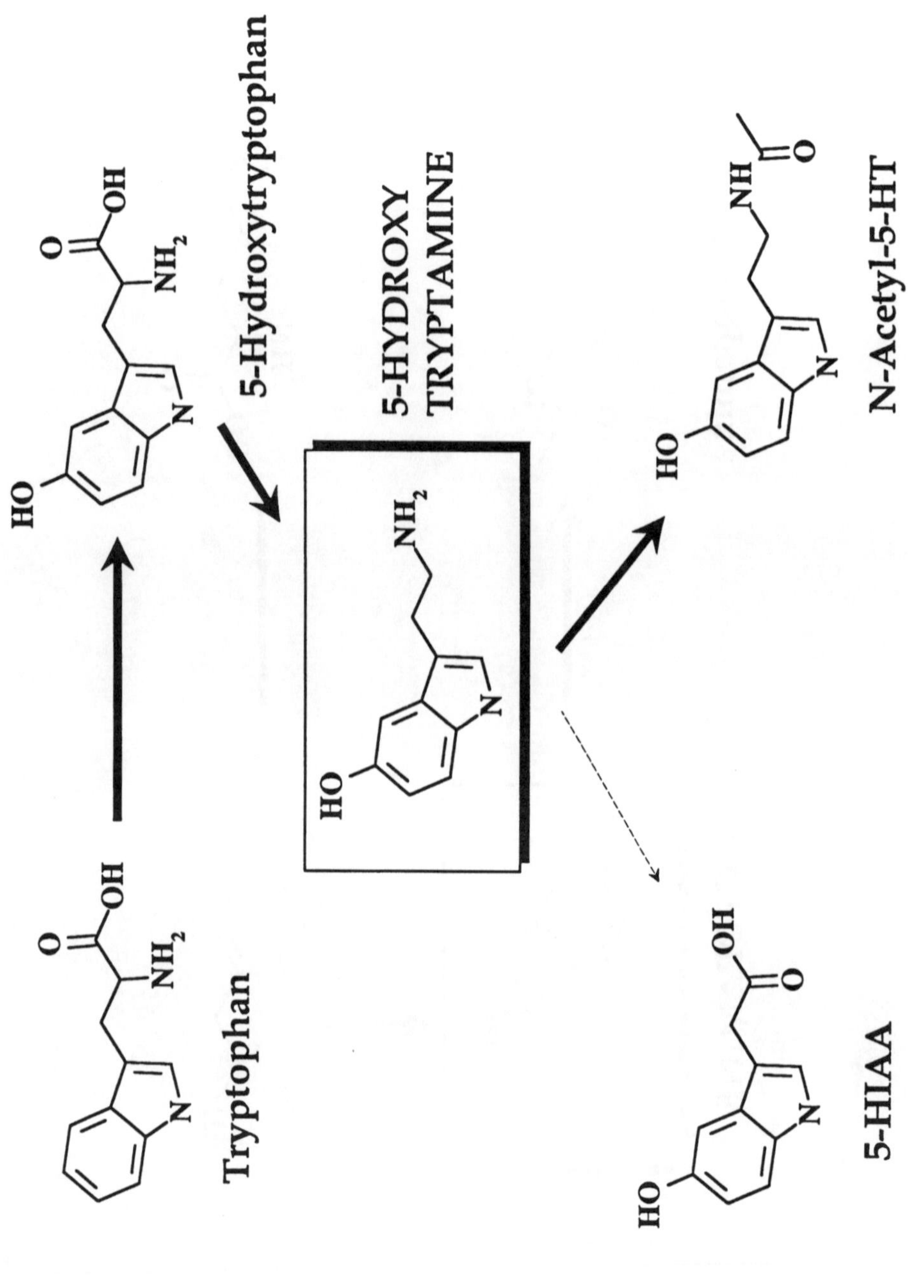

Figure 2. Structures of 5-HT, its probable biosynthetic pathway and two possible metabolites: *N*-acetyl 5-HT which has been positively identified in insects, and 5-HIAA, common mammalian metabolite arising from the action of monoamine oxidases.

1994).

Although insect biogenic amine biosynthesis is similar to that of mammals, the metabolism/degradation of biogenic amines in insect is distinct. In mammalian systems monoamine oxidases and *O*-methyltransferases predominate (Downer, 1990; Vaughan, 1988), whereas biogenic amine metabolism in insect systems primarily involves *N*-acetylation (Downer and Hiripi, 1994; Vaughan, 1988) and in some cases, conjugation (Downer, 1990; Downer and Hiripi, 1994).

SAMPLE PREPARATION AND CHROMATOGRAPHIC STANDARDS

Insect sample preparation for n-EC-HPLC is relatively easy and quite similar to that used for typical single or dual-channel HPLC-EC (Evans *et al.*, 1985). Following dissection, tissues are typically rinsed in saline, and blotted dry (to remove the possibility of hemolymph contamination with its associated biogenic amines) and homogenized in a small volume (e.g., 100μL) of perchloric acid (e.g., 0.1M, 4^oC) (Table 1). A freeze-thaw cycle to facilitate biogenic amine extraction is often performed at this time. Hemolymph samples are collected and then typically diluted in an equal volume of perchloric acid (e.g., 0.4M) or hydrochloric acid (Table 1). Both homogenized tissues and diluted hemolymph are then typically centrifuged (e.g., 5min., 12,000g, 4^oC) to pellet cellular debris (Table 1). In most studies, a volume (e.g., 40-60μL) of the clear supernatant is analyzed directly on the n-EC-HPLC system.

Table 1.
Sample Preparation Methods for n-EC-HPLC

	References					
	Tissues				Hemolymph	
	a	b	c,d,e	f*	b,g	Table 3
Sample Solution						
HCl	X				X	
Perchloric acid		X	X	X		X
Homogenization	X	X	X			
Freeze/thaw						X
Centrifugation	X	X	X	X	X	X
Filtration	X					
Rotary evaporation	X					
Dilution						
Perchloric acid	X	X			X	

*Preceeded by extraction in NaCl and centrifugation. a) Shimizu *et al.*, (1991a); b) Takeda *et al.*, (1991); c) Geng *et al.*, 1993; d) Sparks and Geng (1992); e) Sparks *et al.*, (1994); f) Shimizu *et al.*, (1991b); g) Shimizu and Takeda, (1991).

As already mentioned, a real advantage of n-EC-HPLC over single and dual channel HPLC-EC is the availability of two additional parameters beyond retention

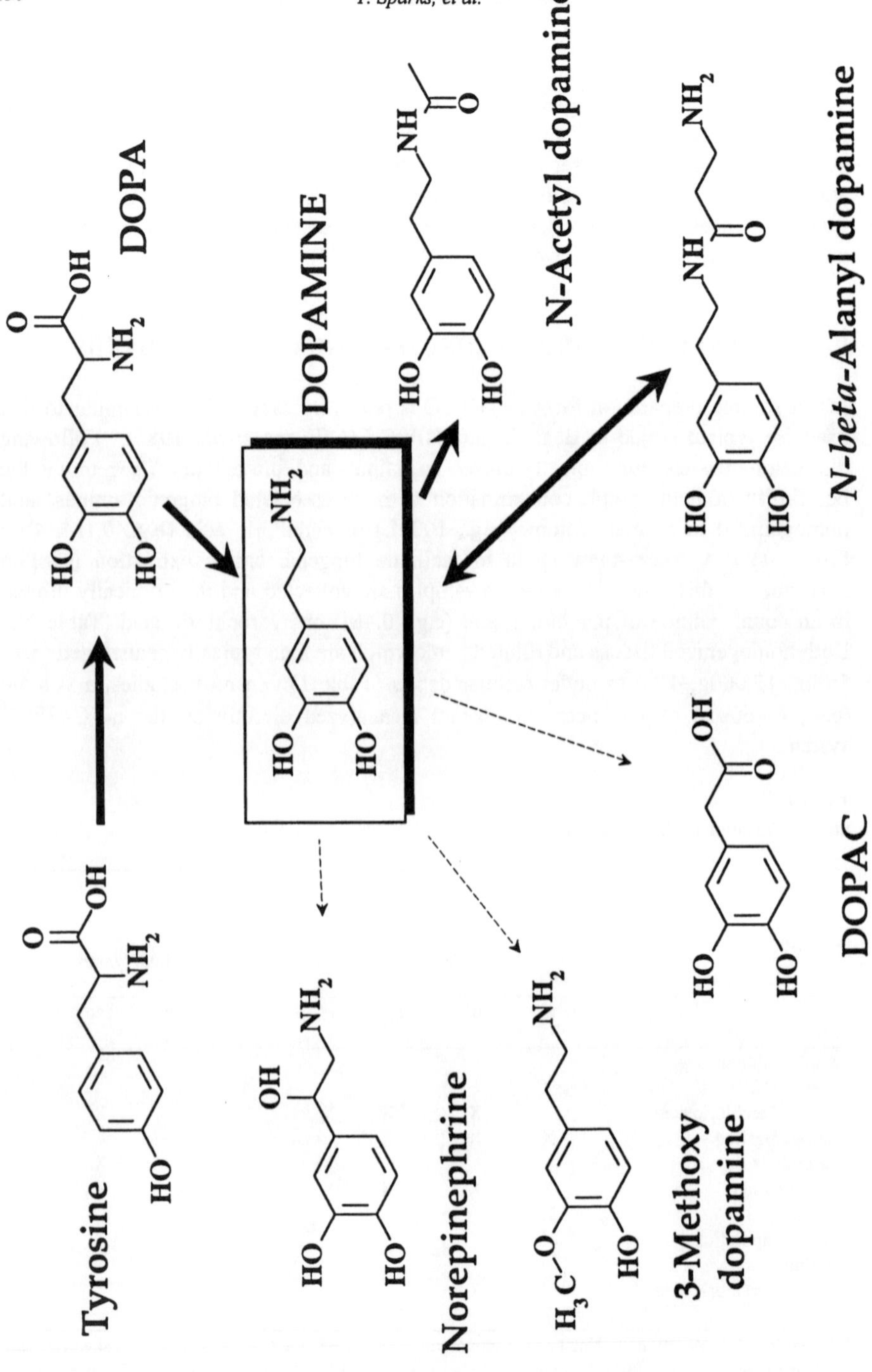

Figure 3. Structures of dopamine and norepinephrine, the biosynthetic pathway and several dopamine metabolites: *N*-acetyldopamine and *N*-β-alanyldopamine, both present in insects, and two metabolites arising from the action of monoamine oxidases (DOPAC) and catecholamine methyltransferases (3-methoxytyramine).

time (peak height ratio and maximum oxidation potential) for use in identifying the biogenic amines present (Beal *et al.*, 1990; Matson *et al.*, 1987). The 3-dimensional-like character of the resultant chromatograms allows the clear separation and identification of 20 to 25, or more, reference standards from which the peaks in a sample will be identified. This capability allows a variety of biogenic amines, precursors and metabolites to be identified in a single sample. However, implicit is this process is that the standards used must be appropriate for the intended samples. Whenever possible, preliminary studies investigating a wide array of potential precursors and metabolites should be conducted to determine the potential biogenic amine composition in the samples and the most appropriate set of standards to be used. This is especially important for samples from insects such as hemolymph, where a large number of peaks may be present. As is shown in Table 2, a wide array of standards have been used in studies of insect derived samples. As an examination of the standards used in the different studies (Table 2) will reveal, their compositions can flavor the nature of what is, or is not, found in the samples analyzed (see below).

BIOGENIC AMINES IN INSECTS AND MITES AS DETERMINED BY n-EC-HPLC

Biogenic amines present in insects

Neural tissues. Although numerous studies have been conducted on the nature of the biogenic amines in insects, information on lepidopterans is somewhat limited, especially for indolalkylamines (Brown and Nester, 1985; Evans, 1985). Studies utilizing n-EC-HPLC have recently been conducted on several insect species, including several lepidopterans. In n-EC-HPLC based studies of the silkworm, *Bombyx mori,* tyramine, dopamine and 5-HT were present in the neural tissue in almost all stages examined while octopamine was only detected in neural tissue of fourth stadium (but not last stadium) larvae (Takeda *et al.*, 1991). In addition, norepinephrine was also found to be present in the brain and subesophageal ganglion of late last stadium larvae of *B. mori* (Takeda *et al.*, 1991). In contrast to *B. mori*, n-EC-HPLC based analysis of neural tissues from last stadium larvae of several other lepidopterans including *M. sexta, T. ni, H. virescens* and *Anticarsia gemmatalis* (velvetbean caterpillar) all revealed the presence of octopamine as well as tyramine, dopamine and 5-HT, but not norepinephrine or epinephrine (Geng *et al.*, 1993; Sparks and Geng, 1992) (Table 3). Previous studies of *M. sexta* neural tissue have also found octopamine, dopamine and 5-HT (see Brown and Nester, 1985; Evans, 1985; Krueger *et al.*, 1990).

Although previous studies utilizing other techniques reported the presence of norepinephrine in neural tissues of *M. sexta* larvae (Taylor and Newburgh, 1979), n-EC-HPLC analysis of a variety of *M. sexta* neural tissues has failed to confirm its presence (Sparks and Geng, 1992). n-EC-HPLC analysis of *M. sexta* neural tissue does detect a peak potentially corresponding to norepinephrine; however careful analysis demonstrated that this peak was not norepinephrine (Geng *et al.*, 1993). Thus, neither epinephrine nor norepinephrine are present in detectable levels in neural tissues of most lepidopterans examined to date using n-EC-HPLC (Geng *et al.*, 1993) (Table 3). In general, a comparison of the concentrations of biogenic amines detected by n-EC-HPLC in several neural tissues of *M. sexta* corresponds well with the values reported in studies using other methodologies (Sparks and Geng, 1992).

Table 2.
Standards used for n-EC-HPLC

Compound	Code	References					
		a	b	d,e	c	g	Table 3
Tyrosine	TYR	x	x	x	x	x	x
Tyramine	TA	x	x	x	x	x	x
Octopamine	OA	x	x	x	x	x	x
Synephrine	SYN	-	-	x	x	-	x
N-Acetyl octopamine	NAOA -	-	x	x	-	x	
Vanillic acid	VA	x	x	-	-	x	-
DOPA	DOPA	x	x	x	x	x	x
Dopamine	DA	x	x	x	x	x	x
Norepinephrine	NE	x	x	x	x	x	x
Epinephrine	E	x	x	x	x	x	x
Metanephrine	MN	x	x	x	x	x	x
Normetanephrine	NMN	x	x	x	x	x	x
Epinine	EPI	x	x	x	x	x	x
3-Methoxytyrosine	3MTYR x	-	-	-	x	x	
3-Methoxytyramine	3MTA	x	x	-	-	x	x
3-Methyoxy-4-hydroxy-phenyl ethylene glycol	MHPG x	x	-	-	x	-	
4-Hydroxy-3-methoxy-mandelic acid	VMA	x	x	-	-	x	-
4-Hydroxy-mandelic acid	HMA	-	-	x	x	-	x
3,4-Dihydroxyphenyl acetic acid	DOPAC x	x	-	-	x	-	
Homovanillic acid	HVA	x	x	-	-	x	x
N-Acetyl dopamine	NADA -	-	x	x	-	x	
N-$-Alanyl dopamine	NBDA -	-	x	-	-	x	
Tryptophan	TRP	x	x	x	x	x	x
5-hydroxytryptophan	5HTRP x	x	x	x	x	x	
5-hydroxytryptamine	5HT	x	x	x	x	x	x
5-hydroxyindoleacetic acid	5HIAA x	x	x	x	x	x	
N-methyl-5-hydroxy-tryptophan	NM5HTRP	-	x	-	-	x	-
N-methyl-5-hydroxy-tryptamine	NM5HT x	-	x	x	x	x	
Melatonin	MET	x	x	-	-	x	x
N-acetyl-5-hydroxy-tryptamine	NA5HT x	-	x	x	x	x	

a) Shimizu *et al.*, (1991a); b) Takeda *et al.*, (1991); c) Geng *et al.*, (1993); d) Sparks and Geng (1992); e) Sparks *et al.*, (1994); f) Shimizu *et al.*, (1991b); g) Shimizu and Takeda (1991).

Hemolymph and non-neural tissues. An analysis of hemolymph from *B. mori* detected only dopamine and norepinephrine; octopamine and 5-HT were absent (Takeda *et al.*, 1991). n-EC-HPLC-based studies of *Mamestera brassicae* hemolymph also detected norepinephrine (in treated larvae) and dopamine, as well as octopamine, but not 5-HT (Shimizu and Takeda, 1991) (Table 3). Unlike the neural tissues, and unlike *B. mori* or *M. brassicae*, the only biogenic amine detected in the hemolymph, corpora allata and all other non-neural tissues of last stadium larvae of *M. sexta* was dopamine (Geng *et al.*, 1993; Sparks and Geng, 1992) (Table 3). Likewise, dopamine was the only biogenic amine present in non-neural tissues (hemolymph and midgut) from larvae of *T. ni* (Table 3). In contrast, octopamine, dopamine and 5-HT were absent in the venom glands of the hymenopteran, *Apanteles kariyai*; only tyramine was present (Shimizu *et al.*, 1991b).

Although extensive data is available on the biogenic amines present in insects, information on biogenic amines in acarines is confined to only a few studies (Knowles *et al.*, 1989; Megaw and Robertson, 1974; Scott *et al.*, 1985). The application of n-EC-HPLC to the study of biogenic amines in mites provided an opportunity to greatly expand our knowledge in this economically important group of plant pests. The biogenic amine profile of whole body homogenates of the two-spotted spider mite, *Tetranychus urticae*, was examined in different life stages using n-EC-HPLC (Sparks *et al.*, 1994). As with most of the insects examined using n-EC-HPLC, octopamine and dopamine were both present in homogenates of most stages of *T. urticae* (Sparks *et al.*, 1994). Tyramine was detected in adult female mites, but was absent (or below the limit of detection) in all other stages examined. Likewise, 5-HT was present in adult female mites and large nymphs, but absent in all other stages examined (Sparks *et al.*, 1994) (Table 3).

Biogenic amine biosynthesis

Based on n-EC-HPLC analysis, the proposed biosynthetic pathways in the *B. mori* nervous system included tyrosine - DOPA - dopamine (Figure 3), tyrosine - tyramine - octopamine (Figure 1); and tryptophan - 5-HT (Figure 2) (Takeda *et al.*, 1991). n-EC-HPLC-derived data for nervous tissues from other lepidoterans (*M. sexta, H. virescens, T. ni, A. gemmatalis*) (Geng *et al.*, 1993; Sparks and Geng, 1992) (Table 3) also support the presence of these three biosynthetic pathways for the primary biogenic amines. Although the data is much less complete, these same three pathways also appear to exist in some stages of the spider mite, *T. urticae* (Sparks *et al.*, 1994) (Table 3).

In most of the n-EC-HPLC studies, DOPA and 5-hydroxytryptophan were absent (Table 3). In *M. sexta* DOPA was not detected in neural tissue until just prior to pupation (Geng *et al.*, 1993). The presence of high levels of decarboxylase for these precursors in *M. sexta* neural tissue (Geng and Sparks, 1991) suggests that they are turned over very quickly thereby preventing accumulation and, hence, detection. n-EC-HPLC studies designed to evaluate the utility of DOPA and tyramine as substrates for the production of dopamine clearly demonstrated that DOPA is the only substrate leading to dopamine (Geng and Sparks, 1991) (Figure 4). Thus, the results of n-EC-HPLC-based studies generally support the results of previous studies concerning the likely pathways for biogenic amine synthesis.

Among all of the insects studied using n-EC-HPLC, dopamine appears to be the predominate biogenic amine in the hemolymph and non-neural tissues (Table 3). In addition to dopamine, tyrosine and DOPA were also present in the hemolymph of *B. mori* (Takeda *et al.*, 1991), *M. brassicae* (Shimizu and Takeda, 1991) and *M. sexta* (Table 3). The precursors for octopamine synthesis (tyrosine - tyramine - octopamine) also appear to be present in the hemolymph of treated larvae of *M. brassicae* (Shimizu and Takeda, 1991). Like the hemolymph, non-neural tissues of *M. sexta* (corpora

Table 3.
Insect biogenic amines identified using n-EC-HPLC

Species / Tissue [ref]	TYR	TA	OA	NAOA	SYN	NE	DOPA	DA	NADA	NBAD	TRP	5OHTR	5HT	NA5HT	Other
LEPIDOPTERA															
Manduca sexta - 4th															
Brain [c]	X	X	X	*	*	*	*	X	X	--	X	*	X	X	*
Manduca sexta - 5th															
Brain [c]	X	X	X	*	*	*	X	X	X	X	X	*	X	X	*
Subesophageal ganglion [d]	X	X	X	*	*	*	*	X	X	--	X	*	X	X	*
Thoracic ganglion [d]	X	X	X	*	*	*	*	X	X	--	X	*	X	X	*
Abdominal ganglion [d]	X	X	X	*	*	*	X	X	X	--	X	*	X	X	*
Terminal ganglion [d]	X	X	X	*	*	*	*	X	X	--	X	*	X	X	5-HIAA
Corpora allata (V6) [h]	X	*	*	*	*	*	X	X	*	X	X	*	*	*	*
Hemolymph [h]	X	*	*	*	*	*	X	X	X	X	*	*	*	*	*
Fat body [h]	--	*	*	*	*	*	*	X	X	--	*	*	*	*	*
Malpighian tubules [h]	--	*	*	*	*	*	--	X	--	--	*	*	*	*	*
Epidermis [h]	--	*	*	*	*	*	--	X	--	--	*	*	*	*	*
Midgut [h]	--	*	*	*	*	*	--	X	--	--	*	*	*	*	*
Muscles [h]	--	*	*	*	*	*	--	X	--	--	*	*	*	*	*
Salivary gland [h]	--	*	*	*	*	*	--	X	--	--	*	*	*	*	*
Trichoplusia ni - 5th															
Brain [h]	X	*	X	X	*	*	*	X	X	X	X	*	X	?	*
Hemolymph [h]	*	*	*	*	*	*	X	X	X	--	*	*	*	*	*
Midgut [h]	X	*	*	*	*	*	X	X	X	X	*	*	*	*	*

Species / Tissue	TYR	TA	OA	NAOA	SYN	NE	DOPA	DA	NADA	NBAD	TRP	5OHTR	5HT	NA5HT	Other
Anticarsia gemmatalis															
Brain (adult) [h]	X	*	X	X	*	*	X	X	X	X	X	*	X	X	*
Brain (larva - 5th) [h]	X	?	X	?	*	*	*	X	X	X	X	*	X	X	*
Heliothis virescens - 5th															
Brain [h]	X	?	X	*	*	*	*	X	X	--	X	*	X	*	*
Bombyx mori - 4th															
Brain [b]	X	X	X	--	--	*	X	X	--	--	X	*	X	--	DOPAC, VMA
Subesophageal ganglion [b]	X	*	X	--	--	X	X	X	--	--	X	*	X	X	DOPAC, 5HIAA
Hemolymph [b]	X	X	*	--	--	*	X	X	--	--	X	*	*	--	3-MT, DOPAC, MEL, MHPG
Bombyx mori - 5th															
Brain [b]	X	X	*	--	--	X	X	X	--	--	X	*	X	--	HVA, DOPAC, MN, NMN, 5-HTP
Subesophageal ganglion [b]	X	X	*	--	--	X	X	X	--	--	X	*	X	X	HVA, DOPAC, NMN
Hemolymph [b]	X	*	*	--	--	X	X	X	--	--	X	*	*	--	3-MT, DOPAC
Mamestera brassicae															
Hemolymph [g]	X	X	?	--	--	X	X	X	--	--	*	*	*	*	HVA, epinine
ORTHOPTERA															
Perplaneta americana															
Corpus cardicum [a]	X	X	X	--	--	*	*	X	--	--	X	*	*	--	VA
HYMENOPTERA															
Apanteles kariyai															
Venom sac [f]	X	X	*	--	*	*	*	*	--	--	X	*	*	--	NMN, VA, HVA, KYN, HPAC

Species / Tissue	TYR	TA	OA	NAOA	SYN	NE	DOPA	DA	NADA	NBAD	TRP	5OHTR	5HT	NA5HT	Other
TETRANYCHIDAE															
Tetranychus urticae															
Whole body (adult females) [e]	X	X	X	X	*	*	?	X	?	--	X	*	X	?	*
Whole body (large nymphs) [e]	X	*	X	X	*	*	?	X	?	--	X	*	X	?	*
Whole body (eggs) [e]	X	*	?	X	*	*	?	X	?	--	X	*	*	?	*

Table codes: X = biogenic amine is present, * = biogenic amine is absent, -- = biogenic amine was not assayed for, ? = biogenic amine may be present. All biogenic amine codes are listed in Table 2 with the exception of KYN - kynurenine and HPAC - hydroxyphenyl acetic acid.
a) Shimizu *et al.*, (1991a); b) Takeda *et al.*, (1991); c) Geng *et al.*, (1993); d) Sparks and Geng (1992); e) Sparks *et al.*, (1994); f) Shimizu *et al.*, (1991b); g) Shimizu and Takeda (1991); h) *Manduca sexta* (tobacco hornworm), *Heliothis virsecens* (tobacco budworm) *Trichoplusia ni* (cabbage looper), and *Anticarsia gemmatalis* (velvetbean caterpillar) larvae were reared as described previously (Dowd *et al.*, 1987; Geng *et al.*, 1993; Sparks *et al.*, 1979; Fescemeyer *et al.*, 1986). Selected tissues (see Table 1) were dissected from last stadium larvae (or adults for *A. gemmatalis*) in ice cold saline (0.15M NaCl), rinsed in saline, blotted dry and placed in ice-cold 0.1M perchloric acid (100μL) in a polypropylene microfuge tube. The tissues were homogenized using a glass pestle and then frozen on dry ice. Hemolymph was collected via a clipped anal horn (*M. sexta*) or clipped proleg, and diluted in an equal volume of 0.4M perchloric acid. Hemolymph or thawed tissue homogenates were centrifuged at 12,000g (5min, 4°C) and 40-60μL of the clear supernatant was analyzed directly using the n-EC-HPLC system. The n-EC-HPLC system (Coulochem Electrode Array System, ESA Inc., Chelmsford, MA) consisted of a two-pump computer control gradient system coupled to a 16-electrode electrochemical detection unit and a refrigerated autosampler unit, and a Supelcosil LC-18 column (5μm, 20 x 0.46cm, Supelco Inc.) protected by a guard column (RP18 NewGuard, 7μm, Munhall Co.). The detectors were set from 0 to +750mV at increments of +50mV. Samples were analyzed as described previously (Geng *et al.*, 1993, Sparks and Geng, 1992).

allata) and *T. ni* (midgut) both possessed the necessary precursors (tyrosine and DOPA) for dopamine synthesis. As with all of these studies, the mere presence of these components does not necessarily imply that synthesis is occurring in a specified tissue.

Biogenic amine metabolism

The proposed metabolic/degradative products of dopamine identified using n-EC-HPLC in the nervous system of *B. mori* include normetanephrine and DOPAC (Figure 3), while the primary metabolite for 5-HT was found to be 5-HIAA (Takeda *et al.*, 1991) (Figure 2). n-EC-HPLC analysis of biogenic amine metabolites in the hemolymph included DOPAC and 3-methoxytyramine (Figure 3) for dopamine and 3-methoxy-4-hydroxyphenyl ethylene glycol for norepinephrine (Takeda *et al.*, 1991). These metabolites are consistent with degradative pathways in *B. mori* that involve systems that are in common with mammals, specifically, catecholamine methyltransferases and monoamine oxidases (Vaughan 1988). Likewise, homovanillic acid and vanillic acid, which also arise from the action of catecholamine methyltransferases and monoamine oxidases on dopamine, were identified as biogenic amine metabolites in the corpus cardiacum of *P. americana* (Shimizu *et al.*, 1991a) and hemolymph of *M. brassicae* (Shimizu and Takeda, 1991), respectively, and both metabolites along with hydroxyphenyl acetic acid were identified in the venom glands of *A. kariyai* (Shimizu *et al.*, 1991b). Thus, the data presented in all of the above studies suggest metabolic pathways that are associated with catecholamine methyltransferases and/or monoamine oxidases.

N-acetylation is generally considered a primary metabolic pathway in insects (Brown and Nester, 1985; Downer, 1990; Evans, 1985; Vaughan, 1988). However, the catabolic pathways identified in insects using n-EC-HPLC are defined, in part, by the reference standards used with the system (see above). In the studies of *B. mori*, *N*-acetyl derivatives of octopamine, dopamine and 5-HT were not included in the reference standards used (Takeda *et al.*, 1991) (Table 2), nor were *N*-acetyloctopamine or *N*-acetyldopamine included in the studies of *P. americana* (Shimizu *et al.*, 1991a) and *M. brassicae* (Shimizu and Takeda, 1991). Thus, their presence in the respective tissues could have been missed.

In contrast to *B. mori, P. americana* and *M. brassicae*, the primary degradative pathway present in *M. sexta*, *T. ni.*, *H. virescens*, *A. gemmatalis* and *T. urticae* all appear to involve *N*-acetylation (Geng *et al.*, 1993; Sparks and Geng, 1992) (Table 3). In addition to *N*-acetylation, *N*-β-alanyl conjugation of dopamine (Figure 3) (standards for *N*-β-alanyl derivatives of octopamine and 5-HT were not available) was also found to be present in *M. sexta, T. ni* and *A. gemmatalis*. In the corpora allata, *N*-β-alanyldopamine was the only metabolite detected (Table 3). In *M. sexta*, the levels of *N*-β-alanyldopamine exceed those of *N*-acetyl dopamine suggesting that *N*-β-alanyl conjugation may be a preferred pathway (Geng *et al.*, 1993). This concept is supported by results of *in vitro* studies indicating a significantly higher rate of *N*-β-alanyldopamine formation compared to *N*-acetyldopamine (Geng *et al.*, 1993; Krueger *et al.*, 1990). However, *N*-β-alanyldopamine may not be a metabolite *per se*; instead it may be functioning as a storage form of dopamine (see below).

Initial n-EC-HPLC studies of *M. sexta* neural tissues and hemolymph included a wider array of standards than indicted in Table 2, including many associated with monoamine oxidases and methyltransferases. However, preliminary studies indicated

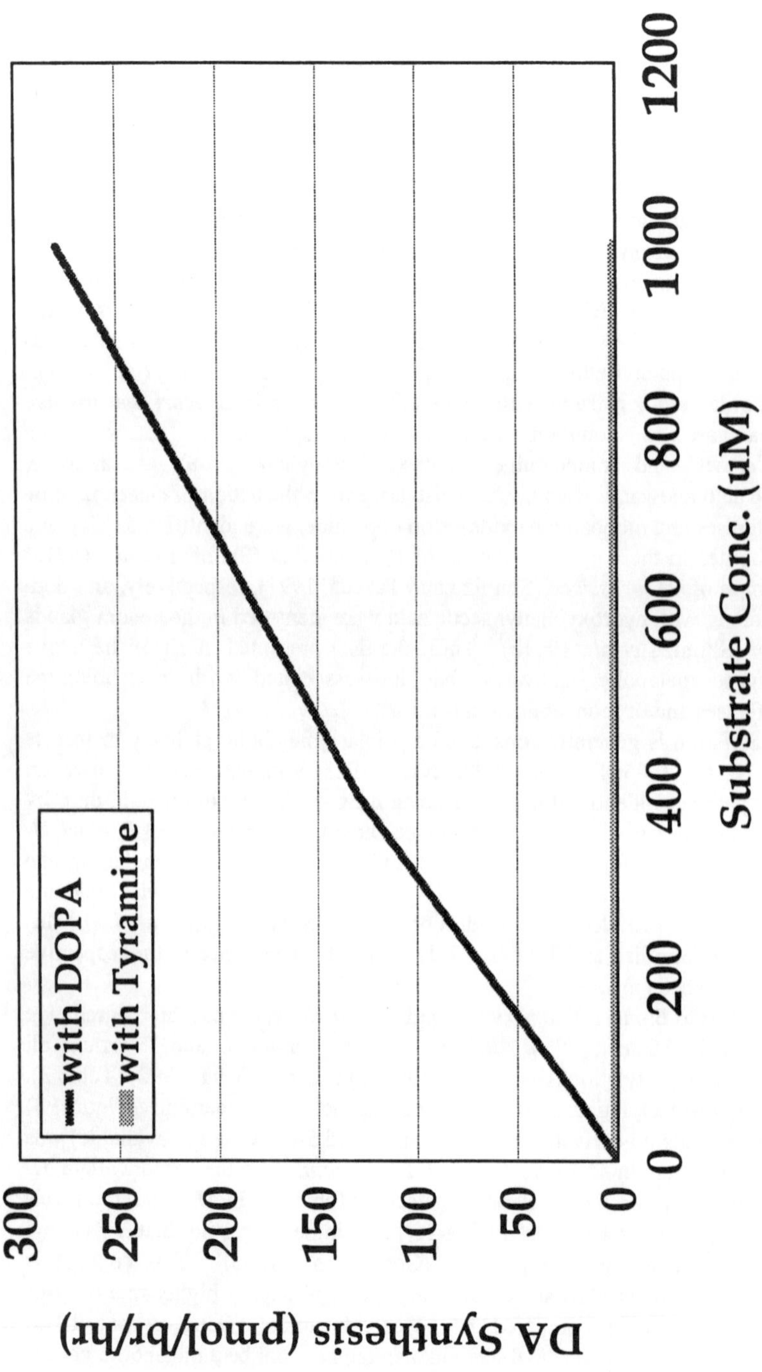

Figure 4. Synthesis of dopamine (DA) *in vitro*, in brain homogenates with DOPA or tyramine as substrate; analyzed using n-EC-HPLC. Dissected brains were homogenized in phosphate buffer, pH6.5, 0.1M, 100μL) incubated with either DOPA or tyramine (22°C). The reaction was terminated by the addition of an equal volume of 0.2M perchloric acid (see Table 3 for details on analysis).

that metabolites resulting from the action of monoamine oxidases or methyl transverses were not present in neural tissues and hemolymph in the insect larvae and mites examined (Sparks and Geng, unpublished data). Therefore subsequent n-EC-HPLC studies focused more on *N*-acetyl and, where possible, *N*-β-alanyl conjugation derivatives. Interestingly, *N*-acetyl octopamine appears to be absent, or at least present in quantities below the detection limit, in the neural tissue of *M. sexta* (Geng and Sparks, 1993; Geng *et al.*, 1993; Sparks and Geng, 1992) (Table 3, unpublished data) and *H. virescens* (Table 3). In contrast, *N*-acetyl octopamine appears to be present in neural tissues of *T. ni* and *A. gemmatalis* and whole body homogenates of *T. urticae* (Table 3, Sparks *et al.*, 1994).

n-EC-HPLC STUDIES OF DEVELOPMENTAL CHANGES IN BIOGENIC AMINE CONTENT

One of the advantages of n-EC-HPLC is that it allows the simultaneous monitoring of numerous biogenic amines as well as their precursors and metabolites. As such, studies aimed at examining developmental changes in these chemical messengers become more enlightening and valuable because shifts in precursor and metabolite profiles can suggest the basis for changes in the biogenic amine content of tissues.

Octopamine, dopamine and 5-HT were present in the neural tissue of *M. sexta* during the late fourth (IV), throughout the fifth (last: V) larval and early pupal (P) stadiums. The neural content of octopamine and 5-HT increased during the pre-wandering phase (V0 to V5) of the last stadium, reaching a peak on day 6 (V6) (Figure. 5) (Geng *et al.*, 1993), and declined and remained low thereafter. At the peak, the brain content of both amines was about double that of their initial values at the beginning of the last stadium (Figure 6). More dramatic were the changes observed in the dopamine content of *M. sexta* neural tissues during the course of the last stadium. The neural dopamine content closely paralleled the octopamine and 5-HT levels in the brains of *M. sexta* until one day post-wandering (V6), where a large peak (more than seven-fold that of the initial value) in the dopamine content occurred, followed by a rapid decline and a final upswing at pupation (P0, Figure 6). A similar pattern had been noted previously for dopamine (Krueger *et al.*, 1990). As already noted, the only biogenic amine identified in the hemolymph of *M. sexta* was dopamine. The hemolymph dopamine content was relatively constant until wandering (V5) where, in sharp contrast to the brain pattern, there was a substantial decrease in the hemolymph dopamine content (Figure 7) reported previously (Hopkins *et al.*, 1984; Krueger *et al.*, 1989). As observed for the brain content, a large increase in hemolymph dopamine content occurred at the time of pupation. Dopamine was undetectable in the corpora allata prior to wandering, but was detected in post wandering larvae (V6) and in pupae (P0), a pattern similar to that observed for neural tissue. Although previous studies suggest relatively high levels of dopamine in the corpora allata/corpora cardiaca (211pg/complex, (Krueger *et al.*, 1990)), n-EC-HPLC analysis suggest that much lower amounts (ca.10pg/CA) are present in the corpora allata (Geng *et al.*, unpublished).

In contrast to *M. sexta*, the profile of biogenic amines in brain tissue from last stadium *B. mori* contained no octopamine, and the dopamine pattern was distinct in that its content was relatively high on day 1, declined to very low levels by day 3,

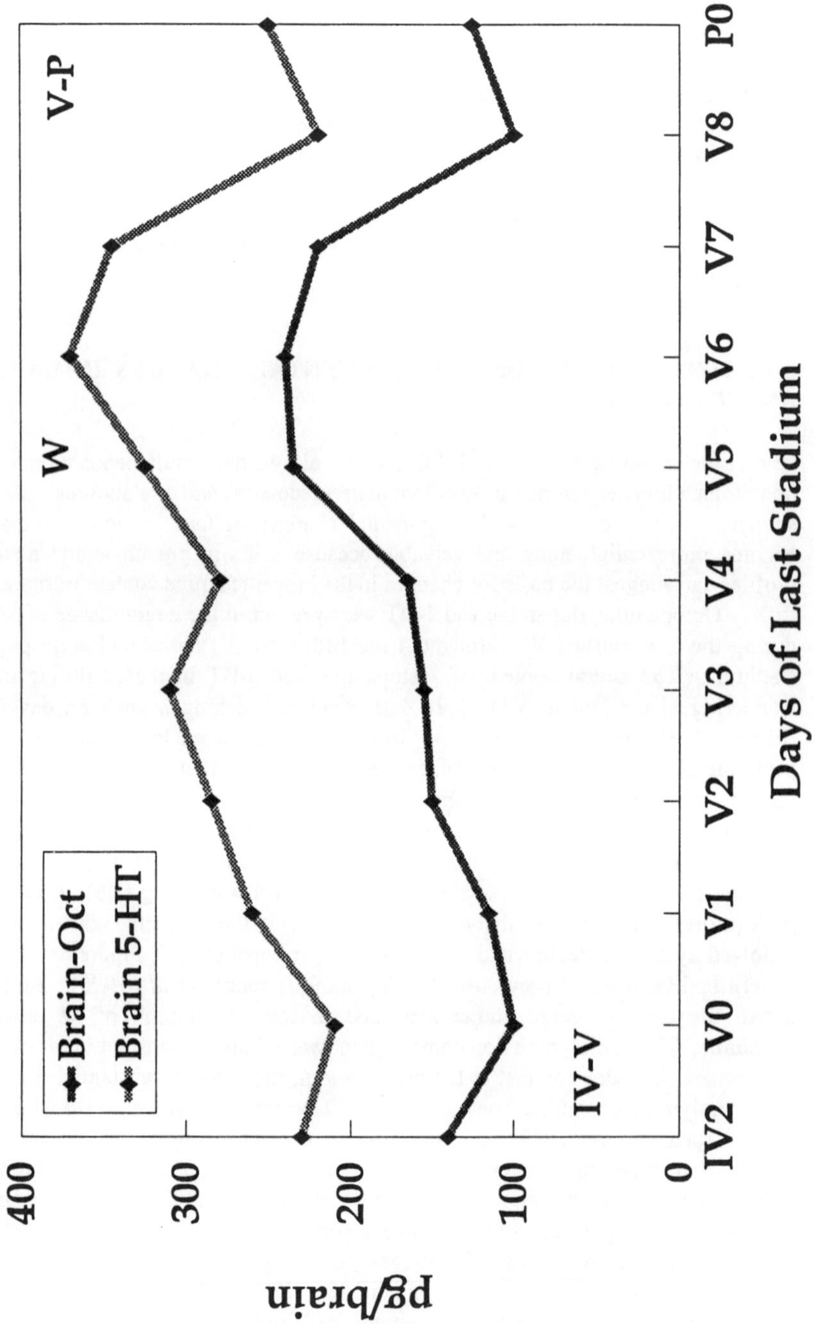

Figure 5. Octopamine and 5-HT levels in neural tissue from last stadium larvae of *M. sexta* (data adapted from Geng *et al.*, (1993)).

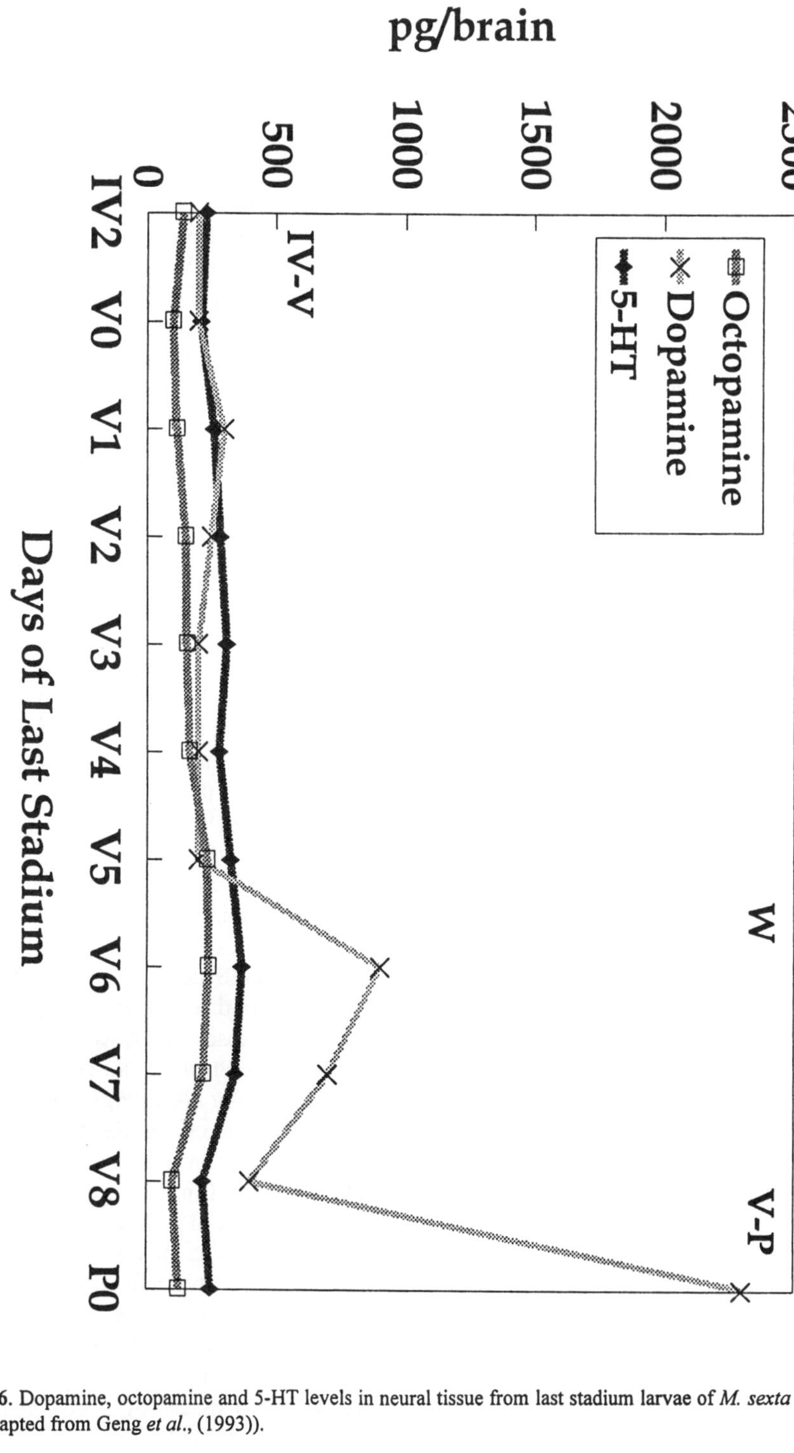

Figure 6. Dopamine, octopamine and 5-HT levels in neural tissue from last stadium larvae of *M. sexta* (data adapted from Geng *et al.*, (1993)).

remained low on day 6 and then increased significantly in post-spinning pharate pupae (Takeda *et al.*, 1991). The 5-HT content in last stadium brains of *B. mori* was about 70-80pg/brain on days 1, 3 and in post-spinning pharate pupae with a small peak on day 7 (135pg/brain) (Takeda *et al.*, 1991). The biogenic amine profiles for the subesophageal ganglion of *B. mori* were similar to those of the brain (Takeda *et al.*, 1991). As with *M. sexta*, the only biogenic amine detected in the hemolymph of *B. mori* was dopamine, which increased from day 1 to day 3, and declined thereafter (Takeda *et al.*, 1991). Thus, either due to differences between the species or in sampling frequency, there appears to be significant differences in the last stadium developmental profiles of biogenic amines for these two lepidopterans.

N-Acetyl 5-HT, the only 5-HT metabolite detected in neural tissue from *M. sexta* (Geng *et al.*, 1993; Sparks and Geng 1992) (Table 3), occurs at levels that are far lower than those of 5-HT and shows little variation throughout the last larval stadium of *M. sexta* (Geng *et al.*, 1993). As already mentioned, no octopamine metabolites were identified in neural tissue of *M. sexta* throughout the last stadium (Geng *et al.*, 1993). In contrast, two metabolites of dopamine were present in the neural tissue of *M. sexta* throughout the last larval stadium; *N*-acetyldopamine and *N*-β-alanyldopamine. Although the *N*-acetyldopamine neural content in *M. sexta* larvae remained fairly constant through the last stadium: the *N*-β-alanyldopamine content was found to peak on the day of wandering (V 5), one day prior to a peak in dopamine content (Fig. 8). As had been suggested previously (Krueger *et al.*, 1990), this observation is consistent with *N*-β-alanyldopamine functioning as a storage form of dopamine (Geng *et al.*, 1993). n-EC-HPLC-based *in vitro* studies clearly demonstrate that *N*-β-alanyldopamine, but not *N*-acetyldopamine, can be hydrolyzed to dopamine by *M. sexta* neural tissue (Geng and Sparks, 1993). Other tissues, such as *M. sexta* fat body or rat neural tissue, were incapable of hydrolyzing *N*-β-alanyldopamine to dopamine (Geng and Sparks 1993). Interestingly, the *N*-β-alanyldopamine hydrolase activity in *M. sexta* neural tissue was higher at the time of wandering (V5 & V6) than during the prewandering phase (V2) (Geng and Sparks, 1993) (Figure 9). Thus, there appears to be a specific enzyme system involved in the hydrolysis of *N*-β-alanyldopamine to dopamine at a time when the dopamine content of neural tissue is increasing.

The presence of a dopamine peak in the neural tissue and in the corpora allata of *M. sexta* near the time of wandering is interesting in that there is also a peak in the level of insect juvenile hormone (JH) synthesis and release from the corpora allata that is critical to successful ecdysis and pupal development (Sparks *et al.*, 1989). Biogenic amines are known to impact JH biosynthesis (Granger *et al.*, 1996; Lafon-Cazal and Baehr, 1988; Pastor *et al.*, 1991; Rachinski, 1994; Thompson *et al.*, 1990; Woodring and Hoffmann, 1994), and the corpora allata of *M. sexta* have been shown to contain dopamine n-EC-HPLC (Granger *et al.*, 1996; Table 3). Thus, the available data may suggest that dopamine may have a role in JH regulation in *M. sexta*. Indeed, recent studies demonstrate that dopamine alters cAMP levels in *M. sexta* corpora allata (Granger *et al.*, 1995), and compounds affecting cAMP typically also affect JH/JH acid biosynthesis (Granger *et al.*, 1994). Indeed, recent studies demonstrate that dopamine stimulates JH biosynthesis during the early last stadium (V0-V2) and then later inhibits JH biosynthesis (V3-V6) (Granger *et al.*, 1996). The ability of the insect to alter the dopamine content in neural (and other?) tissues by a combination of synthesis, metabolism and selective storage and retrieval (via *N*-β-

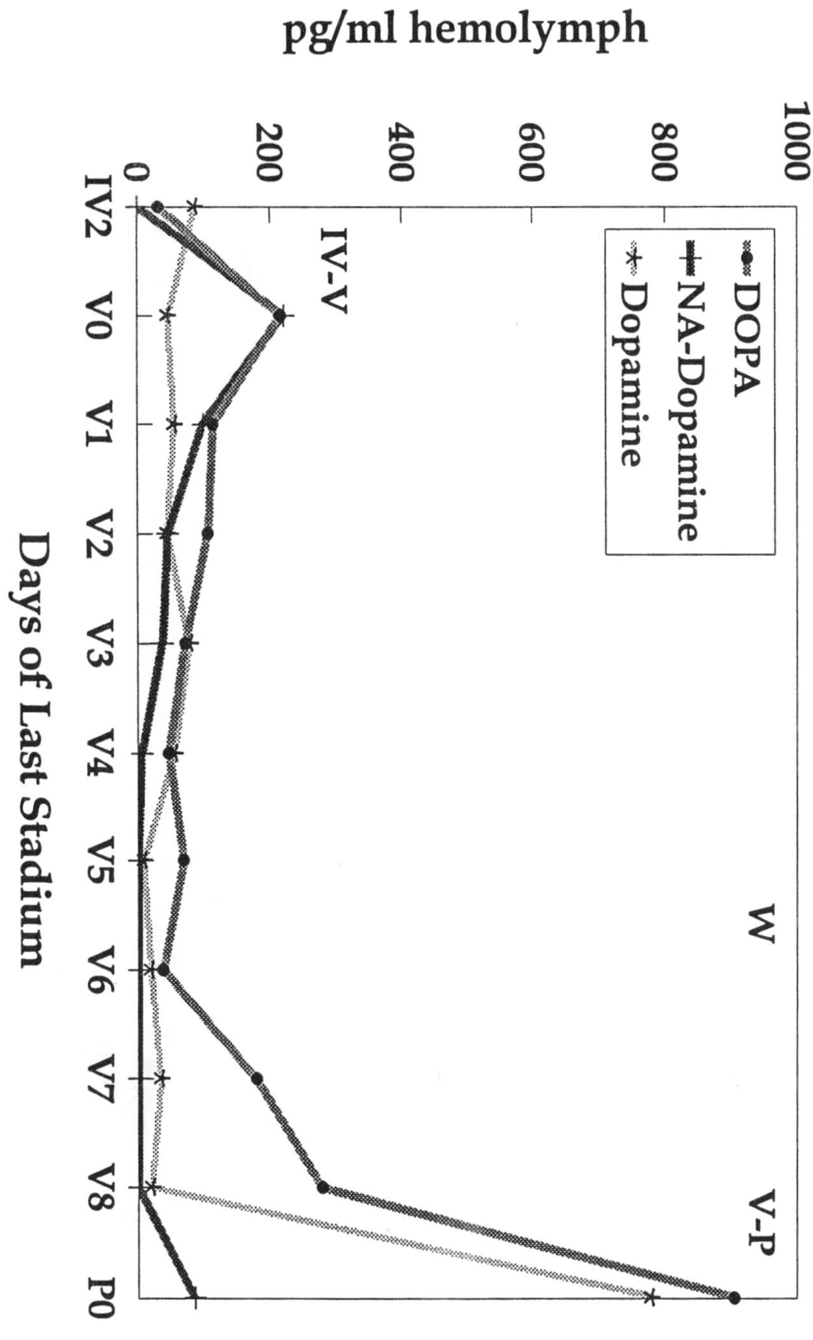

Figure 7. Dopamine, DOPA and *N*-acetyldopamine titers in *M. sexta* hemolymph during last stadium as determined using n-EC-HPLC (see Table 3 for details).

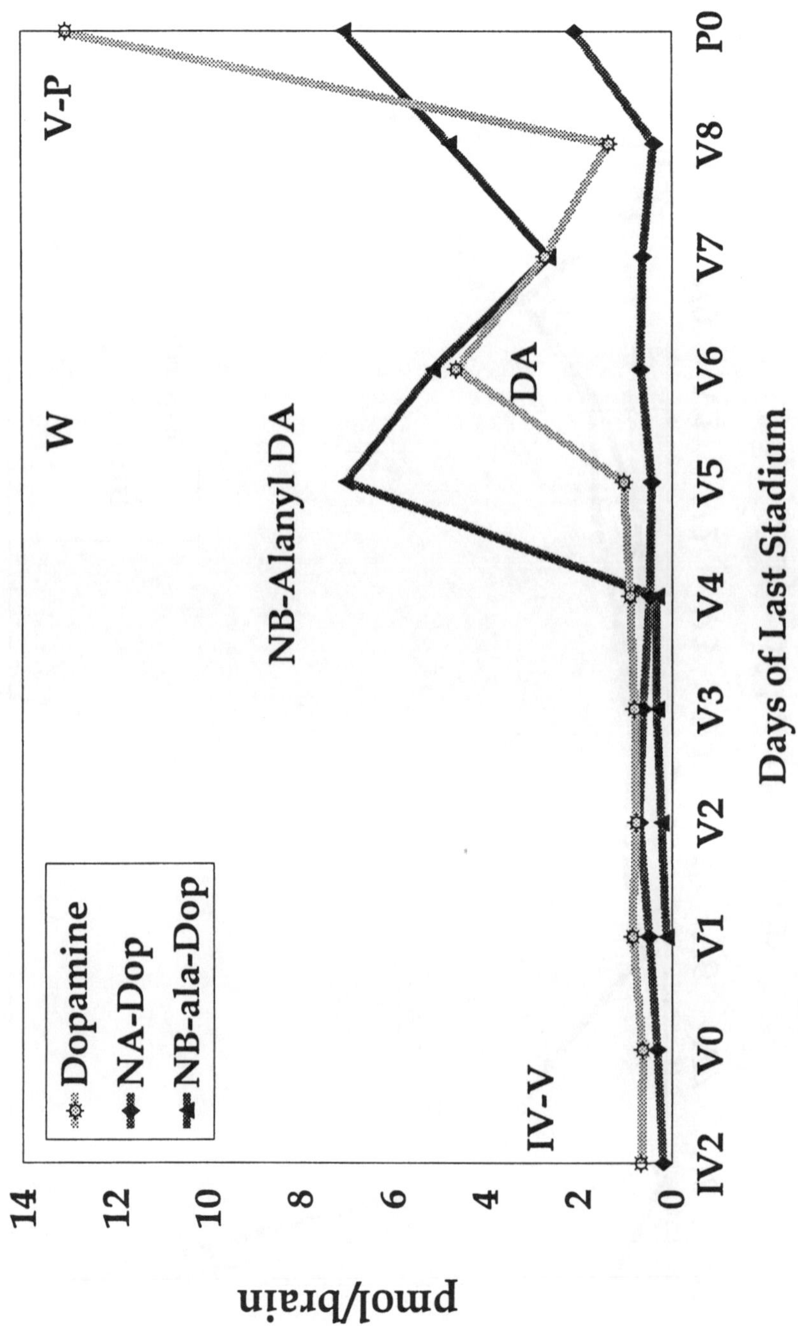

Figure 8. Dopamine, *N*-acetyldopamine and *N*-β-alanyldopamine levels in neural tissue of last stadium *M. sexta* larvae (data adapted from Geng *et al.*, (1993)).

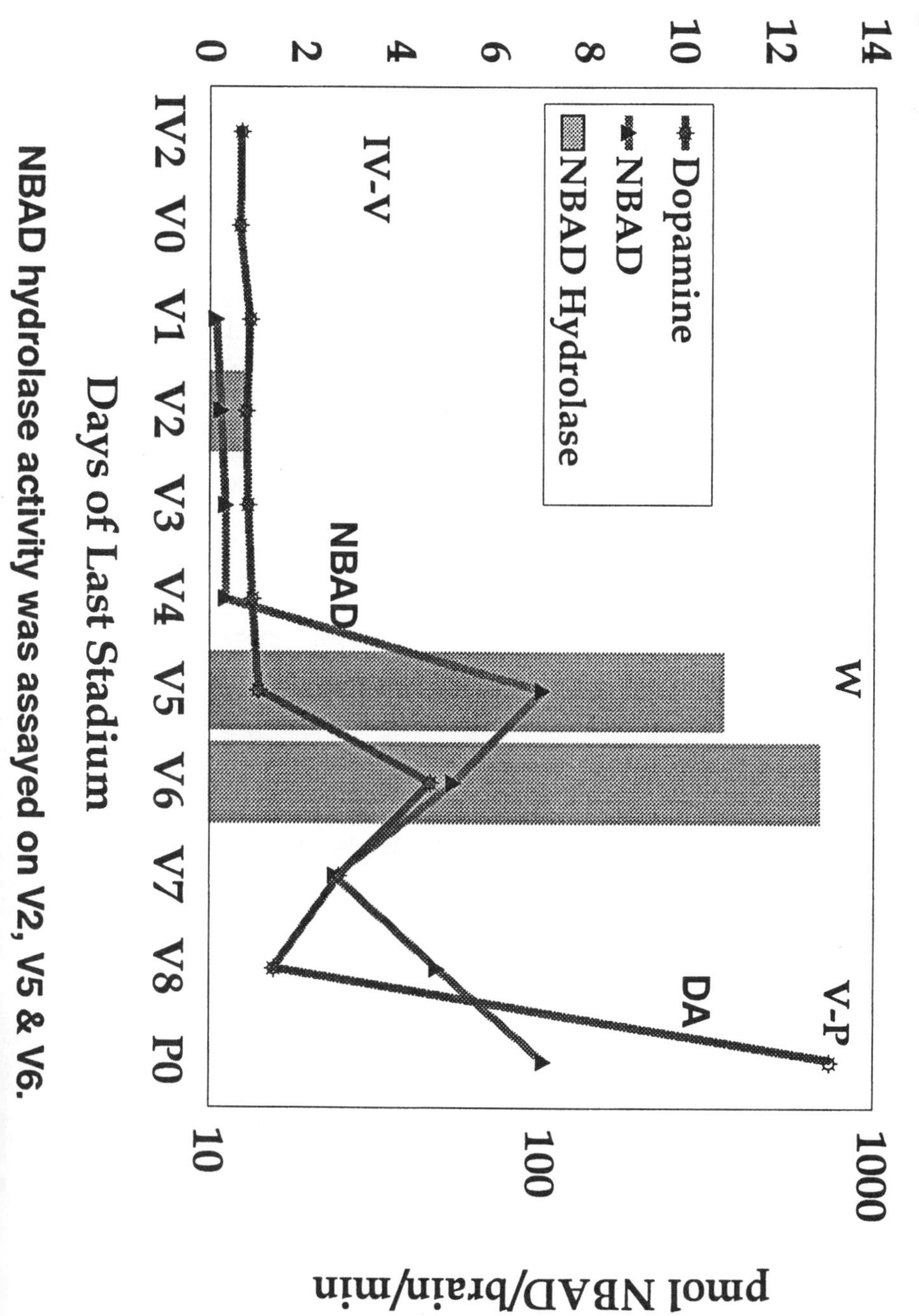

Figure 9. Dopamine, *N*-β-alanyldopamine and *N*-β-alanyldopamine hydrolase (assayed on V2, V5, V6) content of *M. sexta* neural tissue in last (V) stadium larvae. Dopamine and *N*-β-alanyldopamine data is adapted form Geng *et al.* (1993). *In vitro* assay for *N*-β-alanyldopamine hydrolase activity used dissected brains homogenized in phosphate buffer, pH6.5, 0.1M, 100μL) incubated with *N*-β-alanyldopamine (22°C, in the dark, 0.15M NaCl, 0.5M succinic acid). The reaction was terminated by the addition of a equal volume of 0.2M perchloric acid (see Table 3 for details on analysis).

alanyldopamine) certainly is suggestive of a finely controlled regulatory system for a chemical messenger involved in critical developmental processes.

SUMMARY AND CONCLUSIONS

The unique capabilities of n-EC-HPLC can make available much more comprehensive and detailed information concerning the nature of insect biogenic amines, and their precursors and metabolites, than was previously possible. More importantly, this technology has allowed, in some cases (e.g., *M. sexta*), detailed evaluation of the interplay of these processes during critical phases of insect development. Where comparisons have been made, there has generally been good agreement between the results of the n-EC-HPLC studies and studies utilizing other technologies (Sparks and Geng, 1992). Certainly the uncomplicated sample preparation and relative ease of compound identification with n-EC-HPLC facilitates studies that would be difficult to undertake using other technologies.

Acknowledgments
Thanks are due to DowElanco for supporting all of the unpublished studies and much of the published work discussed in this chapter.

REFERENCES

Beal, M. F., Matson, W. R., Swartz, K. J., Gamache, P. H. and Bird. E. D. (1990). Kynurenine pathway measurements in Huntington's Disease Striatum: evidence for reduced formation of kynurenic acid. J. Neurochem. **55**, 1327-1339.

Brown, C. S. and Nestler, C. (1985). Catecholamines and indolalkylamines. In: Comprehensive Insect Physiology, Biochemistry and Pharmacology, **11**. Kerkut, G. A. and Gilbert, L. I. (Eds.). Pergamon Press, New York, USA, pp. 435-497.

Dowd, P. F. and Sparks, T. C. (1987). Comparative hydrolysis of permethrin and fenvalerate in *Pseudoplusia includens* (Walker) and *Heliothis virescens* (F.). Pestic. Biochem. Physiol. **27**, 123-131.

Downer, R. G. H. (1990). Octopamine, dopamine and 5-hydroxytryptamine in the cockroach nervous system. In: Cockroaches as Models for Neurobiology: Applications in Biomedical Research. **2**. I. Huber, E. P. Masler and B. R. Rao (Eds.). CRC Press, Boca Raton, USA, pp. 103-124.

Downer, R. G. H. and Hiripi, L. (1994). Biogenic amines in insects. In: Insect Neurochemistry and Neurophysiology 1993, Borkovec, A. B. and Loeb, M. J. (Eds.). CRC Press, Boca Raton, USA, pp. 23-38.

Evans, P. D. (1985). Octopamine. In: Comprehensive Insect Physiology, Biochemistry and Pharmacology, **11**. Kerkut, G. A. and Gilbert, L. I. (Eds.). Pergamon Press, New York, USA, pp. 499-530.

Evans, P. D., Davenport, A. P., Elias, M. S., Morton, D. B. and Trimmer, B. A. (1985). Assays for biogenic amines in insects. In: Neurochemical Techniques in Insect Research, Breer, H. and Miller, T. A. (Eds.). Springer-Verlag, New York, USA, pp. 25-46.

Fescemeyer, H. W., Rose, R. L., Sparks, T. C. and Hammond, A. M. (1986). Juvenile hormone esterase activity in developmentally synchronous ultimate stadium larvae of the migrant insect, *Anticarsia gemmatalis*. J. Insect Physiol. **32**, 1055-1063.

Geng, C. and Sparks, T. C. (1991). Synthesis and metabolism of biogenic amines in neural tissue of larval *Manduca sexta*: Catecholamine pathways. Entomological Society of America, National Meeting, TMP1382, Reno, NV, Dec. 1991.

Geng, C. and Sparks, T. C. (1993). Synthesis and metabolism of the dopamine metabolite *N*-β-alanyldopamine in *Manduca sexta* central nervous system. Entomological Society of America,

National Meeting, DSP0275, Indianapolis, IN Dec. 1993.

Geng, C., Sparks, T. C., Skomp, J. R. and Gajewski, R. P. (1993). Biogenic amines in the brain of *Maduca sexta* during larval-pupal metamorphosis. Comp. Biochem. Physiol. **106C**, 275-284.

Granger, N. A., MacDonald, J. D., Menold, M., Ebersohl, R., Hiruma, K., Cann, C. L. and Riddiford, L. M. (1994). Evidence of a stimulatory role for cyclic AMP in the regulation of corpus allatum activity in *Manduca sexta*. Mol. Cell. Endocrinol. **103**, 73-80.

Granger, N. A., Allen, L. G., Sturgis, S. L., Comnest, W. and Ebersohl, R. (1995). Corpora allata of larval tobacco hornworm contain a calcium/calmodulin-sensitive adenyl cyclase. Arch. Insect Biochem. Physiol. **30**, 149-164.

Granger, N. A., Sturgis, S. L., Ebersohl, R., Geng, C. and Sparks, T. C. (1996). Dopaminergic control of corpora allata activity in the larval tobacco hornworm, *Manduca sexta*. Arch. Insect Biochem. Physiol. In Press.

Hopkins, T. L. and Kramer, K. J. (1992). Insect cuticle sclerotization. Ann. Rev. Entomol. **37**, 273-302.

Hopkins, T. L., Morgan, T. D. and Kramer, K. J. (1984). Catecholamines in haemolymph and cuticle during larval, pupal and adult development of *Manduca sexta*. Insect Biochem. **14**, 533-540.

Knowles, C. O., McKee, M. J. and Hamed, M. S. (1989). Dopamine and octopamine in whole body extracts of the bulb mite. Experientia **46**, 205-207.

Krueger, R. R., Kramer, K. J., Hopkins, T. H. and Spiers, R. D. (1989). *N*-β-Alanyldopamine levels and synthesis in integument and other tissues of *Manduca sexta* (L.) during the larval-pupal transformation. Insect Biochem. **19**, 169-175.

Krueger, R. R., Kramer, K. J., Hopkins, T. L. and Speirs, R. D. (1990). *N*-β-alanyldopamine and *N*-acetyldopamine occurrence and synthesis in the central nervous system of *Manduca sexta* (L.). Insect Biochem. **20**,: 605-610.

Lafon-Cazal, M. and Baehr, J. C. (1988). Octopaminergic control of corpora allata activity in an insect. Experientia **44**, 895-899.

MacFarlane, R. G., Midgley, J. M., Watson, D. G. and Evans, P. D. (1990). The analysis of biogenic amines in the thoracic nervous system of the locust, Schistocerca gregaria, by gas-chromatography negative ion chemical ionization mass spectrometry (GC-NICIMS). Insect Biochem. **20**, 305-311.

Matson, W. R., Langlais, P., Volicer, L., Gamache, P. H., Bird, E. and Mark, K. A. (1984). n-Electrode three-dimensional liquid chromatography with electrochemical detection for determination of neurotransmitters. Clin. Chem. **30**, 1477-1488.

Matson, W. R., Gamache, P. H., Beal, M. F. and Bird, E. D. (1987). EC array sensor concepts and data. Life Science, **41**, 905-908.

Megaw, M. W. J. and Robertson, H. A. (1974). Dopamine and noradrenaline in salivary glands and brain of the tick, *Boophilus microplus*: Effect of reserpine. Experientia **30**, 1261-1262.

Orchard, I. (1982). Octopamine in insects: neurotransmitter, neurohormone, and neuromodulator. Can. J. Zool. **60**, 659-669.

Pastor, D., M. Piulachs, D., Cassier, P., Andre, M. and Belles, X. (1991). Etude *in vivo* et *in vitro* de l'action de la dopamine sur la croissance des ovocytes et la production d'hormone juvenile chez *Blattela germanica* (L.). C. R. Acad. Sci. Paris **313**, 207-212.

Rachinski, A. (1994). Octopamine and serotonin influence corpora allata activity in honey bee (*Apis mellifera*) larvae. J. Insect Physiol. **40**, 549-555.

Scott, J. A., Johnson, T. L. and Knowles, C. O. (1985). Influence of amidines and anilides on biogenic amine regulatory mechanisms in the bulb mite *Rhizoglyphus echinopus*. Pestic. Sci. **16**, 504-510.

Shimizu, T. and Takeda, N. (1991). Biogenic amine levels in the hemolymph of the cabbage armyworm larvae (*Mamestra brassicae*) following injection of octopaminergic insecticides. Z. Naturforsch. **46c**, 127-132.

Shimizu, T., Mihara, M. and Takeda, N. (1991a). High-performance liquid chromatography of biogenic amines in the corpus cardiacum of the American cockroach, *Periplaneta americana*. J. Chromatogr. **539**, 193-197.

Shimizu, T., Tatsuki, Y. and Takeda, N. (1991b). Aromatic amino acids in the venom of the braconid parasitoid *Apanteles kariyai*. Z. Naturforsch. **48c**, 108-109.

Sparks, T. C. and Geng, C. (1992). Analysis of the biogenic amines in the central nervous system of the tobacco hornworm by high-performance liquid chromatography with 16-sensor electrochemical detection. Anal. Biochem. **205**, 319-325.

Sparks, T. C., Willis, W. S., Shorey, H. H. and Hammock, D. B. (1979). Haemolymph juvenile hormone esterase activity in synchronous last instar larvae of the cabbage looper, *Trichoplusia ni*. J. Insect Physiol. **25**, 125-132.

Sparks, T. C., Allen, L. G., Schneider, F. and Granger, N. A. (1989). Juvenile hormone esterase activity from *Manduca sexta* corpora allata *in vitro*. Arch. Insect Biochem. Physiol. **11**, 93-108.

Sparks, T. C., Skomp, J. R., Babcock, J. and Geng, C. (1994). Biogenic amines in the two-spotted spider mite. Arch. Insect Biochem. Physiol. **25**, 1-8.

Takeda, N., Takaoka, H., Shimizu, T., Yazawa, M. and Yagi, S. (1991). Biogenic amine levels in the central nervous system and hemolymph of the silkworm, *Bombyx mori*. Comp. Biochem. Physiol. **100C**, 677-682.

Taylor D. P. and Newburgh, R. W. (1979). The synthesis and content of neurotransmitters and their effect on cyclic nucleotide accumulation in the CNS of *Manduca sexta*. Insect Biochem. **9**, 265-272.

Thompson, C. S., Yagi, K. J., Chen, Z. F. and Tobe, S. S. (1990). The effects of octopamine on juvenile hormone bisynthesis, electrophysiology and cAMP content of the corpora allata of the cockroach, *Diploptera punctata*. J. comp. Physiol. **160B**, 214-249.

Vaughan, P. F. T. (1988). Amine transmitters and their associated second messenger systems. In: Comparative Invertebrate Neurochemistry, Lunt, G. G. and Olsen, R. W. (Eds.). Cornell Univ. Press., Ithaca, NY, USA, pp. 124-174.

Woodring, J. P. and Hoffmann, K. H. (1994). The effects of octopamine, dopamine and serotonin on juvenile hormone synthesis, *in vitro*, in the cricket *Gryllus bimaculatus*. J. Insect Physiol. **40**, 797-802.

Progress in HPLC-HPCE, Vol. 6, pp. 169–211
I.N. Acworth *et al.* (Eds)

A PHYLOGENETIC ANALYSIS OF BIOGENIC MONOAMINES IN THE CENTRAL NERVOUS SYSTEMS OF VARIOUS ORGANISMS AND APPLICATION TO THE FIELDS OF MEDICAL BIOCHEMISTRY

NAOKUNI TAKEDA

Department of Biotechnology, Research and Development Center, COSMO Research and Development Center, COSMO Research Institute, Satte, Saitama 340-01, Japan

INTRODUCTION

Biogenic amines such as the catecholamines and indolealkylamines are one of a number of neuroactive substances which include the amino acid transmitters, acetylcholine and neuropeptides. The monoamines are thought to act as neurotransmitters, neuromodulators and neurohormones in animals. Catecholamines and indoleamines have been widely detected in the major taxa of both invertebrates and vertebrates (Brown and Nestler, 1985; Klemm, 1985; Parent *et al.*, 1984; Smeets and Reiner, 1995). Phylogenetic analysis of the central nervous system can provide important clues to the evolution underlying the development of brain complexity. With such a perspective, we have been trying to obtain a clearer understanding of the organization of the monoaminergic systems in both invertebrates and vertebrates, and to identify new principles related to the evolution of the central nervous system.

A variety of techniques have been used to localize and quantify specified biogenic monoamines. However, all these approaches have major limitations, such as the detection of only a few biogenic amines. A sensitive histochemical technique was introduced initially. It was based on the reaction of the biogenic amines, such as the monoamines, dopamine (DA), norepinephrine (NE) and serotonin (5-hydroxytryptamine, 5-HT), with formaldehyde vapor which resulted in the formation of highly fluorescent condensation products (Falck *et al.*, 1962). More useful information on the distribution of monoamines in the nervous systems of invertebrates (Klemm, 1985) and vertebrates (Parent *et al.*, 1984; Smeets and Reiner, 1995) has resulted from numerous modifications of this method.

Monoamines are unevenly distributed throughout both invertebrate and vertebrate nervous systems. The early work on the levels of monoamines involved the use of fluorometric methods that unfortunately suffered from a lack of sensitivity and also required very large amounts of nervous tissue. Progress in the quantitation of monoamines was accelerated by the development of radioenzymatic methods (Evans *et al.*, 1985). These methods made it possible to measure picogram amounts of compounds, allowing the accurate quantitation of monoamines in both invertebrates and vertebrates. In addition to such quantitative analysis, immunohistochemical techniques using antibodies against catecholamine-synthesizing enzymes, such as tyrosine hydroxylase (Goldstein *et al.*, 1973), and also against the catecholamines themselves (Steinbusch *et al.*, 1986), yielded better visualization of the monoamine neurotransmitter pathways. A great advance in the analysis of the biogenic monoamines in the nervous system came with the introduction of high-performance liquid chromatography with electrochemical detection (HPLC-ECD). Identification of monoamines with single-electrode HPLC-ECD is based solely on retention time. For qualitative identification of biogenic amines and their metabolites, a dual-electrode HPLC-ECD system is a better tool than the single-electrode HPLC-ECD system. Further improvements in analytical capability were soon realized with the development of a multi-electrode electrochemical detector for use with HPLC (Matson *et al.*, 1983). This system allows monoamines to be separated in an additional dimension on the basis of their characteristic oxidation potential, so that identification is not only based on retention time. Levels of compounds analyzed by conventional single electrode detectors may be over-estimates because the levels of many unidentified compounds can often co-elute with the target compound of interest. Unlike the electrode array the single electrode is not quantitative. The present review summarizes information about the monoamines detected by multiple electrode HPLC-ECD, together with putative metabolic pathways deduced from such information in various organisms. Furthermore, the evolutionary implications of the organization of the biogenic monoamines systems, as well as their potential relevance to the field of medical biochemistry and monoamine-related disorders are discussed.

The animals used for this study are shown in Figure 1 in a phylogenetic tree which illustrates their phylogenetic relationship.

THREE-DIMENSIONAL HPLC WITH COULOMETRIC ELECTRO-CHEMICAL DETECTION (HPLC-ECD)

The analytical apparatus used to obtain most of the results described below was a three-dimensional HPLC system (Coulochem® Electrode Array System, ESA, Chelmsford, MA, USA). The system consisted of a gradient HPLC system and 16 high-sensitivity coulometric electrochemical detectors coupled with a compatible computer. The concept and inherent advantages of multi-electrode HPLC system have been described elsewhere (Matson *et al.*, 1983). Analytes were resolved on a reversed-phase C18 column (4.6 x 150mm; 5μm; NBS column, Niko Bioscience, Tokyo, Japan). Two mobile phases, solvent A and B, were employed with both linear and step gradients. Solvent A consisted of 0.1M sodium phosphate and 10mg/mL sodium dodecylsulfate at pH3.35. Solvent B consisted of methanol and water (1:1,v/v) containing 50mg/L sodium dodecylsulfate (final pH3.35). Our protocol permitted the simultaneous separation of about thirty compounds that were relevant to our analyses (Figure 2). A standard chromatogram, including typical compounds, with their abbreviations and oxidation potentials, is shown in Table 1. The 16 serial electrodes were set in an array that ranged from 0 to +900mV in increments of 60mV. The

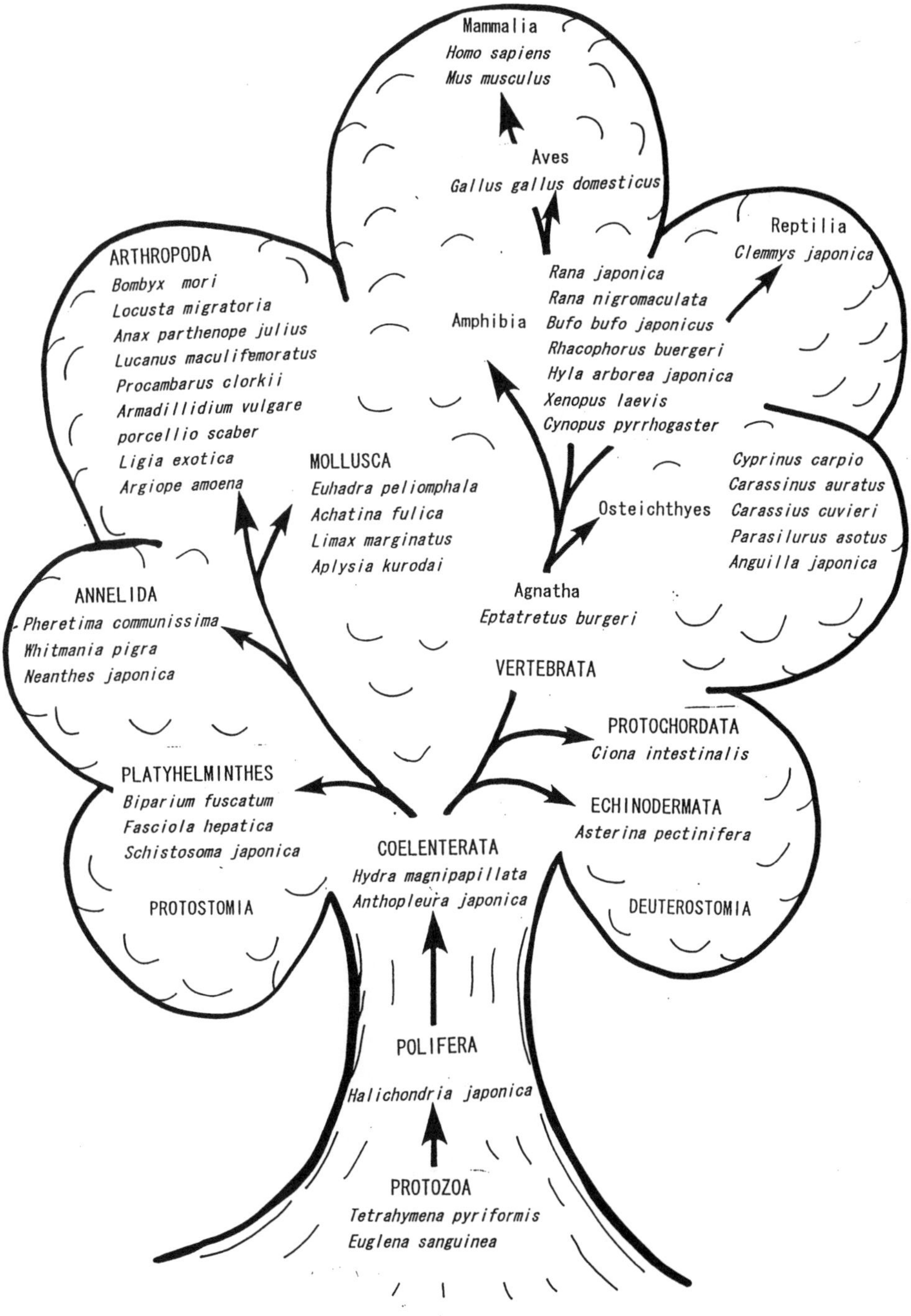

Figure 1. Phylogenetic tree.

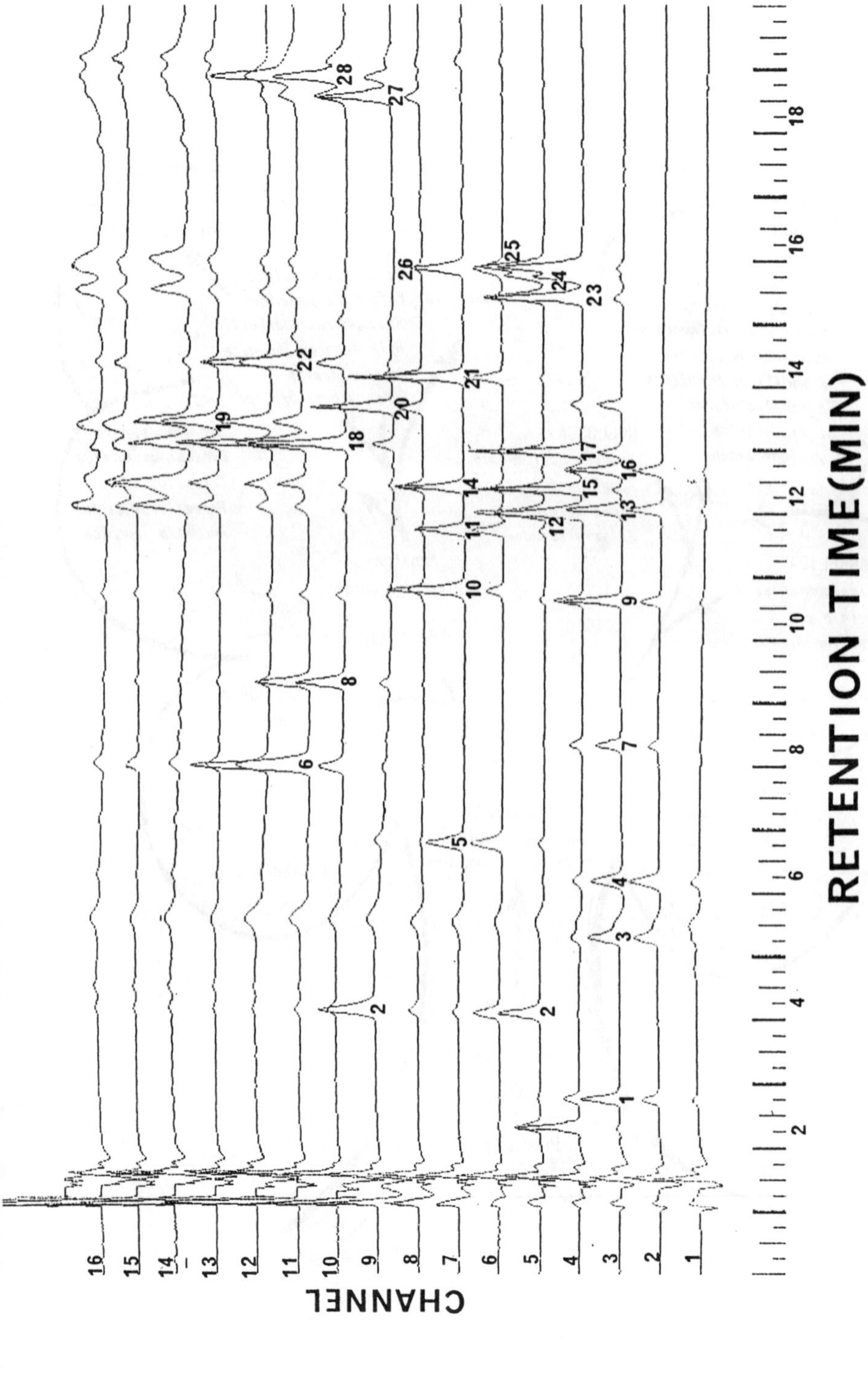

Figure 2. Chromatogram of standards (see Table 1 for peak identification).

column and electrodes were maintained at 34°C. Each compound was characterized not only by its retention time but also the oxidation potential at which a maximum response was observed. Typically, each compound was detected at three electrodes, with an average ratio of peak heights between adjacent electrodes of 1:6:1. However, the exact ratio was found to be specific for each compound. This voltammetric response was used to establish peak purity, to confirm identification and to establish co-elutions. Therefore, using gradient HPLC with coulometric array detection, compound identity could be confirmed by both retention time and voltammetric behavior. A compounds was only identified if it had a similar retention time and similar voltammetric behavior to those of an external standard. A ratio accuracy of 100% means that the unknown compound had exactly the same voltammetric response to that of an external standard. Although in biological samples many compounds do reach ratio accuracies of 100%, some show ratio accuracies lower

Table 1.
List of standards used in these studies (in order of increasing retention time)

Compound	Abbreviation	Oxidation Potential (mV) First	Second
1: Dihydroxyphenylethylene glycol	DOPEG	240	
2: Vanillylmandelic acid	VMA	300	600
3: Norepinephrine	NE	180	
4: Dihydroxyphenylalanine	L-DOPA	150	
5: 3-Methoxy-4-hydroxyphenyl-ethylene glycol	MHPG	450	
6: Octopamine	OCT	620	
7: Epinephrine	E	180	
8: Tyrosine-4 (p-hydroxyphenylalanine)	TYR	650	
9: 3,4-Dihydroxyphenylacetic acid	DOPAC	150	
10: Normetanephrine	NMN	480	
11: 3-0-Methylindoleacetic acid	3-O-MD	450	
12: 5-Hydroxyindoleacetic acid	5-HIAA	180	750
13: Dopamine	DA	150	
14: Metanephrine	MN	480	
15: 5-Hydroxytryptophan	5-HTP	300	650
16: Epinine	EPIN	120	
17: N-Acetyl-5-hydroxytryptophan	N-ACT-5-HT	180	700
18: 4-Hydroxyphenylacetic acid	HPAC-4	650	
19: Kynurenine	KYN	800	
20: Homovanillic acid	HVA	450	
21: Vanillactic acid	VLA	360	
22: Vanillic acid	VA	480	
23: 5-Hydroxytryptamine	5-HT	180	700
24: Bufotenine	BUTN	240	840
25: N-Methyl-5-hydroxytryptamine	N-MET	300	700
26: 3-Methoxytyramine	3-MT	450	
27: Melatonin	MEL	600	
28: Tryptophan	TRP	600	

than this. In general, if compounds reached a ratio accuracy above 70% they were usually regarded as being present in a relatively pure form. Final concentrations were calculated by comparing the size of the dominant peak of the standard to the dominant peak of the unknown compounds in the sample.

DETECTION OF BIOGENIC MONOAMINES IN THE CENTRAL NERVOUS SYSTEM OF VARIOUS ORGANISMS

Invertebrates

Protozoa

The protozoa are unicellular organisms that do not possess a nervous system. Nevertheless, for transmission of information, protozoa use cyclic AMP, cyclic GMP and calcium ions, as do multicellular organisms (Renaud *et al.*, 1991). Biogenic monoamines have been reported in some species such as *Tetrahymena pyriformis* (Goldman *et al.*, 1981), *Crithidia fasciculata* (Janakidevi *et al.*, 1966) and *Entamoeba histolytica* (McCaman *et al.*, 1973). However, biogenic monoaminergic systems have not, as yet, been characterized.

Table 2.
Level of biogenic monoamines and their metabolites in *Tetrahymena pyriformis*

Compound	Non-lyophilized samples (ng/106 cells; n=5) Mean ± SD
TYR	64.84±11.5
L-DOPA	8.39±1.70
DA	15.56±3.27
E	44.09±2.82
DOPEG	221.93±365.81
VMA	65.69±5.41
TYRA	7.30±0.34
HPAC-4	380.89±32.49
TRP	52.54±2.90
5-HT	7.93±1.71
KYN	5.92±0.18
XA	11.43±3.3*

(Mean±SD); *(n=4)

In the sample of *Tetrahymena pyriformis*, a wide variety of catecholaminergic and indolealkylaminergic analytes, precursors and catabolites were found (Table 2). As *Tetrahymena* is known to lack tyrosine hydroxylase, the L-DOPA detected in samples might have been incorporated from the medium into the cell body. Representative biogenic monoamines, such as DA, 5-HT, E, and KYN, were found to be synthesized in the cell body and released into the growth medium. As judged from the released metabolites, systems for the generation of biogenic monoamines in *Tetrahymena* are active during the logarithmic phase of its growth. The following metabolic pathways seem to be operative: L-DOPA → DA → EPIN → E → DOPEG → HPAC-4; TRP → 5-HT; and TRP → KYN → XA (Takeda and Sugiyama, 1993). In *Euglena sanguinea*, biogenic monoamines, their precursors and metabolites were also detected, namely TYR, L-DOPA, DOPEG, VMA, TRP and 5-HIAA. To examine the possible chemical relationship between the eye spot and the flagellum that is used for movement, cells were fractionated. However, no clear relationship between these areas was found. TYR, DOPAC, TYRA, VA and TRP were found in the nuclear fraction, TYR, TRP and 5-HIAA were found in the mitochondrial fraction and TYR, L-DOPA, DOPEG, MN, MHPG, HPAC-4, TRP, 5-HIAA were measured in the microsomal fraction.

Biogenic monoamines have been shown to be involved in various processes in invertebrates, such as cell division (Iwata and Kariya, 1973), regeneration of cilia (Carstroaded *et al.*, 1988) and phagocytosis (Wyroba, 1989).

Porifera

Resembling the life cycles of other multicellular organisms, the life cycle of sponges is characterized by a return to the unicellular condition which is, in turn, followed by redevelopment of the multicellular condition. The presence of a nervous system in members of *Porifera* is uncertain, although cells with a morphological resemblance to neurons have been described. In *Sycon ciliatum*, some vesicles that react positively in histochemical tests for the monoamines were identified in bipolar cells situated in the mesenchyme (Lentz, 1966; Scharrer, 1970).

Table 3.
Levels of biogenic monoamines and their metabolites in *Halichondria japonica*

Compounds	(ng/mg tissue fragment)
TYR	10.0±0.3
EPIN	100.6±8.3*
TYRA	1.6±0.8
OCT	23.3±22.6
TRP	7.6±1.1
5-HT	48.1±31.9*
N-ACT-5-HT	366..6±184.2*
KYN	314.3±226.1*

(Mean±SD; n=6; *; pg/mg tissue)

In *Halichondria japonica,* biogenic monoamines were examined in both the outer and inner parts of a cluster of specimens. The results are shown in Table 3. Although these results were not totally reproducible, a variety of monoamines, their precursors and metabolites were detected. Thus, the pathways: TYR → TYRA

→OCT; TYR → EPIN; TRP → KYN; and TRP → 5-HT → N-ACET-5-HT appear to be operating.

Coelenterata

Together with the sponges, coelenterates are usually regarded as the most primitive living metazoans. Phylogenetically, they are situated at the branching point between Protostomia and Deuterostomia. In the evolution of development of the nervous system, a neural network is first found in coelenterates such as *Hydra*. Therefore, the first nervous system seems to have evolved in this phylum. The presence of monoaminergic cells have been detected by a histochemical method. In a chromatographic study with an electrochemical detection, DA and NE were detected in *Chlorohydra viridissima* (Venturini *et al.*, 1989). Evidence suggests that DOPA, DA and NE are present in *Hydra attenuata, Tubularia indivisa*, the octocorallian *Kophobelemnon stroemi* and the ctenophore *Beeroe* sp. (Carlberg and Rosengren, 1985). In addition, DA, NE and E have been reported to be present in the colonial anthozoan *Renilla kollikeri* and in different parts of the colony (de Waele *et al.*, 1987). Furthermore, DA is thought to be present in the nervous system of the jellyfish *Polyorchis penicillatus* (Chung *et al.*, 1989). The presence of 5-HT had not been proven in spite of suggestive reports in some species (Castano and Rossi, 1978; Venturini *et al.*, 1989). L-DOPA has been reported to be a putative biogenic amine in the tentacles of the sea anemone *Metridium senile* (Carlberg, 1990; Carlberg and Rosengren, 1985; Lenicque *et al.*, 1977). Recently, HPLC-ECD has been used to show that the biogenic monoamines (including NE, E and 5-HT) are present in the colonial tissues, such as the rachis, polyps and peduncle of the sea pansy *Renilla koellikeri* (Pani and Anctil, 1994a, b).

We have examined the role of the nervous system in a from of *Hydra*, called epithelial *Hydra*. In this mutant strain of *Hydra magnipapillata*, an increase in temperature from 18°C to 25°C blocks the proliferation of nearly all the interstitial cells (Marcum *et al.*, 1980). Furthermore, the cells derived from interstitial cells, including nerve cells, gradually decrease in number. Normal polyps were shown to contain the monoamine precursor amino acids TRP and TYR, as well as the biogenic amines, 5-HT, N-MET and DA. Polyps maintained at 25°C lose nearly all of their interstitial cells and more than 90% of the products of their differentiation, namely nerve cells and nematocytes. At 25°C, the ratio of nerve cells to epithelial cells decreases considerably (from 0.121 to 0.014 after 15 days). Under these conditions, levels of monoamines also fell, in particular those of 5-HT and DA. In fact, 5-HT disappeared completely within 15 days. The level of N-MET was, however, seen to increase, suggesting that 5-HT was degraded to N-MET. However, bufotenine (BUTN) was not detected in *Hydra magnipapillata*. Table 4A shows the distribution of monoamines in each region of *Hydra magnipapillata*.

We conclude that the main indolealkylamine in *Hydra magnipapillata* is not 5-HT but N-MET. Furthermore, 5-HT does not seem to act as a neurotransmitter but it may nevertheless act as a venom for the capture of live prey. The pathway from TRP to 5-HT and N-MET is probably present, as is the pathway from TYR to DA (Takeda and Svendsen, 1991).

In the sea anemone *Anthopleura japonica*, L-DOPA, NE, TYRA, OCT, 5-HT and N-MET were detected in each part of the body and DA was detected in the hypostome only (Table 4B). Some metabolites were also detected, namely DOPAC,

3-MT, NMN and VMA. In other species, BUTN has been shown to be present. However at this time BUTN was not routinely measured in our analysis. Thus, in *Anthopleura japonica*, the pathways: TYR → L-DOPA → DA → NE; TYR → TYRA; and TRP → 5-HT → N-MET → BUTN appear to be present.

Table 4A.
Levels of biogenic monoamines and their precursors in each part of the body of *Hydra magnipapillata*

	TYR	DA	TRP	5-HT	NME
Tentacles	388.1±32.5	2.7±0.7	1155.6±157.0	6.8±0.5	50.8±0.5
Hypostome	958. 2±23.5	8.8±0.4	2043.7±58.8	14.4±3.0	106.7±2.5
Body column	454.4±67.0	31.7±0.1	2041.0±173.0	51.3±3.0	412.4±36.5
Peduncle	815.4±17.0	7.7±0.4	772.6±65.5	17.4±1.8	188.4±4.0

(Mean±SD; n=5; pg/mg wet tissue)

Hydrozoan larvae are produced extracellularly to regulate metamorphosis (Kolberg and Martin, 1988). Although the primitive nervous system has been suggested to be peptidergic (Grimmelikhuijzen and Graff, 1991), other conventional monoamines, including L-DOPA, also act as neurotransmitters in these primitive nervous systems. L-DOPA evokes the transitory inhibition of spontaneous contractions of tentacle muscles (Hudman and McFarlene, 1995).

Table 4B.
Levels of main biogenic monoamines in each part of the body of *Anthoplura japonica*

—	Tentacles	Hypostome	Body column	Peduncle
TYP	9.2±4.3*	24.5±12.3*	12.4±9.4*	8.2±7.8*
L-DOPA	1.4±0.7	0.3±0.1	0.8±0.6	1.2±0.8
DA		1.2±1.0		
NE	19.6±10.4	60.6±23.5	83.4±55.5	122.6±98.4
TYRA	399.2±225.6	248.5±18.4	411.1±342.4	401.3±367.3
OCT	1.6±0.1	1.0±0.2	1.3±0.7	2.2±1.0
TRP	3.6±2.5	12.5±9.9	6.7±4.8	4.2±2.6
5-HT	583.6±554.2	659.1±602.5	737.8±546.3	727.6±673.1
N-MET	7.1±5.8	96.6±87.4	97.7±89.3	104.3±91.7

(Mean±SD; n=6; pg/mg tissue: *, ng/mg tissue)

Platyhelminthes

Platyhelminthes are bilaterally symmetrical apparent centralization and cephalization of the nervous system. The phylum *Platyhelminthes* is divided into three classes: *Trematoda*, *Cestoda* and *Turbellaria.* The former two classes are parasitic and the latter is free-living. In free-living species, biogenic monoamines have been studied only in the freshwater species. The presence of 5-HT in the nervous system has been demonstrated in *Platyhelminthes.* DOPA, DA and NE have also been detected in *Dugesia gonocephala* (Algeri *et al.*, 1983), *Diphyllobothrium dendritium* (Gustafsson and Eriksson, 1991), *Stenostonum leucops* (Reuter and Eriksson, 1991), *Bdellocephala angarensis* and *Rimacephalus arecepta* (Eriksson *et al.*, 1993), and *Dugesia tigrina* (Eriksson and Panula, 1994) by HPLC. The distribution of catecholamines and aminergic elements in the nervous system has been examined in some species (Sakharov and Winlow, 1991).

Biogenic monoamines have been detected in terrestrial species such as *Biparium fuscatum*. The central nervous system in *Biparium* consists of the cerebral ganglion and one pair of longitudinal nerves that extend from it to the parenchyma. In this species, 5-HT has been detected immunohistochemically (Fujii and Takeda, 1988). Data from a more in-depth study using HPLC-ECD are shown in Table 5A (Takeda, 1995b).

Table 5A.
Concentrations of biogenic monoamines in the land planarian, *Biparium fuscatum*

Compound	Head region (ng/mg tissue)	Head region (ng/head region)	Tail region (ng/mg tissue)
TYR	17.36±1.50	234.66±91.66	165.06±196.5
L-DOPA	0.05±0.03	0.85±0.86	1.59±1.88
DA	0.22±0.04	2.84±0.67	0.35±0.42
DOPAC	0.01±0.01	0.19±0.08	0.08±0.08
3-O-MD	1.03±0.34	12.63±3.20	7.27±8.14
NE	0.12±0.01	1.66±0.51	
E	0.24±0.11	3.22±1.47	2.64±3.35
HVA	0.09±0.06	1.60±1.51	1.22±1.25
TYRA	0.08±0.03	1.09±0.45	
TRP	1.96±0.08	26.85±11.44	3.54±0.17
5-HT	0.31±0.23	3.74±2.69	0.84±1.14
5-HIAA	0.01±0.01	0.12±0.02	
MEL	0.51±0.36	5.13±3.78	1.03±0.72

(Mean ± SD; n =20)

In the head region, which consists mainly of the cerebral ganglion, evidence for the TYR → L-DOPA → DA → NE → E pathway was obtained. This is interesting as E has not been previously detected in this organism. In this phylum, the presence of tyrosine hydroxylase, required for the synthesis of DOPA, has been demonstrated in *Diphyllobothrium dendriticum* (Wikgren and Lindroos, 1989). Furthermore, the pathways L-DOPA → 3-O-MD → HVA, DA →DOPAC → HVA; and TYR → TYRA also appear to be present. The TRP → 5-HT → 5-HIAA or MEL pathways are also present. The tail region, used as control tissue, included small amounts of nerve tracts. Higher levels of biogenic monoamines were detected here than in the head region. The submuscular plexus has been shown to exhibit immunoreactivity specific for 5-HT and to send 5-HT neurons towards the ventral epithelium (Fujii and Takeda,

1988; Kabotyanski *et al.*, 1991). Although biogenic monoamines are highly localized in the nerve cell bodies themselves, the high levels of monoamines in the tail region might be due to the complicated innervation that gives rise to the peripheral networks (Baguna and Ballester, 1978). A role for biogenic amines in the early stage of regeneration has also been suggested (Martelly and Franquinet, 1984). Administration of 5-HT and NE to regenerate promotes regeneration, while antagonists of 5-HT and DA significantly delay regeneration. The presence of larger amounts of biogenic amines in the peripheral region might, therefore, be involved in the regeneration of flatworms.

The role of biogenic monoamines in regeneration has also been examined in some planarian species (Webb, 1988). Furthermore, 5-HT has a general activating effect on metabolism and excitatory effects on the ciliated cells of the ventral surface and on the pharynx (Kabotyanski *et al.*, 1991; Webb, 1988). In *Dugesia*, several drugs that affect DA receptors alter motor activity (Venturini *et al.*, 1989). In *Dugesia dorotocephala*, a MEL-metabolizing system has already evolved that responds to cycles of light and darkness (Morita *et al.*, 1987). The presence of melatonin provides strong evidence for the existence of active neurotransmitters or neurohormones in planarians.

5-HT and DA may play different but complementary roles in the growth processes of *Planaria.*

Table 5B.
Levels of biogenic monoamines and their metabolites in the whole body of some flukes

—	*Fasciola hepatica* (pg/mg tissue: *ng/mg)	*Schidtodoms isponica* (pg/20μL homogenate: *ng/20μL)
TYR	60.5±23.3*	111.4±34.3*
L-DOPA	21.3±9.3	119.0±47.4
3-O-MD	103.1±39.2	
DA	10.6±4.6	
NE	600.0±143.9	
E	42.0±13.5	374.1±122.6
MN	10.6±5.3	
TYRA	147.0±38.5	196.5±77.3
—		
TRP	29.1±11.4	21.6±6.8
5-HT	9.3±2.8	
5-HIAA		59.0±15.6
MEL		152.0±73.4

(Mean±SD; n=3; Takeda and Kawashima unpublished data)

Motility is an important function of parasitic worms, allowing them to remain in place in spite of the movements of tissue and body fluid of their hosts. Parasitic trematodes and cestodes, which do not express tryptophan hydroxylase, apparently depend on 5-HT released from the host for maintenance of motility. Regulation of ciliary activity and cell motility might be a general function of neurotransmitters in primitive organisms and early invertebrate embryos. The biogenic amines in the whole body of the liver fluke, *Fasciola hepatica,* and blood fluke, *Schistosoma japonica,* were also examined (Table 5B). The former has main amines, such as DA, NE, E, TYRA and 5-HT, and their metabolites. By contrast, DA and NE were not detected in the latter. In these species, 5-HT was present at low levels or was absent, even though 5-HT has been found in free-living species (Terenina, 1991).

Annelida

Biogenic monoamines have been shown to be present in nervous tissue or in immunohistochemically identified neurons of annelids (Gilchrist *et al.*, 1995; Klemm, 1985). These results have not been confirmed by more sensitive techniques, such as HPLC-ECD, except in the case of *Chaetopterus variopedatus* (Polychaeta) (Anctil *et al.*, 1990), *Ophryotrocha puerilis* (Schlawny *et al.*, 1991) and *Hirudo medicinalis* (Lent *et al.*, 1991). Analytical results using HPLC-ECD are shown in Table 6. In *Annelida*, polychaetes are thought to be the most primitive and representative organisms. In the nereis, *Neanthes japonica*, 5-HT, DA and TYRA were detected as the main monoamines. Based upon the level of metabolites, the TYR → L-DOPA → DA → DOPAC; TYR → TYRA and TRP → 5-HTP → 5-HT pathways seem to be present. NE and OCT, which were detected in *Chaetopterus variopedatus*, were not present at detectable levels in *Neanthes japonica.* To our knowledge, no other analysis of biogenic monoamines by HPLC-ECD has yet been reported in earthworms, although some compounds have been analyzed by microspectro-fluorometry. In *Pheretima communissima*, the main monoamines were 5-HT, DA, NE, OCT and TYRA. The levels and types of monoamines were similar in the brain and the subesophageal ganglion and low in the ventral nerve cord. Our data suggest that the TYR → L-DOPA → DA → DOPAC → HVA, DA → NE, TYR → TYRA → HPAC-4 or OCT, and TRP → 5–HT pathways may be present.

The presence of catecholamines has not been unequivocally confirmed in the leech (Klemm, 1985). In contrast, high levels of 5-HT appear to associate with the exhibition of hunger and low levels with satiation in *Hirudo medicinalis* (Groome *et al.*, 1993). The amount of 5-HT in leech ganglia is known to be a potentially useful neurochemical indicator of behavioral state. In addition, OCT immunoreactive neurons have also been found in the same species (Gilchrist *et al.*, 1995). From our analyses the following pathways are thought to be present: TYR → DA → DOPAC → HVA; TYR → TYRA → OCT; TRP → 5-HT → BUTN; and TRP → KYN (Takeda, 1996). In particular, high levels of OCT were detected in the brain and BUTN in the last ventral ganglion. As described below, the presence of BUTN in the nervous system is of great interest, but its exact function in the leech still remains unclear. The results also suggest that DA is the major catecholamine transmitter and that 5-HT might have roles in interneuronal transmission in *Annelida* (Schlawny *et al.*, 1991).

Mollusca

In mollusks, research into the biogenic monoamines in the central nervous system has mainly been performed in gastropods such as *Aplysia*, *Lymnaea*, *Helisoma* and *Helix.* Monoamine neurons are particularly well developed in mollusks. Giant neurons containing 5-HT have yielded significant information on the role of 5-HT. Compared to reports on immunohistochemical detection, however, quantitative analytical reports

Table 6.
Levels of biogenic amines and their metabolites in the central nervous system of annelidan species

	Neanthes japonica	*Pheretima communissima*			*Whitmania pigra*		
	Brain & Sub-esophageal ganglion (pg/mg tissue: * ng: n=15)	Brain	Subesophageal ganglion	Ventral nerve cord (ng/mg tissue: n=12)	Brain & Sub-esophageal ganglion	Ventral nerve cord	Ventral ganglion
TYR	36.0±9.8*	120.4±46.8	25.5±9.2	22.2±15.2	40.6±1.6	5.0±2.0	55.8±13.2
L-DOPA	115.3±29.6	0.13±0.03	0.09±0.02	0.02±0.01	-----	-----	0.06±0.01
DA	221.2±104.9	18.7±11.3	22.2±11.2	4.8±1.9	2.6±1.7	0.5±0.2	2.0±0.8
DOPAC	11.5±3.15	0.8±0.2	0.4±0.1	0.09±0.01	0.3±0.2	0.2±0.1	-----
HVA	-----	-----	0.05±0.01	-----	3.5±1.5	-----	0.3±0.1
NE	-----	4.0±1.9	3.7±2.1	1.9±0.2	-----	-----	-----
TYRA	31.5±13.3	1.3±0.3	0.6±0.3	-----	0.5±0.2	1.2±0.6	1.4±0.4
OCT	-----	5.9±3.2	3.1±2.6	0.9±0.7	25.1±13.5	-----	-----
HPAC-4	-----	2.4±1.1	1.2±0.4	-----	-----	-----	-----
TRP	4.5±1.6*	39.9±22.5	25.5±17.9	6.0±3.3	10.4±1.6	21.4±9.0	18.2±11.2
5-HTP	44.3±17.2	-----	-----	-----	0.15±0.09	0.05±0.02	-----
5-HT	498.1±284.8	58.5±32.4	43.2±33.5	9.6±5.8	15.3±9.5	3.9±1.6	-----
(KYN)	-----	-----	-----	-----	-----	12.7±5.7	-----
(BUTN)	-----	-----	-----	-----	-----	-----	14.0±10.5

(Mean ± SD)

are few in number. Among the monoamines detected by various methods, 5-HT and DA may be involved, for example, in the modulation of locomotion, feeding behavior and learning (Walker, 1982). NE occurs in much smaller amounts and OCT has been identified in some mollusks.

Aplysia has widely been used for research in the neurosciences. Several reports of biogenic monoamines in the central nervous system of *Aplysia* have appeared (Klemm, 1985). DA has been detected biochemically in the visceral ganglion in large quantities (Carpenter *et al.*, 1971). OCT has also been detected in single neurons (Saavedra *et al.*, 1974) and has been shown to occur at high levels in the buccal ganglion (Farnham *et al.*, 1978). Among indolealkylamines, 5-HT has been detected in the central nervous system (McCaman *et al.*, 1973). Catecholamines and 5-HT-containing neurons have been detected by fluorescent histochemical staining (Salimova *et al.*, 1987) and immunohisto-chemical staining (Fujii and Takeda, 1988; Longley and Longley, 1989; Takeda and Mizuno, 1989), but overall there has been little biochemical research on biogenic monoamines in *Aplysia*.

Based on the data presented in Table 7, the following metabolic pathways seem to be present: TYR → DA → DOPAC; TYR → 3-O-MD; DA → 3-MT; and TYR → TYRA → OCT. The last pathway was found only in the buccal ganglion. The buccal ganglion has been reported to develop after hatching (Kriegstein, 1977). Thus, this pathway seems to appear late in the development of the nervous system of *Aplysia*. Moreover, the pathways E → DOPEG and E → MN → VA seemed to be present as minor routes. With respect to indolealkylamines, TRP → 5-HT was shown to be the major metabolic pathway with conversion of 5-HT to N-ACT-5-HT being a minor route. Furthermore, it appeared that the pathway TRP → KYN was also present.

The numbers and levels of detected compounds were large in the cerebral, pedal and parieto-visceral ganglia. This seems reasonable from the viewpoint of ontogeny: the cerebral-pedal ganglion system appears first among the ganglia during early development while the other ganglia only develop after hatching (Kriegstein, 1977). With the progress of morphogenesis, the pedal and parieto-visceral ganglia appear to control mainly the abdominal region. The higher-ranking cerebral ganglion appear to synthesize and transport large amount of transmitters for the control of lower-ranking ganglia, such as the buccal, pleural and parieto-visceral ganglia. As a pathway for neurotransmitters, the nerve tracts, which include a bundle of axons, were found to contain large amounts of monoamines. Similar considerations seem also to apply to lower-ranking ganglia and the nerve tracts in that order. It seems reasonable that the numbers and amounts of neurotransmitters might reflect the numbers of target tissues that are to be controlled.

Biogenic monoamines have also been examined in the snails *Achatina fulica* and *Euhadra peliomphala* and in the slug *Aglioimax reticulatus*. In *Achatina fulica*, the distribution of monoamines in the central nervous system has been examined by immunohistochemistry in the central nervous system (Croll, 1988). Furthermore, the actions of these monoamines have been widely examined using electrophysiological methods (Walker, 1982). Analytical data are shown Table 8. In *Achatina fulica* and *Euhadra peliomphala*, levels of the main monoamines, such as DA, NE, E and 5-HT, were found to be higher in the subesophageal ganglion than in the brain, whereas the opposite was found for the metabolites, suggesting higher rates of metabolism in the

Table 7.
Levels of biogenic mines and their metabolites in the central nervous system of *Aplysia kurodai*

	Buccal ganglion (n = 9)	Cerebral ganglion (n = 10)	Pleural ganglion (n = 16)	Pedal ganglion (n = 14)	Parieto-visceral ganglion (n = 10)	Bag cell clusters (n = 16)	Nerve tract I** (n = 10; pg/mg tissue)	Nerve tract II*** (n = 10; pg/mg tissue)
TYR	230.2±112.0	381.2±44.5	142.2±53.5	433.6±53.4	698.1±132.4	114.2±57.2	7.9±8.7*	5.9±1.6*
DA	19.9±6.5	64.1±18.3	4.2±1.3	277.4±53.0	153.8±11.2	11.3±9.0	581.5±550.2	206.8±80.9
DOPAC	1.1±0.9	12.4±7.1	0.5±0.3	10.2±4.3	18.6±2.3	0.6±0.2	5.5±6.2	19.3±24.5
3-O-MD	-----	-----	-----	4.0±0.8	2.3±0.7	-----	-----	-----
E	-----	1.2±0.1	0.2±0.04	-----	1.9±0.3	0.1±0.07	-----	-----
TYRA	0.6±0.2	1.3±0.4	-----	0.4±0.8	1.5±0.2	-----	-----	-----
OCT	3.4±1.1	-----	-----	-----	-----	-----	-----	-----
TRP	139.9±24.9	248.2±29.2	106.7±2.2	282.9±18.3	392.2±32.0	49.8±14.1	7.9±7.2*	3.8±0.7
5-HTP	0.2±0.03	-----	-----	-----	-----	0.1±0.05	-----	-----
5-HT	8.3±3.4	31.8±9.1	3.6±0.5	106.3±23.9	39.4±6.2	14.9±0.7	5.9±7.9*	367.1±100.8
5-HIAA	-----	0.2±0.1	-----	0.6±0.1	-----	-----	-----	-----
N-ACT-5-HT	-----	-----	-----	-----	-----	0.2±0.08	-----	22.0±11.2
KYN	-----	-----	2.1±0.4	6.6±2.7	27.5±27.0	-----	163.0±175.8	54.1±14.0

(Mean ± SD; ng / organ; * ng; ** A part of the nerve tract which runs from the cerebral ganglion to the pedal ganglion:
*** A part of the nerve tract which runs from the pleural ganglion to the paroeto-visceral ganglion)

Table 8.
Levels of biogenic amines and their metabolites in the central nervous system of terrestrial snails and slugs

	Achatina fulica		*Euhadra peliomphala*		*Agliolimax reticulatus*
	Brain	Subesophageal ganglion	Brain	Subesophageal gagnlion	Brain plus Subesophageal ganglion
	(n = 5)		(n =10)		(n = 12)
TYR	204.7±41.7*	198.5±20.5*	35.0±45.7	12.4±32.8	27.2±10.9
L-DOPA	68.5±11.6	75.7±40.2	53.2±38.5	29.6±21.9	292.7±49.5
DA	390.6±80.4	1.7±0.1*	1.85±0.2*	2.2±0.6*	2.7±1.8*
DOPAC	19.8±6.0	13.6±3.1	-----	-----	-----
3-O-MD	68.7±55.4	75.7±67.9	32.4±2.7	26.5±27.4	512.2±226.0
NE	79.9±20.5	47.4±7.9	67.9±54.2	-----	66.7±31.7
E	120.1±27.6	191.0±28.7	43.8±23.1	-----	-----
HVA	45.3±15.9	105.0±27.2	-----	-----	-----
3-MT	20.7±7.3	15.3±4.7	-----	-----	18.3±5.9
MHPG	65.1±15.7	31.6±7.0	36.8±22.6	-----	84.8±4.6
MN	-----	-----	155.2±120.9	128.3±32.4	-----
TYRA	-----	-----	213.6±29.3	84.9±66.3	27.8±10.9
OCT	-----	-----	-----	37.9±11.6	116.9±72.0
HPAC-4	213.7±119.6	157.7±121.4	-----	5.3±4.2	-----
TRP	3.5±0.5*	5.6±0.6*	1.2±0.1*	1.0±0.1*	3.2±0.8*
5-HT	745.9±180.7	1.7±0.3*	382.7±299.6	785.8±294.7	520.9±463.3
5-HIAA	1.6±0.4*	654.0±77.5	4.8±2.1	4.4±3.6	-----
KYN	240.2±128.6	156.0±114.2	-----	-----	-----

(Mean ± SD; pg / mg tissue; * ng)

brain. In addition, the level of DA was higher in *Euhadra peliomphala* and *Agliolmax reticulatus* than that in *Achatina fulica*. Levels of 5-HT and E were far higher in these pulmonates than in *Aplysis*. The pathways TYR → L-DOPA → DA → NE → E and TRP → 5-HT → 5-HIAA appear to be the main metabolic pathways, while L-DOPA → 3-O-MD → HVA, DA → DOPAC → HVA and TYR → TYRA pathways are also operating in pulmonate slugs and snails.

Arthropoda

A considerable amount of information has been reported on the detection and role of biogenic monoamines in Arthropods (Brown and Nestler, 1985; Klemm, 1985).

Insects. A variety of biogenic monoamines their precursors and metabolites have been identified in the nervous systems of insects (Brown and Nestler, 1985; Evans, 1980; Evans *et al*., 1985; Klemm, 1985; Vaughan, 1988). Such measurements have been improved with the development of technology that allows accurate quantification of these compounds (Barreteau *et al*., 1991; Goudey-Perrire *et al*., 1991; Linn *et al*., 1994; Puiroux *et al*., 1992; Pyza *et al*., 1991; Shimizu *et al*., 1991; Sparks and Geng, 1992; Takeda *et al*., 1991). It has been suggested that in insects these monoamines function as neurohormones, neurotransmitters and neuromodulators (Brown and Nestler, 1985; Evans, 1980; Klemm, 1985; Orchard *et al*., 1993).

In the brain, biogenic monoamines have been identified by HPLC-ECD. DA and 5-HT were detected in *Periplaneta americana* (Owen *et al*., 1987). In *Pieris*, DA, DOPAC, N-acetyldopamine (N-ACT-DA) and 5-HT were detected (Puiroux *et al*., 1990). In *Manduca sexta*, N-ACT-DA, DA, DOPA and 5-HT were measured (Krueger *et al*., 1990). In the ventral nerve cord of *Rhodnius prolixus*, DA and NE were analyzed (Orchard, 1990).

Several metabolic pathways were shown to be present in the central nervous system of *Bombyx mori* (Takeda *et al*., 1991) and the data are shown in Table 9A. The putative neurochemical pathways appear to be TYR → L-DOPA → DA → NMN; DA → DOPAC; TYR → TYRA → OCT; and TRP → 5-HT → 5-HIAA. The levels of DA and sometimes those of L-DOPA were particularly high. From the fourth instar to the post-spinning stage, generally, the levels of biogenic amines and precursors rose with development both in the brain and in the subesophageal ganglion. However, levels of DA and L-DOPA at the late stage of the fifth instar and of TYRA and 5-HT in post-spinning stage were reduced. 5-HT was not detected in the subesophageal ganglion at the early fourth instar or at the post-spinning stage. 5-HIAA was detected only in the subesophageal ganglion at the fourth instar. At some stages, VMA, MN, OCT and NE were occasionally detected. However, they were not detected consistently. DOPAC was detected in the subesophageal ganglion at the fourth instar, and in both the brain and subesophageal ganglion at the post-spinning stage. NMN was detected at the late stage of the fifth instar and post-spinning stage. HVA was only detected at the post-spinning stage.

In hoppers, biogenic monoamines, including OCT, DA, 5-HT and NE, have been detected in *Locusta migratoria* (Hiripi and S-Rozsa, 1973; Orchard *et al*., 1986), in the desert locust, *Schistocerca gregaria* (Klemn and Axelesson, 1973; Robertson, 1976) and in *Schistocerca americana gergaria* (Morton and Evans, 1983).

In the brain and corpus cardiacum of the migratory locust, *Locusta migratoria,* we found evidence of the following metabolic pathways (Table 9B):

Table 9A.
Levels of biogenic amines and metabolites in the central nervous system during fifth-instar of *Bombyx mori*

	Brain				Subesophageal ganglion			
	(Fifth-Instar)							
	Early (n = 19)	Middle (n = 21)	Late (n = 19)	Post-spinning (n = 20)	Early (n =19)	Middle (n = 20)	Late (n = 26)	Post-spinning (n= 8)
TYR	14.4±2.1*	11.8±0.9*	24.5±0.1*	62.2±5.3*	9.9±1.0*	13.1±1.1*	15.7±2.0*	28.8±2.9*
L-DOPA	87.4±22.2	1.1±0.1*	223.4±42.7	3.1±0.9*	79.9±7.0	1.0±0.2*	326.7±3.8	2.7±0.1*
DA	119.5±2.5	1.6±0.1*	1.6±0.4*	27.6±6.8*	82.6±11.1	2.9±0.1*	1.8±0.7*	11.1±0.7*
DOPAC	-----	-----	-----	121.6±29.9	-----	-----	9.3±0.1	63.0±30.8
HVA	-----	-----	-----	97.6±0.8	-----	-----	-----	71.5±0.1
NE	-----	-----	105.5±0.1	54.3±0.1	-----	-----	102.7±0.1	-----
TYRA	19.8±2.2	34.3±9.5	32.9±6.1	21.4±1.6	23.0±3.6	36.5±11.8	19.1±15.5	368.3±286.0
TRP	1.2±0.3*	2.5±0.1*	6.8±1.4*	12.4±1.7*	863.8±40.7	2.7±0.2*	3.8±0.6*	6.8±0.4*
5-HT	70.6±2.7	87.7±0.1	135.4±40.3	71.8±18.0	51.2±0.8	87.2±0.9	118.9±14.6	-----

(Mean ± SD; pg / organ; * ng)

TYR → L-DOPA → DA → NE → E → VMA; DA → 3-O-MD → HVA; DA → DOPAC → HVA; TYR → TYRA → HPAC-4; TYR → OCT; TRP → 5-HT → 5-HIAA; and TRP → KYN. The KYN pathway was found for the first time in the neuroendocrine system of an insect. In particular, the clearest difference between the albino and control locusts was associated with KYN. KYN disappeared from both the brain and the corpus cardiacum of the fifth instar nymphs on day 0. The results are shown in Table 9B (Tanaka and Takeda, 1996).

Table 9B.
Levels of biogenic monoamines and their metabolites in the brain of normal and albino fifth-instar nymphs of *Locusta migratoria.*

—	Normal		Albino	
	Male	Female	Male	Female
TYR	67.3±6.2	86.2±5.5	82.8±5.9	77.3±17.2*
L-DOPA	605.0±72.4	885.8±232.6	664.2±80.9	524.0±31.9
DA	484.3±182.4	454.3±229.4	757.1±12.6	448.8±48.0
DOPAC	29.7±2.6		33.6±12.1	24.4±14.1
3-O-MD	20.1±5.7	29.3±10.0		30.0±12.6
EPIN			36.7±11.2	12.4±3.9
NE		124.7±49.8		116.1±47.5
VMA	0.4±1.2*	1.0±0.4	0.3±0.1	0.8±0.4
TYRA	723.8±158.2	701.8±102.4	619.2±305.3	668.8±424.2
OCT				711.7±699.8
HPAC-4	73.4±0.5		23.0±32.5	43.5±47.7
TRP	0.9±1.4*	10.5±2.9*	20.2±4.8*	23.4±16.3
5-HT		14.8±1.4		38.4±36.3
KYN	809.6±68.2	838.2±504.1		

(Mean±SD; n=15 in each sample; pg/organ; *ng)

Levels of biogenic monoamines in hoppers are altered by changes in breeding conditions. In *Locusta migratoria*, levels of OCT in the brain and other nervous tissues have been shown to be higher in isolated locusts than in those reared under crowded condition (Fuzeau-Braesch and David, 1980). Similar results were also obtained with *Schistocerca gregaria* (Morton and Evans, 1983). In contrast, levels of DA in the brain were found to be higher in gregarious, than in the solitary species of *Locusta migratoria* (Fuzeau-Braesch, 1977).

Biogenic monoamines have also been examined in the dragonfly, *Anax parthenope julius*, and beetle, *Lucanus maculifemoratus* (Table 9C). In the former, the following pathways appeared to be present: TYR → 3-O-MD or DA → HVA; TYR → TYRA → OCT or HPAC-4; and TRP → 5-HT or KYN. In the latter, the pathways were TYR → L-DOPA → DA → DOPAC; TYR → L-DOPA → 3-O-MD; TYR → TYRA → OCT; and TRP → 5-HT.

Three-dimensional HPLC with ECD has also been used to examine monoamines in other insects, such as *Periplaneta americana* (Shimizu *et al.*, 1991), *Mamestra brassicae* (Shimizu and Takeda, 1991), *Manduca sexta* (Geng *et al.*, 1993), *Apanteles kariyai* (Shimizu *et al.*, 1993), *Mythimna separata* (Shimizu and Takeda,

1994) and *Trichoplusia ni* in relation to photoperiod (Linn *et al.*, 1994) and in whole-body extracts of the two-spotted spider mite, *Tetranychus urticae* (Sparks *et al.*, 1994).

Table 9C.
Levels of biogenic monoamines and their metabolites in the central nervous system of the dragonfly and beetle

	Anax parthenope julius	*Lucanus maculifemoratus*
TYR	1.0±0.5*	104.9±45.6
L-DOPA		1.1±0.1
3-0-MD	2.0±0.6	0.2±0.1
DA	4.6±0.2	1.4±0.3
DOPAC		83.9±74.9
TYRA	0.8±0.2	0.8± 0.4
OCT	13.0±0.9	2.3±0.1
HPAC-4	2.1±0.3	
TRP	48.9±0.9	13.0±3.5
5-HT	10.3±1.5	1.1±0.2
KYN	0.9±0.4	

(Mean±SD; n=12; ng/organ; μg/organ)

Crustacea. Analysis of various biogenic monoamines by HPLC has been performed in crustaceans that include *Procambarus clorkii* (Fingerman and Kulkarni, 1993) and *Pacifastacus leniusculus* (Elofsson *et al.*, 1982; Laxmyr, 1984). Biogenic amines in crustaceans were reviewed in terms of identification, location and roles by Fingerman and Nagabhushanam (1994).

The presence of aggregation pheromone and the neural mechanism of aggregation have been reported from the viewpoint of land adaptation in the terrestrial isopods *Ligia exotica, Porcellio scaber* and *Armadillidium vulgare* (Takeda, 1984; Takeda and Mizuno, 1989). To clarify the neural basis of aggregation behavior, a wide range of monoamines and related metabolites have been examined (Takeda, 1990). Among biogenic amines, 5-HT-immunoreactive neurons have been detected in the nervous system of some species of isopods (Fujii and Takeda, 1988; Thompson *et al.*, 1994). The analytical results are shown in Table 10A.

In the three species mentioned above, high levels of TYR and TRP were detected. In *Ligia exotica*, L-DOPA was not present at detectable levels. However, 3-O-MD was found and evidence for the pathways TYR → DA → NE and TYR → TYRA → OCT was observed. Similarly, in *Porcellio scaber,* 3-O-MD, TYRA and OCT were detected. In *Armadillidium vulgare*, DA was at an undetectable level. In all species, 5-HT was found to be present. In *Porcellio scaber*, not only 5-HT but also N-MET was found to be present with high peak purity. So, the pathway TRP → 5-HT → N-MET seems to be present.

In *Procambarus clarkii*, the major monoamines were DA, EPIN, 5-HT and KYN and high levels of 3-O-MD and MN were also detected. E was detected in the nervous system by two-dimensional methods (Elofsson *et al.*, 1982; Fingerman and Kulkarni, 1993). The results obtained by the three-dimensional method were, however, inconsistent (i.e., the peak purity was under 0.5). This suggests that identification of E by the 2-dimensional methods may be in error. Instead of E, we found that its precursor EPIN was clearly present. Generally, EPIN is not abundant in the nervous system. Results are presented in Table 10b. The main metabolic pathways seem to be TYR → DA → EPIN → MN; TRP → 5-HT and TRP → KYN. The TYR → TYRA → HPAC-4 pathway was found to be particularly active in the thoracic and ventral ganglia.

Table 10A.
Levels of biogenic monoamines and their metabolites in the central nervous system of some isopods

	Ligia exotica	*Porcellio scaber*	*Armadillidium vulgare*
TYR	441.7±65.2*	351.7±0.07*	159.7±3.3*
L-DOPA		683.3±8.3	743.0±1.3
3-O-MD	709.1±219.3	8.1±0.1*	3.7±0.6*
DA	366.6±67.1	125.8±3.3	
DOPAC		290.1±16.8	
NE	9.39±59.3	81.2±3.7	251.0±1.0
TYRA	100.6±31.3	103.7±2.9	770.0±133.0
OCT	688.4±229.6	2112.2±20.4	127.6± 1.0
TRP	132.1±21.6*	165.7±0.8*	36.0±1.5*
5-HTP			28.8±2.0
5-HP	189.0±40.1	55.2±3.2	30.4±0.8
N-MET		396.6±3.3	

(Mean±SD; n=50 in each sample; pg/mg tissue; *ng/mg)

Table 10B.
Levels of biogenic monoamines and their metabolites in the central nervous system of the crayfish *Procambarus clarkii*

	Brain	Subesophageal ganglion	Thoracic ganglion	Abdominal ganglion
TYR	85.2±24.1	78.1±13.0	29.4±7.3	31.1±4.8
3-O-MD	0.4±0.08	0.4±0.05	0.3±0.07	0.1±0.04
DA	0.3±0.08	0.4±0.01	0.1±0.01	0.08±0.01
EPIN	0.3±0.1	0.2±0.1	0.1±0.03	
TYRA			0.2±0.1	0.3±0.08
OCT	1.9±0.6	2.0±0.07	1.4±0.1	1.4±0.09
HPAC-4	0.03±0.02	0.03±0.01	0.02±0.01	0.02±0.01
TRP	25.6±4.5	24.7±0.5	8.9±1.4	9.0±1.5
5-HT	0.3±0.1	0.1±0.02	0.06±0.01	0.01±0.02
KYN	1.4±0.2	0.8±0.2	0.6±0.4	0.4±0.2

(Mean±SD; n=8 in each sample; ng/mg tissue)

Arachnidae. In Arachnidae, the presence of monoamine oxidase has been examined in spiders. Evidence for the presence of catecholamines was reported by Meyer and Johnen (Meyer and Jehnen, 1980). Recently, the existence of OCT immunoreactive neurons was suggested. OCT was detected in the hemolymph by HPLC (Seyfarth *et al.*, 1993). Results are shown in Table 11. In the present analysis, 5-HT, DA, OCT and NE were clearly shown to be present in the nervous system of the spider, *Argiope amoena.* Thus, the TYR → DA → DOPAC → NE, TYR → OCT and TRP → 5-HT pathways appear to be present in the central nervous system of spiders.

Table 11.
Levels of biogenic monoamines and their metabolites in the central nervous system of the spider *Argiope amoena*

Compound	(ng/mg tissue)
TYR	60.0±12.2
DA	1.6±1.5
DOPAC	0.3±0.1
NE	0.3±0.3
OCT	1.2±0.4
TRP	20.0±9.2
5-HT	1.5±1.3

(Mean±SD; n=8)

Echinodermata

In the Echinodermata, biogenic monoamines have been examined in the starfish, such as *Pisaster ochraceus*, *Solaster stimpsoni* and *Pycnopodia helianthiodes* by HPLC-ECD (Sloley and Juorio, 1990). High levels of DA and NE, TYRA and MEL have been detected in the radial nerves (Table 12).Our current data suggests that the TYR → L-DOPA → DA → NE, TYR → TYRA, TRP → 5-HTP → 5-HT → 5-HIAA and 5-HT → MEL pathways are to be found in the starfish, *Asterina pectinifera.* It is unusual for MEL to be detected in the nervous system.

Table 12.
Levels of biogenic monoamines and their metabolites in the radial nerve of the starfish *Asterina pectinifera*

Compounds	(pg/mg tissue: *ng/mg)
TYR	60.6±20.1*
L-DOPA	182.5±110.3
DA	11.1±2.9*
NE	1.2±0.3*
TYRA	401.2±156.0
TRP	16.6±2.2*
5-HTP	14.8±18.4*
5-HT	101.0±81.8
5-HIAA	13.9±12.9
MEL	619.1±259.9

(Mean±SD; n=12)

Protochordata

The ascidian is a representative protochordate which is thought to be a phylogenetic offshoot from the deuterostome line from invertebrates to vertebrates.

The neural complex of ascidians consists of the cerebral ganglion and a neural gland. The latter is situated ventral to the former and consequently is called the hypoganglionic gland. Since its first description by Hancock (1868), the neural complex of ascidians has attracted considerable interest because of its proposed homology to the vertebrate pituitary. At the present time, however, the endocrine function of the neural complex remains to be established. Nonetheless, the classical monoamine neurotransmitters have been reported in the neural complex of these animals (Osborne *et al.*, 1979; Welsh and Loveland, 1968).

Levels of detected biogenic monoamines in the neural complex of *Ciona intestinalis* are presented in Table 13. Based on our data, the following metabolic pathways are thought to occur in the neural complex: TYR → L-DOPA → DA → NE, TYR → DA → DOPAC, TYR → TYRA, and TRP → 5-HTP → 5-HT.

Table 13.
Levels of biogenic monoamines and their metabolites in the neural complex of the ascidian, *Ciona intestinalis*

Compounds	Concentration (ng / neural complex)
TYR	125.07±0.09
L-DOPA	0.71±0.30
DA	14.46±10.85
NE	3.05±2.85
NMN	0.28±0.19
DOPAC	1.40±0.67
VA	0.12±0.07
TYRA	0.20±0.11
TRP	24.56±16.35
5-HTP	0.07±0.03
5-HT	0.38±0.31

(Mean±SD: n=4)

Compared to the compounds detected in protostomian species, the metabolic pathways involved in the synthesis of monoamines in *Ciona interstinalis,* a deuterostome, seem to be rather simple. For example, E, which is usually present in higher-ranking protostomes, was absent. 5-HIAA, a metabolite of 5-HT, was also absent. This profile might be due to the fixed lifestyle, without motility, of adults. 5-HT is known to be present in the endostyle of *Ciona, Ascidia, Ascidiella* and *Phallucia* and in the digestive tract of *Ciona* (Fritsch *et al.*, 1982). In *Halocynthia roretzi,* immunoreactivity specific for 5-HT has been detected in cells in the central nervous system (Fujii and Takeda, 1988).

Vertebrates

Compared to those of invertebrates, vertebrate nervous systems are very highly organized and numerous physiologically active substances are involved in their function. Vast numbers of reports on neurotransmitters have appeared (e.g., Anwyl, 1990; Smeets *et al.*, 1995; Winberg and Nilsson, 1993). Much research has focused on the histochemical identification of monoamine neurons. The distribution and destinations of monoamine-containing neurons of various vertebrates have been extensively investigated. To gain a better insight into the development of monoamine systems in representative vertebrates, phylogenetic analysis has been performed. In this section, levels of monoamines, precursors and metabolites in vertebrate brains are discussed in each phylum, systematically.

Agnatha

Hagfish and lampreys are the only living species in this very primitive class of vertebrates. Data on catecholamines in the brain of hagfish are limited to a few histochemical studies on the distribution of monoamine oxidase, tyrosine hydroxylase and dopamine (Pierr *et al.*, 1984). In the inshore hagfish, *Eptatretus burgeri*, high levels of 5-HT, NE and TYRA were detected in the whole brains. These results correspond well to those obtained using immunohistochemical analysis (Kadota, 1991). Our results are shown in Table 14. The monoamine systems, in particular, the DA system, are well developed in these brains. The following pathways seem to be present: TYR → DA → DOPAC, TYR → DA, TYR → TYRA → HPAC-4, TRP → 5-HTP → 5-HT → 5-HIAA and TRP → KYN.

Table 14.
Levels of biogenic monoamines and their metabolites in the brain of the hagfish *Eptatretus burgeri.*

Compounds	(pg/mg tissue: *ng/mg)
TYR	26.7±7.2*
DA	73.0±17.1
DOPAC	11.2±6.1
VMA	246.1±273.2
TYRA	64.1±114.7
HPAC-4	1.6±3.1*
TRP	5.3±2.2
5-HTP	166.1±124.8
5-HT	717.2±193.6
5-HIAA	13.8±6.6
KYN	98.5±205.4

(Mean±SD ; n= 15)

Osteichthyes

Fish are generally subdivided into two major groups, cartilaginous fish and bony fish. Teleosts (bony fish) are the most abundant group of fish. The role of serotonin has been found in the brain of bony fish (Winberg and Nilsson, 1993). Immuno-

Table 15.
Levels of biogenic amines and their metabolites in the brain of some bony fish

	Anguilla japonica	*Parasilurus asotus*	*Carassius cuvieri*	*Carassius auratus*	*Cyprinus carpio* **
TYR	5.5±10.3*	6.6±2.2*	3.1±1.5*	4.6±0.4*	3.4±0.7*
3-O-MD	174.6±321.4	——	——	58.4±45.6	——
DA	199.2±389.0	96.2±20.2	98.7±51.2	110.6±13.7	92.9±17.7
DOPAC	48.3±87.1	16.4±3.8	——	5.1±0.9	11.7±1.0
HVA	——	3.4±0.6	3.0±1.6	——	1.2±0.2
3-MT	——	——	6.0±2.1	5.1±2.1	2.9±1.1
NE	——	212.3±38.4	296.4±152.8	179.6±40.4	255.3±66.3
E	70.5±141.6	25.8±13.0	17.5±7.3	28.6±6.7	38.9±23.6
VA	——	——	2.2±0.9	2.3±1.2	——
OCT	——	——	6.4±2.2	——	13.8±3.4
TRP	1.6±2.8*	3.7±0.7*	1.2±0.6*	1.8±0.4*	1.6±0.4
5-HT	214.5±426.1	80.9±20.1	90.3±46.1	99.4±9.0	69.2±16.0
KYN	——	27.6±6.7	25.8±5.2	20.4±0.9	26.5±2.0

(Mean ± SD; n = 6 in each sample; pg / mg tissue; * ng; ** Juvenile) histochemical localization of catecholamines has also been examined in the central nervous system (Meek, 1994). Monoamine levels were examined in whole brains of the bony fishes *Anguilla japonica, Parasilurus asotus, Carassius cuvieri, Cyprinus carpio* and *Carassius auratus*. Results are shown in Table 15. DA, E and 5-HT were detected in all species. NE and KYN were also detected except in *Anguilla japonica*. The following pathways appear to be present: TYR → DA → NE, DA → DOPAC → HVA, TRP → 5-HT and TRP → KYN.

The above mentioned species have a so-called caudal neurosecretory system in the caudal region of the spinal cord. The neurosecretory cells send their axons mainly to the vascularized urophysis. We also examined the monoamine, precursor and metabolite levels in the urophysis. Based on our data we have found evidence for the following pathways: TYR → DA → EPIN → E and, TYR → DA → DOPAC. The TRP → 5-HT → 5-HIAA pathways were found only in *Cyprinus carpio*. The TYR → TYRA pathway was only detected in *Carassius cuiviri,* the TYR → TYRA → OCT pathway only in *Carassius auratus* and the TRP → KYN pathway only in *Parasilurus asotus*.

Thus, the catecholamine systems are more highly developed in the more advanced bony fishes. Immunohistochemical techniques have yielded detailed information on the distribution of catecholamine neurons. However, the data on the distribution are too limited for general consideration.

Amphibia

Amphibians belong to three different orders: the Anura, which includes frogs and toads; the Urodela, which includes salamanders and newts; and the Apoda. Most of our knowledge of the details of monoamine systems comes from studies of the anuran. Anuran amphibians undergo striking changes, not only during the embryonic period but also during post-embryonic development. To gain better insight in the development of monoamine systems during metamorphosis, biogenic monoamines were analyzed in embryos at early stages and in tadpoles of *Xenopus laevis*. Table 16 shows the development of metabolic pathways from the precursor amino acids TYR and TRP to the monoamines and metabolites. During embryogenesis, TYR → 3-O-MD → HVA is the main metabolic pathway. 5-HT was barely detectable until stage III. Then the TYR → DA → 3-MT → HVA, DA → DOPAC, DA → E, TRP → 5-HT → 5-HIAA → N-MET and TRP → KYN pathways developed during the premetamorphic stages of the whole embryo. From the premetamorphic stage to prometamorphic stage, these pathways also advanced in the nervous system itself. The TYR → TYRA pathway also developed. At metamorphic climax, levels of DA and 5-HT were also maximal and the TRP → 5-HT → N-MET and TYR → OCT pathways were also introduced. At the early juvenile stage, each monoamine was at its highest level. These results coincided with the results of immunochemical staining of 5-HT and catecholamines (Gonzalez *et al*., 1994; Pollack and Harold, 1988). More details of the changes in levels of monoamines during metamorphosis in *Bufo* and *Rana* were published elsewhere (Takeda and Takaoka, 1991; Takeda, 1996). The results were basically similar to those obtained with *Xenopus laevis*.

Levels of biogenic monoamines and their metabolites were also analyzed in the whole brains of adult frogs, such as *Rana japonica*, *Rana nigromaculata*, *Rhacophorus buergeri* and *Hyla arborea japonica*, and of the newt such as *Cynopus pyrrhogaster*. High levels of 5-HT and E were detected in all frogs examined.

Table 16.
Levels of biogenic monoamines and their metabolites during metamorphosis in *Xenopus laevis*

Stage Compound	I (St. 1)	II (St. 15)	III (St. 25)	IV (St. 39)	V (St. 45)
TYR	1859.5±384.0	2453.7±456.4	5272.7±456.4	8602.9±1626.7	1174.4±515.4
3-O-MD	9.0±2.4	9.0±1.0	10.4±3.8	17.0±3.2	14.8±6.0
HVA			0.6±0.3	2.3±2.8	2.2±1.3
DA					11.9±3.0
3-MT				17.3±2.3	21.4±4.7
DOPAC					1.4±0.4
E					35.7±6.8
VA					2.2±1.1
TRP	605.0±135.1	1104.0±246.2	2326.8±950.2	3508.2±709.3	752.7±209.2
5-HT				19.9±10.8	10.0±2.2
5-HIAA				55.4±7.9	21.3±6.5
NMET					13.3±4.7
KYN				83.5±19.0	30.1±7.1

Stage Compound	VI (St. 47)	VII (St. 54)	VIII (St. 59)	IX (St. 65)	X
TYR	3771.7±3416.5	2510.2±870.5	4537.1±639.8	1705.7±378.9	6553.5±257.2
3-O-HT	5.3±4.1	10.0±4.4	15.7±1.7	27.0±3.3	
HVA	8.2±6.5	15.8±5.1	18.4±4.3	10.8±0.9	22.6±8.7
DA	55.2±44.9	69.5±5.0	117.2±42.2	145.5±57.2	1595.6±355.2
3-MT					
DOPAC	10.7±8.8	15.6±5.0	19.7±4.2	15.0±1.1	63.1±11.9
E	80.2±57.4	146.8±52.7	148.0±30.9	141.6±14.8	329.3±99.6
VA	10.8±9.5	3.2±0.3	4.3±1.9		
TYRA		29.0±6.8	29.9±12.4	22.4±3.3	40.5±23.4
OCT				7.3±1.4	22.1±9.8
TRP	891.5±779.6	1059.2±390.2	1381.3±225.7	675.8±135.3	1690.0±776.7
5-HT	103.3±87.0	133.0±45.5	178.3±32.2	396.4±159.8	5206.8±982.8
5-HIAA	102.4±88.3	163.2±66.7	151.3±25.8	90.2±1.2	223.7±721.3
NMET				11.7±7.4	442.5±342.1
KYN	24.3±19.5	20.5±5.9	18.9±6.8	17.6±8.1	26.4±13.7

(pg/mg Embryo; Mean±SD; n=25)
Stages were determined by the developmental table of Nieukoop and Faber (1956)

In general, the following pathways appeared to be present: TYR → DA → DOPAC → HVA, DA → E, TYR → TYRA → HPAC-4, TRP → KYN, and TRP → 5-HTP → 5-HT → 5-HIAA. In *Rhacophorus buergeri,* we found the serotonin-degradation pathway: 5-HT → N–ME → BUTN, as described below for *Bufo bufo japonicas*. In the newt, high levels of DA were detected. Other pathways were similar but fewer in number for those in frogs. Amphibians may be unique among the vertebrates in that E appears to be the predominant brain catecholamine in this group.

Reptilia

The reptiles occupy a crucial position in phylogeny, since they gave rise to both birds and mammals. Catecholamine systems in the central nervous system have mainly been studied immunohistochemically (Smeets, 1994).

Analytical results from *Clemmys japonica* showed that 5-HT, DA, E, and NE are the predominant monoamines. In addition, EPIN, 5-HIAA, DOPAC, HVA were detected, as shown in Table 17. Our data suggest that the following pathways are operating in reptiles: TYR → DA → DOPAC → HVA, DA → NE → EPIN or NE → E, and TRP → 5-HT → 5-HIAA.

Table 17.
Levels of biogenic monoamines and their metabolites in the brain of *Clemmys japonica*

Compounds	(ng/mg of whole brain; *ng/mg)
TRY	3.2±1.1*
DA	380.9±102.2
DOPAC	63.7±83.8
HVA	9.9±8.5
NE	250.4±157.2
EPIN	2.7±1.1
E	321.4±189.1
TRP	1.0±0.3*
5-HT	558.4±487.7
5-HIAA	56.4±47.2

(Mean±SD ; n=6)

Aves

Our knowledge of monoamine systems in the avian brain is still meager. Data are largely based on histofluorescence investigations in the brain of the chicken, *Gallus gallus domesticus* (Pulles and Medina, 1994). Levels of biogenic monoamines in *Gallus gallus domesticus* during neurogenesis until just before hatching are shown in Table 18 (Takeda, 1992c). Just after oviposition, the TYR → TYRA → HPAC-4 pathway first appears in the embryonic area. 5-HT was markedly elevated at stage 12. In contrast, DA first appeared in the whole brain at stage 37. Generally, at early embryonic stages, the TRP → 5-HT, and TYR → L-DOPA → 3-O-MD → HVA pathways first appeared. The TRP → 5-HTP → 5-HT, TYR → DA → DOPAC → HVA, DA → 3-MT → HVA and NE → NMN pathways then gradually developed.

Mammalia

The central monoamine systems are particularly widespread and highly developed. The results of numerous studies reveal that the monoamine systems in the mammalian brain are organized according to a pattern that is basically similar in all species (Kitahama *et al.*, 1994).

In mammals, such as the mouse (*Mus musculus)*, as in birds, the brain monoamine systems are more complex than those in reptiles. However, the overall pattern of organization in mammals does not differ significantly from the fundamental

Table 18.
Levels of biogenic amines and metabolites during embryogenesis in the chicken *(Gallus gallus domesticus)*

Stages / Compounds	Just after oviposition (embryo region) (n = 4)	Stage 12 (n = 6) (Developing whole embryo)	Stage 20 (n = 6) (Developing whole embryo)	Stage 29 (n = 6) (Developing head region)	Stage 31 (n = 6) (Developing brain)	Stage 37 (n = 6) (Whole brain)	Stage 42 (n = 6: each) (Whole brain just before hatching) Forebrain	Hindbrain
TYR	1.3±0.9*	48.7±33.6*	44.4±32.2*	9.9±1.6*	11.1±3.5*	1.5±0.5*	2.2±0.7*	2.3±0.6*
L-DOPA	0.9±0.7	68.6±66.5	59.7±81.8	-----	-----	-----	3.9±0.6	6.0±3.6
3-O-MD	-----	7.1±5.2	20.3±16.7	11.1±4.9	10.6±3.0	58.0±56.8	114.5±92.8	127.1±108.2
DA	-----	-----	-----	-----	-----	3.7±2.1	66.3±32.5	17.9±8.3
DOPAC	-----	-----	-----	-----	-----	2.2±1.4	5.5±1.7	5.9±1.5
3-MT	-----	-----	-----	-----	-----	-----	16.8±9.0	6.4±3.0
HVA	-----	-----	4.6±3.7	1.1±0.2	1.2±0.3	3.5±3.3	22.1±7.1	16.3±4.8
NE	-----	-----	1.4±0.5	5.4±3.2	40.0±16.2	16.1±9.2	50.9±21.9	104.3±37.5
EPIN	-----	5.4±4.8	12.6±5.8	-----	-----	-----	-----	-----
E	-----	-----	-----	33.6±20.6	53.2±5.1	6.7±3.3	-----	-----
NMN	-----	-----	-----	-----	2.5±1.7	1.1±0.9	11.6±7.9	31.2±20.3
MN	-----	-----	-----	-----	-----	-----	30.8±31.2	29.1±28.7
TYRA	29.6±11.4	356.6±257.4	204.3±148.6	48.9±20.6	101.3±102.4	18.5±10.7	56.8±10.2	41.2±9.0
HPAC-4	1.0±1.2	23.3±18.9	18.1±8.5	19.7±7.4	24.1±15.7	6.0±3.4	26.5±8.4	23.6±5.9
TRP	450.8±423.7	12.7±10.0*	15.8±11.8*	2.9±0.7*	2.6±1.7*	0.6±0.2*	0.9±0.3*	1.0±0.1*
5-HTP	-----	-----	-----	1.7±0.3	1.6±0.4	1.2±0.9	5.0±2.6	5.8±2.1
5-HT	-----	29.8±21.6	26.0±19.2	10.5±2.6	13.3±3.3	15.3±13.1	69.2±22.5	79.6±21.3
5-HIAA	-----	-----	-----	-----	-----	-----	-----	12.0±4.0

(Mean ± SD: Stages were determined by the normal table of Hamburger and Hamilton (1951) (41) arrangement of monoamine neurons in reptiles.

APPLICATIONS TO THE FIELD OF MEDICAL BIOCHEMISTRY

Monoamines in some drinks and herbs

In Japan, tea made from powdered mulberry leaves, *Morus alba*, which is also the major diet of the silkworm, *Bombyx mori*, is drunk by some people as a traditional health-promoting drink (Nomura, 1988). A wide range of biogenic amines and related compounds were detected in the mulberry leaves, namely, DA, DOPAC, HPAC-4, L-DOPA, NE, and TYRA. DA occurred at ten-fold higher levels than other compounds and NE was present at the lowest level (Shimizu *et al.*, 1992).

Tea made of the dried plant, *Houttuynia cordata*, and its leaves and roots is also used as a health-promoting drink. Although the main medicinal compounds were flavonoids and aldehydes, high levels of DA, E, TYRA and KYN were also detected.

The aromatic amino acids in two kinds of oolong tea, green tea and coffee were also examines. One oolong tea contained TYRA and high levels of KYN, while TYRA, KYN and high levels of NE were found in the other oolong tea sample and in the green tea preparation. TYRA and high levels of NE were detected in black tea. Interestingly, DA, NE, TYRA, OCT and high levels of 5-HT were found in coffee. Although the fate of these ingested monoamines in the body is not clearly understood, these drinks clearly have an effect on the brain.

Table 19.
Levels of biogenic monoamines and their metabolites in some herbs.

	Datura metal	*Lycoris radiata*	*Phytolacca americana*
TYR	19.9±1.1	14.5±7.5*	3.9±2.2*
L-DOPA		56.9±51.1	1.7±1.4*
3-O-MD			1.6±1.8*
DA	466.8±403.0	236.4±293.3	2.1±1.3*
DOPAC		120.1±89.1	1.3±1.2*
3-MT	1.1±1.1*		
HVA			511.9±419.4
NE	83.3± 40.3		1.5±1.1*
E	61.2±77.4	210.0±247.8	
MN		4.1±1.1*	
NMN			652.7±380.4
MHPG	1.6±1.5*		
VMA	384.3±232.9	48.6±43.4	
DOPEG	160.5±90.5		374.3±31.3
TYRA			7.1±3.6*
OCT	1.5±0.4*		
HPAC-4			1.3±1.4*
TRP	4.3±1.9*	2.7±1.1*	3.9±2.2
5-HTP	8.9±7.3*		

(Mean±SD : n=12 : ng/mg tissue : *µg/mg)

The levels of biogenic monoamines and related compounds were also measured in some medical herbs as shown in Table 19. In three species of popular herbs, monoamines were shown to be present in addition to the main medicinal compounds. In *Datura metel*, DA, NE, E and high levels of OCT were detected in addition to alkaloids. In *Lycoris radiata*, DA, NE, and some metabolites were detected in addition to alkaloids. In *Phytolacca americana*, OCT, TYRA and high levels of DA, NE and many metabolites were detected in addition to the toxin phytolacca.

Disorders

Psychiatric disorders

The methylation theory of psychiatric disorders suggests that the human body is capable of producing toxic methylated substances that induce psychotic behavior. We examined this theory from the viewpoint of comparative biology.

In most animals, 5-HT and DA are synthesized continuously. In other animals, metabolic pathways other than those necessary for maintaining homeostasis have developed, including different degradation routes. In *Hydra magnipapillata*, as described above, the main indolealkylamine in the primitive nervous system is not 5-HT but N-MET. This compound acts not as a neurotransmitter but as a venom. It is interesting to note that many such unusual metabolic pathways are associated with the production of venoms (Habermehl, 1981). Indeed, we have detected bufotenine (BUTN) and N-MET in the poison glands of some bees (Shimizu *et al.*, 1993), in the nematocytes of some sea anemones, in glandular secretions of some toads and in *Amanitae mushrooms*, and so on (Takeda, unpublished observations). In the course of our phylogenetic analysis of monoamines and their metabolites, we detected N-MET and BUTN in the brain of the toad *Bufo bufo japonicus*. Furthermore, both N-MET and BUTN were detected in the blood and urine of *Bufo bufo japonicas*, which suggests their passage from the brain (Takeda, 1994a and b). BUTN is a dimethylated derivative of 5-HT and is known to be a hallucinogen. The unusual metabolism of 5-HT in the brain of *Bufo bufo japonicus* reminded us of the relationship between the metabolism of 5-HT and human psychiatric disorders that originate from functional diseases of the central nervous system, such as schizophrenia.

The first evidence, obtained by paper chromatography, that urine from schizophrenic patients contains a BUTN-like substance, in contrast to the urine of healthy controls, was reported by Fisher *et al.* (Cottreall *et al.*, 1977; Fischer *et al.*, 1961). In the 1960's, there were many attempts to determine whether BUTN or a BUTN-like substance might be present in urine from normal controls or schizophrenic patients. At that time, however, methods for detection of BUTN were too insensitive to allow the routine measurements of minute quantities. In spite of the positive results obtained with urine from schizophrenics, little further progress was made until Karkkainen and coworkers reported the detection of BUTN by gas chromatography-mass spectrometry (Karkkainen *et al.*, 1988). Now, three-dimensional HPLC with ECD permits the simultaneous measurement of indolealkylamines (including BUTN), catecholamines, precursors and metabolites, easily and with sensitivity and qualitative accuracy.

Table 20A.
Levels of BUTN and related compounds in urine specimens of patients with various psychiatric disorders

Disorders	Serotonin	N-MET ng/mg Creatinine (no. of patients)	BUTN
Autism			
a. with mental retardation (N=47) [20±4]	10.8±4.8 (47)	3.0±3.3 (8)	3.3±1.5 (32)
b. with epilepsy and mental retardation (N=18) [22±3]	12.1±9.6 (18)	2.5±1.7 (7)	3.6±4.3 (18)
Mental retardation with epilepsy or cerebral palsy (N=6) [21±4]	7.2±3.9 (6)	2.0±0.8 (3)	2.5±2.0 (6)
Epilepsy (N=6) [26±5]	118.7±78.0 (6)	102.3±85.9 (4)	13.7±15.6 (5)
Depression (N=18) [26±14]	94.7±116.3 (18)	43.8±48.4 (17)	46.0±48.2 (15)
Schizophrenia (N=15) [32±11]	446.4±693.8 (15)	16.4±377.8 (7)	287.9±487.6 (13)
Non-psychiatric disorders (N=30) [34±14]	69.4±97.7 (30)	(0)	(0)
Healthy control (N=200) [37±9]	84.6±26.8 (200)	(0)	10.9 ± 8.6 (2)

(Mean±SD); [age in years]

In a pilot study, BUTN was detected in urine from all autistic patients with mental retardation and epilepsy (n=18) and in many autistic patients (32/47) with mental retardation. BUTN was detected in the urine of 15 of 18 patients with depression. Thirteen of 15 schizophrenic patients were also positive for BUTN. N-MET was also detected in some cases of each disorder. Urine from patients with other disorders, such as liver disease and hyperthyroidism, was negative for both compounds. Only two of 200 urine specimens from healthy controls were positive for BUTN. Characteristically, levels of 5-HT, N-MET and BUTN were all high in patients with schizophrenia and low in patients with autism, as compared to controls. The main results are summarized in Table 20A (Takeda, 1993b; Takeda *et al.*, 1995). These results indicate that the degradation pathway from 5-HT and BUTN might be associated with each of the various psychiatric disorders. Metabolites other than BUTN and N-MET were not detected, with the exception of 5-HIAA which was present at normal levels. The activity of the BUTN pathway was high in patients with schizophrenia, depression and epilepsy, in that order, and low in patients with autism. Since the brain of *Bufo bufo japonicus* has been suggested to provide a useful pharmacological model for the metabolism of BUTN, many drugs have been tested for their ability to reduce levels of BUTN in the brain of this toad. In *Bufo bufo japonicus*, *p*-chlorophenylalanine and ritanserin have been shown to reduce brain levels of BUTN (Takeda, 1995a). Our observation should encourage research into the development of new and potent drugs that reduce the levels of these compounds. The old methylation theory of mental disorders now seems to have some validity.

Several researchers have found BUTN at low concentrations in the urine of apparently normal controls. This is surprising, as even though the limit of detection of our HPLC multi-electrode electrochemical detector system for BUTN was about 50pg/mL, we only detected BUTN in urine from two of 200 healthy normal individuals. One reason for this discrepancy might be the racial differences among the subjects who were analyzed since our subjects were exclusively Japanese. BUTN has been shown to be capable of producing a schizophrenia-like syndrome. When administered to humans as a hallucinogen, or when present in excess in the body, BUTN can cause psychotic behavior. Recent research has shown that BUTN has an effect on 5-HT neurons similar to that of *p*-chloroamphetamine in rats (Monroe *et al.*, 1994).

At the present time, we have no chemical markers to assist in the diagnosis of psychiatric disorders. Therefore, we propose that levels of BUTN in urine should be determined in large-scale primary screenings of psychiatric patients in an attempt to establish whether BUTN is a useful marker for the diagnosis of psychiatric disorders.

Neuroblastoma

Analysis of urine for the diagnosis of neuroblastoma in infants has been successfully performed using one particular assay that has been applied under the aegis of the Japanese Ministry of Social Welfare. The assay quantifies HVA and VMA in urine specimens (Sawada *et al.*, 1991). Increases in levels of these metabolites of DA and NE, respectively, indicate the presence of neuroblastoma which is a highly malignant tumor of the sympathetic nervous system. Therefore, widespread screening for HVA and VMA in infant urine has been performed in Japan to facilitate an early diagnosis. If a large number of similar metabolites, as well as a whole range of other precursors and unrelated compounds, could be accurately quantitated at once, and if such compounds included those known to be indicative of specific tumors, more information of diagnostic value could rapidly be accumulated, as well as further details about the biochemistry of these devastating disorders.

Table 20B.
Compounds identified in urine from patients with neuroblastoma and in control specimens

Compounds	Neuroblastoma (n=23)	Control (N = 10)
TYR	17.0±17.6	64.3±11.8
3-O-MD	0.7±0.3	2.0±1.2
DA	3.4±2.2	0.5±0.9
DOPAC	10.5±13.0	2.0±0.7
3-MT	1.9±2.4	8.8±0.2
HVA	28.5±20.0	9.4±3.8
MHPG	1.0±1.6	0.2±0.5
MN	0.7±0.8	0.1±0.1
VMA	61.5±66.5	8.9±0.5
VA	2.0±0.6	0.2±0.2
HPAC-4	56.7±43.1	4.5±6.3
TRP	10.0±11.2	50.3±9.8
5-HTP		0.1±0.1
5-HT	0.1±0.1	0.7±0.2
5-HIAA	2.8±2.4	10.0±2.9
KYN	0.3±0.3	2.2±1.4

(Mean±SD; μg/mg creatinine)

Table 20B provides a summary of the compounds detected in urine from both patients with neuroblastoma and controls. A large amount of data on urine chemistry, which would normally require many assays and a lot of time, can be rapidly obtained by HPLC with multi-electrode ECD. Among catecholamines, elevated levels of HVA, VMA and HPAC-4 were seen in urine from patients with neuroblastoma. A significant elevation in levels of DOPAC and MHPG were also seen. These levels provide evidence for the operation of the following pathways in neuroblastoma: TYR → DA → DOPAC → VMA, DA → MHPG → VMA → VA and TYR → HPAC-4. No evidence for either the L-DOPA → HMPPA → VLA or NE → NMN → VMA pathways could be found. Furthermore, the conversion of DA → 3-MT → HVA and E → MN → VMA appeared to be low. No conversion of N-ACT–DA into VMA could be found, even though the latter has been reported to occur at a high level in neuroblastoma. Indolealkylamines, by contrast, appeared to be present at lower than control levels.

The pathways TRP → 5-HTP → 5-HT → 5-HIAA and TRP → KYN were apparently not operative in neuroblastoma. In urine of neuroblastoma patients, levels of detected compounds were generally higher in the case of catecholamines and lower in the case of indolealkylamines when compared to normal control urine specimens.

The biosynthesis of biogenic monoamines is under the control of complicated enzymatic systems, and metabolic pathways seem to differ between tumors and/or even within the same tumor as the rate of proliferation or extent of differentiation of the tumor changes.

CONCLUSION

The present review emphasizes the importance of utilizing different approaches for a complete understanding of the biogenic amines in the central nervous system. While monoamine levels are poor sources of information on the actual activity of the monoaminergic system, the monoamines and their metabolites do provide information about monoaminergic activity in the central nervous system. Moreover, the levels of biogenic monoamines have been shown to vary with age, physiological condition and environmental changes.

The results summarized here show that conventional monoamines and their metabolites have been conserved throughout evolution, from the earliest animals with a nervous system to mammals. The ubiquity of monoamines in living animals suggests that these neurochemicals have a very long and intricate phylogenetic history.

Let us consider the phylogenetic development of monoamine systems in the central nervous system, which control many physiological phenomena and act to maintain homeostasis. In unicellular organisms, the cell itself responds to environmental stimuli. In multicellular organisms, receptor cells (sensory cells) and effector cells (muscle cells) are differentiated and nerve cells, acting as interneurons, connect these cells via the action on transmitters with ascending and descending function. Biogenic monoamines are always associated with these interneurons. The nervous system provides rapid communication allowing, among other things, systematic contraction of muscles during locomotion. Complicated chains of interconnected neurons are necessary for the transmission of transient impulses, together with the highly localized production of chemicals. The classic monoamines, 5-HT and DA, seem to be the first that appeared in the nervous system.

In the nervous system, neurotransmitters act as chemical mediators of intercellular communication, activating specific receptors and second messengers in post-synaptic cells. This specialized role may have evolved from more primitive functions in lower organisms where these substances were used as both intra- and intercellular signaling devices.

Monoamines are synthesized and act in primitive unmyelinated neurons. They are small, chemically stable molecules. By contrast, neuroactive amino acids, such as γ-amino butyric acid, act in myelinated neurons. Acetylcholine acts in both types of neurons. Monoamines are also known to coexist with neuroactive peptides in some neurons (Takayanagi and Takeda, 1988).

In diploblastic animals, which are sedentary or slow-moving and basically symmetrical, the nervous system retains its peripheral position just beneath the surface epithelium, where it forms a diffuse network that covers the entire body. In coelenterates, for example, the diffuse nervous system is dominant for movement. When the body developed bilateral symmetry, a central nervous system developed. From the viewpoint of phylogeny, the central nervous system evolved in two directions. In the more active triploblasts, it tended to move towards a more protected internal position and become aggregated to from a central nervous system. In deuterostomes, the central nervous system formed an invagination of neuroectoderm, the neural tube, which is different from the ladder-like nervous system in protostomes.

With the development of organ and complicated body systems, further compounds became necessary for the transmission of more complicated information.

In the nervous system, sensory afferent nerves and efferent motor neurons appeared. So, as the next step, neurosecretory cells appeared in the nervous system as a result of transformation of nerve cells. These cells included many neuropeptide-producing cells and formed systems through neurohemal areas such as the pars cerebrovascular complex in nereis, the vena cave neurosecretory system in cephalopods, the X organ-sinus gland system in Crustacea, the pars intercerebrailis-corpus cardiacum in insects, all in the invertebrates, and the hypothalamo-neurohypophysial system and hypothalamo-median eminence system in vertebrates. In the less highly organized invertebrates, neurotransmitters and neurosecretory substances act as coordinators of the body system. The next evolutionary step was the appearance of epithelial endocrine organs, such as the infracerebral gland in Annelida, the dorsal bodies and optic gland in Mollusca, the Y-organ and androgenic glands in Crustacea, and the corpora allata and prothoracic glands in insects, among invertebrates. In vertebrates, these organs include the classic endocrine glands, such as the pituitary, the parathyroid, the thyroid, the ultimobranchial bodies, the adrenal medulla, and the pancreatic islets of Langerhans.

As briefly mentioned above, the most primitive and important neuroactive substances in physiology are neurotransmitters such as 5-HT and DA. Therefore, these compounds can be found in all phyla of the animal kingdom.

Some details of the evolution of monoamines in the central nervous system of invertebrates can be postulated. Precursor amino acids, such as tyrosine for catecholamines and tryptophan for indolealkylamines, are detectable in the most primitive organisms. In Porifera, the various compounds occur not systematically but apparently randomly. 5-HT and DA are consistently present in the nervous system. These biogenic monoamines are even present in Protozoa. The pathway from L-DOPA to 3-O-MD is found in nearly all nervous systems and appears first in early embryogenesis, as for example, in *Xenopus laevis* and *Gallus gallus domesticus*. Therefore, this pathway seems to have made its first appearance early in evolution. Soon afterwards, the TRP → 5-HT and DA → DOPAC pathways seem to have evolved. With respect to the timing of their initial appearance, the TRP → 5-HT pathway developed remarkably early and became distributed at a rapid rate as found in *Gallus gallus domesticus*. Thus, from the ontogenic viewpoint, it is necessary to estimate the evolution process of the neurochemicals in the nervous system. The next major putative monoamines to evolve are OCT and KYN. With the development of the complexity of the body system, biosynthetic pathways tended to advance, with the appearance of TYR → TYRA → OCT or HPAC-4, DA → NE → E and TRP → KYN pathways. In particular, NE and E were clearly detected only in Arthropoda and Mollusca in the Protostomia. NE appears to play, at most, a very minor role as a transmitter in invertebrates. Relatively high levels of NE have been demonstrated in echinoderms and chordates. This could be significant, as these phyla belong to the Deuterostomia which differ from other invertebrates that belong to the Protostomia. Furthermore, in Deutrostomia, the DA → 3-MT → HVA and 5-HT → MEL pathways developed. From these basic pathways, further more sophisticate pathways slowly developed.

In invertebrates, monoamines are known to act as hormones or neuromodulators. In addition, the timing of appearance of monoamines suggests that early embryos use these substances as intracellular signals to regulate cell proliferation and the morphogenetic movements of cells during cleavage and gastrulation (Lauder, 1993), which is important for closure of the neural tube in amphibians and avians.

We often ingest monoamines in drinks such as oolong tea. Although the fate of orally ingested monoamines still remains unclear, they may have a positive effect on human health. Some monoamines in herbs may even act as drugs.

Biogenic monoamines, in particular, DA and 5-HT, are distributed throughout the animal kingdom. These compounds seem to be indispensable for animal life. They are not only the most primitive substances in terms of phylogeny but also the most basic of the neuroactive substances. Other neuroactive substances such as the neuropeptides seem to have developed with the progress of evolution. Monoamines transfer information from a neuron across a synapse, either exciting the post-synaptic cell into action or inhibiting its activity. Acting as neuromodulators they are also released from neurons to amplify or dampen the activity of other neurons. Some biogenic monoamines also serve as neurohormones in the hemolymph of invertebrates and in the blood of vertebrates. Altered circulating levels of monoamines and metabolites may be useful markers of abnormal metabolism associated with diseases (e.g., autism). There are high levels of metabolites, such as VMA and HVA, in urine from patients with neuroblastoma. 5-HT is degraded to BUTN in patients with some psychiatric disorders. Loss of DA in the striatum induces Parkinson's disease, and so on.

We have reviewed the evolutionary development of the neurochemicals in the central nervous system phylogenetically, from protozoa to humans. The development of the nervous system from the viewpoint of phylogeny leads not only to an enhanced understanding of neurogenesis, but also to a new perspective on the present development state of the nervous system, including abnormalities such as tumors and psychiatric disorders.

REFERENCES

Algeri, S., Caroleo, A., Ferretti, P., Gallone, C., Pallandini, G. and Venrurini, G. (1983). Effects of dopaminergic agents on monoamine levels and motor behaviour in plannaria. *Com. Biochem. Physiol.* **74c**, 27-29.

Antcil, M., de Waele, J.-P., Miron, M.-J. and Pani, A.K. (1990). Monoamines in the nervous system of the tube-worm Chaetopterus variopedatus (Polychaeta): Biochemical detection and serotonin immunoreactivity. Cell Tissue Res. **259**, 81-92.

Anwyl, R. (1990). Neurophysiological actions of 5-hydroxytryptamine in the vertebrate nervous system. Prog Neurobiol. **35**, 4521-468.

Baguna, J. and Ballester, R (1978). The nervous system structure and the relationship between central nervous system structure and the accelomate organization. J. Morph. **155**, 237-252.

Barreteau H., Perriere, C., Brousse-Gaury, P., Gayraly, P., Jacquot. C. and Goudey-Perriere, A. (1991). Indolamines in the cockroach *Blaberus cranifer burm*. Nervous system-1. Fed and crowded young females. Comp. Biochem. Physiol. **99C**, 567-571.

Brown, C. S. and Nestler, C. (1985). Catecholamines and indolealkylamines. In: Comprehensive Insect Physiology, Biochemistry & Pharmacology, Vol. Pharmacology. Kerkut, G.A. and Gilbert, L.I. (Eds.). Pergamon Press, Oxford, pp. 435-497.

Carlberg, M. (1990). 3,4--Dihydroxyphenylethylamine, L-3, 4- dihydroxyphenylalanine and 3,4,5-trihydroxyphenylalanine: Oxidation and binding to membranes. A comparative study of a neurotransmitter, a precursor and a neurotransmitter candidate in primitive nervous systems. J. Neural. Transm. **81**, 111-119.

Carlberg. M. and Rosengren, E. (1985). Biochemical basis for adrenergic neurotransmissions in coelenterates. Toxicon **15**, 177-196.

Carpenter, D., Breese, G., Schanberg, S. and Kopin. I. (1971). Serotonin and dopamine: Distribution and accumulation in *Aplysia* nervous and non- nervous tissues. Int. J. Neurosci. **2**, 49-56.

Carstroaded, F.A., Renaud F.L. and Phillips, D.M. (1988). Biogenic amines stimulates cilia regeneration *Tetrahymena thermophila*. J. Protozool. **35**, 260-264.

Castano, P. and Rossi, S. (1978). Cytochemical, ultrastructural and fluorescence study of the nervous net of *Hydra* sp. J. Submicro. Cytol. **10**, 381-395.

Chung, J.M., Spencer, A.N. and Gahm, K.H. (1989). Dopamine in tissues of the hydrozoan jellyfish, *Polyorchis pencillatus* as revealed by HPLC and GC/MS. J. Comp. Physiol. B. **159**, 173-181.

Cottreall, A.C., McLeod, M.F. and McLeod, W. (1977). A bufotenine-like substance in the urine of schizophrenics. Am. J. Psychiatry **134**, 322-323.

Croll, R. P (1988). Distribution of monoamines within the central nervous system of the juvenile pulmonate snail, *Achatina fulica*. Brain Res. **460**, 29-49.

de Waele, J.-P., Anctil, G. and Carlberg, M. (1987). Biogenic catecholamines in the cnidercarian *Renilla kollikeri*: Radioenzymatic and chromatographic detection. Can. J. Zool. **65**, 2458-2465.

Elofsson, R., Laxmyr, L., Rosengren, E. and Hansson, C. (1982). Identification and quantitative measurements of biogenic amines and DOPA in the central nervous system and haemolymph of the crayfish, Pacifastacus leniusculus. Comp. Biochem. Physiol. **71C**, 195-201.

Eriksson, K., Reuter, M. and Timoshkin, O. (1993). Analysis of catecholamines in tubellarians of lake baikal. Comp. Biochem. Physiol. **105C**, 509-511.

Erikson, K. and Panula P. (1994). Gamma-aminobutyric acid in the nervous system of planarian. J. Comp. Neurol. **345**, 528-536.

Evans, P.D. (1980). Biogenic amines in the insect nervous system. Adv. Insect Physiol. **15**, 313-473.

Evans, P.D., Davenport A.P., Wlias, M.S., Morton, D.B. and Trimmer, B.A. (1985). Assay for biogenic amines in insect nervous tissue. In: Neurochemical Techniques in Insect Research. Breer, H. and Miller, T.A. (Eds.). Springer-Verlag, New York, pp. 25-46.

Falck, B., Hillarp, N.A., Thieme, G. and Torp, A. (1962). Fluorescence of catecholamines and related compounds with formaldehyde. J. Histochem. Cytochem. **10**, 348-354.

Farnham, P.J., Novak, R.A. and McAdoo, D.J. (1978). A re-examination of the distribution of octopamine and phenylethanolamine in *Aplysia* nervous system. J.Neurochem. **30**, 1173-1176.

Fingerman, M. and Kulkarni, G.K. (1993). Quantitative measurement by reverse phase high performance liquid chromatography of norepinephrine in the central nervous system of the red swamp crayfish, *Procambarus clarkii* and physiologically and pharmacologically induced alterations. Comp. Biochem. Physiol. **104C**, 117-123.

Fingerman, M. and Nagabhushanam, R. (1994). Biogenic amines in crustaceans: Identification, localization, and roles. J. Crustacean Biol. **14**, 413-437.

Fischer, E., Fernandez Lagravere, T.A., Vazquez, A.J. and Di Stefano, A.O. (1961). Bufotenine-like substance in the urine of schizophrenics. J. Nervous Metal Disease **133**, 441-446.

Fritsch, H.A.R., van Noorden, S. and Pearse, A.D.E. (1982). Gastro-intestinal and neurohormonal peptides in the alimentary tract and cerebral complex of *Ciona intestinalis* (Ascidiaceae). Cell Tissue Res. **223**, 369-402.

Fujii, K. and Takeda, N. (1988). Phylogenic detection of serotonin immunoreactive cells in the central nervous system of invertebrates and membrane permeability. Bull. Soc. Zool. Fr. **102**, 327-328.

Fuzeau-Braessch, S. (1977). Aspect neurophysiologique de la differentiation phasaire chez le criquet migrateur: Amines biogenis et permeabilite membranaire chez Locusta migratoria. Bull. Soc. Zool. Fr. **102**, 327-328.

Fuzeau-Braesch, S. and David, J.C. (1980). Taux d'octopamine, mutation et differentitation phasaire chez Locusta migratoria. C. R. Soc. Biol. **174**, 6-9.

Geng, C., Sparks, T., Skomp, J.R. and Gajewski, R.P. (1993). Biogenic amines in the brain of Manduca sexta during laveal-pupal metamorphosis. Comp. Biochem. Physiol. **106C**, 275-284.

Gilchrist, L.S., Klukas, K.A., Jellies, J., Rapus, J., Eckert, M. and Mesce, K.A. (1995). Distribution and developmental expression of octopamine-immunoreactive neurons in the central nervous system of the leach. J. Comp. Neurol. **353**, 451-463.

Goldman, M.E., Gunderson, R., Erikson, C.K. and Thompson, G.A. (1981). High performance liquid chromatography analysis of catecholamines in growing Tetrahymena. Biochem. Biophys. Acta **676**, 221-225.

Goldstein, M., Anagnoste, B., Freedman, L. S., Roffman, M., Ebstein, R.P., Prak, D.H., Fuxe, K. and Hokfelt, T. (1973). Characterization, localization and regulation of catecholamine synthesizing enzymes. In: Frontiers in Catecholamine Research, Usdin, E. and Snyder, S. (Eds.). Pergamon Press, New York, pp. 69-78.

Gonzalez, A., Marin, O., Tuinhof, R. and Smeets, W.J.A.J. (1994). Developmental aspects of catecholamines systems in the brain of anuran amphibians. In: Phylogeny and Development of Catecholamine Systems in the CNS of Vertebrates, Smeets, W.J.A.J. and Reiner, A. (Eds.). Cambridge University Press, Cambridge, pp. 343-360.

Goudey-Perrire, F., Barreteau, H., Perriere, C, Gayral, P., Jacqout C. and Brousse-Gaury, P. (1991). Biogenic amine levels in the cockroach *Blaberus cranifer Burm.* nervous system. Comp. Biochem. Physio. **100C**, 451-455.

Grimmelikhuijzen, C.J.P. and Graff, D. (1991). Isolation of Glu-Gly-Arg-phe-NH2 (antho-RFamide), a neuropeptide from sea anemones. Proc. Natl. Acad. Sci. **83**, 9817-9821.

Groome, J.R., Clark, M.C. and Lent, C.M. (1993). The behavioral state of satiation in the leech is regulated by body distention and mimicked by serotonin deletion. J. Exp. Biol. **182**, 265-270.

Gustafsson, M.K.S. and Eriksson, K. (1991). Localization and identification of catecholamines in the nervous system of *Diphyllobothrium dendritium* (Cestods). Parasitol. Res. **77**, 498-502.

Habermehl, G.G. (Ed.). (1981). Venomous Animals and their Toxins. Springer, Berlin.

Hamburger, V. and Hamilton, H.L. (1951). A series of normal stages in the development of chick embryo. J. Morph. **88**, 49-92.

Hancock, A. (1868). On the anatomy and physiology of the Tunicata. J. Lin Soc. **9**, 309-346.

Hiripi, L. and S-Rozsa, K.C. (1973). Fluorometric determination of 5-hydroxytryptamine and catecholamines in the central nervous system and heart of Locust amigratorides. J. Insect Physiol. **19**, 1481-1485.

Hudman, D. and McFarlene, I.D. (1995). The role of L-dopa in the nervous system of sea anemones: A putative inhibitory transmitter in tentacles. J. Exp. Biol. **198**, 1045-1050.

Iwata, H. and Kariya, K. (1973). Changes in catecholamine content of *Tetrahymena pyriformis* during growth. Jap. J. Pharmacol. **23**, 751-752.

Janakidevi, K., Deway, V.C. and Kidder, G.W. (1966). The biosynthesis of catecholamines in the two genera of protozoa. J. Biol. Chem. **241**, 2576-2578.

Kabotyanski, E.A., Nezlin, L.P. and Sakharov, D.A. (1991). Serotonin neurons in the planarian pharynx. In: Simpler Nervous Systems. Sakharov, D.A. and Winlow, W. (Eds.). Manchester University Press, Manchester, pp. 138-152.

Kadota, T. (1991). Distribution of 5-HT (serotonin) immunoreactivity burgeri. Cell Tissue. Res. **266**, 107-116.

Karkkainen, J., Raisanen, M., Naukkarinen, H., Spoov, J. and Rimon, R. (1988). Urinary excretion of free bufotenin by psychiatric patients. Biol. Psychiatry. **24**, 441-446.

Kitahama, K., Nagatsu, I. and Pearson, J. (1994). Catecholamine systems in mammalian midbrain and hindbrain: theme and variations. In: Phylogeny and Development of Catecholamine Systems in the CNS of Vertebrates. Smeets, W.J.A.J. and Riner, A. (Eds.). Cambridge University Press, Cambridge, pp. 343-360.

Klemm, N. (1985). The distribution of biogenic monoamines in invertebrates. In: Neurobiology. Gilles, R. and Balthazart, J. (Eds.). Springer-Verlag, Heidelberg. pp. 280-296.

Klemm, N. and Axelesson, S. (1973). Detection of dopamine and noradrenalin in nervous structures of the insect brain. Brain Res. **26**, 459-464.

Kolberg, K.J.S. and Martin, V.J. (1988). Morphological, cytological and neuropharmacological evidence for the presence of catecholamines in hydrozoan planulae. Development **103**, 249-258.

Kriegstein, A.R. (1977). Development of the nervous system of *Aplysia californica*. Proc. Natl. Acad. Sci. **74**, 375-378.

Krueger, R.R., Kramer, K.J., Hopkins, T.L. and Speirs, R.D. (1990). *N-β*-Alanyldopamine and N-acetyldopamine occurrence and synthesis in the central nervous system of *Manduca sexta* (L). Insect Biochem. **20**, 605-610.

Kulkarni, G.K. and Figerman, M. (1992). Quantitative analysis by reversed phase high performance liquid chromatography of 5-hydroxytryptamine in the central nervous system of the red swamp crayfish, *Procambarus clarkii*. Biol. Bull. **182**, 341-347.

Laxmyr, L. (1984). Biogenic amines and DOPA in the central nervous system of decapod crustaceans. Comp. Biochem. Physiol. **77C**, 139-143.

Lauder, J.M. (1993). Neurotransmitters as growth regulatory signals: roles of receptors and second messengers. TINS **16**, 233-240.

Lenicque, P.M., Toneby, M.I and Doumenc, D. (1977). Demonstration of biogenic amines and localization of monoamine oxidases in the sea anemone *Metridium seile* (Linne). Comp. Biochem. Physiol. **56C**, 31-34.

Lent, C.M., Zundel, D., Freedman, E. and Groome, J.R. (1991). Serotonin in the leech central nervous system: anatomical correlates and behavioral effects. J. Comp. Physiol. A **168**, 191-200.

Lentz, T.L. (1966). Histochemical localization of neurohumors in a sponge. J. Exp. Zool. **162**, 171-180.

Linn, Jr. C.E., Poole, K.R. and Roelofs, W.L. (1994). Studies on biogenic amines and their metabolites in nervous tissue and hemolymph of adult male cabbage looper moths-1. Quantitation of photoperiod changes. Comp. Biochem. Physiol. **108C**, 73-85.

Longley, R.D. and Longley, A.J. (1989). Serotonin immunoreactivity of neurons in the gastropod *Aplysis californica*. J. Neurobiol. **17**, 339-358.

Marcum, B.A., Fujisawa, T. and Sugiyama, T. (1980). A mutant hydra strains (SF-1) containing temperature sensitive interstitial cells. In: Development and Cellular Biology of Coelenterates. Tradent, P. and Tradent, R. (Eds.). Elsevier, Amsterdam, pp. 429-434.

Martelly, I. and Franquinet, R. (1984). Planarian regeneration as a model for cellular activation studies. TIBS **11** 468-471.

Matson, W.R., Langlais, P., Volicer, L., Gamache, P.H., Bird, E. and Mark, K.A. (1983). *n* -Electrode three-dimensional liquid chromatography with electrochemical detection for determination of neurotransmitters. Clin Chem. **30**, 1447-1488.

McCaman, M.W., Weinreich, D. and McCaman, R.E. (1973). The determination of picomol levels of 5-hydroxytryptamine and dopamine in *Aplysia, Tritonia* and leech nervous tissues. Brain Res. **5**, 129-137.

McGoowan, K., Kane, A., Asarkov, N., Wicks, J., Guerina, V., Kellum, J., Baron, S., Gintzler, A. and Dowitz, M. (1983.) Entamoeba histolytica causes intestinal secretion: role of serotonin. Science **221**, 762-764.

Meek, J. (1994). Catecholamines in the brain of osteichthyes (bony fishes). In: Phylogeny and Development of Catecholamine Systems in the CNS of Vertebrates. Smeets, W.J.A.J. and Reiner, A. (Eds.). Cambridge University Press, Cambridge, pp. 49-76.

Meyer, W. and Jehnen, R. (1980). The distribution of monoamine oxidase and biogenic monoamines in the central nervous system of spiders (Archinda: *Araneida*). J. Morph. **164**, 69-81.

Monroe, P.J., Smith, D.L,. Williams, G.M. and Smith, D.J. (1994). Bufotenine has a para-chloroamphetamine-like action on the storage and release of serotonin in rat spinal cord synaptosomes. Biogenic Amines. **10**, 273-284.

Morita, M., Hall, F., Best, J.B. and Gern, W. (1987). Photoperiodic modulation of cephalic melatonin in planarians. J. Exp. Zool. **241**, 383-388.

Morton, D.B. and Evans, P.D. (1983). Octopamine distribution in solitary and gregarious forms of the locust, *Schistocerca americana gregaria*. Insect Biochem. **13**, 177-183.

Nieukoop, P.D. and Faber, J. (Eds.). (1956). Normal Table of Xenopus laevis (Daudin). North-Holland Publishing Company Co., Amsterdam.

Nomura, T. (1988). Phenolic compounds of the mulberry tree and related plants. In: Progress in the Chemistry of Organic Natural Products. **53**. Herz, W., Grisebach, H.,. Kirby, G.W. and Tamm, C.H. (Eds.). Springer-Verlag, New York, pp. 188-201.

Orchard, I. (1990). Tyrosine hydroxylase-like immunoreactivity in previously described catecholamine-containing neurons of *Rhodinus prolixus*. J. Insect Physiol. **36**, 593-600.

Orchard, I., Martin, R.J., Sloley, B.D. and Downer, R.G.H. (1986). The association of 5-hydroxytryptamine, octopamine, and dopamine with the intrinsic (glandular) lobe of the corpus cardiacum of *Locusta migratoria*. Can J. Zool. **64**, 271-274.

Orchard, I., Ramirez, J.-M. and Lanfe, A.B. (1993). A multifunctional role of octopamine in locust flight. Ann. Rev. Entomol. **38**, 227-249.

Osbone, N.N., Neuhoff, V., Ewers, E., and Robertson, H A. (1979). Putative neurotransmitters in the cerebral ganglia of the tunicate *Ciona intestinalis*. Comp. Biochem. Physiol. **63C**, 209-212.

Owen, M.D., Pfaff, L. and Sloley, B.D. (1987). The absence of diet changes in the concentration of dopamine. 5-hydroxytryptamine and their metabolism in the cerebral ganglion of the cockroach, *Periplaneta americana*. Insect Biochem. **17**, 723-729.

Pani, A.K. and Anctil, M., (1994a). Evidence for biosynthesis and catabolism of monoamines in the sea pansy *Renilla koellikeri* (Cnidaria). Neurochem. Int. **25**, 465-474.

Pani, A.K. and Anctil. M. (1994b). Quantitative survey of biogenic monoamines, their precursors and metabolites in the coelenterate *Renilla koellileri*. Biogenic Amines **10**, 161-180.

Parent, A., Poitras, D. and Dube, L. (1984). Comparative anatomy of central monoaminergic systems. In: Handbook of Chemical Neuroanatomy.· 2. Classical Transmitters in the CNS, Part 1. Bjorklund, A. and Hokfelt, T. (Eds.). Elsevier, Amsterdam, pp. 409-439.

Pierr, J., Rio, J.P., Mahouche, M. and Reperant, J. (1994). Catecholamine systems in the brain of cyclostomes, the lamprey, *Lampetra fluviatilis*. In: Phylogeny and Development of Catecholamine Systems in the CNS of Vertebrates.·Smeets, W.J.A.J. and Reiner, A. (Eds.). Cambridge University Press, Cambridge, pp. 7-19.

Pollack, E.D. and Harold, D. B. (Eds.). (1988). Developmental Neurobiology of the frog, Alan R. Liss, Inc., New York.

Puiroux, J., Moreau, R., and Gourdoux, L. (1990). Variation of biogenic amine levels in the brain of Pieris brassicae pupae during non-diapausing and diapausing development. Arch. Insect Biochem. Physiol. **14**, 57-69.

Puiroux, J., Moreau, R. and Gourdox, L. (1992). Simultaneous estimation of dopamine, serotonin and related compounds in the brain of *Pieris brassicase* pupae: Degradation pathways of biogenic amines in insect nervous tissue. Biogenic Amines **8**, 391-400.

Pulles, L. and Medina, L. (1994). Development of neurons expressing tyrosine hydroxylase and dopamine in the chicken brain: A comparative segmental analysis. In: Pylogeny and Development of Catecholamine Systems in the CNS of Vertebrates. Smeets, W.J.A.J. and Reiner, A. (Eds.). Cambridge University Press, Cambridge, pp. 381-404.

Pyza, E., Golembiowsla, K. and Antkiewicz-Michaluk, L. (1991). Serotonin, dopamine, norepinephrine and their metabolites: levels in the brain of the house cricket (Acheta domesticus L.) during a 24 hour period and after administration of quipazine, a 5-HT2 receptor agonist. Comp. Biochem. Physiol. **100C**, 367-371.

Renaud, F.L., Chieesa, R., de Jesus, J.M., Lopez, A., Mirands, J. and Tomassini N. (1991). Hormones and signal transduction in protozoa. Comp. Biochem. Physiol. **100A**, 41-45.

Reuter, M. and Eriksson, K. (1991). Catecholamines demonstrated by glyoxylic-acid-induced fluorescence and HPLC in some microtubellarians. Hydrobiologia **227**, 209-219.

Robertson, H. A. (1976). Octopamine, dopamine and noradrenaline content of the brain of the locust, *Schistocerca gregaria*. Experientia **32**, 552-553.

Saavedra, J.M., Brownstein, M.J., Carpenter, D.O. and Axelrod, J. (1974). Octopamine: Presence in single neurons of *Aplysia* suggests neurotransmitter function. Science **185**, 364-365.

Sakharov, D.A. and Winlow, W. (Eds.). (1991). Simpler Nervous Systems. Manchester University Press, Manchester.

Salimova, N.B., Sakharov, D.A., Milosevic, I., Turpaev, T.M. and Raki, L. (1987). Monoamine containing neurons in the *Aplysia* brain. Brain Res. **400**, 285-299.

Sawada, T., Matsumae, T., Kawakatsu, H., Sugimoto, T., Kuroda, H., Hosoi, H., Michihata, T., Saida, T. and Hino, T. (1991). Long term effects of mass screening for neuroblastoma in infancy. Am. J. Pediatric Hematol. Oncol. **13**, 3-7.

Seyfarth, E.-A., Hammer, K., Sporhase-Eichmann, U. Horner and Vullings, H.G.B. (1993). Octopamine immunoreactive neurons in the fused central nervous system of spiders. Brain Res. **611**, 197-206.

Scharrer B. (1970). Current problems in invertebrate neurosecretion. In: Aspects of Neuroendocrinology. Bargmann, W. and Scharrer, B. (Eds.). Springer-Verlag, Berlin, pp. 1-23.

Schlawny, A., Hamann, T., Muller, M.A. and Pfannenstiel, H-D. (1991). The catecholaminergic system of an annelid (Ophrptrocha puerilis, Polychaeta). Cell Tissue Res. **265**, 175-184.

Shimizu, T., Mihara, M. and Takeda, N. (1991). High performance liquid chromatography of biogenic amines in the corpus cardiacum of the American cockroach, *Peliplaneta americana*. J. Chromatogr. **539**, 193-197.

Shimizu, T. and Takeda, N. (1991). Biogenic monoamine levels in the haemolymph of the cabbage armyworm, *Mamestra brassicae*, following injection of octopaminergic insecticides. Z. Naturforsch. **46C**, 127-132.

Shimizu, T., Yazawa, M. and Takeda, N. (1992). Aromatic amino acids in the leaves of Morus alba and their possible medicinal value. Sericologia **32**, 633-636

Shimizu, T., Yazawa, M. and Takeda, N., (1993). Aromatic amino acids in the venom of the Braconid parasitoid *Apanteles kariyai*. Z. Naturforsch. **48C**, 108-109.

Shimizu, T. and Takeda, N. (1994). Aromatic amino acid and amine levels in the haemolymph of parasitized and unparasitized larvae of Mythimna separata. Z. Naturforsch. **49C**, 693-695.

Sloley, B.D. and Juorio, A.V. (1990). Biogenic amines in the nervous and other tissues of several species of starfish: presence of relatively high levels of tryptamine and low levels of 5-hydroxytryptamine. Biogenic Amines 7, 341-349.

Smeets, W.J.A.J. (1994). Catecholamine systems in the CNS of reptiles: Structure and functional correlations. In: Phylogeny and Development of Catecholamine Systems in The CNS of Vertebrates. Smeets, W.J.A.J. Reiner, A. (Eds.). Cambridge University Press, Cambridge, pp. 103-133.

Smeets, W.J.A.J. and Reiner, A. (Eds.). (1995).·Phylogeny and Development of Catecholamine Systems in the CNS of Vertebrates. Cambridge University Press, Cambridge.

Sparks, T.C. and Geng, C. (1992). Analysis of the biogenic amines in the central nervous system of the tobacco hornworm by high-performance liquid chromatography with 16-sensor electrochemical detection. Anal. Biochem. **205**, 319-325.

Sparks, T.C., Skomp, J.R., Babocock, J. and Geng, C. (1994). Biogenic amines in the two-spotted spider mite. Arch. Insect Biochem. Physiol. **25**, 1-8.

Steinbusch, H.W.M., de Vevte, J. and Schipper, J. (1986). Immunohistochemistry of monoamines in the central nervous system. In: Neurohistochemistry Today, Neurology and Neurobiology. **20**. Panula, P., Paivarinta, H. and Soinila, S. (Eds.). Alan Liss, New York, pp. 75-105.

Takayanagi, H. and Takeda, N. (1988). Co-existence of FMRF-amide, met-enkephalin and serotonin in molluscan neurons. Comp. Biochem. Physiol. **91A**, 613-620.

Takeda, N. (1984). The aggregation pheromone in terrestrial isopods. In: The Biology of Terrestrial Isopods.·Holdich, D.M. and Sutton, S.L. (Eds.). Academic Press, London. pp. 381-404.

Takeda, N. and Mizuno, J. (1989). Land adaptation and neuropeptides in terrestrial isopods. *Montore* Zool. Ital. Monogr. **4**, 249-270.

Takeda, N. (1990). Dynamics of monoamine metabolism in the central nervous system of some terrestrial isopods. In: The Biology of Terrestrial Isopods.·Juhault, P. and Mocquard, J.P. (Eds.). University of Poitier Press, Poitier, France, pp. 195-200.

Takeda, N., Takaoka, H., Shimizu, T., Yazawa, M. and Yagi, S. (1991). Biogenic monoamine levels in the central nervous system and haemolymph of the silkworm, *Bombyx mori*. Comp. Biochem. Physiol. 99C, 588-593.

Takeda, N. and Takaoka, H. (1991). Dynamics of neurotransmitter metabolism in the central nervous system of frogs. Soc. For Neurosci. **17**, 745.

Takeda, N. (1992a). Biogenic monoamine system detected simultaneously in the neural complex of the ascidian, *Ciona intestinalis*. Comp. Biochem. Physiol. **102C**, 489-493.

Takeda, N. (1992b). Biogenic monoamine levels in the central nervous system of the sea hair, Aplysia kurodai. Comp. Biochem. Physiol. **103C**, 511-519.

Takeda, N. (1992c). Development of neurotransmitter, metabolism in the brain of the chick, *Gallus gallus domesticus*. Soc. for Neurosci. **18**, 1472.

Takeda, N. (1993a). Simultaneous analysis of monoamines and their metabolites in urine specimens of patients with neuroblastoma. Comp. Biochem Physiol. **105C**, 373-377.

Takeda, N. (1993b). Serotonin-degradative pathways in the toad brain and human psychiatric disorders. Soc. for Neurosci. **19**, 1171.

Takeda, N. and Sugiyama, K. (1993). Metabolism of biogenic monoamines in the ciliated protozoan, *Tetraphymena pyriformis*. Comp. Biochem Physiol. **106C**, 63-70.

Takeda, N. (1994a). Serotonin-degradative pathways in the toad (*Bufo bufo japonicus*) brain: Clues to the pharmacological analysis of human psychiatric disorders. Comp. Biochem. Physiol. **107C**, 275-281.

Takeda, N. (1194b). Dynamics of serotonin-degradative pathways in the brain of some frogs. Soc. for Neurosci. **20**, 1165.

Takeda, N. (1995a). Effects of serotonin reagents on bufotenine levels in the brain of *Bufo*. Soc. for Neurosci. **21**, 1370.

Takeda, N. (1995b). Biogenic monoamines in the land planarian, *Biparium fuscatum*. Biogenic Amines (in press).

Takeda, N., Ikeda, R., Ohba, K. and Kondo, M. (1995). Bufotenine reconsidered as a diagnostic indicator of psychiatric disorders. Neuro-Report. **6**, (in press).

Takeda, N. (1996a). Biogenic monoamine system in the central nervous system of some annelidan species. Neuro-Report. (submitted).

Takeda, N.(1996b). Biogenic monoamine system in the central nervous system during metamorphosis of frogs. Neuro-Report. (submitted).

Tanaka, S. and Takeda, N. (1996). Comparison of biogenic monoamines in the brain and the corpus cardiacum between albino and normal strains of the migratory locust, *Locusta migratoria*. Comp. Biochem. Physiol. (in press).

Terenina, N. B. (1991). Biogenic amines in helminths. In: Simpler Nervous System. Sakharov, D.A. and Winlow, W. (Eds.). Manchester University Press, Manchester, pp. 153-172.

Thompson, K.S.J., Zeidler, M.P. and Bacon, J.P. (1994). Comparative anatomy of serotonin-like immunoreactive neurons in isopods: Putative homologues in several species. J. Comp. Neurol. **347**, 553-569.

Vaughan, P.F.T. (1988). Amine transmitters and their associated second messenger system. In: Comparative Invertebrate Neurochemistry. Lunt, G.G. and Olsen, R.E. (Eds.). Cornell University Press, New York, pp. 124-174.

Venturini, G., Stocchi, F., Margotta, V., Ruggieri, S., Bravi, D., Bella, P. and Palladini, G. (1989). A pharmacological study of dopaminergic receptors in planaria. Neuropharmacology **28**, 1377-1382.

Walker, R.J. (1982). Current trends in invertebrate neuropharmacology. Verh. Dtsch. Zool. Ges. **23**, 31-59.

Webb, R.A. (1988). Endocrinology of acoelomates. In: Endocrinology of Selected Invertebrate Types. Laufer, H. and Downer, R.G.H. (Eds.). Alan R. Liss Inc., New York. pp. 31-62.

Welsh, J.H. and Loveland, R.E. (1968). 5-Hydroxytryptamine in the ascidian, *Ciona intestinalis L.* Comp. Biochem. Physiol. **27**, 719-722.

Wikgren, M. and Lindroos, P. (1989). Histamine and catecholamines in flatworms. Gen. Comp. Endocrinol. **74**. 295-296.

Winberg, S. and Nilsson, G.E. (1993). Roles of brain monoamine neurotransmitters in agonistic behaviour and stress reactions, with particular reference to fish. Comp. Biochem. Physiol. **106C**, 597-614.

Wyroba, E. (1989). Beta-adrenergic stimulation of phagocytosis in the unicellular eukaryote *Paramecium aurelia*. Cell Biol. Int. Reps. **13**, 667-678.

Progress in HPLC-HPCE, Vol. 6, pp. 213–225
I.N. Acworth *et al.* (Eds)

MULTICOMPONENT QUANTITATIONS OF MONOAMINES AND RELATED COMPOUNDS IN BIOLOGICAL FLUIDS AND CNS TISSUES BY HPLC COUPLED WITH MULTI-ELECTRODE ELECTROCHEMICAL DETECTION

PETER HUETTL and GREG A. GERHARDT*

Rocky Mountain Center for Sensor Technology, Departments of Psychiatry, Pharmacology, and the Neuroscience Training Program*, University of Colorado Health Sciences Center, 4200 East Ninth Avenue, Denver, CO 80262*

INTRODUCTION

Over the last two decades, the technique of high performance liquid chromatography coupled with electrochemical detection (HPLC-EC) has seen dramatic improvements in design and function (Bird *et al.*, 1990; Godefroy *et al.*, 1990; Hall *et al.*, 1989; Matson *et al.*, 1984; Mefford, 1981; Rizzo *et al.*, 1991; Turk and Blank, 1990). These developments have resulted in the preferred use of HPLC-EC methods for analyses of the monoamine neurotransmitters and related compounds in tissue samples, microdialysis samples, cerebrospinal fluid, urine, plasma and saliva (Drebing *et al.*, 1989; Gerhardt *et al.*, 1986; Gerhardt *et al.*, 1989; Godefroy *et al.*, 1990; Hall *et al.*, 1989; Langlais *et al.*, 1991; Mefford, 1981; Ogawa *et al.*, 1992; Ogawa *et al.*, 1990; Suzuki and Mori, 1992; Swartz *et al.*, 1990; Volicer *et al.*, 1985; Yamamoto, 1991). In addition, the recent advances in the coupling of HPLC to arrays of electrochemical detectors, termed Coulochem Electrode Array System (CEAS; ESA Inc.), allows for the quantitation of over 20 monoamine-related neurochemicals from a single biological sample (Bird *et al.*, 1990; Godefroy *et al.*, 1990; Issa *et al.*, 1994a,b; Matson *et al.*, 1984; Ogawa *et al.*, 1990; Rizzo *et al.*, 1991; Swartz *et al.*, 1990). The ability to separate and quantitate numerous inter-related neurochemicals in tissues and biological fluids may be of use in studies of psychiatric or neurodegenerative disorders (Matson *et al.*, 1990), such as schizophrenia (Issa *et al.*, 1994a,b), Parkinson's Disease (LeWitt *et al.*, 1992; Naoi *et al.*, 1993; Ogawa *et al.*, 1992; Tohgi *et al.*, 1991), and Alzheimer's Disease (Volicer *et al.*, 1985).

A significant area of interest for our research group is the neurochemistry of schizophrenia. Our work has focused on the study of plasma and cerebrospinal fluid (CSF) samples from patients diagnosed with schizophrenia that are on and off neuroleptic medication (Adler *et al.,* 1990; Baker *et al.,* 1990; Baker *et al.,* 1988; Freedman *et al.,* 1987; Issa *et al.,* 1994a,b; Waldo *et al.,* 1992; Waldo *et al.,* 1994). The need to analyze a variety of compounds having different elution properties and oxidation potentials from a single sample is made possible by use of gradient and isocratic HPLC separations in conjunction with a series of electrochemical detectors as found in the CEAS. Compounds of interest not only include the catecholamines, indoleamines, and their metabolites, but parent compounds and a variety of amino acids. Recent sample preparation methodologies that we and others have developed allow samples to be processed using a minimum number of steps prior to injecting into the CEAS (Naio *et al.,* 1993). In addition, custom modifications made to the CEAS by our group have served to improve the reproducibility and subsequent analyses of biological samples. The purpose of this chapter is to summarize some improvements in CEAS methods that our research group has developed, and applications of these methods for the measurements of neurochemicals and related compounds in biological fluids and tissue samples from humans and animals.

METHODOLOGY

Sample preparation

Aliquots (100-500μL) of CSF were centrifuged at 14,000 x g for 5 min. at 5°C. The supernatants were transferred to a 10,000MW cut-off micro-filtration tube (Microcon™ 10, Amicon, Inc., Beverly, MA.) and centrifuged at 14,000 x g for 50 min. at 5°C. The filtrates of CSF were diluted 2X with mobile phase and 50μL samples were injected into the CEAS or the isocratic HPLC system (see below). For plasma studies, 250μL of each sample is diluted 2X with mobile phase. In studies of tissue samples, 300-500μL of mobile phase were added to 1-60mg pieces of tissue. Plasma and tissue samples were then sonicated for ~5 sec., (output = 55 - Vibra Cell, Sonics & Materials Inc., Danbury, CT.) and centrifuged at 14,000g for 5 min. at 5°C. Supernatants were transferred to Micropure™ 0.45μm inserts and Microcon™ 10 microconcentrators and centrifuged at 14,000 x g for 45-120min. at 5°C. Filtrates were collected and 50μL samples were injected into the HPLC system. Once processed, all filtered samples were maintained at ~5°C.

Instrumentation and chromatography conditions

The separations and quantitations of neurochemicals in the samples were carried out using a gradient HPLC coupled to sixteen coulometric electrochemical detectors arranged in series set at increasing potentials from 0 mV to +950mV (CEAS; ESA Inc., Chelmsford, MA; see Figure 1). For some of our studies, we have used the detector potential settings, eluents, columns, and chromatography conditions analogous to Naoi and co-workers (Issa *et al.,* 1994a,b; Naio *et al.,* 1993; Stromberg *et al.,* 1991). For our more current studies we have modified the voltages of the detectors (D) 1 to 16 as follows: D1 (0mV), D2 (+20mV) , D4 (+40mV), D5 (+200mV), D6 (+250mV), D7 (+300mV), D8 (+350mV), D9 (+450mV), D10 (+570mV), D11 (+680mV), D12 (+700mV), D13 (+750 mV), D14 (+800mV), D15 (+850mV) and D16 (+950mV).

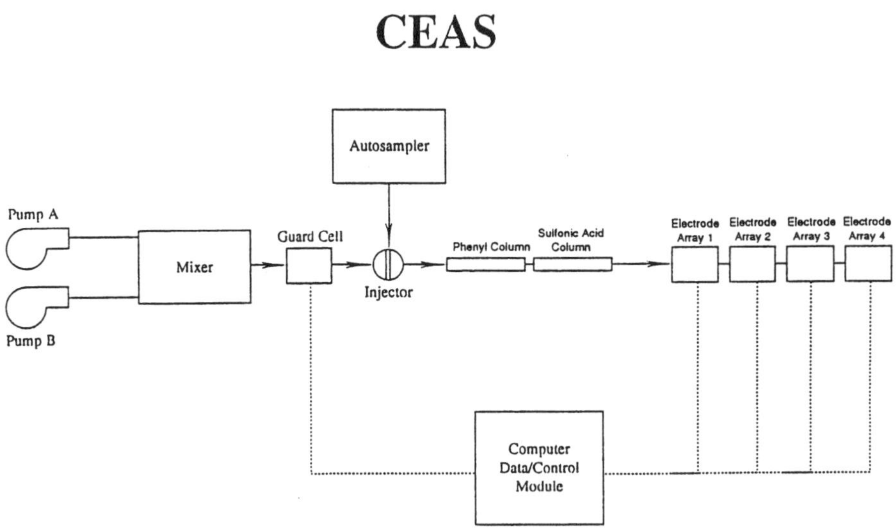

Figure 1. Schematic diagram of the 16-channel CEAS system that is used for our studies. Note the inclusion of the guard cell and the use of both phenyl and sulfonic acid HPLC columns.

The modifications that we have made to the commercially available CEAS are summarized in Figure 1. First, a Model 5020 guard cell (ESA Inc.), which is controlled by an additional 5100A Coulochem (ESA Inc.) system, was placed following the mixer and set at a potential of +150mV above the highest detector setting (+1100mV). This pretreatment and screening of the eluent resulted in decreased baseline noise. Secondly, in contrast to using an ODS column and adding ion pairing agents to the eluent, a sulfonic acid cation exchange column is inserted immediately following a phenyl column. This combination of columns allowed for excellent resolution of the different compounds of interest without the equilibration problems of using pairing agents with an alcohol gradient elution. A phenyl column (4.6mm x 150mm, 5µm particles, BDS. Hypersil Phenyl; Keystone Scientific Inc., Bellefonte, PA) and sulfonic acid column(4.6mm x 30mm, 5µm particles, 100SCX; Keystone) were coupled together and analytes eluted using gradient elution with a pH2.95 phosphate buffer containing 0.5% - 17.5% ethanol (v/v). The use of ethanol and the BDS Phenyl column permitted decreases in retention times thereby improving peak symmetry and reduced tailing as compared to methanol and more traditional C18 columns. The compositions of the buffers were as follows: Buffer A - 0.5% ethanol, 4.26g/liter sodium phosphate (dibasic), 0.275g/liter, nitrilotriacetic acid (trisodium) and the pH was adjusted to 2.95 with perchloric acid; Buffer B - 25.0% ethanol, 4.26g/liter sodium phosphate (dibasic), 0.275g/liter, nitrilotriacetic acid (trisodium) and the pH was adjusted to 2.95 with perchloric acid. The modified gradient profile for our studies is seen in Figure 2. Over 20 neurochemicals and related compounds can be reliably quantitated using this separation protocol. The retention times and the dominant channels/voltages that were used to quantitate the compounds of interest are seen in Table 1. In addition, a 16 channel chromatogram showing 26 compounds is seen in Figure 3. It should be noted that we have optimized our system for the determination of these compounds in CSF, plasma and tissues. Due to the low levels of dopamine, norepinephrine, epinephrine and serotonin that exist in CSF and plasma, we also used an isocratic HPLC coupled with a dual-coulometric electrochemical detector to quantitate the low levels of these compounds in the CSF and plasma samples (see isocratic system below).

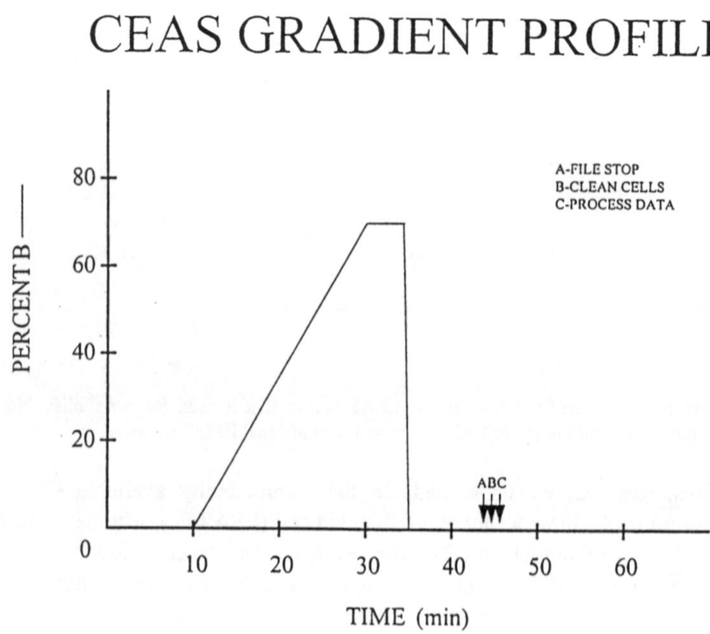

Figure 2. Gradient profile of our mobile phase that is used for separation and quantitation of over 26 compounds.

Isocratic HPLC system

The low level detection of 3,4-dihydroxyphenylacetic acid (DOPAC), dopamine (DA), epinephrine (E), serotonin (5-HT), norepinephrine (NE), metanephrine, 3-methoxytyramine (3-MT) and normetanephrine were performed using an isocratic HPLC system (Beckman, Ins., Fullerton, CA.) coupled to a dual-channel electrochemical array detector (ESA, Inc., 5100A, E_1 +0.35mV and E_2 -0.25mV using a 5011 dual analytical cell.). Separation of compounds of interest was performed using a C18 reverse phase column (4.6mm x 100mm, 3μm particles, ODS Hypersil; Keystone Scientific) with a pH 4.0 citrate-acetate mobile phase containing ~5.5% methanol and ~0.35mM 1-octane-sulfonic acid. We have previously published a more thorough description of this method (Hall *et al.*, 1989).

Table 1.
Compounds analyzed by the CEAS

COMPOUND	ABBREVIATION	RETENTION TIME (Minutes)	DOMINANT CHANNEL
L-3,4-DIHYDROXYPHENYLALANINE	L-DOPA	2.6	5
GUANOSINE	GUA	4.2	14
HOMOVANILLYL ALCOHOL	HVOL	17.8	9
3-HYDROXYANTHRANILIC ACID	3-OHAN	6.8	5
p-HYDROXYBENZOIC ACID	p-HBAC	8.3	14
5-HYDROXYINDOLE-3-ACETIC ACID	5-HIAA	17.1	6
3-HYDROXY-DL-KYNURENINE	3-OHKY	10.7	5
4-HYDROXY-3-METHOXYPHENYLACETIC ACID	HVA	21.0	10
4-HYDROXY-3-METHOXYPHENYLGLYCOL	MHPG	4.4	10
p-HYDROXYPHENYLACETIC ACID	p-HPAC	12.2	11
p-DL-HYDROXYPHENYLLACTIC ACID	p-HPLA	7.7	11
5-HYDROXYTRYPTAMINE	5-HT	21.6	6
5-HYDROXY-DL-TRYPTOPHAN	5-HTP	6.6	6
5-HYDROXYTRYPTOPHOL	5-HTOL	14.5	6
L-KYNURENINE	KYN	22.2	14
MELATONIN	MEL	36.7	11
L-METHIONINE	L-MET	3.4	15
3-METHOXYTYRAMINE	3-MT	23.2	9
TRYPTAMINE	TRPMN	34.6	11
L-TRYPTOPHAN	L-TRP	17.1	11
TRYPTOPHOL	TRPOL	28.8	11
TYRAMINE	TYRAMINE	15.2	11
L-TYROSINE	L-TYR	6.5	11
XANTHINE	XANTHINE	2.8	13
XANTHURENIC ACID	XANA	10.3	10

APPLICATIONS OF THE CEAS

Tissue samples

One of the more standard uses of HPLC-EC methods is the quantitation of monoamines and related species in brain tissue samples from rats, mice, monkeys and man. These measures are used to determine the effects of drugs on CNS systems and to confirm the extent of depletion produced by neurotoxins (Hall *et al.,* 1989; Kirch *et al.,* 1990; Stromberg *et al.,* 1991). Using the CEAS, a much greater understanding of the neurochemistry of CNS tissues can be determined. Table 2 summarizes the numerous compounds that can be detected in the rat striatum.

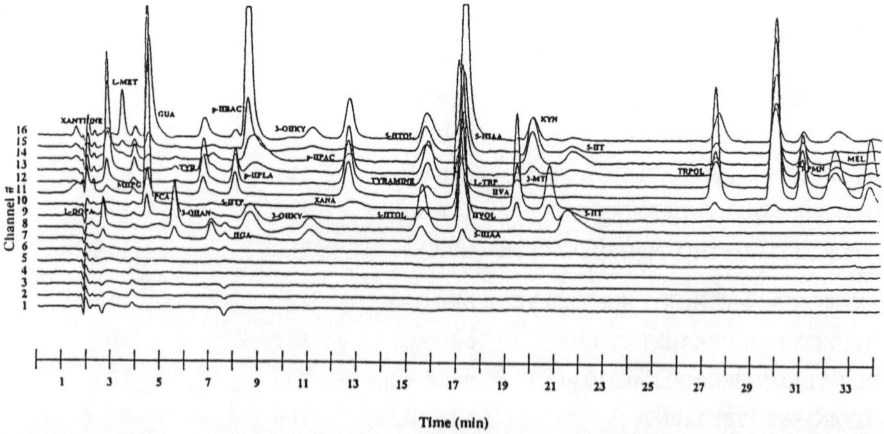

Figure 3. CEAS chromatogram of a 10^{-6} M standard (50 µL injection) containing 26 compounds of interest. Refer to Table 1 for retention times and abbreviations. Please note that an internal standard, PCA (protocatechuic acid), is seen in this chromatogram and doesn't appear as a detectable compound in Table 1.

Recently, we have used the CEAS to investigate the neurochemical characteristics of both rat tissues and transplanted human tissues (Stromberg *et al.*, 1991) for studies relating to the use of transplanted fetal brain tissues for patients with

Table 2.
CEAS analysis of rat striatum

Compound	ng/g wet wt.*	Compound	ng/g wet wt.*
3-OHKY	18±7	HVA	1617±578
3-OHAN	18±3	HVA-4-SULF	1915±739
3-MT	37±1	HVOL	3±1
p-HBAC	13±2	KYN	206±34
p-HPAC	62±26	L-DOPA	22±4
p-HPLA	406±92	L-MET	8785±882
5-HTOL	5±2	MHPG	17±8
5-HIAA	865±254	Metanephrine	7±1
5-HT	1341±131	NE	95±18
Ascorbate	2332024±17983	Normetanephrine	5±1
DA	15365±803	Tryptophol	3±1
DOPAC	2107±168	L-TRP	4706±298
E	3±1	L-TYR	17423±2618
GUA	3572±656	Xanthine	6618±561

* All values n = 6 animals. See Table 1 for abbreviations. Variance is expressed as standard error of the mean (S.E.M.)

Parkinson's disease (Hoffer *et al.,* 1992). Table 3 contains data from rat substantia nigra, rat striatum and fetal human mesencephalic tissues that were transplanted into the striatum of immunosuppressed rats. Using the unique capabilities of the CEAS, we were able to detect numerous neurotransmitters and related compounds. In particular, the human tissues contained very high levels of 5-HT, HVA, DA and the sulfated metabolite of HVA. In contrast, very low levels of DOPAC were measured in the human tissues as compared to the relative high ratios of DA:DOPAC that are seen in rat tissues (Hall *et al.,* 1989; Kirch *et al.,* 1990; Stromberg *et al.,* 1991). Thus, the human tissues, although transplanted to a rat host, are maintaining some of their inherent neurochemical

Table 3.
CEAS analysis of rat and human brain tissues

Compound	Normal Rat Striatum (n = 6)	Rat Substantia Nigra (n = 5)	Human Xenografts (n = 4)
DA	15,356 ± 803	682 ± 111	11,308 ± 3,324
DOPAC	2,107 ± 168	121 ± 30	112 ± 17
L-Dopa	22 ± 4	61 ± 32	11 ± 4
HVA	1,617 ± 578	152 ± 49	915 ± 143
HVA-4-sulfate	1,915 ± 739	nd	3,382 ± 1,672
L-TYR	17,423 ± 2,618	12,184 ± 2,122	14,966 ± 3,990
L-TRP	4,706 ± 298	3,953 ± 828	3,404 ± 448
5-HT	1,341 ± 131	2,849 ± 1,531	10,652 ± 4,122
5-HIAA	865 ± 254	1,575 ± 275	2,499 ± 903
Ascorbate	2,332,024 ± 17,983	212,248 ± 38,155	401,601 ± 128,663

nd = not detectable. See Table 1 for abbreviations. All values are in units of ng/g wet wt. Variance is expressed as standard error of the mean (S.E.M.)

properties. The unique capabilities of the CEAS allows for thorough neurochemical investigations to be readily performed on such tissues.

Human CSF samples

Our recent studies have capitalized on the unique multicomponent assay capabilities of the CEAS in order to perform measures of 20 neurochemicals and related compounds in the lumbar CSF from normal subjects and patients diagnosed with schizophrenia (Issa *et al.,* 1994a,b). The biogenic amines, their metabolites, and other compounds were simultaneously measured in the CSF from 24 medication-free schizophrenic patients and 12 normal control subjects. The results from these measurements are seen in Table 4. In addition, a representative chromatogram of a human CSF sample is seen in Figure 4. After covariation for storage time, results of a stepwise discriminate function analysis comparing the control and patient groups identified tryptophan, tryptophol, and epinephrine as discriminating variables. Hotelling's paired T^2 test from a subgroup of schizophrenic patients studied while they were and were not receiving neuroleptic treatment did not yield any significant differences between subgroups. Thus, these studies did not support changes in CSF monoamines that support a role for DA systems in schizophrenia. We are currently investigating a large number of CSF samples (>200 per group) from normal controls and schizophrenic patients with and without neuroleptic treatment to further investigate the neurochemistry of schizophrenia. Thorough descriptions of our recent results have been recently published (Issa *et al.*, 1994a,b).

Table 4.
CEAS analysis of human lumbar csf samples from normal subjects and patients diagnosed with schizophrenia

Compound	Control Subjects (n = 12)	Patients with Schizophrenia (n = 24)
L-TYR	9139±1749	9052±1885
HVOL	17.7±7.7	15.3±5.3
HVA	229.3±58.4	227.7+77.0
MHPG	39.1±9.3	42.7±8.3
L-TRP	1841±265	2134±441
TRPOL	48.9±2.9	48.8±4.1
5-HTP	6.4+1.3	7.5+2.0
5-HT	1.5±0.1	1.5±0.1
5-HIAA	100.2±18.7	103.7±31.5
5-HTOL	7.3±1.5	6.4±1.8
KYN	57.1±61.7	27.8±11.9
3-OHKY	6.2±4.6	5.9±4.2
Xanthine*	2.8±0.8	2.6±0.6

All values are in united of nmol/L. * = units in µmol/L, Mean±SD. See Table 1 for abbreviations. Variance is expressed as standard deviation (SD).

Nonhuman primate CSF

A recent application of our CEAS procedures has involved studies of cisternal CSF from young monkeys. In these studies, samples of CSF were obtained weekly and the monkeys, after a suitable baseline period, received chronic infusion of monoamine oxidase (MAO) inhibitors. Data seen in Figure 5 are single channel chromatograms (Detector #10, +570mV) obtained over eight weeks from a young monkey which received clorgyline (2mg/kg per day for 4 weeks), a selective MAO A inhibitor, at week 4 through week 8. A significant decline in CSF levels of MHPG and HVA were seen. However, tryptophan levels, which are one indication of the food intake of this animal, did not change. The effects of clorgyline on the CSF levels of MHPG and HVA are consistent with its pharmacological actions. These studies demonstrate the capabilities of the CEAS for such longitudinal studies of monoamine metabolites to investigate the effects of drugs on CSF function in nonhuman primates.

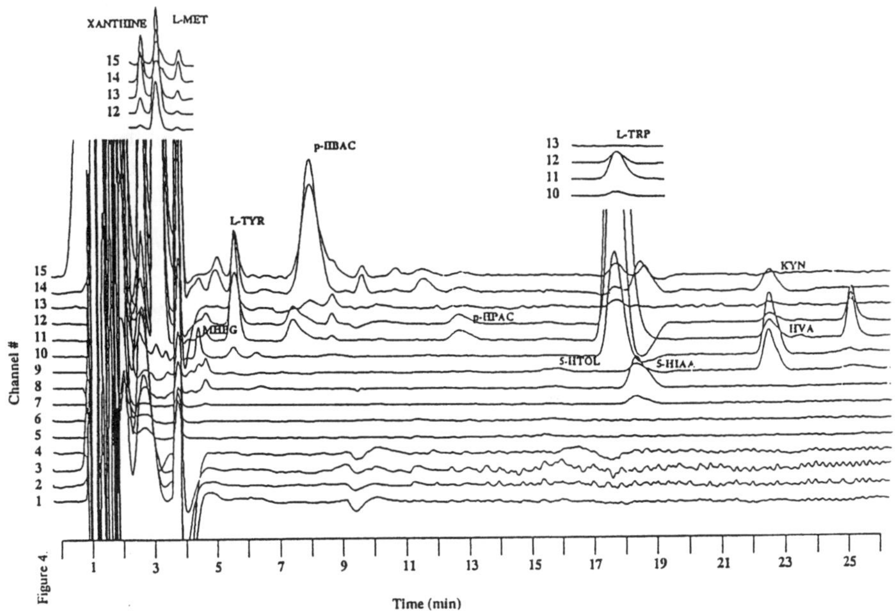

Figure 4. CEAS chromatogram of a human CSF sample. The insets show the re-scaled responses for L-TRP, xanthine, and L-MET.

Human plasma and saliva

Previous studies from our group have concentrated on the use of monoamine metabolite measures in human plasma to investigate the neurochemistry of schizophrenia and mania (Adler *et al.*, 1990; Baker *et al.*, 1990; Baker *et al.*, 1988; Freedman *et al.*, 1987; Waldo *et al.*, 1992; Waldo *et al.*, 1994). Our older HPLC procedures involved extraction of plasma with ethyl acetate, evaporation of the sample, and then the sample was resuspended in mobile phase for injection into a gradient HPLC system coupled to a dual-electrode coulometric detector (Gerhardt *et al.*, 1986). This was a time consuming procedure that required the use of standard addition protocols for optimum quantitation of the monoamine metabolites. Using the CEAS and our modified plasma filtration procedures (Gerhardt *et al.*, 1989), we are now capable of quantitating numerous neurochemicals and related compounds in filtrates of human plasma (see Methodology). We are currently using the CEAS methods to investigate plasma levels of monoamine metabolites in studies relating to the neurochemistry of schizophrenia.

A relatively new avenue of research involves the potential use of saliva samples to track changes in the CNS (Drebing *et al.,* 1989). Saliva is more of an ultrafiltrate of plasma and can be collected from patients without the stress and pain of the plasma and CSF collection procedures. We have previously demonstrated that levels of HVA and MHPG, are lower in saliva as compared to plasma. However, a good correlation between plasma and saliva was seen (Drebing *et al.,* 1989). More recently me have used the CEAS to investigate salivary levels of MHPG, HVA and a variety of other neurochemicals (data not shown). In contrast to our previous work, the use of the CEAS allows for the processing of saliva samples by filtration only, and there is not the need to use the extraction procedures that we have previously employed for such studies. We are

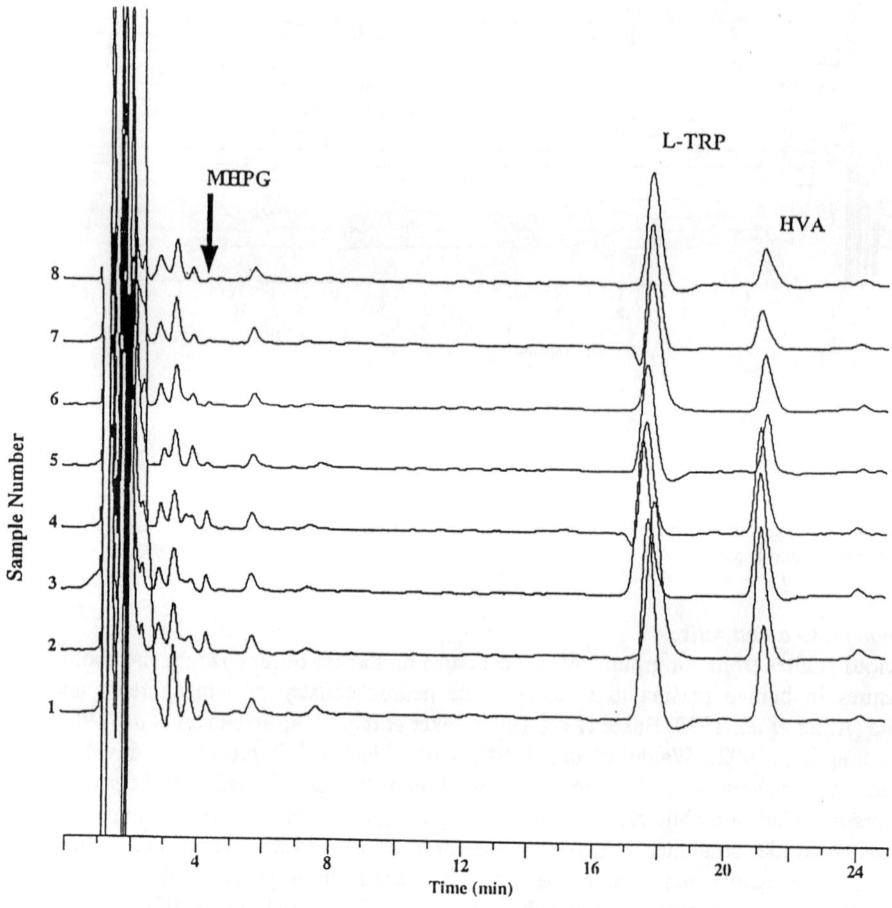

Figure 5. Series of CEAS chromatograms of cisternal CSF samples from young monkeys. Each trace represents data from channel #10 (+570mV) showing the detection of MHPG, HVA and L-TRP every week for eight weeks. Clorgyline (2mg/kg per day) was administered to this animal on week 4 for a total of 4 weeks. Note the reductions in the MHPG and HVA levels.

currently using the CEAS to investigate monoamine metabolites and related compounds in saliva as a noninvasive method to study changes in CNS function. Such procedures could be used for longitudinal studies of CNS function in a variety of patient populations.

CONCLUSION

The unique capabilities of the CEAS allow for the determination of greater than 20 compounds in single samples of tissue, plasma, CSF and saliva. This ability to measure numerous inter-related compounds has greatly increased our ability to investigate neurochemicals that are or are not involved in schizophrenia. In addition, the CEAS is a powerful tool for basic research that facilitates a more complete evaluation of the neurochemistry of CNS tissues and biological fluids. It is hoped that such technology can eventually be used to aid in the diagnosis and treatment of psychiatric disorders and neurodegenerative diseases.

Acknowledgments
This work was supported by the United States Public Health Service from grants NS-09199 and AG-06434 . We would like to thank Mr. Paul Gamache (ESA, Inc.) for technical support. In addition, we would like to thank Mr. Adam Coles for help with the artwork in this chapter.

REFERENCES

Adler, L. E., Gerhardt, G. A., Franks, R., Baker, N., Nagamoto, H., Drebing, C. and Freedman, R. (1990). Sensory physiology and catecholamines in schizophrenia and mania. Psychiatry Res., **31**, 297-309.

Baker, N. J., Staunton, M., Adler, L. E., Gerhardt, G. A., Drebing, C., Waldo, M., Nagamoto, H. and Freedman, R. (1990). Sensory gating deficits in psychiatric inpatients: Relation to catecholamine metabolites in different diagnostic groups. Biol. Psychiatry, **27**, 519-528.

Baker, N .J., Adler, L. E., Waldo, M., Gerhardt, G., Drebing, C., Cox, B., Berry, S., Macdonald, M., Phillips, W. and Freedman, R. (1988). Reproducibility of the measurement of plasma noradrenergic and dopaminergic metabolites in normal subjects. Psychiatry Res., **23**, 119-130.

Bird, E. D., Matson, W. R., Beal. M. F. and Ogawa, T. (1990). Patterns of tyrosine and tryptophan metabolites in controls and various degenerative disorders, in: Basic, Clin., and Therap. Aspects of Alzheimer's and Parkinson's Diseases, Plenum Press, New York, **1**. pp. 529-535.

Drebing, C. J., Freedman, R., Waldo, M. and Gerhardt, G.A. (1989). Unconjugated methoxylated catecholamine metabolites in human saliva: Quantitation methodology and comparison with plasma levels. Biomed. Chromatog., **3**, 217-220.

Freedman, R., Adler, L. E., Gerhardt, G. A., Waldo, M., Baker, N., Rose, G.M., Drebing, C., Nagamoto, H., Bickford-Wimer, P. and Franks, R. (1987). Neurobiological studies of sensory gating in schizophrenia. Schizophrenia Bulletin, **4**, 669-678.

Gerhardt, G. A., Drebing, C. J. and Freedman, R. (1986). Simultaneous determination of free homovanillic acid, 3-methoxy-4-hydroxyphenylethylene glycol, and vanilmandelic acid in human plasma by high performance liquid chromatography coupled with dual-electrode coulometric electrochemical detection. Anal. Chem., **58**, 2879-2883.

Gerhardt, G. A., Drebing, C. J., Stephen, C. and Freedman, R. (1989). Direct determination of unconjugated HVA in human plasma filtrates by HPLC coupled with dual-electrode coulometric electrochemical detection. Biomed. Chromatog., **3**, 105-109.

Godefroy, F., Matson, W. R., Gamache, P. H., Weil-Fugazza, J. (1990). Simultaneous measurements of tryptophan and its metabolites, kynurenine and serotonin, in the superficial layers of the spinal dorsal horn. A study in normal and arthritic rats, Brain Res., **526,** 169-172.

Hall, F., Hoffer, B. J. and Gerhardt, G. A. (1989). Rapid and sensitive determination of catecholamines in small tissue samples by high performance liquid chromatography coupled with dual-electrode coulometric electrochemical detection. LC-GC, **7,** 258-265.

Hoffer, B. J., Leenders, K. L., Young, D., Gerhardt, G., Zerbe, G.O., Bygdeman, M., Seiger, Å., Olson, L., Strömberg, I. and Freedman, R. (1992). Eighteen-month course of two patients with grafts of fetal dopamine neurons for severe Parkinson's disease. Exp. Neurol., **118,** 243-252.

Issa, F., Gerhardt, G. A., Bartko, J. J., Suddath, M. L., Gamache, P. H., Freedman, R., Wyatt, R. J. and Kirch, D. G. (1994a). A multidimensional approach to analysis of cerebrospinal fluid biogenic amines in schizophrenia: I. Comparisons with healthy control subjects and neuroleptic-treated/unmedicated pairs analyses. Psychiatry Res., **52,** 237-249.

Issa, F., Kirch, D. G., Gerhardt, G. A., Bartko, J. J., Suddath, R. L., Freedman, R. and Wyatt, R. J. (1994b). A multidimensional approach to analysis of cerebrospinal fluid biogenic amine analysis in schizophrenia: II. Correlations with psychopathology. Psychiatry Res., **52,** 251-258.

Kirch, D. G., Taylor, T. R., Gerhardt, G. A., Benowitz, N. L., Stephen, C. and Wyatt, R. J. (1990). Effect of chronic caffeine administration on monoamine and monoamine metabolite concentrations in rat brain. Neuropharmacology, **29,** 599-602 .

Langlais, P. J., Wardlow, M. L., and Yamamoto, H. (1991). Changes in CSF neurotransmitters in infantile spasms. Pediatr. Neurol., **7,** 440-445.

LeWitt, P. A., Galloway, M. P., Matson, W., Milbury, P., McDermott, M., Srivastava, D. K., Oakes, D., and the Parkinson Study Group. (1992). Markers of dopamine metabolism in Parkinson's disease. Neurol., **42,** 2111-2117.

Matson, W.R., Bouckoms, A., Svendsen, C., Beal, M.F., and Bird, E.D. (1990). Generating and controlling multiparameter data bases for biochemical correlates of disorders. In: Basic, Clin., and Therap. Aspects of Alzheimer's and Parkinson's Diseases, Plenum Press, New York, **1.** pp. 513-516.

Matson, W. R., Langlais, P., Volicer, L., Gamache, P. H., Bird, E., and Mark, K. A. (1984). n-Electrode three-dimensional liquid chromatography with electrochemical detection for determination of neurotransmitters. Clin. Chem., 30, 1477-1488.

Mefford I. N. (1981). Application of high performance liquid chromatography with electrochemical detection to neurochemical analysis: measurement of catecholamines, serotonin and metabolites in rat brain. J. Neurosci. Meth., **3,** 207-224.

Naoi, M., Maruyama, W., Acworth, I. N., Nakahara, D., and Parvez, H. (1993). Multi-electrode detection systems for determination of neurotransmitters. In: Methods in Neurotransmitter and Neuropeptide Research, Part 1. **11.** Techniques in The Behavioral and Neural Sciences. Parvez, S.H., Naoi, M., Nagatsu, T. and Parvez, S. (Eds.). Elsevier Science Publishers, New York, pp. 1-39.

Ogawa, T., Matson, W. R., Beal, M. F., Myers, R. H., Bird, E. D., Milbury, P., and Saso, S. (1992). Kynurenine pathway abnormalities in Parkinson's disease. Neurol., **42,** 1702-1706.

Ogawa, T., Saso, S-i, Beal, F., Kenton, S., Matson, W. and Bird, E. D. (1990). Tryptophan metabolites in Parkinson's brain. In: Basic, Clin., and Therap. Aspects of Alzheimer's and Parkinson's Diseases. **1.** Plenum Press, New York, pp. 329-332.

Rizzo, V., D'Eril, G.M., Achilli, G. and Cellerino, G. P. (1991). Determination of neurochemicals in biological fluids by using an automated high-performance liquid chromatographic system with a coulometric array detector. J. Chromatogr., **536,** 229-236.

Strömberg, I., van Horne, C., Bygdeman, M., Weiner, N. and Gerhardt, G. A. (1991). Function of intraventricular human mesencephalic xenografts in immunosuppressed rats: An electrophysiological and neurochemical analysis. Exp. Neurol., **112,** 140-152.

Suzuki, S. and Mori, A. (1992). Regional distribution of tyrosine, tryptophan, and their metabolites in the brain of epileptic El mice. Neurochem. Res., 17, 693-698.

Swartz, K. J., Matson, W. R., MacGarvey, U., Ryan, E. A. and Beal, M. F. (1990). Measurement of kynurenic acid in mammalian brain extracts and cerebrospinal fluid by high-performance liquid chromatography with fluorometric and coulometric electrode array detection. Anal. Biochem., **185,** 363-376.

Tohgi, H., Abe, T., Takahashi, S., Nozaki, Y. and Kikuchi, T. (1991). Concentrations of tyrosine, L-dihydroxyphenylalanine, dopamine, and 3-O-methyldopa in the cerebrospinal fluid of Parkinson's disease. Neurosci. Lett., **127,** 212-214.

Turk, D. L. and LeRoy Blank, C. (1990). NEUBA: a multicolumn, multidetector liquid chromatograph with electrochemical detection for use in the identification and determination of neurochemicals and related species. In: Basic, Clinical, and Therapeutic Aspects of Alzheimer's and Parkinson's Disease. **1.** Plenum Press, New York, pp. 517-520.

Volicer, L., Langlais, P. J., Matson, W. R., Mark, K. A. and Gamache, P. H. (1985). Serotoninergic system in dementia of the Alzheimer type: Abnormal forms of 5-hydroxytryptophan and serotonin in cerebrospinal fluid. Arch. Neurol., **42,** 1158-1161.

Waldo, M., Gerhardt, G. A., Baker, N., Drebing, C., Adler, L. and Freedman, R. (1992). Auditory sensory gating and catecholamine metabolism in schizophrenic and normal subjects. Psychiatry Res., **44,** 21-32.

Waldo, M. C., Cawthra, E., Adler, L. E., Dubester, S., Staunton, M., Nagamoto, H., Baker, N., Madison, A., Somon, J., Scherzinger, A., Drebing, C., Gerhardt, G. A. and Freedman, R. (1994). Auditory sensory gating, hippocampal volume, and catecholamine metabolism and schizophrenics and their siblings. Schizophrenia Res., **12,** 93-106.

Yamamoto, H. (1991). Studies on CSF tryptophan metabolism in infantile spasms. Pediatr. Neurol., **7,** 411-414.

Progress in HPLC-HPCE, Vol. 6, pp. 227–246
I.N. Acworth *et al.* (Eds)

THE APPLICATION OF THE MULTI-ELECTRODE DETECTION SYSTEM TO NEUROCHEMISTRY: *IN VIVO* MICRODIALYSIS STUDIES

MAKOTO NAOI[1], WAKAKO MARUYAMA[2], DAIICHIRO NAKAHARA[3] and HASAN PARVEZ[4]

[1]Department of Biosciences, Nagoya Institute of Technology, Nagoya, Japan. [2]Laboratory of Biochemistry and Metabolism, Department of Basic Gerontology, National Institute of Longevity Sciences, Obu, Japan. [3]Department of Psychology, Hamamatsu University School of Medicine, Hamamatsu, Shizuoka, Japan. [4]Neuroendocrinolodie and Neuropharmacologie, du Developpement, Universite de Reims-UFR Sciences, Reims Cedex, France

INTRODUCTION

The analysis of the neurotransmitters is the basic method for the studies of the metabolism and function of neurotransmission in the brain and other neuronal tissues. For quantitation of the biogenic compounds, sensitive methods are required since their concentrations are normally very low in the brain and the sample volume prepared from specific brain regions are often limited. High-performance liquid chromatography (HPLC) with electrochemical detection (ECD) is at present the most sensitive, accurate and simplest method for quantitative analysis of neurotransmitters. Among neurotransmitters, catecholamines and indoleamines can be easily analyzed by their electrochemical reactivity. As reviewed previously (Naoi *et al.*, 1993b), the multi-electrode detection system [CEAS, (NEUROCHEM in Japan), ESA, Chelmsford, MA] can analyze not only the biogenic amines, but also their precursor amino acids and metabolites. Compounds with electrochemical reactivity can be oxidized across the sixteen electrode array (typically with applied potentials between 0 to >900mV) yielding distinct redox profiles. Compounds are then identified both by their retention time and by their redox behavior across the array. This accuracy in compound identification and quantitation is most important when the volume of the sample is limited, when the concentration of analytes are low, and when many compounds are being simultaneously quantitated in complex biological matrices. This is truly the case with samples obtained using *in vivo* microdialysis perfusion, where only 1-2μL/minute. of sample is usually obtained. Until our application on the CEAS, researchers typically only measured a few compounds in microdialysis samples due to the inadequacies of their single electrode ECD system. By using the CEAS we can analyze many compounds in microdialysis perfusates obtained from the rat brain and we have used this approach to study the metabolism and function of neurotoxins involved in neurodegenerative diseases, alcoholism and other neuro-psychiatric disorders.

1-Methyl-6,7-dihydroxy-1,2,3,4-tetrahydroisoquinoline (salsolinol, Sal) is identified in the mammalian brain as one of the naturally occurring alkaloids (Sandler *et al.*, 1973; Sjöquest *et al.*, 1981, 1982). It is endogenously synthesized by condensation of dopamine (DA) with pyruvic acid, followed by enzymatic decarboxylation and reduction to yield a stereospecific enantiomer, (*R*)Sal (Dostert *et al.*, 1990). In the urine of healthy subjects, only the (*R*)-enantiomer was detected (Strolin Benedetti *et al.*, 1989), while (*S*)Sal was found in the urine of chronic alcoholics (Dostert *et al.*, 1991) and in the urine from patients on L-DOPA therapy (Dostert *et al.*, 1989). Recently a novel enzyme was isolated from the human brain which can synthesize enantio-selectively (*R*)Sal from dopamine and acetaldehyde (Naoi *et al.*, 1996b). Previous papers suggest that ethanol intake increases the biosynthesis and metabolism of Sal (Ung-Chhun *et al.*, 1985), and that Sal may be involved in clinical features and pathogenesis of alcoholism. In alcoholics, perturbation of brain monoamines may be involved in the alcohol dependence, and ethanol itself or one of its metabolites like Sal may also play a role in the perturbation of function and metabolism of monoamines in the brain.

Furthermore, 1-methyl-4-phenyl-1,2,3,6-tetrahydropyridine (MPTP) was found to induce parkinsonism in humans, and its clinical features and pathological findings are very similar to those of idiopathic Parkinson's disease (Davis *et al.*, 1979; Langston *et al.*, 1984). This finding indicates that Parkinson's disease may be caused by naturally-occurring MPTP-like compounds, and a series of 6,7-dihydroxy-1,2,3,4-tetrahydroisoquinolines (DHTIQs), such as Sal, were examined for the cytotoxicity to dopaminergic neurons. When Sal is N-methylated into 1,2-dimethyl-6,7-dihydroxy-1,2,3,4-tetrahydroisoquinoline [N-methyl-salsolinol, NMSal], the structure becomes similar to MPTP, as shown in Figure 1. By gas chromatography-mass spectrometry the occurrence of NMSal was confirmed in the human brain (Niwa *et al.*, 1991).

This article reviews our recent results obtained using *in vivo* microdialysis and analysis performed on the CEAS and examines the biosynthesis of the isoquinoline neurotoxins and their mechanism of cytotoxicity. The first part of this article reviews the N-methylation of (*R*)Sal into NM(*R*)Sal in the rat brain proven by an *in vivo* microdialysis study (Maruyama *et al.*, 1992 and 1993b). In addition, the effects of DHTIQ perfusion on monoamine levels in the brain were examined and the release of monoamines from the nerve terminals was confirmed (Maruyama *et al.*, 1993a; Nakahara *et al.*, 1994). The catechol isoquinolines were found to inhibit the activities of the related enzymes, monoamine oxidase [monoamine: oxygen oxidoreductase (deaminating), EC 1.4.3.4., MAO] and catechol-O-methyltransferase (S-adenosyl-L-methionine, catechol-O-methyltransferase, EC 2,1,1,6, COMT). Finally, the application of the CEAS to the analysis of hydroxyl radicals in the brain is reviewed (Maruyama *et al.*, 1995b). The results are discussed in relation to possible involvement of these isoquinolines in the pathogenesis and clinical symptoms of some neuro-psychiatric disorders, such as Parkinson's disease (PD) and alcoholism.

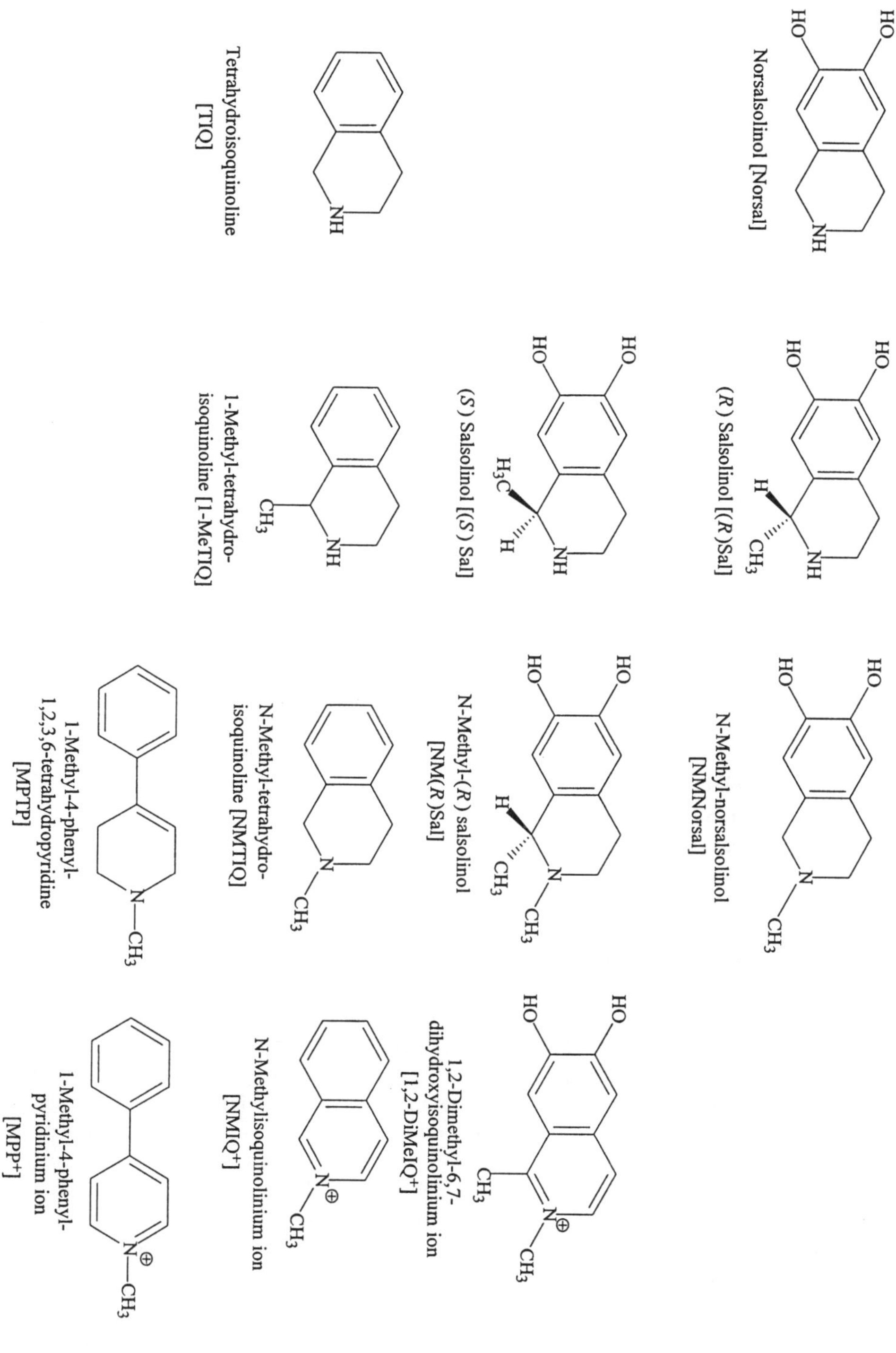

Figure 1. Structure of various analytes discussed in text.

N-METHYLATION OF SALSOLINOL

MPTP is transported into the brain through the blood-brain barrier and oxidized to 1-methyl-4-phenylpyridinium ion (MPP^+) by type B MAO. The oxidation produces a truly potent neurotoxin with specificity for DA neurons (for a review see, Tipton and Singer, 1993), and indicates that neurotoxins are synthesized by metabolism in the brain from less toxic precursors. This "metabolic bioactivation" usually involves two enzymatic reactions, N-methylation and oxidation, and this has also proven to be the case for 6,7-dihydroxy-1,2,3,4-tetrahydroisoquinolines (DHTIQs) [Maruyama *et al.*, 1993c, 1995a; Naoi *et al.*, 1994, 1995a, b), 1,2,3,4-tetrahydroisoquinolines (TIQs) (Naoi *et al.*, 1993a) and the β-carbolines (Matsubara *et al.*, 1995a, b). In general, N-methylation produces a MPTP-like compound, which is further oxidized into a cytotoxic cation structurally similar to MPP^+. N-Methylation was for the first time confirmed with TIQs by *in vitro* experiments using enzyme samples prepared from the human brain (Naoi *et al.*, 1989 and 1993a). This section reviews our results demonstrating for the first time the *in situ* synthesis of a neurotoxin in the brain, as proved by *in vivo* microdialysis and CEAS analysis (Maruyama *et al.*, 1992, 1993a).

(*R*)Sal, (*S*)Sal, NM(*R*)Sal and 2(N)-methyl-6,7-dihydroxy-1,2,3,4-tetrahydro-isoquinoline (N-methylnorsalsolinol, NMNorsal) were synthesized according to Teitel *et al.,* (1972) and were kindly donated by Dr. P. Dostert (Pharmacia-Upjohn, Milan, Italy). 1-Methyl-1,2,3,4-tetrahydroisoquinoline (1-MeTIQ) was kindly donated by Dr. S. Ohta (University of Tokyo, Tokyo, Japan); 2(N)-methyl-1,2,3,4-tetrahydroisoquinoline (NMTIQ) and 2(N)-methyl-isoquinolinium ion ($NMIQ^+$) were from Dr. S. Matsuura (Fujita Health University, Toyoake, Japan). TIQ and sodium octanesulfonate (SOS) were purchased from Aldrich (Milwaukee, WI) and 6,7-dihydroxy-1,2,3,4-tetrahydroisoquinoline (norsalsolinol, Norsal) from Jannsen (Beerse, Belgium). The standard monoamines and their metabolites were purchased from Sigma (St. Louis, MO).

Male adult Wistar rats weighing 260-340g were anesthetized with sodium pentobarbital. The animal experimental protocols were approved by the Ethical Committee of Department of Biosciences, Nagoya Institute of Technology. Microdialysis in the rat brain was carried out as reported previously (Maruyama *et al.*, 1992; Nakahara *et al.*, 1993). The guide cannula was implanted above the left striatum (coordinates [in mm]: A 0.5 to the bregma, L-3.0 to the midline, V-3.0 to the dura), the substantia nigra (A -5.3, L 2.2, V 6.4), the hypothalamus (A -3.0, L 1.0, V 7.2) or the hippocampus (B -5.5, L 5.0, D -3.0), according to the stereotaxic atlas of Paxinos and Watson (1986). Dummy probes were inserted into the guide cannula until the start of the experiment. Twenty four to forty eight hours following surgery the dummy probe was removed, a dialysis probe was inserted through the cannula and was perfused with Ringer's solution (147mM NaCl, 2.3mM $CaCl_2$, 4mM KCl, pH 6.0) at a flow rate of 2μL/min. Dialysis samples were collected every 20 minutes after a 2 hour stabilization period. The first 3 samples were used to determine the baseline levels of compounds in the dialysate. Then the perfusion of 1mM (*R*)Sal or other DHTIQs in Ringer's solution started and continued for 2 hours (6 samples). All samples were frozen immediately and stored at -80°C until analysis. (*R*)Sal, NM(*R)*Sal, monoamines and their metabolites in the dialysate were quantitatively analyzed using an isocratic CEAS. The ECD electrodes were set from 0 to +900mV in 60mV increments. Analytes were eluted using a flow rate of 0.6mL/minute and a temperature of 25°C. The column used was a pre-packed reversed-phase column, Innertosil, (GL Sciences, ODS-2, 4.6mm i.d. x 250mm). The mobile phase consisted of 90mM sodium acetate-35mM citric acid buffer, pH 4.35, containing 130μM disodium EDTA and 230μM sodium octanesulfonate; methanol (10.5%. v/v). The identity of each peak was verified by retention time and by the relative ratio of the peak heights measured across 2 or 3 electrodes at different voltages (ratio accuracy). Quantitation of the compound was made with the detector to give the best resolution of the peak and by comparison of the peak height with that of the standard.

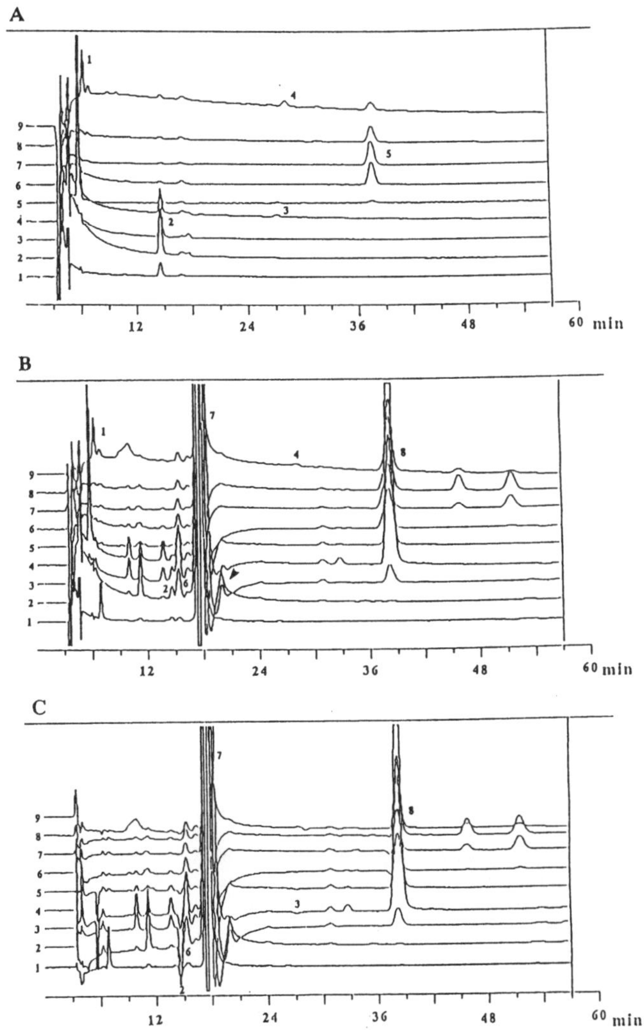

Figure 2. Chromatograms of the dialysate of the rat striatum. The striatum was perfused with 1mM (*R*)Sal and the dialysate was analyzed using HPLC-ECD, as described in the text. **A** - Before perfusion; **B** - after 120 minutes perfusion with 1mM (*R*)Sal; and **C** - subtraction of **A** from **B**. Peak 1 - L-tyrosine; peak 2 - 3,4-dihydroxyphenylacetic acid (DOPAC); peak 3 - 5-hydroxyindole-3-acetic acid (5-HIAA); peak 4 - L-tryptophan; peak 5 - homovanillic acid (HVA); peak 6 - DA, peak 7 - (*R*)Sal, peak 8 - 5-hydroxytryptamine (5-HT); arrow - NM(*R*)Sal.

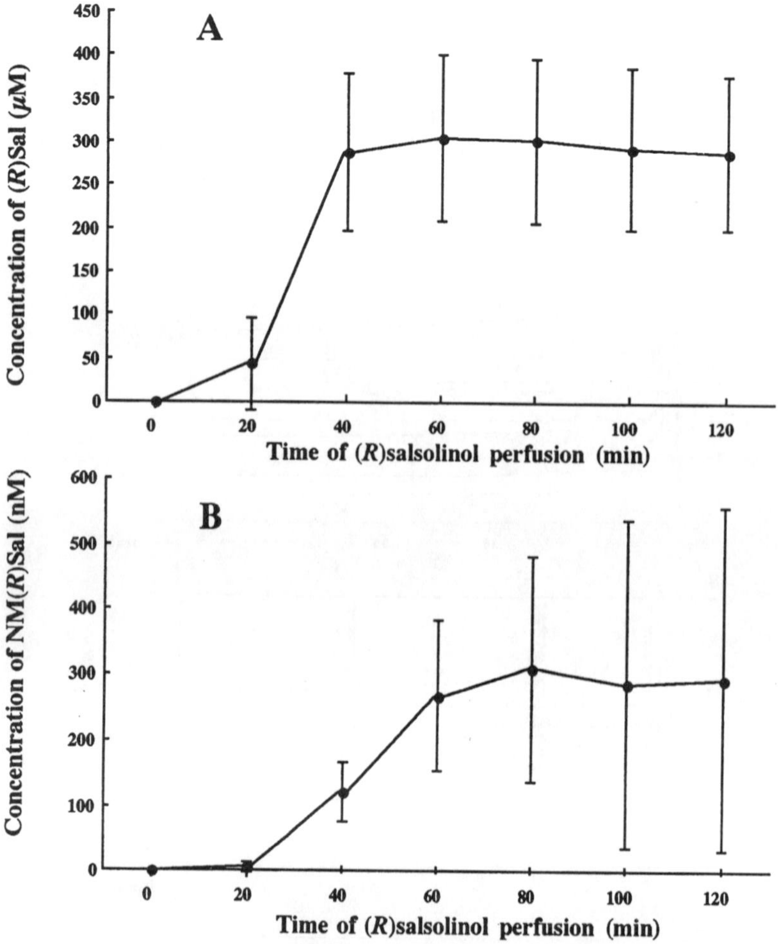

Figure 3. Time dependent change in the concentrations of (*R*)Sal and NM(*R*)Sal in the dialysate samples obtained from the rat striatum. The rat striatum was perfused with 1mM (*R*)Sal for 120 minutes. The dialysate samples were collected at 20 minute intervals and analyzed quantitatively by HPLC-ECD. The circle and bar represents mean±SD of the values obtained from 4 experiments. **A** - The concentration of (*R*)Sal in the dialysate (μM). **B** - The concentration of NM(*R*)Sal in the dialysate (nM).

Data from the samples obtained by *in vivo* microdialysis from the rat striatum before and after perfusion of (*R*)Sal are presented in Figure 2 (Figure 2A shows a basal sample and Figure 2B shows the effect of 120 minutes of Sal perfusion). The changes in metabolite levels (Figure 2B - Figure 2A) is presented as a difference chromatogram in Figure 2C. The voltages of the detectors 1, 2, 3 and 4 were set to 0, +60, +120 and +180 mV, respectively, and this array permitted the most efficient measurement of (*R*)Sal and NM(*R*)Sal. After 40 minutes perfusion of (*R*)Sal, both it and other new peaks were detected in the dialysate. One of the new peaks was identified as NM(*R*)Sal from its HPLC characteristics. The retention time of the peak in the dialysate was 19.11 minutes and the relative ratio of the peak height detected with the detector 3/2 was 1.56, which corresponded well to those of the standard, 19.13 minutes and 1.58, respectively. The (*R*)Sal concentration of the striatal dialysate was increased in a time-dependent fashion and reached a plateau after 40 minutes perfusion, as shown in Figure 3. NM(*R*)Sal concentration also increased with the perfusion time and a plateau was obtained after 80 minutes of perfusion. In the other three regions of the rat brain, namely the substantia nigra, hypothalamus and hippocampus, N-methylation of (*R*)Sal was also examined by *in vivo* microdialysis. The concentrations of (*R*)Sal and NM(*R*)Sal in the dialysate after 80 minutes of perfusion are summarized in Table 1. The relative ratio of NM(*R*)Sal/(*R*)Sal was statistically significant and higher in the substantia nigra when compared to the other three brain regions, as calculated using a one-way ANOVA followed by a Scheffe F-test ($p<0.01$).

Table 1.
Concentration of (*R*)Sal and NM(*R*)Sal in dialysate samples.

	Striatum (n=4)	Substantia nigra (n=4)	Hypothalamus (n=2)	Hippocampus (n=2)
Concentration of (*R*)Sal (μM)	290.8 ± 28.9	358.2 ± 57.4	178.9 176.6	190.5 171.4
Concentration of NM(*R*)Sal (nM)	289.2 ± 163.5	1011 ± 127.4*	160.8 199.2	162.5 144.8
Relative ratio of concentration [(*R*)NMSal/(*R*)Sal] x 10^3	0.863 ± 0.398	2.97 ± 0.536*	0.899 1.13	0.853 0.845

The concentration of (*R*)Sal and NM(*R*)Sal in the dialysate after 80 minutes perfusion of 1mM (*R*)Sal. The samples were collected at 20 minutes intervals and flow rate was 2μL/min. The data of the striatum and the substantia nigra were represented as mean±SD of 4 experiments. *$p<0.01$ by Scheffe F- test.

In situ biosynthesis of NM(*R*)Sal from (*R*)Sal does occur in the brain. (*R*)Sal, in turn, is synthesized from DA either inside and/or around the dopaminergic neurons and is easily and specifically accumulated by them. Recently *in vitro* experiments have shown that NM(*R*)Sal can be oxidized to 1,2-dimethyl-6,7-dihydroxyisoquinolinium ion (1,2-DMDHIQ$^+$) in a spontaneous non-enzymatic process (Maruyama *et al.*, 1995b) and enzymatically (Naoi *et al.*, 1995c). In Parkinson's disease, specific and localized degeneration of dopaminergic neurons in the substantia nigra is a characteristic pathological feature. The neurotoxin that causes idiopathic Parkinson's disease should have affinity to the DA neurons in the substantia nigra. Our data indicate that in the rat substantia nigra a significantly larger amount of N-methylation of (*R*)Sal occurs than in the other 3 regions examined. In addition, NM(*R*)Sal is specifically taken up by the DA transporter (Takahashi *et al.*, 1994) and is accumulated in DA neurons by specific binding to melanin (Naoi *et al.*, 1994a).

The possible involvement of NM(*R*)Sal in the pathogenesis of Parkinson's disease has been further suggested by our recent results. Following injection of DHTIQs into the striatum, only NM(*R*)Sal could induce parkinsonism in rats (Naoi *et al.*, 1996a). The behavioral, biochemical and histological changes were very similar to those observed in Parkinson's disease. In the cerebrospinal fluid of untreated parkinsonian patients, this NM(*R*)Sal was found to increase significantly (Maruyama *et al.*, 1996a). Considering that this catechol neurotoxin is synthesized in the brain by a (*R*)salsolinol synthase and an N-methyltransferase, the enzymes related to its biosynthesis might play an important role in cell death of DA neurons in the normal aging process and in Parkinson's disease.

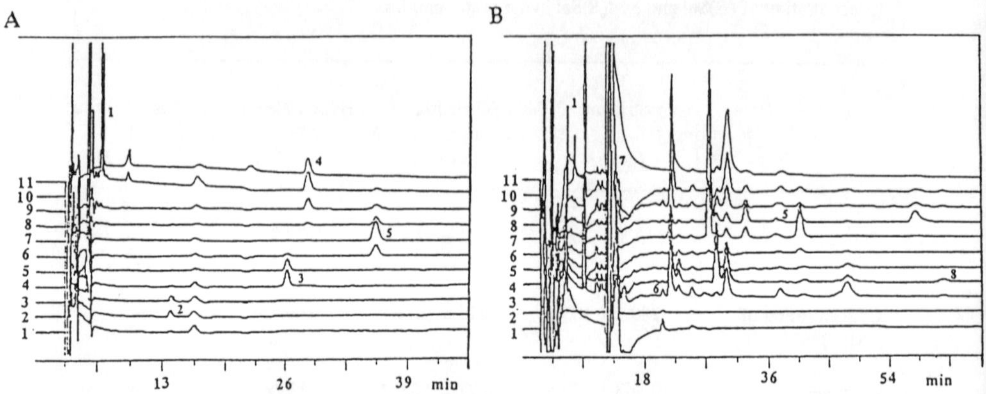

Fig. 4. Chromatograms of rat striatal microdialysis perfusate. The rat striatum was perfused with 1mM TIQ or Norsal, and the dialysate was analyzed using HPLC-ECD. **A** and **B**, after 120 minutes perfusion of 1mM TIQ and Norsal, respectively. Peak 1 - L-tyrosine; peak 2 - DOPAC; peak 3 - 5-HIAA; peak 4 - L-tryptophan; peak 5 - HVA; peak 6 - DA; peak 7 - Norsal; peak 8 - 5-HT.

This is the first report demonstrating the biosynthesis of a neurotoxin in the brain using an *in vivo* microdialysis method. Using this method the metabolism of toxins can be examined in a selected brain region, and by use of multi-channel electro-chemical detectors, precise identification and quantitation of a compound is possible without time-consuming sample preparation.

RELEASE OF MONOAMINES BY SALSOLINOL INFUSION

As shown in Figure 2 perfusion of (*R*)Sal into the striatum causes the monoamine and metabolite levels to change markedly. The most significant change was an enormous increase in 5-HT levels, as shown by the difference chromatogram, Figure 2C. The 5-HT concentration increased from an undetectable level to 2.53±0.12μM (mean±SD).

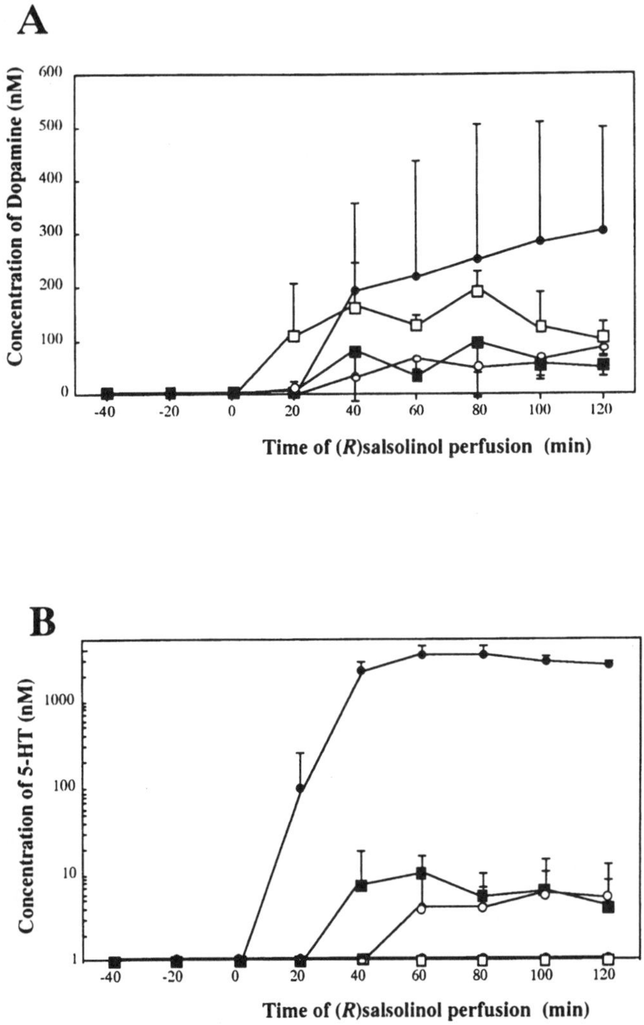

Figure 5. The time-dependent change of DA and 5-HT concentrations in striatal dialysates after perfusion. The striatum was perfused with 1mM DHTIQs and dialysate samples were collected at 20 minute intervals and analyzed on the CEAS. The point and bar represent mean±SD of 4 [(*R*)Sal] or 2 [(*S*)Sal, NM(*R*)Sal, Norsal, NMNorsal] experiments. Filled circles - (*R*)Sal perfusion; filled triangles - (*S*)Sal perfusion open circles - NM(*R*)Sal perfusion; filled squares - Norsal perfusion; open squares - NMNorsal perfusion. **A** - The DA concentration (nM) in the striatal dialysate. **B** - The 5-HT concentration (nM) in the striatal dialysate.

The DA concentration also increased (Figure 2C, peak 6), but this increase was not so marked as for 5-HT; 303.7±96.5nM (mean±SD). In contrast, DOPAC, HVA and 5-HIAA were all reduced (Figure 2, peaks 2 and 3).

The influence of the isoquinoline structure on monoamine release is illustrated by comparison of the effects of Norsal and TIQ on monoamine release and metabolism. The only difference between Norsal and TIQ is the presence or absence of a catechol structure. As shown in Figure 4, perfusion of Norsal caused a marked increase in the levels of DA (Figure 4B, peak 6) and 5-HT (peak 8) and produced a number of new peaks. TIQ perfusion had much less effect (Figure 4A). Differences were also seen between (*R*)Sal or (*S*)Sal and 1-MeTIQ, NMSal and NMTIQ, and the 1-methyl-6,7-dihydroisoquinolinium ion and 2(N)-methylisoquinolinium ion ($NMIQ^+$). These results suggest that the presence of the catechol structure is a prerequisite for the release of DA and 5-HT to occur.

The structure-relationship and ability to release DA and 5-HT was examined between DHTIQs, (*R*) and (*S*)Sal, NM(*R*)Sal, Norsal and NMNorsal (Figure 5). DA and 5-HT levels in the dialysate began to increase after the first 20 minutes of perfusion, reaching a plateau at 40 minutes which continued for a further 120 minutes (Figure 5). DA's concentration also increased with all five DHTIQs, and the increase was the most marked with (*R*)Sal which increased DA's level by about 60 fold above baseline. The increase in DA level by (*S*)Sal was only about one fifteenth of that of (*R*)Sal (Figure 5A). (*R*) and (*S*)Sal both released 5-HT, whereas NM(*R*)Sal and Norsal had much less effect than Sal, and NMNorsal had little effect on 5-HT levels at all (Figure 5B). These results indicate that 5-HT is released selectively by Sal, and N-methylation at C2 of DHTIQs reduces this effect, as shown by comparison of (*R*)Sal with NM(*R*)Sal, and Norsal with NMNorsal. (*R*)Sal was also perfused into other regions of the rat brain including the striatum, hippocampus and hypothalamus. In the striatum 5-HT and DA levels released by (*R*)Sal were highest among the 4 brain regions. Release of 5-HT was much more marked than that of DA in all of these regions.

The mechanism of the release of monoamines by (*R*)Sal perfusion was studied by the same technique. The release of 5-HT in the rat striatum was dose-dependent from 1μM to 1mM in the perfusion medium, but DA levels could only be detected at the 1mM level (Figure 6). Based upon the recovery in the dialysate of 1μM (*R*)Sal perfusion medium, the (*R*)Sal concentration in the extracellular space is estimated to be approximately 0.4μM, which is comparable to the intracellular concentration of salsolinol which has previously been reported (Sjöquest *et al.*, 1981). These data suggest that 5-HT may be released from nerve terminals by (*R*)Sal under physiological conditions. The effect of tetrodotoxin, a blocker of nerve-firing activity through inhibition of voltage-dependent sodium channels, was also examined. Pretreatment with perfusion of 2μM tetrodotoxin through the microdialysis probe could not prevent 5-HT release. This suggests that 5-HT release may be via a carrier-mediated exchange-diffusion mechanism.

Several compounds have been reported to increase extracellular levels of the monoamines. A dopaminergic neurotoxin, MPP+, has previously been reported to release DA in the rat striatum, but was without effect on 5-HT (Ozaki *et al.*, 1987). Amphetamine and methamphetamine also increased 5-HT level in the striatum in addition to DA and noradrenaline, but 5-HT release was less marked than for DA (Hernandez *et al.*, 1991). Our results demonstrate that both (*R*) and (*S*)Sal induce

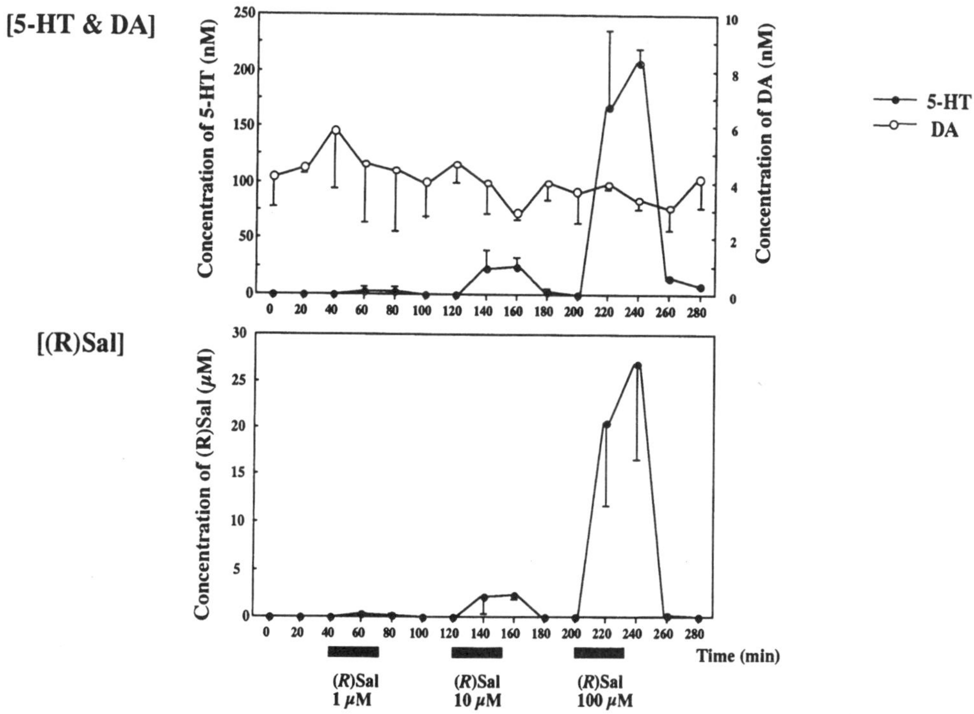

Figure 6. Effects of various concentrations of (*R*)Sal (1, 10 and 100μM) in the perfusion medium on the levels of 5-HT, DA, their metabolites and (*R*)Sal in dialysate samples. Symbols and bars represent mean±S.E.M. Each dose of (*R*)Sal was perfused for a period of 20 minutes after a washout period of 60 minutes. The maximal concentration of (*R*)Sal recovered from the dialysate was 0.38±0.09, 5.4±1.9 and 49.9±17.1μM, respectively, when the (R)Sal dose was increased in the order, 1, 10 and 100μM in the perfusate. * Significant beyond $p<0.05$ compared with the respective pre-drug values.

efflux of a large amount of 5-HT into the extracellular fluid of the brain. The massive release of 5-HT from the serotonergic nerve terminals may be followed by 5-HT depletion in the terminals, which might be further enhanced by reduction of 5-HT biosynthesis through inhibition of tryptophan hydroxylase [L-tryptophan, tetrahydropteridine: oxygen oxidoreductase (3-hydroxylating), EC 1.14.16.4] (Ota *et al.*, 1992; Matsubara *et al.*, 1994). 5-HT is now proposed to be involved in some psychiatric symptoms such as illusion and delirium, and sleep disturbance, and our results suggest that Sal may be involved in such symptoms.

EFFECTS ON MONOAMINE METABOLISM: INHIBITION OF CATABOLISM

As described above, the monoamine metabolites were also affected by (*R*)Sal perfusion. The concentration of DA, DOPAC and HVA (Figure 7), and of 5-HT and 5-HIAA (Figure 8) were analyzed in the dialysate obtained from the striatum. Mean basal values of DA, DOPAC, HVA, 5-HT and 5-HIAA in basal striatal dialysate samples were as follows (nM in the dialysate; mean±S.D): DA - 1.73±0.49; DOPAC - 1238±611; HVA - 659±194; 5-HT not detect; 5-HIAA 66.6±93.4. Following perfusion of (*R*)Sal, DOPAC, a metabolic product of MAO, and HVA, a product of MAO and COMT, were decreased significantly to 24.3±26.7% and 33.5±8.98 % of their basal level, respectively, and 5-HIAA could not be detected in the dialysate sample.

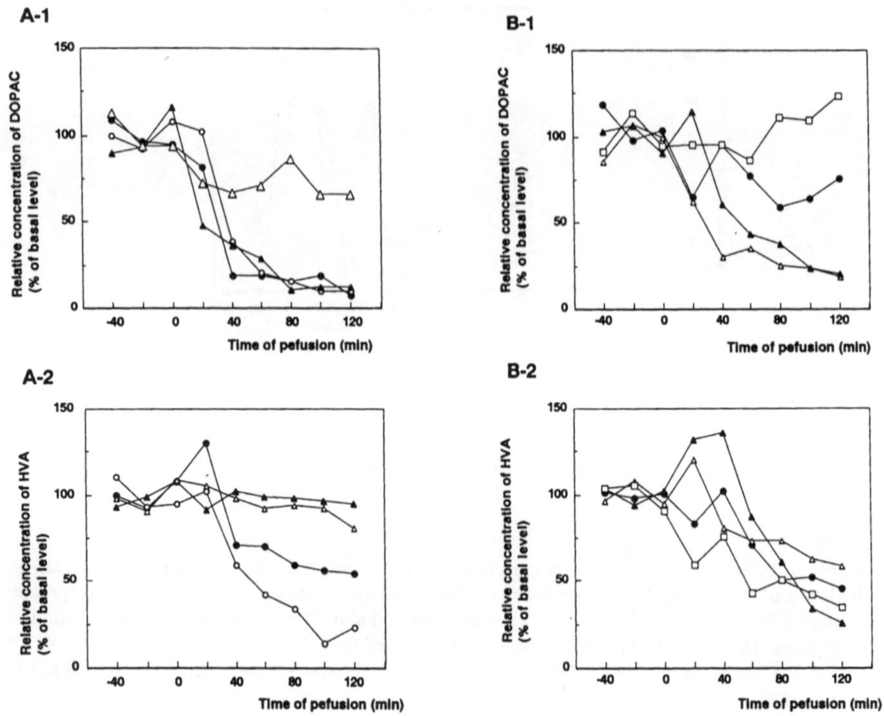

Figure 7. The time-dependent changes in the relative concentration of DOPAC and HVA in striatal dialysate following perfusion with DHTIQs and TIQs. The dialysate samples were collected at 20 minute intervals and were analyzed by HPLC-ECD. Each point represents the mean of data obtained from 2 experiments. **A-1**: The relative concentration of DOPAC after perfusion of DHTIQs. **A-2**: The relative concentration of HVA after perfusion of DHTIQs. Filled circles - Norsal; open triangles - NMNorsal; filled triangles - (*R*)Sal; open circles - NM(*R*)Sal. **B-1**: The relative concentration of DOPAC after perfusion of TIQs. **B-2**: The relative concentration of HVA after perfusion of TIQs: filled circles - TIQ; open triangles - NMTIQ; filled triangles - 1-MeTIQ; open squares - $NMIQ^+$.

The effects of the chemical structure of DHTIQs and TIQs on DA catabolism were examined. As shown in Figure 7, NMNorsal and (*R*)Sal (2-methylated and 1-methylated derivatives of DHTIQ) reduced the extracellular level of DOPAC, but neither NM(*R*)Sal (1,2-dimethyl-DHTIQ) nor Norsal (DHTIQ without any methyl group) had any affect on DOPAC levels (A-1). However, all of these DHTIQs reduced HVA level (A-2). Among the TIQs, the isoquinolines without catechol structure (TIQ, 1-MeTIQ, and $NMIQ^+$) reduced DOPAC levels (B-1). In addition, TIQ and $NMIQ^+$ decreased striatal HVA, but less markedly so than with DOPAC (B-2). The effects of DHTIQs and TIQs on the DA metabolites in the rat striatum after 100 minutes of perfusion are summarized in Table 2. The ratio of (HVA+DOPAC)/DA (i.e. DA turnover) was reduced with all isoquinolines except for NMTIQ. DHTIQs were found to markedly reduce DA turnover from 344.7±32.7 to 2.69-43.0 after 100 minutes of perfusion.

Table 2.
The effect of THIQs on the concentration of DA metabolites in the rat striatum.

	Relative concentration (% of basal value)		Relative concentration ratios	
	DOPAC	HVA	(DOPAC+HVA)/DA	HVA/DOPAC
Norsal	75.7*	45.6*	5.24*	0.25
NMNorsal	19.2*	58.9	3.02*	2.51*
(R)Sal	17.8*	23.0*	4.90*	0.26
(S)Sal	95.4*	61.8	41.5*	0.25
NM(R)Sal	105.0*	44.4*	24.2*	0.14*
TIQ	19.0*	51.6*	83.0*	3.08*
NMTIQ	88.1*	87.3	231.0	1.82
1-MeTIQ	12.7*	95.2	94.0*	2.78*
$NMIQ^+$	9.46*	23.0*	48.1*	0.77
Basal Control	1591±201.9	609±35.8	344.7±32.7	0.368±0.029

The concentrations of DA, DOPAC, and HVA were quantitatively analyzed using HPLC-ECD in the striatal dialysate after 100 minutes of perfusion of the THIQs. Each value represents the mean of 2 experiments. * The difference is statistically significant from the basal value by one-sample runs test on the samples obtained from 40 minutes to 100 minutes after isoquinoline perfusion ($p<0.05$).

The effects of isoquinolines on 5-HT metabolism were also examined. Figure 8 shows the effects of isoquinolines on 5-HT catabolism. (*R*)- and (*S*)Sal released a large amount of 5-HT while there was a marked reduction in 5-HIAA levels. As in the case of DA, DHTIQs were shown to inhibit *in vivo* 5-HT catabolism in the rat brain more strongly than TIQ. The order of the potency of DHTIQs to reduce 5-HT catabolism was (*R*)- and (*S*)Sal, NMNorsal, NM(*R*)Sal and Norsal. Amongst the TIQs, the 5-HIAA level was reduced by $NMIQ^+$ most strongly, followed by TIQ, 1-MeTIQ and NMTIQ.

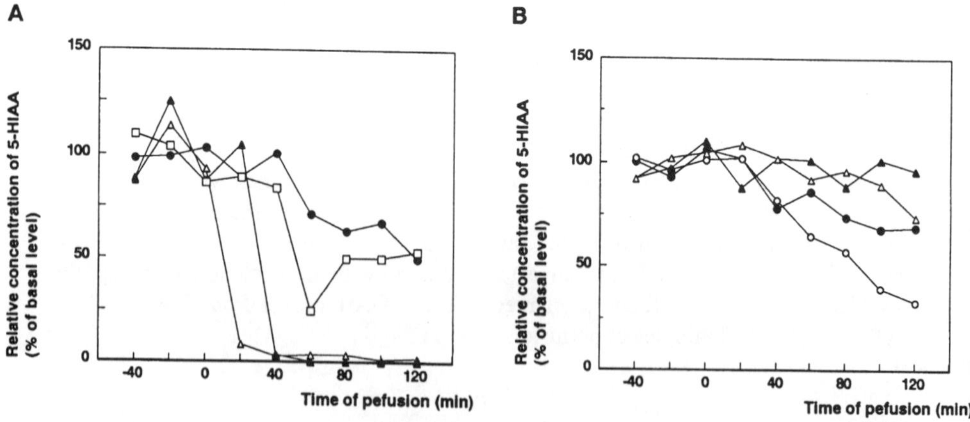

Figure 8. The time-dependent change of the relative concentration of 5-HIAA levels in the striatal dialysate after perfusion of TIQs and DHTIQs. The dialysate samples were collected at 20 minute intervals and analyzed by HPLC-ECD. Each point represents the mean of data obtained from 2 experiments. **A**: The relative concentration of 5-HIAA after perfusion of DHTIQs. Filled circles - Norsal; open triangles - NMNorsal; filled triangles - (*R*)Sal; open squares - NM(*R*)Sal. **B**: The relative concentration of 5-HIAA after perfusion of TIQs. Filled circles - TIQ; open triangles - NMTIQ; filled triangles - 1-MeTIQ; open circles - $NMIQ^+$.

DHTIQs were found to reduce both DOPAC and HVA levels, indicating inhibition of both MAO and COMT activity. This is comparable to *in vitro* results that show that DHTIQs inhibit type A MAO (Minami *et al.*, 1993; Naoi *et al.*, 1994c), and also COMT activity (Sanft and Thomas, 1989). *In vitro* experiments showed that the two enantiomers of Sal inhibited the activity of the rate-limiting enzyme in the synthesis of monoamines - tyrosine hydroxylase [tyrosine, tetrahydropteridine: oxygen oxidoreductase (3-hydroxylating), EC 1.14.16.2] (Minami *et al.*, 1992), and tryptophan hydroxylase (Ota *et al.*, 1992; Matsubara *et al.*, 1994). These results suggest that DA-derived DHTIQs affect the monoamine metabolism in the brain, by the inhibition of enzymes involved in their biosynthesis and catabolism. Perturbation of monoamine metabolism may be related to some psychiatric symptoms induced by L-DOPA therapy, where Sal levels are expected to be increased in the brain.

HYDROXYL RADICALS IN THE BRAIN

One factor to account for the vulnerability of the dopaminergic neurons in Parkinson's disease might be a high production of activated oxygen radicals resulting from DA catabolism. In the brain DA is oxidized by MAO or can autooxidize non-enzymatically (Maruyama *et al.*, 1995a). Hydrogen peroxide, a product of the oxidation of DA by MAO, produces hydroxyl radicals by the Fenton reaction in the presence of iron ions. Furthermore, the substantia nigra is rich in iron, which can increase the radical formation (Riederer *et al.*, 1989; Youdim *et al.*, 1989). Oxidation of DA may contribute to the oxidative stress seen with aging and with certain pathological conditions of the human brain.

The mechanism of the MPTP-induced neuronal death has received great attention (Tipton and Singer, 1993). A biologically active metabolite of MPTP, MPP^+, inhibits oxidative phosphorylation (Mizuno *et al.*, 1987) and the resulting energy crisis causes neuronal death. It has recently been reported that MPP^+ administered directly into the striatum induced a large DA efflux (Chiueh *et al.*, 1992; Ozaki *et al.*, 1987). The autooxidation of the released DA produces hydroxyl radicals resulting in neuronal damage (Chiueh *et al.*, 1992; Maruyama *et al.*, 19995b). Thus, oxidative stress may be involved in the cell death induced by neurotoxins, in addition to the energy crisis (Maruyama et al., 1995a and b).

NM(*R*)Sal and 1,2-$DMDHIQ^+$ have been found to have bioactivity similar to MPTP and MPP^+. Using an animal model of Parkinson's disease with NM(*R*)Sal and 1,2-$DMDHIQ^+$, 1,2-$DMDHIQ^+$ was found to be more cytotoxic than the reduced NM(*R*)Sal (Naoi *et al.*, 1996a). The catechol isoquinolinium ion induced massive necrosis in the striatum, but this cytotoxicity was non-selective. The mechanism of the potent cytotoxicity of 1,2-$DMDHIQ^+$, and oxidation of NM(*R*)Sal should be examined in relation to hydroxyl radical generation and to their possible inhibition of oxidative phosphorylation in mitochondria.

As described above, DHTIQs increased free monoamine levels in the brain and reduced DA catabolism in the rat brain. These results suggest that DHTIQs may increase or reduce hydroxyl radical levels in the brain, either by the release of DA followed by autooxidation or by the inhibition of oxidation of DA. In this section of our review, we summarize the effects of (*R*)Sal, NM(*R*)Sal, and 1,2-$DMDHIQ^+$ on hydroxyl radical production in the rat striatum using the *in vivo* microdialysis technique.

2,3- Dihydroxybenzoic and 2,5-dihydroxybenzoic acid (2,3- and 2,5-DHBA) were purchased from Sigma (St. Louis, MO), and salicylic acid was from Nacalai Tesque (Kyoto, Japan). The procedure for brain microdialysis in the striatum is described above. Twenty-four to 48 hours after the surgery, the dialysis probe was inserted, and 1mM salicylic acid solution dissolved in Ringer's solution was perfused at a flow rate of 2μL/minute. Hydroxyl radicals were trapped with salicylic acid as 2,3- and 2,5-DHBA (Halliwell *et al.*, 1991). The concentrations of 2,3- and 2,5-DHBA, monoamines and their metabolites were measured until their levels stabilized. Then, perfusion of the catechol isoquinolines was started. Dialysis samples were collected every 20 minutes and 20μL of dialysate was used for analysis. Monoamines, their metabolites, 2,3- and 2,5-DHBA were quantitatively analyzed using the CEAS. The conditions of the analysis are described above. Basal levels of the compounds in the dialysate were determined for a 1-hour period, thereafter Ringer's solution containing 1mM salicylic acid and 40 or 200μM of (*R*)Sal, NM(*R*)Sal, or 1,2-$DMDHIQ^+$ was perfused for 2 hours.

When Ringer's solution containing salicylic acid was perfused through the rat striatum, 2,3-DHBA and 2,5-DHBA were detected in the dialysate. After the stabilization period, the levels of 2,3- and 2,5-DHBA and DA metabolites reached a plateau; and the basal values obtained were 26.9±5.7 nM (mean±SD) for 2,3-DHBA and 40.9±8.9nM for 2,5-DHBA. Then, the catechol isoquinolines were perfused. (*R*)Sal, NM(*R*)Sal, and 1,2-DMDHIQ$^+$ were found to affect the level of 2,3-DHBA in the dialysate, and the time-dependent change in its level is shown in Figure 9. (*R*)Sal (Figure 9, I-A and II-A) and 1,2-DMDHIQ$^+$ (Figure 9, I-C and II-C) reduced the 2,3-DHBA levels significantly ($p<0.0156$), and the reduction was greater with the 200μM isoquinolines (II-A and II-C) than with 40μM (I-A and I-C). On the other hand, 40μM NM(*R*)Sal significantly ($p<0.0156$) increased the 2,3-DHBA level (I-B), whereas perfusion with 200μM NM(*R*)Sal (II-B) reduced the level markedly ($p< 0.0156$).

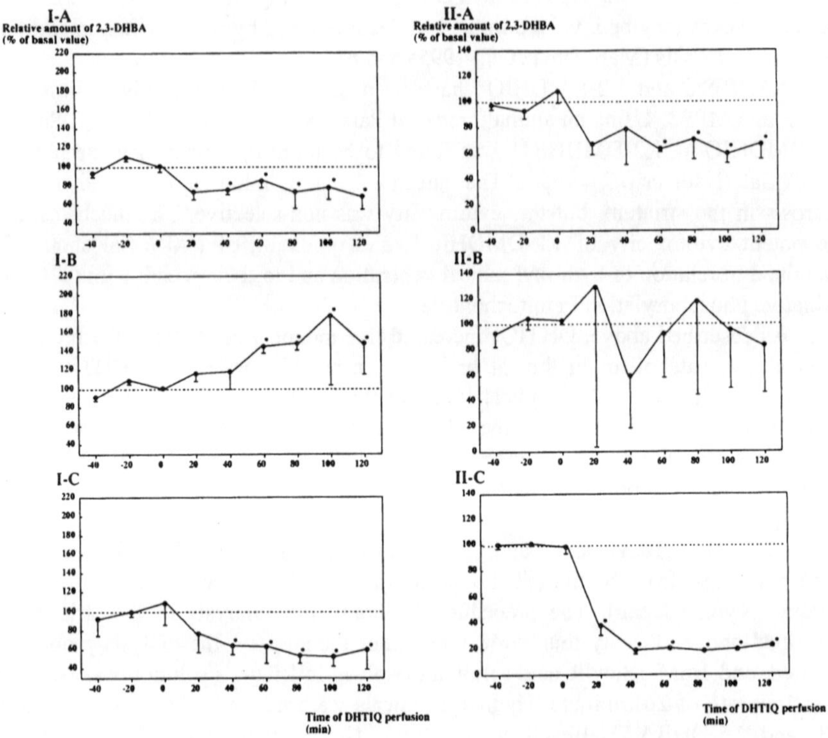

Figure 9. Time-dependent changes in 2,3-DHBA concentration in the striatal dialysate after perfusion with (*R*)Sal, NM(*R*)Sal or 1,2-DMDHIQ$^+$. Ringer's solution containing 1mM salicylic acid was perfused through the rat striatum, and hydroxyl radicals were trapped as 2,3-DHBA and analyzed by the CEAS. The relative amount of 2,3-DHBA was expressed as a percentage from the basal levels. The rat striatum was perfused with 40μM (*R*)Sal (I-A), 40μM NM(*R*)Sal (I-B), 40μM 1,2-DMDHIQ$^+$ (I-C), 200μM (*R*)Sal (II-A), 200μM NM(*R*)Sal (II-B), or 200μM 1,2-DMDHIQ$^+$ (II-C) for 120 minutes. Each point and bar represent the mean±SD of 4 experiments.

Using the *in vivo* microdialysis technique, we showed that 1,2-DMDHIQ$^+$ reduced the levels of hydroxyl radicals markedly. The reduction in hydroxyl radical levels by the catechol isoquinolines is thought to be due to the inhibition of MAO activity. Maruyama *et al.*, (1993a and b) showed that these catechol isoquinolines reduce its catabolism in the brain. As we confirmed earlier by *in vitro* experiments, the catechol isoquinolines are potent inhibitors of type A MAO (Minami *et al.*, 1992; Naoi *et al.*, 1993c and 1994c). The K_i value of 1,2-DMDHIQ$^+$ towards type A MAO was calculated to be 8μM, whereas the concentration recovered in the striatal dialysate was much higher; 22.3±4.5μM for the perfusion of 40μM 1,2-DMDHIQ$^+$. The inhibition of MAO activity by catechol isoquinolines reduces H_2O_2 and hydroxyl radical generation. These results indicate that 1,2-DMDHIQ$^+$ is not, or only poorly, autooxidized, but that it also behaves as a radical scavenger *in vitro*. The effects of (*R*)Sal on hydroxyl radical formation *in vivo* was much weaker than that of 1,2-DMDHIQ$^+$.

NM(*R*)Sal was found to increase hydroxyl radical formation during perfusion of the 40μM solution into the rat striatum. The increased formation of hydroxyl radicals might reflect the production of the reactive oxygen species by autooxidation of NM(*R*)Sal, as proved by *in vitro* experiments (Maruyama *et al.*, 1995c). The autooxidation NM(*R*)Sal produces 1,2-DMDHIQ$^+$ and hydroxyl radicals, simultaneously.

CONCLUSION

Our results on the biosynthesis of neurotoxins and their effects on monoamine metabolism were obtained by use of the CEAS and *in vivo* microdialysis perfusion. The data reported here were obtained from minimal sample volume (10 to 50μL) which were injected directly into the CEAS. We could identify approximately 30 compounds from a single HPLC chromatogram, including monoamines, their metabolites, and dihydroxyisoquinoline derivatives. The conjugated form of monoamines can also be identified. The multi-electrode system can measure most aromatic amino acids directly, and following derivatization with *o*-phthalaldehyde, most naturally occurring amino acids can also be measured in only 1μL of sample and with excellent limits of detection (Naoi *et al.*, 1993b). With the availability of more standard compounds, the range of species measured will be even further increased. In such a case, the data analysis system of the CEAS can store the real HPLC chromatograms and by comparing them to the new standards compounds, other peaks in the chromatogram can be identified and quantitated at any time in the future. Our results demonstrate how the application of a new, sensitive and accurate analytical method can further develop research in neurochemistry and neuropharmacology.

Acknowledgment

This work was supported by Grant-In-Aid for Scientific Research on Priority Area and (C) from the Ministry of Education, Science and Culture, Japan. We thank Dr. K. Murayama and other researchers at MC Medical (Tokyo, Japan) for their skillful help with the CEAS.

REFERENCES

Chiueh, C. C., Krishna, G., Tulsi, P., Obata, T., Lang, L., Huang, S. and Murphy, D. L. (1992). Intracranial microdialysis of salicylic acid to detect hydroxyl radical generation through dopamine autooxidation in the caudate nucleus: effects of MPP^+. Free Rad. Biol. Med. **13**, 581-583.

Davis, C. G., Williams, A. C., Markey, S. P., Ebert, M. H., Caine, E. D., Reichert, C. M. and Kopin, I. J. (1979). Chronic parkinsonism secondary to intravenous injection of meperidine analogues. Psychiat. Res. **1**, 249-254.

Dostert, P., Strolin Benedetti, M., Dordain, G. and Vernay, D. (1989). Enantiometric composition of urinary salsolinol in Parkinsonian patients after Madopar. J. Neural Transm. [P-D Sect] **1**, 269-278.

Dostert, P., Strolin Benedetti, M., Bellotti, V., Allievi, C. and Dordain, G (1990). Biosynthesis of salsolinol, a tetrahydroisoquinoline alkaloid, in healthy subjects. J. Neural Transm. **81**, 215-223.

Dostert, P., Strolin Benedetti, M., Dordain, G. and Vernay, D. (1991). Urinary elimination of salsolinol enantiomers in alcoholics. J. Neural Transm. [GenSect] **85**, 51-59.

Halliwell, B., Kaur, H. and Ingelman-Sundberg, M. (1991). Hydroxylation of salicylate as an assay for hydroxyl radicals: a cautionary note. Free Rad. Biol. Med. **10**, 439-441.

Hernandez, L., Lee, F. and Höbel, B. G. (1987). Simultaneous microdialysis and amphetamine infusion in the nucleus accumbens and striatum of freely moving rats: increase in extracellular dopamine and serotonin. Brain Res. Bull. **19**, 623-628.

Langston, J. W., Ballard, P., Tetrud, J. W. and Irwin, I. (1984). Chronic parkinsonism in humans due to a product of meperidine-analog synthesis. Science **219**, 979-980.

Maruyama, W., Nakahara, D., Ota, M., Takahashi, T., Takahashi, A., Nagatsu, T. and Naoi, M. (1992). N-Methylation of dopamine-derived 6,7-dihydroxy-1,2,3,4-tetrahydroisoquinoline, (R)-salsolinol, in the rat brains: *in vivo* microdialysis study. J. Neurochem. **59**, 395-400.

Maruyama, W., Nakahara, D., Dostert, P., Hashiguchi, H., Ohta, S., Hirobe, M., Takahashi, A., Nagatsu T. and Naoi, M. (1993a). Selective release of serotonin by endogenous alkaloids, 1-methyl-6,7-dihydroxy-1,2,3,4-tetrahydroisoquinolines, (R)- and (S)salsolinol, in the rat striatum; *in vivo* microdialysis study. Neurosci. Lett. **149**, 115-118.

Maruyama, W., Nakahara, D., Dostert, P., Takahashi, A. and Naoi, M. (1993b). Naturally-occurring isoquinolines perturb monoamine metabolism in the brain: studied by *in vivo* microdialysis. J. Neural Transm. [GenSect] **94**, 91-102.

Maruyama, W., Takahashi, T., Minami, M., Takahashi, A., Dostert, P., Nagatsu, T. and Naoi, M. (1993c). Cytotoxicity of dopamine-derived 6,7-dihydroxy-1,2,3,4-tetrahydroisoquinolines. Adv. Neurol. **60**, 224-230.

Maruyama, W., Nakahara, D., Dostert P., Takahashi, T. and Naoi, M. (1995a). Dopamine-derived isoquinolines as dopaminergic neurotoxins and oxidative stress. In: Alzheimer's and Parkinson's Diseases. Recent Developments. Hanin I., Yoshida M. and Fisher A. (Eds.). Plenum Press, New York and London, pp. 575-581.

Maruyama, W., Dostert, P. and Naoi, M. (1995b). Dopamine-derived 1-methyl-6,7-dihydroxyisoquinolines as hydroxyl radical promoters and scavengers in the rat brain: *in vivo* and *in vitro* studies. J. Neurochem. **64**, 2635-2643.

Maruyama, W., Dostert P., Matsubara, K. and Naoi, M. (1995c). N-Methyl(*R*)salsolinol produces hydroxyl radicals: involvement to neurotoxicity. Free Rad. Biol. Med. **19**, 67-75.

Maruyama, W., Abe, T., Tohgi, H., Dostert, P. and Naoi, M. (1886a). A dopaminergic neurotoxin, (*R*)-N-methylsalsolinol, increases in parkinsonian CSF. Ann. Neurol. in press.

Matsubara, K., Collins, M.A. and Neafsey, E.J. (1992a). Mono-N-methylation of 1,2,3,4-tetrahydro-β-carbolines in brain cytosol: Absence of indole methylation. J. Neurochem. **59**, 505-510.

Matsubara, K., Neafsey, E.J. and Collins, M.A. (1992b). Novel S-adenosyl methionine-dependent indol-N-methylation of β-carbolines in brain particulate fractions. J. Neurochem. **59**, 511-518.

Matsubara, K., Ota, M., Takahashi T., Maruyama, W. and Naoi, M. (1994). Structural studies of condensation products of biogenic amines as inhibitors of tryptophan hydroxylase. Brain Res. **655**, 121-127.

Minami, M., Takahashi, T., Maruyama, W., Takahashi, A., Dostert, P., Nagatsu, T. and Naoi, M. (1992). Inhibition of tyrosine hydroxylase by R and S enantiomers of salsolinol, 1-methyl-6,7-dihydroxy-1,2,3,4-tetrahydroisoquinoline. J. Neurochem. **58**, 2007-2101.

Minami, M., Maruyama, W., Dostert, P., Nagatsu, T. and Naoi, M. (1993). Inhibition of type A and type B monoamine oxidase by 6,7-dihydroxy-1,2,3,4-tetrahydroisoquinolines and their N-methylated derivatives. J. Neural Transm. [GenSect] **92**, 125-135.

Mizuno, Y., Sone, N. and Saitoh, T. (1987). Effects of 1-methyl-4-phenyl-1,2,3,6-tetrahydropyridine and 1-methyl-4-phenylpyridinium ion on activities of the enzymes in the electron transport system in mouse brain. J. Neurochem. **48,** 1787-1793.

Nakahara, D., Ozaki, N. and Nagatsu, T. (1993). *In vivo* microdialysis of neurotransmitters and their metabolite. In: Methods in Neurotransmitter and Neuropeptide Research, Parvez, S.H., Naoi, M., Nagatsu, T. and Parvez, S. (Eds.). Elsevier, Amsterdam, Part 1, pp. 219-248.

Nakahara, D., Maruyama, W., Hashiguchi, H. and Naoi, M. (1994). Characterization in the *in vivo* action of (*R*)-salsolinol, an endogenous metabolite of alcohol, on serotonin and dopamine metabolism: A microdialysis study. Brain Res. **644,** 226-232.

Naoi, M., Matsuura, S., Takahashi, T. and Nagatsu, T. (1989). A N-methyltransferase in human brain catalyses N-methylation of 1,2,3,4-tetrahydroisoquinoline into N-methyl-1,2,3,4- tetrahydro-isoquinoline, a precursor of a dopaminergic neurotoxin, N- methylisoquinolinium ion. Biochem. Biophys. Res. Commun. **161,** 1213-1219.

Naoi, M., Dostert, P., Yoshida, M. and Nagatsu, T. (1993a). N-Methylated tetrahydroisoquinolines as dopaminergic neurotoxins. Adv. Neurol. **60,** 212-217.

Naoi, M., Maruyama, W., Acworth, I. N., Nakahara, D. and Parvez, S. H. (1993b). Multi-electrode detection system for detection of neurotransmitters. In: Methods in Neurotransmitter and Neuropeptide Research, Parvez, S.H., Naoi, M., Nagatsu, T. and Parvez, S. (Eds.). Elsevier, Amsterdam, Part 1., pp. 1-39.

Naoi, M., Maruyama, W., Minami, M., Dostert, P., Parvez, S. H. and Nagatsu, T. (1993c). Salsolinols and other 6,7-dihydroxy-1,2,3,4-tetrahydroisoquinolines as inhibitors of monoamine oxidase: *in vivo* and *in vitro* study. Biogenic Amines **9,** 367-379.

Naoi, M., Maruyama, W. and Dostert, P. (1994a). Binding of 1,2(N)-dimethyl-6,7-dihydroxyisoquino-linium ion to melanin: effects of ferrous and ferric ion on the binding. Neurosci. Lett. **171,** 9-12.

Naoi, M., Maruyama, W., Niwa, T. and Nagatsu, T. (1994). Novel toxins and Parkinson's disease: N-methylation and oxidation as metabolic bioactivation of neurotoxin. J. Neural Transm. [Suppl.] **41,** 197-205.

Naoi, M., Maruyama, W., Sasuga, S., Deng, Y., Dostert, P., Ohta, S. and Takahashi, T. (1994b). Inhibition of type A monoamine oxidase by 2(N)-methyl-6,7-dihydroxyisoquinolinium ions. Neurochem. Int. **25,** 475-481.

Naoi, M., Maruyama, W., Dostert, P., Nakahara, D., Takahashi, T. and Nagatsu, T. (1995a). Metabolic bioactivation of endogenous isoquinolines as dopaminergic neurotoxins to elicit Parkinson's disease. In: Alzheimer's and Parkinson's Diseases. Recent Developments. Hanin, I., Yoshida, M. and Fisher, A. (Eds.). Plenum Press, New York and London, pp. 553-559.

Naoi, M., Maruyama, W. and Dostert, P. (1995b). Dopamine-derived 6,7-dihydroxy-1,2,3,4-tetrahydroisoquinolines: Oxidation and neurotoxicity. Progress Brain Res., **106,** 227-239.

Naoi, M., Maruyama, W., Zhang, J.H, Takahashi, T., Deng, Y. and Dostert, P. (1995c). Enzymatic oxidation of the dopaminergic neurotoxin, 1(*R*),2(*N*)-dimethyl-6,7-dihydroxy-1,2,3,4-tetrahydro-isoquinoline, into 1,2(*N*)-dimethyl-6,7-dihydroxyisoquinolinium ion. Life Sci. **57,** 1061-1066.

Naoi, M., Maruyama, W., Dostert, P., Hashizume, Y., Nakahara, D., Takahashi, T. and Ota, M. (1996a). Dopamine-derived endogenous 1(*R*),2(*N*)-dimethyl-6,7-dihydroxy-1,2,3,4-tetrahydro-isoquinoline, *N*-methyl-(*R*)-salsolinol, induced parkinsonism in rat: biochemical, pathological and behavioral studies. Brain Res. **709,** 285-295.

Naoi, M., Maruyama, W., Dostert, P., Kohda, K. and Kaiya, T. (1996b). A novel enzyme enantio-selectively synthesizes (*R*)salsolinol, a precursor of a dopaminergic neurotoxin, N-methyl(*R*)salsolinol. Neurosci. Lett. in press.

Niwa, T., Takeda, N., Yoshizumi, H., Tatematsu, A., Yoshida, M., Dostert, P., Naoi, M. and Nagatsu, T. (1991). Presence of 2-methyl-6,7-dihydroxy-1,2,3,4-tetrahydroisoquinoline and 1,2-dimethyl-6,7-dihydroxy-1,2,3,4-tetrahydroisoquinoline, novel endogenous amines, in parkinsonian and normal brains. Biochem. Biophys. Res. Commun. **177,** 603-609.

Ota, M., Dostert, P., Hamanaka, T., Nagatsu, T. and Naoi, M (1992). Inhibition of tryptophan hydroxylase by (R)- and (S)-1-methyl-6,7-dihydroxy-1,2,3,4- tetrahydroisoquinolines (salsolinols). Neuropharmacol. **31,** 337-341.

Ozaki, N., Nakahara, D., Kaneda, N., Kiuchi, K., Okada, T., Kasahara, Y. and Nagatsu, T. (1987). Acute effects of 1-methyl-4-phenylpyridinium ion (MPP^+) on dopamine and serotonin metabolism in rat striatum as assayed *in vivo* by a micro-dialysis technique. J. Neural Transm. **70**, 241-250.

Paxinos, G. and Watson, C. (1986). The rat brain in stereotaxic coordinates. 2nd edition. Academic Press, Sydney.

Riederer, P., Sofic, E., Rausch, W-D., Schmidt, B., Reynolds, G. P., Jellinger, K. and Youdim, M. B. H. (1989). Transition metals, ferritin, glutathione, and ascorbic acid in Parkinsonian brains. J. Neurochem. **52**, 515-520.

Sandler, M., Carter, S.B., Hunter, K. R. and Stern, G. M. (1973). Tetrahydroisoquinoline alkaloids: *in vivo* metabolites of L-DOPA in man. Nature **241**, 439-437.

Sanft, K. and Thomas, H. (1989). Competitive inhibition of catechol-O-methyltransferase by the tetrahydroisoquinoline alkaloids salsolidine and 1-carboxysalsoline. J. Biosc. **44**, 173-176.

Strolin Benedetti, M., Bellotti, V., Pianeola, E., Moro, E., Carminati, P. and Dostert, P. (1989). Ratio of the R and S enantiomers of salsolinol in food and human urine. J. Neural Transm. 77, 47-53.

Sjöquest, B., Borg, S. and Kvande, H. (1981). Salsolinol and methylated salsolinol in urine and cerebrospinal fluid from healthy volunteers. Subst. Alcohol Actions Misuse **2**, 73-77.

Sjöquest, B., Erisson, A. and Winblad, B. (1982). Salsolinol and catecholamines in human brain and their relation to alcoholism. Prog. Clin. Biol. Res. **90**, 57-67.

Takahashi, T. Deng, Y., Maruyama, W., Dostert, P. Kawai, M. and Naoi, M. (1994). Uptake of a neurotoxin-candidate, (R)-1,2-dimethyl-6,7-dihydroxy-1,2,3,4-tetrahydroisoquinoline into human dopaminergic neuroblastoma SH-SY5Y cells by dopamine transport system. J. Neural Transm. [GenSect] **98**, 107-118.

Teitel, S., O'Brien, J. and Brossi, A. (1972). Alkaloids in mammalian tissue. J. Med. Chem. **15**, 845-846.

Tipton, K. F. and Singer, T. P. (1993). Advances in our understanding of the mechanism of the neurotoxicity of MPTP and related compounds. J. Neurochem. **61**, 1191-1206.

Ung-Chhun, N., Cheng, B.Y., Pronger, D. A., Serrano, P., Chavez, B., Perez, R. F., Morales, J. and Collins, M. A. (1985). Alkaloid adducts in human brain: coexistence of 1-carboxylated and noncarboxylated isoquinoline and ß-carbolines in alcoholics and nonalcoholic. Prog. Clin. Biol. Res. **183**, 125-136.

Youdim, M. B. H., Ben-Shachar, D. and Riederer, P. (1989). Parkinson's disease, a progressive siderosis of substantia nigra resulting in iron and melanin induced neurodegeneration. Acta Neurol. Scand. **126**, 47-54.

Progress in HPLC-HPCE, Vol. 6, pp. 247–258
I.N. Acworth *et al.* (Eds)

HPLC-MULTI-ELECTROCHEMICAL DETECTION SYSTEM: APPLICATION TO NEUROCHEMICAL STUDIES IN PEDIATRICS

YOSHIRO TAKEUCHI, SATOSHI MATSUO, AKIRA NISHIMURA, ICHIRO YAMAZOE, SYOUJI KAWASE and TADASHI SAWADA

Department of Pediatrics, Kyoto Prefectural University of Medicine, Kawaramachi-Hirokoji, Kamikyo-ku, Kyoto 602, Japan

INTRODUCTION

The levels and interrelationships of neurotransmitters, precursors and metabolites are considered to determine the function and integration of the central nervous system. Thus, analyses of these molecules can provide useful information as to the neurochemical mechanisms underlying the pathogeneses of neurological disorders, including degenerative and convulsive disorders (Naoi *et al.*, 1993). The simultaneous measurement of neurotransmitters, precursors and metabolites by means of the Coulochem Electrode Array System (CEAS) provides an overview of their synthesis, release, reuptake and catabolism (Matson *et al.*, 1987 and 1984; Naoi *et al.*, 1993). In the field of neuropediatrics, a few studies involving the CEAS have been reported (Langlais *et al.*, 1991; Yamamoto, 1991; Yamamoto *et al.*, 1994). Consequently, we have applied the CEAS to investigation of monoamine metabolism in cerebrospinal fluid (CSF) in order to clarify the neurochemical abnormalities underlying several pediatric neurological disorders and the mechanism of a new antiepileptic drug.

The levels of monoamine metabolites in CSF generally reflect the turnover of biogenic amines in the central nervous system. Age-related changes in monoamine metabolite levels have been reported in children. Maturational changes in storage pools, rates of turnover, transport systems, and the rate of CSF production may alter the CSF concentrations of such metabolites (Seifert *et al.*, 1980; Silverstein and Hohnston, 1984; Silverstein *et al.*, 1984 and 1986; Yamamoto, 1991). The metabolite levels are also influenced by other factors, such as the time of lumbar puncture and administered drugs. Even considering that a number of factors may influence the levels of monoamine metabolites, CSF still remains one of the most useful tissues for examining monoamine metabolism in the human central nervous system.

MATERIALS AND METHODS

CSF was collected by means of lumbar puncture, while children were in the lateral decubitus position, between 10:00 a.m. and noon. Each CSF sample was immediately frozen at –80°C for later analysis. The CSF was collected in 2 aliquots, the second aliquot (1.0mL) being used for neurochemical analysis. We were careful when sampling to ensure that the same volume of CSF was collected and that the same portion was used for each analysis. This prevented any variation due to the rostrocaudal gradient of the monoamine metabolites in the CSF (Hyland *et al.*, 1993).

For analyses of monoamine metabolites, CSF (50μL) was injected into a computer-controlled, high performance liquid chromatography with 16 high sensitivity electrochemical detectors (CEAS, ESA, Chelmsford, MA). The data for the following compounds are presented: tryptophan (TRP), 5-hydroxyindoleacetic acid (5-HIAA), kynurenine (KYN), homovanillic acid (HVA), and 3-methoxy-4-hydroxyphenyl glycol (MHPG).

MONOAMINE METABOLISM IN INFANTILE SPINAL MUSCULAR ATROPHY

The childhood form of spinal muscular atrophy (SMA) is one of the motor neuron diseases and is clinically subdivided into three types: acute Werdnig-Hoffmann disease (type I), intermediate Werdnig-Hoffmann disease (type II), and Kugelberg-Welander disease (type III). Both the acute and chronic SMA loci were recently mapped to chromosome 5q.11.2-13.3 (Gilliam *et al.*, 1990; Munsat *et al.*, 1990) but the underlying neurochemical defects, which may determine the clinical heterogeneity of this devastating disease, remain unknown. From a neuroanatomical viewpoint, we proposed to re-evaluate the significance of the serotonin (5-HT) neuron system in motor neuron diseases (Takeuchi *et al.*, 1983). Except for the extraocular muscle nuclei, the motor nuclei of the cranial nerves and the motor neurons of the spinal cord receive rich input from 5-HT neurons in all species (Takeuchi *et al.*, 1983). In addition, the coincidence between the motor nuclei rich in 5-HT and the lesions in motor neuron disease is noteworthy from a neuropathological aspect. However, little is known of monoamines in motor neuron disease, especially in SMA.

CSF was obtained from 4 SMA type I infants and 5 control infants. Cases 1, 2, 3 and 5 comprised the SMA group and case 4 was included to compare the data for triplets (cases 2-4). All the SMA infants exhibited muscle weakness within 1 month after birth. Control samples were obtained from infants who required a lumbar puncture for the diagnosis of possible meningitis and peripheral neuropathy, and were later proved to be neurologically intact. The significance of differences between SMA type I patients and controls was determined by means of the Student t-test (Takeuchi *et al.*, 1994).

The CSF levels of monoamine metabolites in each case are presented in Figure 1. The CSF level of TRP in the SMA type I patients (746.4±48.0ng/mL) was not significantly different from the control level (782.0±121.0ng/mL). The level of 5-HIAA in the SMA type I patients (92.3±34.4ng/mL) was significantly lower than that in the controls (131.6±12.0ng/mL, $p<0.05$). The level of HVA in the SMA type I patients (85.3±17.6ng/mL) was also significantly lower than that in the controls

(152.2±18.2ng/mL, p<0.001). Although the level of KYN in the SMA type I patients (29.5±32.1ng/mL) did not differ significantly from the control level (40.0±10.2ng/mL), the level of KYN was extremely high in case 4, whose clinical course was atypical and in whom the disease progressed very slowly, while the level of KYN in cases 1-3 was apparently lower than that in the controls (Figure 1).

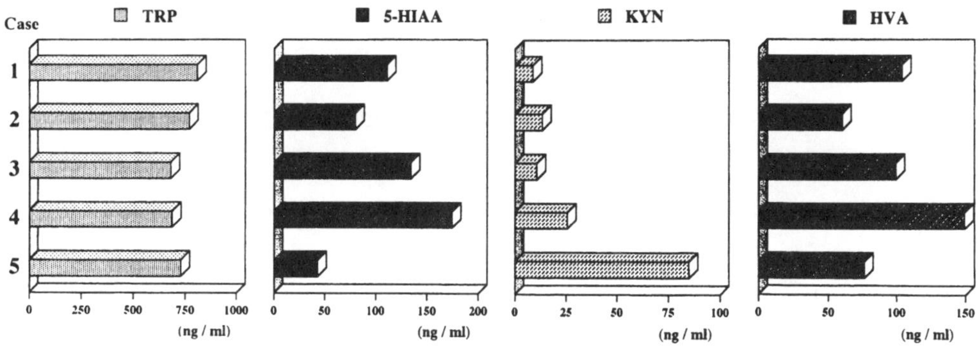

Figure 1. CSF monoamine metabolite levels (ng/mL) in patients with SMA type I, and case 4. Case Nos. are in order of decreasing severity.

In general, the significance of CSF monoamine metabolite levels should be evaluated with caution, and interpretation of the data depends on the comparability of patients and controls (Hyland *et al.*, 1993). In this study, CSF samples from SMA patients were obtained at around 1 month after birth and strictly age-matched controls were used. Therefore, the differences in the levels of 5-HIAA and HVA, between the SMA and control groups demonstrated in this study, are considered to be valid, suggesting the existence of abnormalities in the monoamine neuron system in SMA type I. Moreover, the strong relationship between motor activity and 5-HT neuronal discharges also indicates that the reduction of 5-HT metabolites in the CSF of SMA type I cases is a primary change (Jacobs and Fornal, 1993). The reason for the decreased dopamine metabolites cannot be discussed from a neuroanatomical viewpoint, but it is likely that 5-HT turnover has a regulatory action on dopamine (DA) turnover. The decreases in the concentrations of 5-HIAA and KYN appear to be related to the severity of SMA type I (Figure 1). TRP is catabolized to 5-HT and KYN through different metabolic pathways (Figure 2). In addition, kynurenic acid, a metabolite of KYN, is the only known endogenous antagonist of excitatory amino acid receptors, which plays an important role in neuronal death in the developing brain (Freese *et al.*, 1990; Stone and Connick, 1985). The reduced levels of CSF 5-HIAA and KYN may provide an insight into the nature of the neurochemical abnormalities in SMA type I patients (Takeuchi *et al.*, 1994). Therefore, it is concluded that SMA should be re-evaluated in the light of tryptophan metabolism, especially the kynurenine pathway.

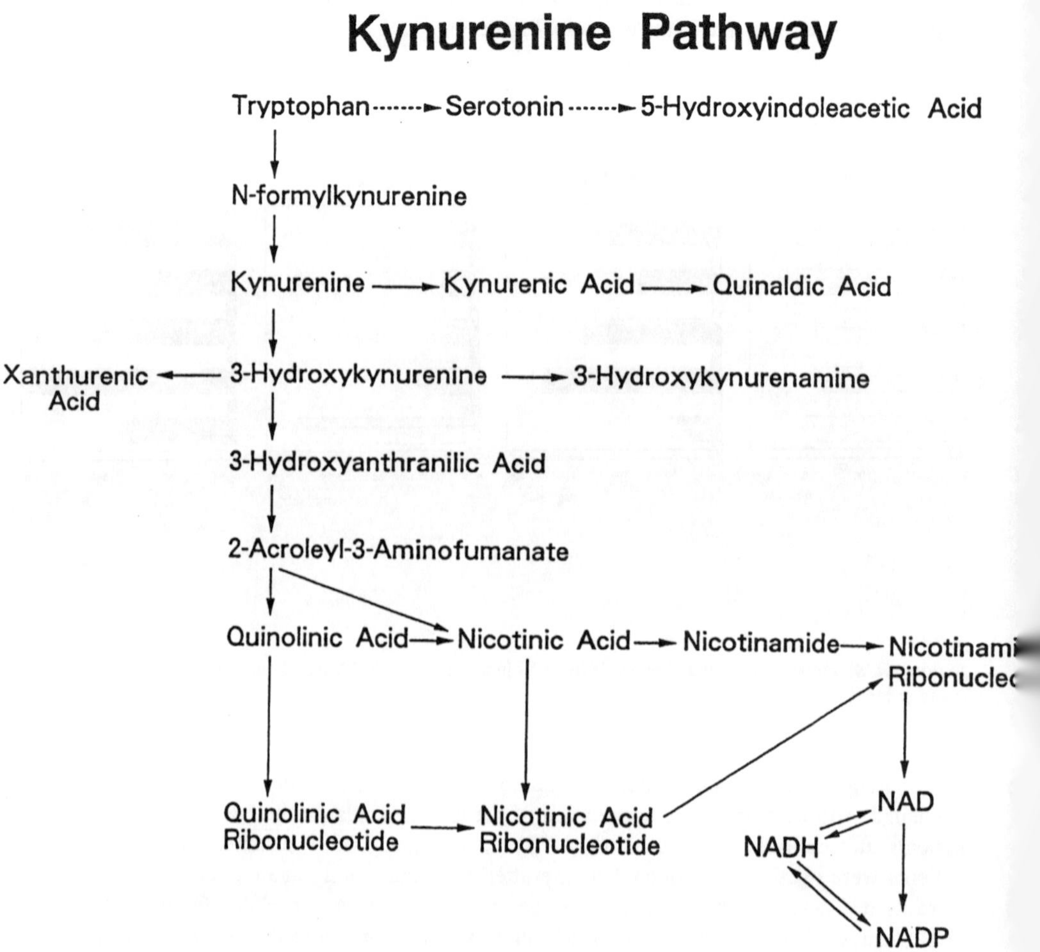

Figure 2. A summary of the metabolic pathways from tryptophan, showing details of the kynurenine pathway.

MONOAMINE METABOLISM IN THE NEUROLEPTIC MALIGNANT SYNDROME

The causes of rhabdomyolysis, that is defined as a clinical syndrome resulting from skeletal muscle injury with release of the muscle cell contents into the plasma, are varied, the neuroleptic malignant syndrome (NMS) being one of the main causes of drug-induced rhabdomyolysis (Levenson, 1985). The levels of monoamine metabolites in CSF were measured serially in a case which exhibited all the diagnostic criteria of NMS (Komatsu *et al.*, 1992).

The patient suffered from spastic quadriplegia, involuntary movements, and difficulty in walking and speaking from 10 years old. On January 1991, he suddenly developed an extremely high fever, delirium, severe opisthotonus and muscle rigidity. The hyperthermia lasted for 4 days and the serum level of creatine kinase increased with time. Thus, we diagnosed this case as having rhabdomyolysis. The concentrations of

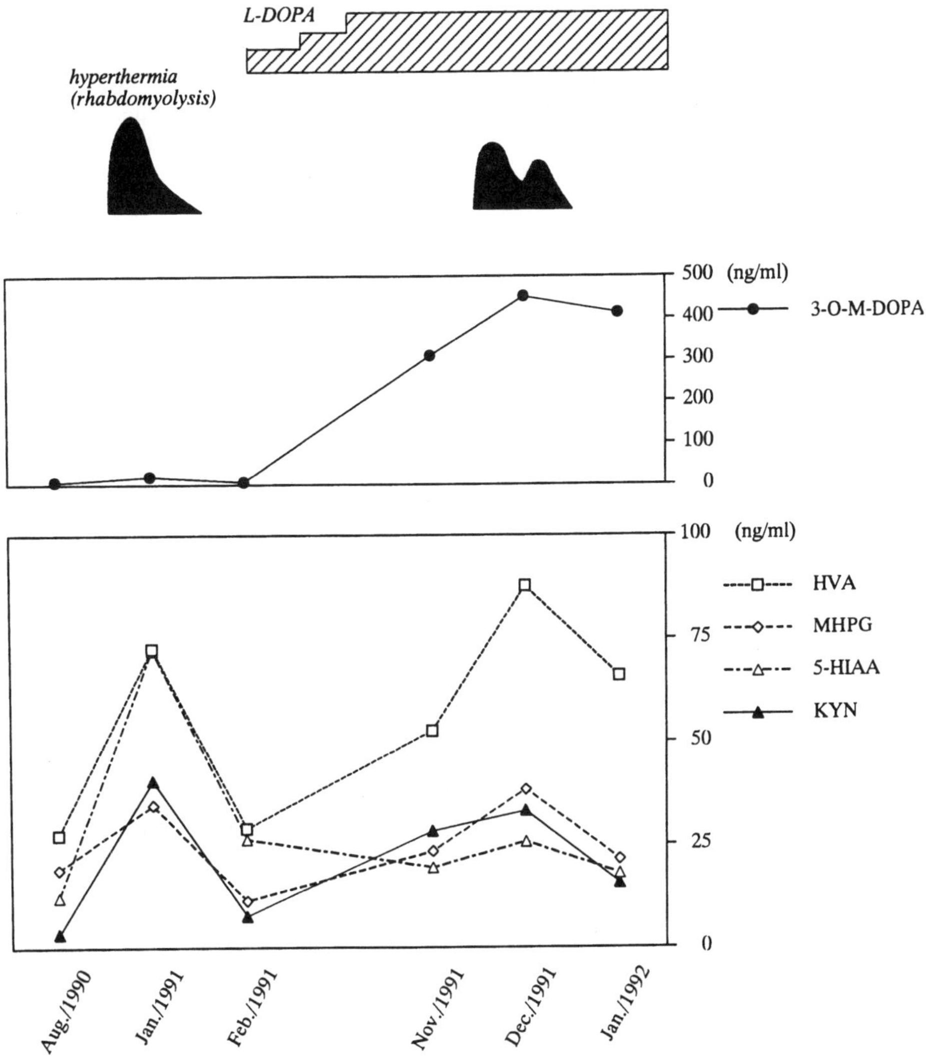

Figure 3. Clinical course of a patient with the neuroleptic malignant syndrome and changes in monoamine metabolites in CSF (ng/mL).

monoamine metabolites in CSF were determined serially, with parental consent. The concentrations of MHPG, 5-HIAA, HVA and KYN were all normal on August 1990, when he exhibited no symptoms of rhabdomyolysis, but remarkable elevation was observed on January 1991, when he developed rhabdomyolysis (Figure 3). These elevated CSF levels of MHPG, 5-HIAA, HVA and KYN returned to within normal ranges on February 1991. From March 1991, he received L-DOPA for treatment of the extrapyramidal symptoms. After the L-DOPA treatment, a marked amount of 3-O-methyl-DOPA (O-M-DOPA) was detected in CSF (Figure 3). 3-O-Methyl-DOPA may compete with L-DOPA and interfere with the uptake of TRP by 5-HT neurons, possibly resulting in worsening of the neurological symptoms.

The most commonly accepted model of the pathogenesis of NMS is a DA receptor blockade (Ansseau *et al.*, 1986; Yamawaki *et al.*, 1986). An excessive dopamine receptor blockade may cause extrapyramidal symptoms, and a hypothalamic blockade may account for disturbance of the autonomic and thermoregulatory functions. Hyperthermia in NMS induced by a hypothalamic DA receptor blockade is considered to be mediated by secondary activation of 5-HT mechanisms. The increased CSF MHPG, 5-HIAA, HVA and KYN levels suggest that NMS in this case may be related to central monoaminergic dysfunctions, including the dopaminergic, noradrenergic and serotonergic systems.

MONOAMINE METABOLISM IN INTRACTABLE EPILEPSY AFTER THYROTROPIN THERAPY AND THYROTROPIN-RELEASING HORMONE ADMINISTRATION

In spite of rapid advances in the treatment of epilepsy with the introduction of new drugs, a number of children with epilepsy remain uncontrolled, so a potent, but safer therapy is still required. Thyrotropin-releasing hormone (TRH), the smallest known peptide hormone, has central effects unrelated to its endocrine action. Recently, TRH was successfully used for treating children with such neurological disorders as intractable epilepsy, spinal muscular atrophy and cerebellar ataxia (Matsuishi *et al.*, 1983, 1987 and 1989; Takeuchi *et al.*, 1994, 1989 and 1995). A large number of animal experiments were performed to study the mechanism underlying the antiepileptic action of TRH (Agarwal *et al.*, 1977; Kerwin and Pycock, 1979; Sato *et al.*, 1984; Sharp *et al.*, 1982; Yamashita *et al.*, 1990), and it has been suggested that TRH regulates the release and turnover of monoamines, and may exert an antiepileptic action through a monoamine mechanism. However, there have been no reports of changes in monoamines induced by TRH therapy in human epilepsy.

We investigated 18 children, 10 males and 8 females, aged 6 months to 12 years, with intractable epilepsy, that remained uncontrolled even after therapy with ACTH and several antiepileptic drugs, including clonazepam, valproate, phenytoin, phenobarbital and vitamin B6.

The details of the TRH therapy were previously reported (Takeuchi *et al.*, 1995). Briefly, TRH-tartrate (TRH-T) was administered at a dose of 0.05mg/kg/day once daily at about 10:00 a.m., for 4 weeks. During the study, the patients continued to receive the previous doses of their antiepileptic agents. The efficacy of serial TRH-T administration, in the terms of the frequency of clinical seizures and changes in

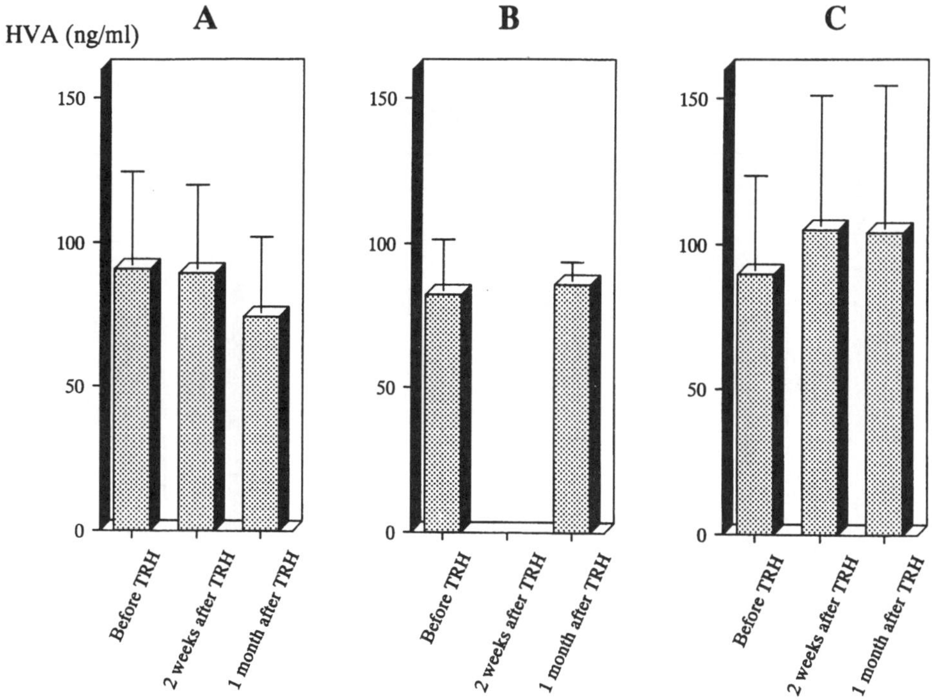

Figure 4. CSF HVA levels before, and 2 weeks and 1 month after TRH therapy in intractable epilepsy (mean ± SD). Group A: very effective; group B: effective, group C; not effective.

electroencephalographic discharges, was evaluated at the end and 1 year after the TRH treatment. Before the TRH therapy, and 2 weeks and 1 month after initiation of the therapy, neurological examination and a lumbar puncture were carried out, between 10:00 am and noon, in most of the patients, with parental consent.

The subjects were classified into three groups based on the effects of TRH therapy: cessation of seizures and seizure discharges for more than 1 year (very effective; group A), moderate reduction of seizures and/or seizure discharges (effective; group B), and no changes in the frequency of seizures or discharges (not effective; group C). There were 6 cases in group A, 3 in group B, and 9 in group C. In each group (A-C), there were no significant changes in monoamine metabolites after the TRH therapy. The data are reported as mean±SD (Figures 4-7).

Despite the "negative" findings, these results concerning the CSF concentrations of monoamine metabolites are important for understanding how TRH works in the case of intractable epilepsy. Although the results of this study do not completely rule out the possibility of an antiepileptic effect of TRH through monoamine mechanisms, other possibilities should be considered. The relatively prolonged duration of the efficacy of the TRH therapy may indicate that the mechanism underlying the antiepileptic action of TRH differs from that of other antiepileptic drugs (Takeuchi *et al.*, 1995). For example, TRH may contribute to the plasticity of the seizure network, analogous to its known trophic effect on spinal motoneurons (Kubek *et al.*, 1993; Takeuchi *et al.*, 1994).

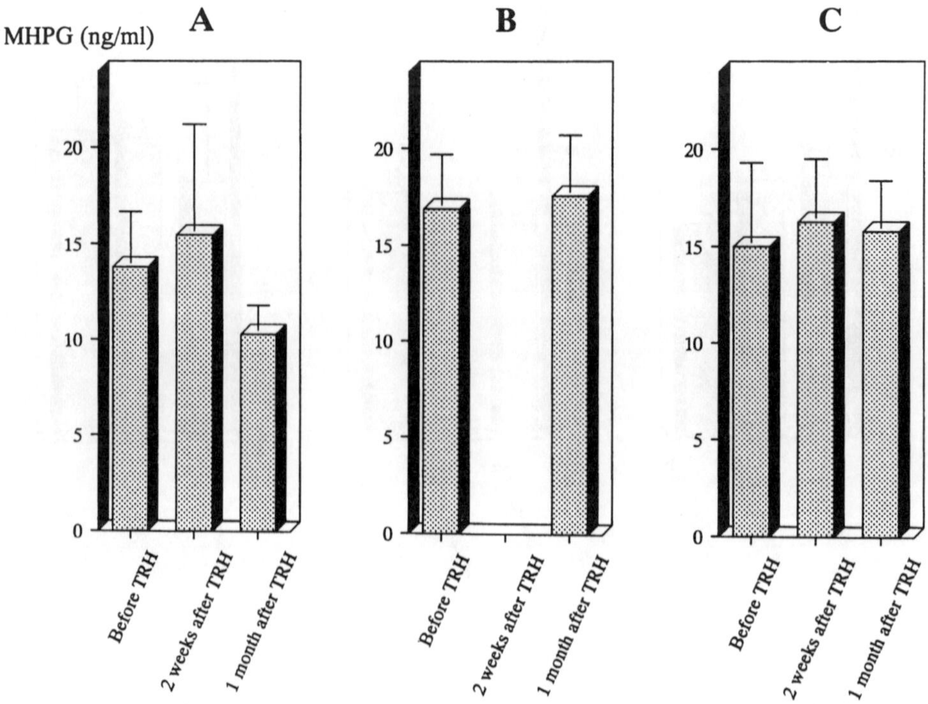

Figure 5. CSF MHPG levels before, and 2 weeks and 1 month after TRH therapy in intractable epilepsy (mean ± SD). Group A: very effective; group B: effective, group C; not effective.

FUTURE TRENDS IN NEUROPEDIATRICS

The multi-electrode detector system has many advantages over other conventional analytical systems. The simultaneous measurement of neurotransmitters, precursors and metabolites, including monoamines and peptides, with this system requires only a small amount of a sample, such as CSF and blood. This point is clinically very valuable, especially in the field of neuropediatrics, for studying the metabolism of neurotransmitters.

For the application of the CEAS to clinical neuroscience and for understanding of the pathogeneses of neurological disorders, it is crucial to combine the knowledge of all fields of neuroscience, i.e., neurochemistry, neuropathology, neuroanatomy, neuropharmacology, neurophysiology and clinical neurology. For example, owing to the progress in chemical neuroanatomy in the past decade, knowledge concerning the precise and detailed distribution of most neurotransmitters in the brain is available at present. Thus, we should utilize the knowledge of chemical neuroanatomy to analyze and understand the results concerning the dynamics of neurotransmitters in CSF, demonstrated with the CEAS.

Regarding monoamine metabolism, the kynurenine pathway, which is a metabolic pathway leading from TRP to nicotinamide adenine dinucleotide and includes both quinolinic and kynurenic acids, has become the subject of great interest. The kynurenine group is worth serious consideration as a crucial factor in degenerative and convulsive disorders. Both in causing cell death and producing convulsions, quinolinic acid is the most potent member of the kynurenine group, and is the only one to produce neuronal excitation (Stone and Connick, 1985). On the other hand, kynurenic acid is the only known endogenous antagonist of excitatory amino acid receptors and is considered to be neuroprotective, therefore it may also influence important physiologic and pathologic processes in neurological disorders. In neuropediatrics, it would be of considerable value to use the CEAS to measure the compounds in the kynurenine pathway with known neuroactive properties (such as quinolinic and kynurenic acids) as well as those with no currently reported neurological roles (such as kynurenine and 3-hydroxykynurenine).

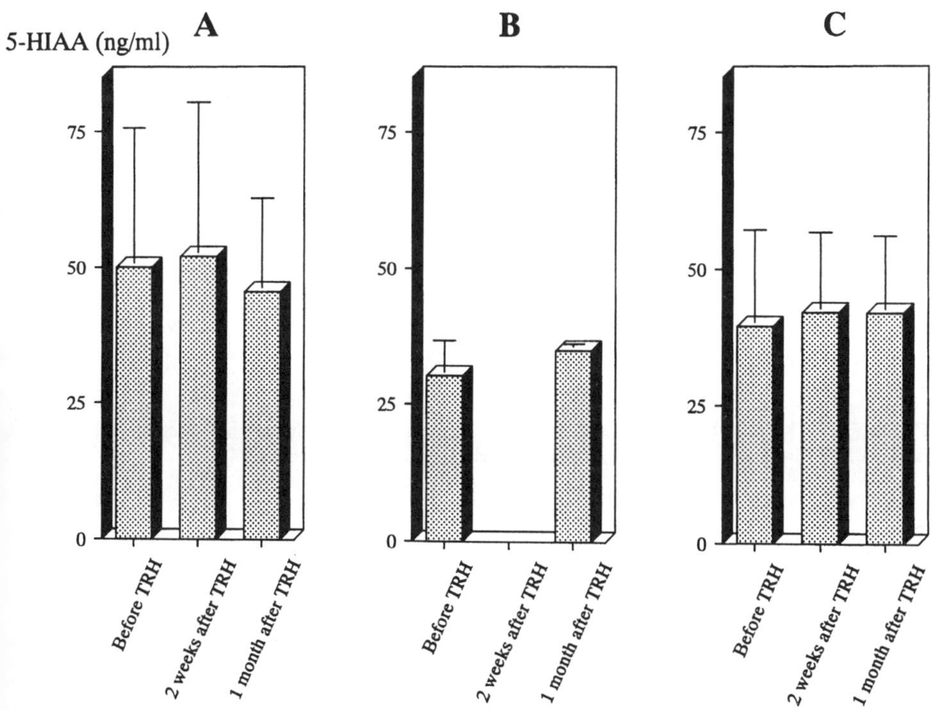

Figure 6. CSF 5-HIAA levels before, and 2 weeks and 1 month after TRH therapy in intractable epilepsy (mean ± SD). Group A: very effective; group B: effective, group C; not effective.

Other than monoamines and their metabolites, analyses of amino acids, such as glutamate, aspartate, glycine and gamma-amino butyric acid (GABA), in the CSF are very useful for understanding the pathogeneses of neurological disorders, especially degenerative and convulsive disorders. Simultaneous analyses of monoamines and amino acids in CSF are at present being performed using the CEAS, and the findings will provide a substantial insight into the nature of neurochemical abnormalities in the neurological disorders in children.

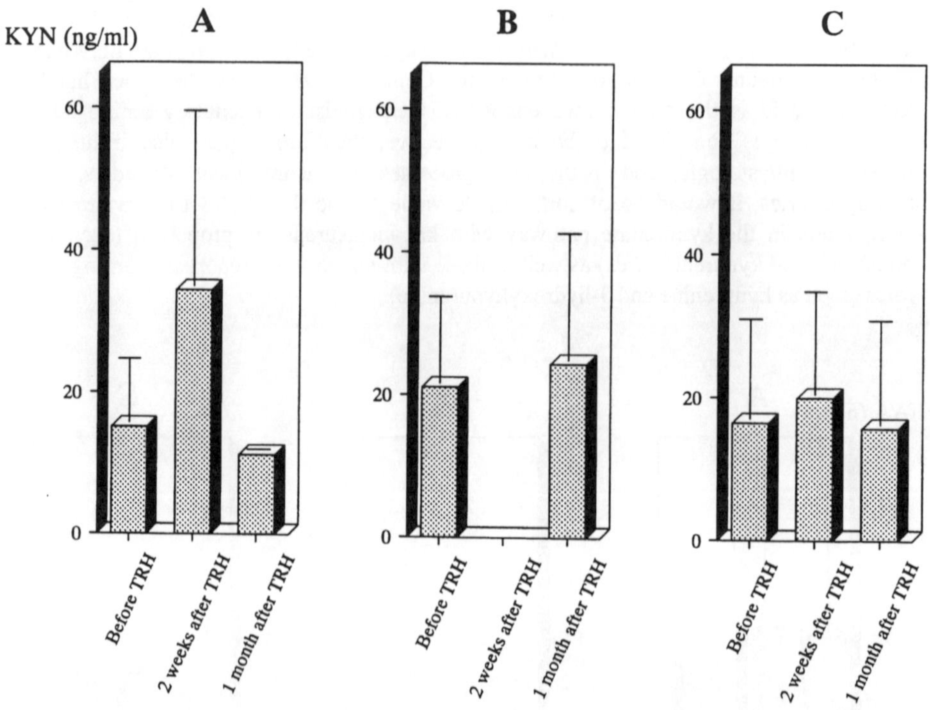

Figure 7. CSF KYN levels before, and 2 weeks and 1 month after TRH therapy in intractable epilepsy (mean ± SD). Group A: very effective; group B: effective, group C; not effective.

Acknowledgment
This study was supported by grants No. 07457181 and 08877126 from the Ministry of Education, Science and Culture, Japan.

REFERENCES

Agarwal, R.A., Rastogi, R. B. and Singhal, R. L. (1977). Enhancement of locomotor activity and catecholamine and 5-hydroxytryptamine metabolism by thyrotropin releasing hormone. Neuroendocrinol. **23**, 236-247.

Ansseau, M., Reynolds, C., Kupfer, D. J., Poncelet, P. F., Franck, G., Dresse, A. E. and Reznik, M. (1986). Central dopaminergic and noradrenergic receptor blockade in a patient with neuroleptic malignant syndrome. J. Clin. Psychiatry **47**, 320-321.

Freese, A., Swartz, K. J., During, M. J. and Martin, J. B. (1990). Kynurenine metabolites of tryptophan: implications for neurologic diseases. Neurology **40**, 691-695.

Gilliam, T. C., Brzustowicz, L. M., Castilla, L. H., Lehner, T., Penchaszadeh, G. K., Daniels, R. J., Byth, B. C., Knowles, J., Hislop, J. E., Shapira, Y., Dubowitz, V., Munsat, T. L., Ott, J. and Davies, K. E. (1990). Genetic homogeneity between acute and chronic forms of spinal muscular atrophy. Nature **345**, 823-825.

Hyland, K., Surtees, R. A., Heales, S. J., Bowron, A., Howells, D. W. and Smith, I. (1993). Cerebrospinal fluid concentrations of pterins and metabolites of serotonin and dopamine in a pediatric reference population. Pediatr. Res. **34**, 10-14.

Jacobs, B. L. and Fornal, C. A. (1993). 5-HT and motor control: a hypothesis. Trends Neurosci. **16**, 346-352.

Kerwin, R. W. and Pycock, C. J. (1979). Thyrotrophin releasing hormone stimulates release of [3H]-dopamine from slices of rat nucleus accumbens in vitro. Br. J. Pharmacol. **67**, 323-325.

Komatsu, H., Nishimura, A., Okano, S., Miyanomae, Y., Takeuchi, Y. and Sawada, T. (1992). Neuroleptic malignant syndrome-like state in a patient with Down syndrome and basal ganglia calcification. Brain Dev. **14**, 400-403.

Kubek, M. J., Knoblach, S. M., Sharif, N. A., Burt, D. R., Buterbaugh, G. G. and Fuson, K. S. (1993). Thyrotropin-releasing hormone gene expression and receptors are differentially modified in limbic foci by seizures. Ann. Neurol. **33**, 70-76.

Langlais, P. J., Wardlow, M. L. and Yamamoto, H. (1991). Changes in CSF neurotransmitters in infantile spasms. Pediatr. Neurol. **17**, 440-445.

Levenson, J. L. (1985). Neuroleptic malignant syndrome. Am. J. Psychiatry **142,** 1137-1145.

Matson, W. R., Gamache, P. G., Beal, M. F. and Bird, E. D. (1987). EC array sensor concepts and data. Life Sci . **41,** 905-908.

Matson, W. R., Langlais, P., Volicer, L., Gamache, P. H., Bird, E. and Mark, K. A. (1984). n-Electrode three-dimensional liquid chromatography with electrochemical detection for determination of neurotransmitters. Clin. Chem. **30**, 1477-1488.

Matsuishi, T., Yano, E., Inanaga, K., Terasawa, K., Ishihara, O., Shiotsuki, Y., Katafuchi, Y., Aoki, N. and Yamashita, F. (1983). A pilot study on the anticonvulsive effects of a thyrotropin-releasing hormone analog in intractable epilepsy. Brain Dev. **5**, 421-8.

Matsumoto, A., Kumagai, T., Takeuchi, T., Miyazaki, S. and Watanabe, K. (1987). Clinical effects of thyrotropin-releasing hormone for severe epilepsy in childhood: a comparative study with ACTH therapy. Epilepsia **28**, 49-55.

Matsumoto, A., Kumagai, T., Takeuchi, T., Miyazaki, S. and Watanabe, K. (1989). Factors influencing effectiveness of thyrotropin-releasing hormone therapy for severe epilepsy in childhood: significance of serum prolactin levels. Epilepsia **30**, 45-49.

Morimoto, K. and Goddard, G. V. (1985). Effects of thyrotropin-releasing hormone on evoked responses and long-term potentiation in dentate gyrus of rat. Exp. Neurol. **90**, 401-410.

Munsat, T. L., Skerry, L., Korf, B., Pober, B., Schapira, Y., Gascon, G. G., Al-Rajeh, R. S., Dubowitz, V., Davies, K., Brzustowicz, L.M., Penchaszadeh, G.K. and Gilliam, T.C. (1990). Phenotypic heterogeneity of spinal muscular atrophy mapping to chromosome 5q11.2-13.3 (SMA 5q). Neurology **40**, 1831-1836.

Naoi, M., Maruyama, W., Acworth, I. N., Nakahara, D. and Parvez, H. (1993). Multi-electrode detection system for determination of neurotransmitters. In: Methods in Neurotransmitter and Neuropeptide Research, Part 1. **11.** Techniques in The Behavioral and Neural Sciences.Parvez, S.H., Naoi, M., Nagatsu, T. and Parvez, S. (Eds.). Elsevier Science Publishers BV, Amsterdam, pp. 1-39.

Sato, M., Morimoto, K. and Wada, J. A. (1984). Antiepileptic effects of thyrotropin-releasing hormone and its new derivative, DN-1417, examined in feline amygdaloid kindling preparation. Epilepsia **25,** 537-544.

Seifert, W. J., Fox, J. L. and Butler, I. J. (1980). Age effect on dopamine and serotonin metabolite levels in cerebrospinal fluid. Ann. Neurol. **8,** 38-42.

Sharp, T., Bennett, G. W. and Marsden, C. A. (1982). Thyrotrophin-releasing hormone analogues increase dopamine release from slices of rat brain. J. Neurochem. **39**, 1763-6.

Silverstein, F. and Johnston, M. V. (1984). Cerebrospinal fluid monoamine metabolites in patients with infantile spasms. Neurology **34**, 102-105.

Silverstein, F. S., Donn, S., Buchanan, K. and Johnston, M. V. (1984). Concentrations of homovanillic acid and 5-hydroxyindoleacetic acid in cerebrospinal fluid from human infants in the perinatal period. J. Neurochem. **43**, 1769-1772.

Silverstein, F. S., Hutchinson, R. J. and Johnston, M. V. (1986). Cerebrospinal fluid biogenic amine metabolites in children during treatment for acute lymphocytic leukemia. Pediatr. Res. **20**, 285-291.

Stone, T. W. and Connick, J. H. (1985). Quinolinic acid and other kynurenines in the central nervous system. Neuroscience **15**, 597-617.

Takeuchi, Y., Fujiwara, K., Ishimura, K. and Sawada, T. (1994). Efficacy of thyrotropin-releasing hormone in the treatment of cerebellar ataxia. J. Child Neurol. **9**, 287-289.

Takeuchi, Y., Fujiwara, K., Ishimura, K., Shimada, Y., Ochi, M., Yoshioka, H. and Sawada, T. (1989). Efficacy of thyrotropin-releasing hormone in the treatment of cerebellar ataxia. Pediatr. Neurol. **5**, 107-110.

Takeuchi, Y., Kojima, M., Matsuura, T. and Sano, Y. (1983). Serotonergic innervation on the motoneurons in the mammalian brainstem. Light and electron microscopic immunohistochemistry. Anat. Embryol. **167**, 321-333.

Takeuchi, Y., Komatsu, H., Matsuo, S., Hirai, K., Kawase, S., Nishimura, A. and Sawada, T. (1994). Monoamine metabolites in the cerebrospinal fluid in infantile spinal muscular atrophy. Neuroreport **5**, 898-900.

Takeuchi, Y., Tominaga, M., Mitsufuji, N., Yamazoe, I., Kawase, S., Nishimura, A., Matsuo, S. and Sawada, T. (1995). Thyrotropin-releasing hormone in the treatment of intractable epilepsy-neurochemical analysis of cerebrospinal fluid monoamine metabolites. Pediatr. Neurol. **12**, 139-145.

Yamamoto, H. (1991). Changes in CSF neurotransmitters during the first year of life. Pediatr. Neurol. **7**, 406-410.

Yamamoto, H., Shindo, I., Egawa, B. and Horiguchi, K. (1994) Kynurenic acid is decreased in cerebrospinal fluid of patients with infantile spasms. Pediatr. Neurol. **10**, 9-12.

Yamashita, K., Mori, A. and Otsuki, S. (1990) Changes in brain thyrotropin-releasing hormone (TRH) of seizure-prone El mice. Exp. Neurol. **108**, 71-75.

Yamawaki, S., Yanagawa, K., Morio, M. and Mori, K. (1986). Possible central effect of dantrolene sodium in neuroleptic malignant syndrome [letter]. J. Clin. Psychopharmacol. **6**, 378-379.

Progress in HPLC-HPCE, Vol. 6, pp. 259–271
I.N. Acworth *et al.* (Eds)

NEUROTRANSMITTER DISTURBANCES IN ALZHEIMER'S DISEASE WITH LEWY BODIES

PHILIP J. LANGLAIS

Department of Neurosciences, Univ. of California at San Diego, School of Medicine, Neurology Research Service, Veterans Affairs Medical Center, San Diego, CA and Department of Psychology, San Diego State University, San Diego, CA, 9218.

INTRODUCTION

There are several pathophysiological substrates implicated in the cognitive, affective, and motor disturbances observed in dementing illnesses among the elderly. Changes in the neurotransmitters acetylcholine, dopamine, serotonin, norepinephrine, and γ-aminobutyric acid (GABA) and their receptors have been observed in many degenerative disorders presenting with dementia. Among the best documented are significant reductions in biochemical markers of cholinergic activity in Alzheimer's disease (AD) (Bowen *et al.*, 1976; Rossor *et al.*, 1984) and of dopaminergic activity in Parkinson's disease (PD) (Hornykiewicz, 1973).

The San Diego Alzheimer's Disease Research Center (ADRC) has observed that 30-40% of clinically diagnosed and pathologically confirmed cases of AD contain cortical and subcortical Lewy Bodies (LBs) (Hansen *et al.*, 1990). Lewy bodies are intracytoplasmic eosinophilic neuronal inclusions that have been historically linked with Parkinson's disease (PD). Although PD was originally described as a movement disorder characterized by tremor, flexed posture and shuffling gait, many neurologist now recognize that PD is often accompanied by dementia. Neuropathologically the link between PD and Lewy bodies has also become muddled by the observation of Lewy bodies in AD and plaques and tangles in PD. Furthermore, recent studies have demonstrated that a high percentage of PD cases have neocortical Lewy bodies (reviewed in Hansen & Galasko, 1992). These findings make the clinical and pathological distinctions between PD and AD less clear than in the past.

Could these AD cases with subcortical and cortical Lewy represent the coexistence of AD and PD? Several features suggest not. Clinically, these cases have all of the features diagnostic of AD but also demonstrate more severe deficits of attention, fluency (letter but not category) and visuospatial processing than the pure AD cases. In addition, features associated with PD such as masked facies and bradykinesia are uncommonly mild, none have resting tremor, and about one-half have an action tremor. Neuropathologically, these cases are also different from pure AD. Although these cases contain the usual density of neocortical senile plaques, the plaques are diffuse and there are fewer neocortical neurofibrillary tangles than pure AD cases, and the neuronal loss in LC, substantia nigra (SN) and substantia innominata (SI) is far greater than observed in the pure AD cases. These cases differ neuropathologically from typical PD in that the pattern and distribution of neuronal loss in the substantia nigra (SN) is unusually mild compared to that observed in typical PD (Gibb & Lees, 1991).

On these grounds, it was suggested that the LBV cases represent a clinically and pathologically distinct subset of AD and labeled the Lewy Body variant (LBV) of AD (Hansen *et al.*, 1990). This view has been challenged by another investigator who proposes that the LBV cases are similar to "diffuse Lewy Body disease" (DLBD) and not a variant of AD (Dickson *et al.*, 1987). Another group has argued that these LBV cases probably represent PD with additional involvement of cerebral cortex and the presence of AD changes (Fearnley *et al.*, 1991). They point out that without a careful quantitative cell count in the various regions of SN, the actual loss of neurons may have been underestimated. A third group has described yet another disorder, i.e., senile dementia of Lewy body type (SDLT), in a group of psychiatric cases in which mild extrapyramidal features were apparent but most did not display the classical triad of tremor, rigidity and akinesia (Perry *et al.*, 1990). These cases also resemble the LBV cases in that they contain Lewy body formations in cortex and basal ganglia but at densities well below those typically observed in diffuse Lewy body disease. However, in contrast to the LBV cases, these SDLT cases do not meet the clinical criteria for AD and often present initially with a confused state, often fluctuating and associated with visual hallucinations and behavioral disturbances.

Several important questions are unanswered regarding the pathological-behavioral relationships in this group of patients. Finding the answers is significant to both their treatment and diagnosis. Are the motor disturbances related to dopamine deficiency in basal ganglia or to some unknown metabolic defect? The mild extrapyramidal disturbances of these patients is relatively unresponsive to levodopa treatment, suggesting that dopamine deficiency is not the underlying cause. Is the greater decline in cognitive functions due to a greater disruption of neocortical cholinergic activity or to additional loss of other neurotransmitter systems implicated in attention and cognitive processing? To provide answers to these questions and the issue of whether LBV is simply the coexistence of PD and AD, a fairly extensive neurochemical study was conducted. This study has been reported in detail elsewhere (Langlais *et al.*, 1993) and is briefly presented here to illustrate the growing need for multiple transmitter analyses in the study of neurodegenerative disorders. To conduct such studies requires sophisticated analytical tools that provide comprehensive yet cost-effective measurements of the entire metabolic pathway of several neuroactive compounds in samples of very limited size. For several years our laboratories have used a computerized, gradient HPLC with series coulometric EC detectors (Model CEAS, ESA Inc.) to simultaneously measure over 30 compounds in the monoamine synthetic and metabolic pathways in post-mortem brain and CSF. As will become evident from our findings, this approach is essential to uncovering unsuspected metabolic abnormalities in an ever increasing array of heterogeneous neurodegenerative disorders such as AD and PD.

NEUROTRANSMITTER DISTURBANCES IN AD & LBV

The levels of dopamine, norepinephrine, serotonin, kynurerine, their precursors tyrosine, DOPA, tryptophan, 5-hydroxytryptophan, and metabolites homovanillic acid (HVA), 3-methoxy,4-hyroxphenylglycol (MHPG), and 5-hydroxyindoleacetic acid (5-HIAA) were measured in frozen post-mortem tissue from caudate, putamen, and Brodman areas 8 (frontal cortex), 21 (temporal cortex), and 39/40 (posterior parietal cortex) with a single chromatographic assay (Langlais *et al.*, 1991) using a gradient, series coulometric-electrochemical detector HPLC system (Model CEAS, ESA Inc.). In addition, the level of tyrosine hydroxylase (TYR-OH) enzyme was measured with an immunoblotting technique (Girault *et al.*, 1989) and the activity of choline acetyltransferase (ChAT) with a radioenzymatic assay (Fonnum *et al.*, 1969). All subjects were diagnosed clinically (McKhann *et al.*, Neurol. 1984) and neuropathologically (Khachaturian, 1985) as having AD. The LBV cases were identified neuropathologically by the presence of Lewy bodies in neocortex and substantia nigra. The AD and LBV cases were matched for age of onset, sex, and postmortem interval. The age of onset, duration of illness, and age at death were similar in the AD and LBV groups. A similar proportion of patients in these groups had reached a floor on the Mini-Mental Status Exam and Blessed Information-Memory-Concentration (IMC) test (Blessed *et al.*, 1968) within the year prior to death, i.e., 2/6 for AD group and 3/7 for the LBV group. Extrapyramidal symptoms (EPS) developed early in 5 of the 7 LBV cases, at a stage when they were relatively mildly demented and preceding neuroleptic exposure. By contrast, none of the pure AD cases had EPS relatively early in the course of illness. Two pure AD cases did develop EPS but only after neuroleptic treatment. Post-mortem tissue from control subjects was obtained from the Brain Tissue Resource Center, McLean Hospital/Harvard Medical School and general autopsies conducted at the UCSD Medical Center.

Severe reductions of biochemical markers of dopamine activity were observed in the basal ganglia of the LBV cases, suggesting a biochemical resemblance to typical PD. In the LBV brains, dopamine levels were reduced to 5% of the control value in the putamen (Figure 1) and to 19% of controls in the caudate (Figure 2). Previous studies have shown that the loss of dopamine in idiopathic PD is likewise greater in the putamen (<10% of control value) than caudate (10-20% of control value) (Fahn et al., 1971; Kish *et al.*, 1988). The observations of a significant reduction of the dopamine precursor, DOPA, the major dopamine metabolite, HVA (Table 1), and TOH enzyme and normal levels of tyrosine within the putamen of the LBV cases are also consistent with a severe functional disruption of nigral cells and loss of terminals. These findings demonstrate that dopaminergic innervation of the basal ganglia of the LBV cases is as severely impaired as that observed in typical PD. Furthermore, reductions of 5-HT (Figure 1) and its metabolite 5-HIAA (Table 1) were observed in both caudate and putamen of the LBV cases that are similar to those observed in typical PD brains (Scatton *et al.*, 1983; Arai *et al.*, 1984; Hornykiewicz, 1982; Dubois *et al.*, 1983; Birkmayer *et al.*, 1974; Fahn *et al.*, 1971). In the only other study of brain dopamine levels in cases equivalent to LBV, i.e., senile dementia of the Lewy body type (SDLT), which contains AD pathology, nigral degeneration and neocortical Lewy bodies (Perry *et al.*, 1990) the dopamine level in caudate was only moderately reduced (42% of control levels) and congruent with the mild EPS observed in those patients.

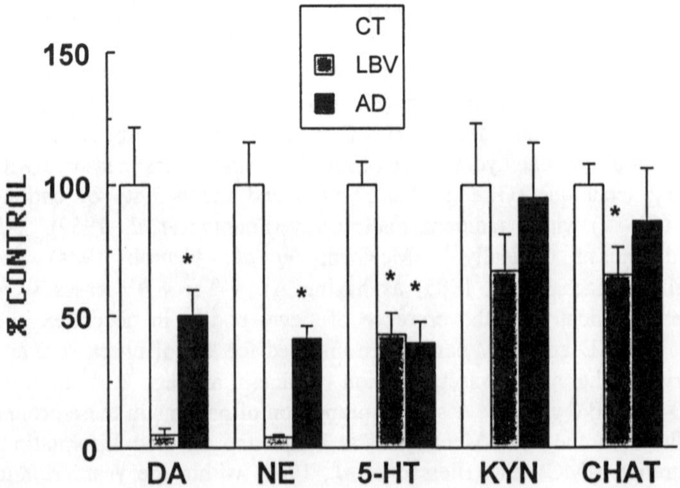

Figure 1. Concentration of dopamine (DA), norepinephrine (NE), serotonin (5-HT), kynurenine (KYN), and the activity of choline acetyltransferase (CHAT) in putamen of subjects diagnosed pathologically as pure Alzheimer disease (AD), Lewy Body variant of AD (LBV) and normal control subjects (CT). Values are the mean ± S.E.M. *p<.05 compared to controls, post hoc Scheffe test; ‡p<.05 compared to controls and AD, post hoc Scheffe test.

The severe dopamine depletion but relatively mild extrapyramidal symptoms (EPS) observed in these LBV cases suggest that additional pathophysiological changes exist in LBV which attenuate parkinsonism, particularly resting tremor. One possible site of damage is the subthalamic nucleus which when destroyed can reverse MPTP-induced PD in monkeys (Bergman *et al.*, 1990). This hypothesis seems unlikely since the subthalamic nucleus was observed to be histologically unremarkable in the few LBV cases examined (Langlais *et al.*, 1993). Damage to other structures within the extrapyramidal motor pathways, such as the globus pallidus (Bravo & Cooper, 1959) and thalamus (Kelly *et al.*, 1987), have been shown to reduce tremor and rigidity in PD. Acetylcholine is a neurotransmitter used in both interneuronal and efferent projections within these basal ganglia structures, and anticholinergic medication improves some of the symptoms of idiopathic PD. On this point, a significant reduction in cholinergic activity was observed within caudate of our LBV cases, averaging only 52% of controls and was less than 40% of the control in four of the seven LBV cases (Figure 2). In contrast, most previous studies have reported very little change in ChAT activity in caudate of PD cases or cases with combined AD and PD pathology (Perry *et al.*, 1990; Gaspar *et al.*, 1980; Rinne *et al.*, 1991; Sirvio *et al.*, 1989; Aubert *et al.*, 1992; Bowen *et al.*, 1976; Perry *et al.*, 1991; Araujo *et al.*, 1988).

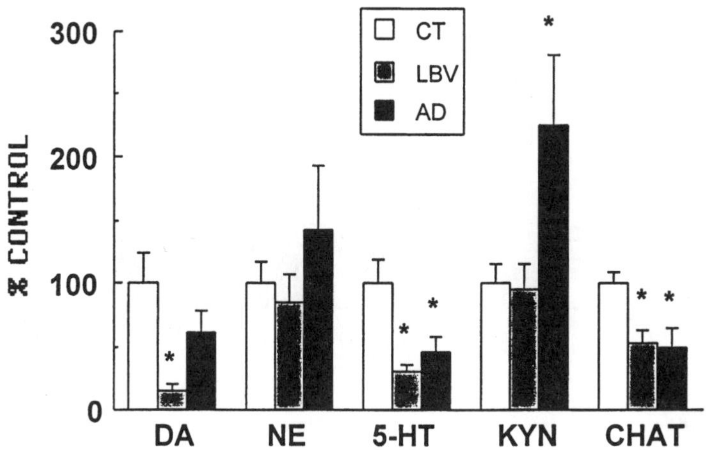

Figure 2. Concentration of dopamine (DA), norepinephrine (NE), serotonin (5-HT), kynurenine (KYN), and the activity of choline acetyltransferase (CHAT) in caudate of subjects diagnosed pathologically as pure Alzheimer disease (AD), Lewy Body variant of AD (LBV) and normal control subjects (CT). Values are the mean ± S.E.M. *$p<.05$ compared to controls, post hoc Scheffe test; ‡$p<.05$ compared to controls and AD, post hoc Scheffe test.

Additional pathophysiological changes were observed within the putamen of LBV, the site of greatest dopamine loss, which are not typically observed in PD nor AD. Within the putamen of the LBV cases, there was a severe reduction of norepinephrine (to 4% of control value) (Figure 1). This decline is much larger than that observed in typical PD (Hornykiewicz, 1982; Fahn *et al*, 1971; Rinne *et al.*, 1974) and suggests a loss of noradrenergic terminals. The presence of a normal level of MHPG in putamen of these LBV cases (Table 1) suggests a compensatory increase in NE turnover within the surviving terminals. It is possible that the reduced cholinergic and noradrenergic activity within basal ganglia contribute to the milder Parkinsonian symptoms of these LBV cases, as some theories of the pathophysiology of PD implicate a cholinergic-dopaminergic imbalance. Unfortunately, this hypothesis could not be tested because of a lack of longitudinal and detailed measures of EPS and other neurologic features. Further studies are needed to conduct similar biochemical analyses of basal ganglia structures in postmortem brain from patients with LBV, PD, and AD in whom careful assessments of motor disturbances, i.e., bradykinesia, rigidity, resting tremor, masked facies, postural instability, shuffling gait, stooped posture, and hypophonic speech are available. Comparison of the distribution and severity of these features with the magnitude of change in dopamine, norepinephrine, and acetylcholine markers within caudate and putamen should help to determine if these transmitter disturbances contribute to the milder Parkinson's symptoms of LBV.

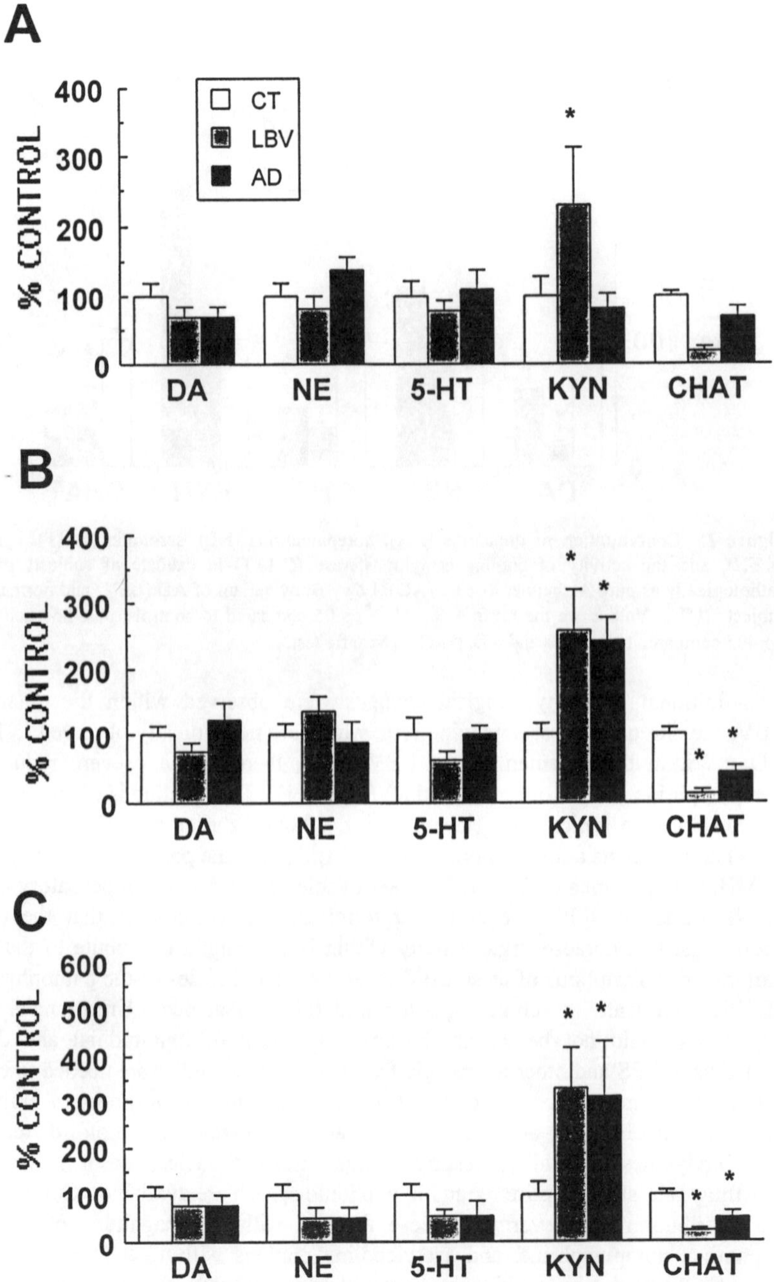

Figure 3. Concentration of dopamine (DA), norepinephrine (NE), serotonin (5-HT), kynurenine (KYN), and the activity of choline acetyltransferase (CHAT) in frontal (A), parietal (B), and temporal (C) neocortex of subjects diagnosed pathologically as pure Alzheimer disease (AD), Lewy Body variant of AD (LBV) and normal control subjects (CT). Values are the mean ± S.E.M. *$p<.05$ compared to controls, post hoc Scheffe test; ‡$p<.05$ compared to controls and AD, post hoc Scheffe test.

Table 1
Concentration of monoamines, metabolites and precursors in postmortem brain regions

REGION/ GROUP	HVA	MHPG	5-HIAA	TYR	DOPA	5-HTP
CAUDATE						
CT (N=6)	32.04±3.6	0.19±.03	1.11±.20	346±37	0.30±.07	0.07±.01
LBV (N=7)	11.38±2.3[a]	0.21±.04	0.33±.07 [a]	400±63	0.19±.07	0.06±.02
AD (N=6)	20.20±4.5	0.40±.03[c]	0.50±.13 [a]	435±28	0.25±.09	0.08±.02
PUTAMEN						
CT N=8)	50.54±7.5	0.35±.06	1.75±.16	296±52	2.12±.49	0.23±.08
LBV (N=7)	12.55±1.9[b]	0.32±.06	0.75±.14 a	376±18	0.69±.18[a]	0.08±.05[a]
AD (N=6)	38.09±2.7	0.39±.04	0.70±.14 [a]	391±24	1.62±.28	0.25±.10
FRONTAL						
CT (N=9)	2.32±0.4	0.17±.04	0.14±.03	339±34	0.11±.02	0.02±.01
LBV (N=6)	1.29±0.2	0.29±.12	0.11±.02	360±67	0.11±.03	0.03±.01
AD (N=6)	1.57±0.3	0.44±.10	0.15±.04	327±47	0.12±.03	0.03±.01
PARIETAL						
CT (N=9)	2.07±0.3	0.69±.12	0.08±.02	616±80	0.37±.04	0.10±.02
LBV (N=7)	1.73±0.4	0.86±.10	0.05±.01	673±79	0.38±.06	0.22±.07
AD (N=6)	2.49±0.5	0.62±.17	0.08±.01	654±83	0.48±..06	0.11±.04
TEMPORAL						
CT (N=12)	1.64±0.3	0.21±.02	0.09±.02	392±50	0.13±.02	0.03±.02
LBV (N=7)	1.10±0.1	0.20±.05	0.05±.01	423±115	0.10±.01	0.04±.03
AD (N=6)	1.53±0.2	0.14±.01	0.05±.02	654±83	0.08±.01	0.10±.03

[a] p<.01, compared to controls, post hoc Scheffe test;[b] p<.01, compared to controls and AD, post hoc Scheffe test. All statistically significant differences remained significant after Bonferroni correction for multiple comparisons. Values represent the mean ± S.E.M in ng/mg protein. HVA = homovanillic acid; MHPG = 3-methoxy-4-hydroxyphenylglycol; 5-HIAA = 5-hydroxyindoleacetic acid; TYR= tyrosine; DOPA = 3,4 Dihydroxyphenylalanine; 5-HTP = 5-hydroxytryptophan; CT = controls matched for age, sex, and postmortem interval to the LBV and AD groups; LBV = Lewy body variant; AD = Alzheimer's disease.

As mentioned previously, the LBV cases studied at our ADRC are different from pure AD in that they have more severe cognitive and memory deficits than AD patients (Hansen *et al.*, 1990). The loss of neocortical ChAT activity is greater in these LBV cases than in pure AD cases, particularly in frontal cortex (Figure 3), and may partially contribute to the suspected greater intellectual decline in these LBV cases. The loss of neocortical ChAT activity in our LBV cases was also greater than that observed in most previous studies of pure AD (Perry *et al.*, 1985, 1990, 1991; Aubert *et al.*, 1992; Bowen *et al.*, 1976; Araujo et al., 1988; Rossor et al., 1984; D'Amato *et al.*, 1987; Whitehouse *et al.*, 1988; Crow *et al.*, 1984). However, the severe reduction of ChAT in the LBV cases closely resemble many but not all studies of patients with combined AD and PD pathology (Perry *et al.*, 1985, 1990, 1991;

Aubert *et al.*, 1992; D'Amato *et al.*, 1987; Whitehouse *et al.*, 1988) and diffuse Lewy body disease (Dickson *et al.*, 1987). Furthermore, the levels of ChAT in the LBV cases are closer to those reported in severely demented PD cases than in most cases of pure AD (Aubert *et al.*, 1992; Sirvio *et al.*, 1989). In many of these PDD cases, AD pathology has been reported in conjunction with the PD lesions. However, in the LBV cases, the severe ChAT depletion appears unrelated to the severity of Alzheimer pathology since neocortical plaque and tangle counts in LBV cases were similar or lower than in pure AD cases (Langlais *et al.*, 1993). These observations suggest that the more severe reduction of ChAT activity may be more related to the presence of Lewy bodies than to numbers of plaques and tangles.

Very little is known about neocortical levels of the monoamines and their role in the behavioral and cognitive impairments of LBV, AD or PD. Several studies have failed to detect any significant differences in neocortical dopamine or homovanillic acid (HVA) levels of control and PD brains, perhaps due to the very low levels of dopamine and their high variability in postmortem brain. One study, however, has reported a significant reduction of dopamine (39% of control) in frontal cortex from PD cases (Scatton *et al.*, 1983). In this study, and a recently published study (Ogawa *et al.*, 1992), dopamine concentration in frontal cortex of LBV cases was reduced but these changes were not significant. However, a reduction of HVA (56% of control) was detected in frontal cortex of our LBV cases (Table) of a similar degree to that reported in PD (62% of control [Scatton *et al.*, 1983]; 58% of control [Perry *et al.*, 1991]) and in cases of combined AD/PD (Perry *et al.*, 1991). In unmedicated PD patients, Ogawa *et al.*, 1992 found even larger reductions (34% of control) of HVA in frontal cortex. The HVA concentration in neocortical regions from the LBV cases was also lower than in the current group of AD cases, suggesting some difference in neocortical dopamine utilization between the LBV and pure AD cases. It is clear, however, that more studies involving a larger number of subjects are needed to determine the integrity of the mesocortical dopamine systems of LBV as well as PD and AD.

We also observed reductions of serotonin (5-HT) concentration in frontal, temporal, and parietal neocortex of the LBV group (Figure 3). The high variance and small number of cases examined probably prevented the demonstration of statistically significant group differences. The absence of PD cases in our study also precluded examining whether these reductions in neocortical 5-HT in the LBV cases resemble typical PD. Two published studies have reported a significant reduction of 5-HT (49-55% of control) and increased 5-hydroxyindoleacetic acid/serotonin (5-HIAA/5-HT) ratio in frontal cortex of PD (Scatton *et al.*, 1983; Adolfsson *et al.*, 1979), suggesting enhanced 5-HT turnover in this region in PD. A third study found no significant changes in 5-HT concentration in frontal cortex of PD either on or off L-DOPA (Ogawa *et al.*, 1992). There was no evidence for this pattern in the frontal cortex of our LBV group since the reductions of 5-HT and 5-HIAA were comparable, and 5-HIAA/5-HT (9.0) was similar to controls (8.4). However, the reduced 5-HT content and near normal levels of 5-HIAA in temporal cortex (Table) suggest a compensatory increase in 5-HT turnover in this area of the LBV group. A similar increase in the 5-HT/5-HIAA ratio was also observed in temporal cortex of the AD group. One previous study (Adolfsson *et al.*, 1979) also observed normal 5-HT levels whereas another study reported a significant decline (39% of control) of 5-HT in frontal cortex of AD patients (D'Amato *et al.*, 1987). A third study (Arai *et al.*, 1984) observed a significant change of 5-HT in frontal but not temporal or parietal cortex and no significant change in 5-HIAA concentration in any of the three regions. It appears, therefore, that changes in serotonergic activity in the LBV and AD groups are highly variable and less severe within neocortical areas compared to basal ganglia structures. Correlational studies are needed to determine if these changes in serotonin activity within cortical regions are related to severity of neurological features and/or neuropsychometric measures of cognition and memory.

Significant elevation of kynurenine was also observed in temporal and parietal cortex of AD and LBV groups (Figure 3) and in the caudate of the AD group (Figure 2). These observations are significant because they suggest a possible mechanism for the loss of neurons and neurotransmitters in AD and LBV. Within the central nervous system, tryptophan is converted to either 5-HT or kynurenine through separate metabolic pathways (Figure 5). Kynurenine is further metabolized to both quinolinic

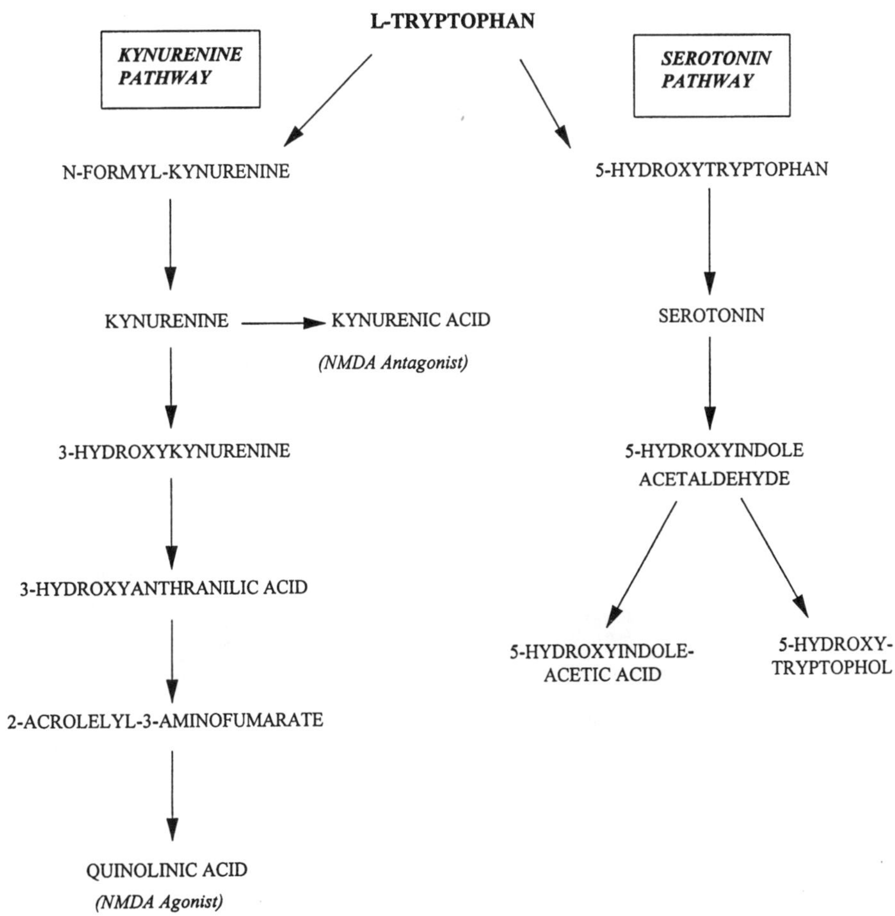

Figure 4. Schematic representation of the metabolism of tryptophan along the serotonin and kynurenine pathways

acid, an n-methyl-d-aspartate (NMDA) receptor agonist (Moroni *et al.*, 1988) and kynurenic acid, an NMDA receptor antagonist (Foster *et al.*, 1984). Increased concentrations of kynurenine and quinolinic acid and decreased levels of kynurenic acid could result in over activation of NMDA receptors, a mechanism implicated in accelerated neuronal death (Olney, 1989).

This type of excitotoxic mechanism involving increased kynurenine and decreased kynurenic acid concentrations has been implicated in the pathogenesis of Huntington's disease (Beal *et al.*, 1990). Our observation of a significant increase of kynurenine/5-HT + 5-HIAA in temporal and parietal cortex of the LBV group (Figure 5) is suggestive of a shift in the utilization of tryptophan from the 5-HT to kyrunenine metabolic pathways in LBV. Levels of the immediate serotonin precursor, 5-HTP, were not significantly altered in these regions (Table 1) but the values were very low and thus may not be a sensitive marker of activity within this pathway. In contrast, reduced levels of kynurenine and kynurenic acid have been reported in frontal cortex of pure PD cases (Ogawa *et al.*, 1992). Measurement of kynurenic acid and/or quinolinic acid in brain tissue are needed to provide a direct assessment of the hypothesis that these tryptophan-derived NMDA receptor ligands are significantly altered in brain regions of LBV and AD and may be strongly correlated with neuronal loss.

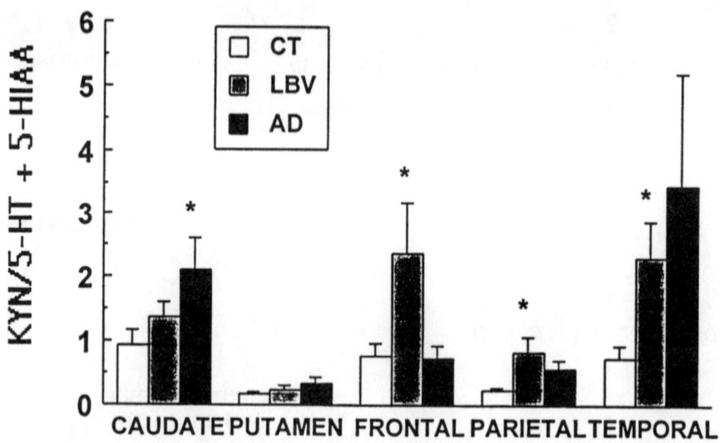

Figure 5. Ratio of the concentration of kynurenine to the combine concentrations of serotonin and its metabolite (KYN/5-HT + 5-HIAA) in basal ganglia and neocortical regions of subjects diagnosed pathologically as pure Alzheimer disease (AD), Lewy Body variant of AD (LBV) and expressed as percentage of value obtained from normal control subjects. Values are the mean ± S.E.M. *p<.05 compared to controls, post hoc Scheffe test.

CONCLUSION

Understanding the relationship between Lewy bodies and disturbances in neocortical and basal ganglia neurotransmitter activity will have a significant impact on the treatment of progressive dementing illnesses. As described above, results from our laboratories suggest that more severe disruptions of cholinergic, noradrenergic, and dopaminergic neurotransmission exist in a relatively large proportion of clinically and neuropathologically diagnosed cases of AD. These cases of AD have diffuse cortical and subcortical Lewy body and severe depletion of dopamine within basal ganglia characteristic of Parkinson's disease. Moreover these cases have additional loss of noradrenergic activity with basal ganglia not typically seen in PD and a decline in cholinergic activity within neocortex that is more pronounced than that observed in pure AD. These observations raise the possibility that imbalances in the interactions of dopaminergic-cholinergic and noradrenergic-cholinergic control of cortical and subcortical brain systems may be involved in the cognitive, mnemonic, and motor disturbances of these progressive dementing subjects. It is important, therefore, to examine a large number of cases of pure AD, PD, and the Lewy Body variant of AD to determine if greater disruptions in dopaminergic, noradrenergic, and or serotonergic activity exist in the LBV cases compared to typical PD and AD. If so, these findings would suggest alternative hypotheses regarding the neurochemical disturbances participating in the clinical symptoms and guide in the development of different and more effective pharmacologic treatments.

Acknowledgements

Supported by funds from the VA Merit Program, NIH/DRR grants S10 RRO 4654 and AG05131, and PHS grant MH NS 31862. This work was conducted in collaboration with Drs. Douglas Galasko, Larry Hansen, Leon Thal, and Eliezer Masliah and with the technical assistance of Michael Alford, Andy Chen, and Mark Wardlow.

REFERENCES

Adolfsson, R., Gottfries, C. G., Roos, B. E. and Winblad, B. (1979) Changes in the brain catecholamines in patients with dementia of Alzheimer type. Brit. J. Psychiat., **135**, 216-223.

Arai, H., Kosaka, K.and Iizuka, R. (1984). Changes of biogenic amines and their metabolites in postmortem brains from patients with Alzheimer's-type dementia. J. Neurochem., **43**, 387-393.

Aubert, I., Araujo, D. M., Cecyre, D., Robitaille, Y., Gauthier, S. and Quiron, R. (1992). Comparative alterations of nicotinic and muscarinic binding sites in Alzheimer's and Parkinson's diseases. J. Neurochem., **58**, 529-541.

Araujo, D. M., Lapchak, P. A., Robitaille, Y., Gauthier, S. and Quirion, R. (1988). Differential alteration of various cholinergic markers in cortical and subcortical regions of human brain in Alzheimer's Disease. J. Neurochem., **50**, 1914-1923.

Beal, M. F., Matson, W. R., Swartz, K. J., Gamache, P. and Bird, E. D. (1990). Kynurenine pathway measurements in Huntington's disease striatum: Evidence for reduced formation of kynurenic acid. J. Neurochem., **55**, 1327-1339.

Bergman, H., Wichman, T. and DeLong, M. R. (1990). Reversal of experimental parkinsonism by lesions of the subthalamic nucleus. Science, **249**, 1436-1438.

Birkmayer, W., Danielczyk, W., Neumayer, E. and Riederer, P. (1974). Nucleus ruber and L-Dopa psychosis: Biochemical post-mortem findings. J. Neural.Trans., **35**, 93-116.

Blessed, G., Tomlinson, B. E. and Roth, M. (1968). The association between quantitative measures of dementia and of senile change in the cerebral gray matter of elderly subjects. Brit. J. Psychiat., **114**, 797-811.

Bowen, D. M., Smith, C. B., White, P. and Davidson, A. N. (1976). Neurotransmitter-related enzymes and indices of hypoxia in senile dementia and other abiotrophies. Brain, **99**, 459-496.

Bravo, G. J. and Cooper, I. S. (1959). A clinical and radiological correlation of the lesions produced by chemopallidectomy and thalamectomy. J. Neurol. Neurosurg. Psychiat., **22**, 1-10.

D'Amato, R. J., Zweig, R. M., Whitehouse, P. J., Wenk, G. L., Singer, H. S., Mayeux, R., Price, D. L. and Snyder, S. H. (1987). Aminergic systems in Alzheimer's disease and Parkinson's disease. Ann. Neurol., **22**, 229-236.

Dickson, D. W., Davies, P., Mayeux, R., Crystal, H., Horoupian, D. S., Thompson, A. and Goldman, J. E. (1987). Diffuse Lewy body disease: Neuropathological and biochemical studies of six patients. Acta Neuropathol., (Berl) **75**, 8-15.

Dubois, B., Ruberg, M., Javoy-Agid, F., Ploska, A. and Agid, Y. (1983). A subcortico-cortical cholinergic system is affected in Parkinson's disease. Brain Res., **288**, 213-218.

Fahn, S., Libsch, L. R. and Cutler, R. W. (1971). Monoamines in the human neostriatum: Topographic distribution in normals and in Parkinson's Disease and their role in akinesia, rigidity, chorea, and tremor. J. Neurol. Sci., **14**, 427-455.

Fearnley, J. M., Revesz, T., Brooks, D. J., Frackowiak, R. S. and Lees, A. J. (1991). Diffuse Lewy body disease presenting with a supranuclear palsy. J. Neurol. Neurosurg. Psychiat., **54**, 159-161.

Fonnum, F. (1969). Radiochemical microassays for the determination of choline acetyltransferase and acetylcholinesterase activities. J. Biochem., **115**, 465-472.

Foster, A. C., Vezzani, A., French, E. D. and Schwarcz, R. (1984). Kynurenic acid blocks neurotoxicity and seizures induced in rats by the related brain metabolite quinolinic acid. Neurosci. Lett., **48**, 273-278.

Gaspar, P., Javoy-Avid, F., Ploska, A. and Agid, Y. (1980). Regional distribution of neurotransmitter synthesizing enzymes in the basal ganglia of human brain. J. Neurochem., **34**, :278-283.

Gibb, W. R. G. and Lees, A. J. (1991). Anatomy, pigmentation, ventral and dorsal subpopulations of the substantia nigra, and differential cell death in Parkinson's disease. J. Neurol. Neurosurg. Psychiat., **54**, 388-395.

Girault, J. -A., Raisman-Vozari, R., Agid, Y. and Greengard, P. (1989). Striatal phosphoproteins in Parkinson disease and progressive supranuclear palsy. Proc. Natl. Acad. Sci. USA, **86**, 2493-2497.

Hansen, L. A. and Galasko, D. (1992). Lewy body disease. Current Opinions Neurol. & Neurosurg., **5**, 889-894.

Hansen, L., Salmon, D., Galasko, D,. Masliah, E., Katzman, R., DeTeresa, R., Thal, L., Pay, M.M., Hofsetter, R., Klauber, M., Rice, V., Butters, N. and Alford, M. (1990). The Lewy body variant of Alzheimer's Disease: A clinical and pathologic entity. Neurology, **40**, 1-8.

Hornykiewicz, O. (1973). Dopamine in the basal ganglia: Its role and therapeutic implications (including the clinical use of L-DOPA). Brit. Med. Bull., **29**, 172-178.

Hornykiewicz, O. (1982). Brain Neurotransmitter Changes in Parkinson's Disease. In: Movement Disorders., Marsden CD, Fahn S, (Eds.). Butterworth Scientific, London, pp.41-58.

Kelly, P. J., Ahlskog, J. E., Goerss, S. J., Daube, J. R., Duffy, J. R. and Kall, B. A. (1987). Computer-assisted stereotactic ventralis lateralis thalamotomy with microelectrode recording control in patients with Parkinson's disease. Mayo Clin. Proc., **62**, 655-664.

Khachaturian, Z. S. (1985). Diagnosis of Alzheimer's disease. Arch. Neurol., **42**, 1097-1105.

Kish, S. J., Shannak, K., and Hornykiewicz, O. (1988). Uneven pattern of dopamine loss in the striatum of patients with idiopathic Parkinson's Disease. N. Engl. J. Med., **318**, 876-880.

Langlais, P. J., Thal, L. J., Hansen, L., Galasko, D., Alford, M. and Masliah, E. (1993). Neurotransmitters in basal ganglia and cortex of Alzheimer's disease with and without Lewy bodies. Neurology, **43**, 1927-1934.

Langlais, P. J., Wardlow, M. L. and Yamamoto, H. (1991). Changes in CSF neurotransmitters in Infantile Spasms. Ped. Neurol., **7**, 440-445.

McKhann, G., Drachman, D., Folstein, M. F., Katzman, R., Price, D. and Stadlan, E. M. (1984). Clinical diagnosis of Alzheimer's disease: Report of the NINCDS-ADRDA work group. Neurology, **34**, 939-944.

Moroni, F. P., Russi, P., Lombardi, G., Beni, M. and Carla, V. (1988) Presence of kynurenic acid in mammalian brain. J. Neurochem., **51**, 177-180.

Ogawa, T., Matson, W. R., Beal, M. F., Myers, R. H., Bird, E. D., Milbury, P. and Saso, S. (1992). Kynurenine pathway abnormalities in Parkinson's disease. Neurology, **42**, 1702-1706.

Olney, J. W. (1989). Excitatory amino acids and neuropsychiatric disorders. Biol. Psychiat., **26**, 505-525.

Perry, E. K., Curtis, M., Dick, D. J., Candy, J. M., Ratack, J., Bloxham, C. A., Blessed, G., Fairbairn, A., Tomlinson, B. E. and Perry, R. H. (1985). Cholinergic correlates of cognitive impairment in Parkinson's disease: Comparisons with Alzheimer's Disease. J. Neurol .Neurosurg. Psychiat., **48**, 413-421.

Perry, E. K., Marshall, E., Perry, R. H., Irving, D., Smith, C. J., Blessed, G. and Tomlinson, B. E. (1990). Cholinergic and dopaminergic activities in Senile Dementia of Lewy Body Type. Alzheimer Dis. Assoc. Disord., **4**, 87-95.

Perry, E. K., McKeith, I., Thompson, P., Marshall, E., Kerwin ,J,. Jabeen, S., Edwardson, J. A., Ince, P., Blessed, G., Irving, D. and Perry, R. H. (1991). Topography, extent, and clinical relevance of neurochemical deficits in Dementia of Lewy Body Type, Parkinson's Disease, and Alzheimer's Disease. Ann. N. Y. Acad. Sci., **640**, 197-202.

Rinne, U. K., Sonninen, V., Riekkinen, P. and Laaksonen, H. (1974). Dopaminergic nervous transmission in Parkinson's Disease. Medical Biol., **52**, 208-217.

Rinne, J. O., Laihinin, A., Lonberg, P., Marjamaki, P. and Rinne, U. K. (1991). A post-mortem study on striatal dopamine receptors in Parkinson's disease. Brain Res., **556**, 117-122.

Rossor, M. N., Iversen, L. L., Reynolds, G. P., Mountjoy, C. Q. and Roth, M. (1984). Neurochemical characteristics of early and late onset types of Alzheimer's disease. Brit. Med. J., **288**, 961-964.

Scatton, B., Javoy-Agid, F., Rouquier, L., Dubois, B. and Agid, Y. (1983). Reduction of cortical dopamine, noradrenaline, serotonin and their metabolites in Parkinson's disease. Brain Res., **275**, 321-328.

Sirvio, J., Rinne, J. O., Valjakka, A., Rinne, U. K., Riekkinen, P. J. and Paljarvi, L. (1989). Different forms of brain acetylcholinesterase and muscarinic binding in Parkinson's disease. J. Neurol .Sci., **90**, 23-32.

Whitehouse, P. J., Martino, A. M., Marcus, K. A., Zweig, R. M., Singer, H. S., Price, D. L. and Kellar, K. J. (1988). Reductions in acetylcholine and nicotine binding in several degenerative diseases. Arch. Neurol., **45**, 722-724.

Progress in HPLC-HPCE, Vol. 6, pp. 273–290
I.N. Acworth *et al.* (Eds)

APPLICATION OF THE MULTI-ELECTRODE DETECTION SYSTEM TO THE MONITORING OF DRUG THERAPY IN NEUROLOGICAL DISEASES

WAKAKO MARUYAMA[1] and MAKOTO NAOI[2]

[1]*Laboratory of Biochemistry and Metabolism, Department of Basic Gerontology, National Institute for Longevity Sciences, Obu, Japan, and* [2]*Department of Biosciences, Nagoya Institute of Technology, Gokiso-cho, Showa-ku, Nagoya 466, Japan*

INTRODUCTION

Changes in brain neurotransmission may underlie the symptoms of some neurological diseases. Drugs which affect neurotransmitter levels or have affinity for specific receptors, are presently being effectively used to treat neurological diseases. In Parkinson's disease (PD), the main pathological change is the selective degeneration of dopamine (DA) neurons in the nigro-striatal system with the subsequent biochemical change, the impairment of DA synthesis in the brain. 3,4-Dihydroxy-L-phenylalanine (L-DOPA), the precursor of DA, is decarboxylated by aromatic L-amino acid decarboxylase (aromatic L-amino acid decarboxylase, EC 4.1.1.28, AADC) producing DA. This precursor is now routinely given to PD patients to alleviated the symptoms of the disease. However, neuropathological studies have revealed that not only DA neurons but also noradrenaline (NA) neurons degenerate in the Parkinsonian brain with the advancement of the disease. Decreases in NA synthesis were more prominent in the PD patients with longer duration of the illness (Narabayashi *et al.*, 1981; Nagatsu *et al.*, 1982) and it has been suggested that NA deficiency may play a role in some of the DA-refractory symptoms associated with PD. L-*threo*-3,4-Dihydroxyphenylserine (L-*threo*-DOPS) is an artificial amino acid which acts as a NA precursor, producing NA following its decarboxylation by AADC. It has been successfully used for supplement therapy of NA (Narabayashi *et al.* 1981). Out of the 4 possible DOPS isomers (L-, D-, *erythro*- and *threo*-) only L-*threo*-DOPS can be metabolized into the biological active isomer of NA, and recently pure L-*threo*-DOPS has been synthesized and used to treat Parkinsonian patients. L-*threo*-DOPS alleviates some DOPA-refractory symptoms, such as freezing phenomenon (Narabayashi *et al.*, 1981). However, the effects on NA metabolism have not been fully examined in the human brain.

To evaluate the *in vivo* effects of the precursor amino acids, L-DOPA and L-*threo*-DOPS, in the brain, the analysis of human tissue is essential. The most commonly used analytical procedures utilize either enzyme-linked immunosorbent assays (ELISA) or high-performance liquid chromatographs (HPLC) coupled to electrochemical detectors (ECD) or fluorometric detectors. However, human brain samples contain a lot of interfering compounds, and many of the compounds of interest occur at very low levels and their accurate identification has been very difficult. Furthermore, complicated sample preparation limits the efficiency of the analysis. Recently HPLC with multi-ECD [CEAS, (NEUROCHEM in Japan), ESA, Chelmsford, MA] was proposed as an effective tool to analyze human tissues (Matson *et al.*, 1984; Naoi *et al.*, 1993). The CEAS contains up to 16 electrochemical detectors set in a series with graded oxidation voltage which gives rise to a three-dimensional HPLC pattern and can be used to differentiate between bioactive compounds such as monoamines. Compounds are identified both by their retention time and their reactivity across the electrode array (Naoi *et al.*, 1993). This method has proven to be extremely useful in the analysis of human tissue, such as brain (Beal *et al.*, 1990), cerebrospinal fluid (CSF) (Maruyama *et al.*, 1996b) and intraventricular fluid (IVF) (Maruyama *et al.*, 1996a,c). However, there have been no systematic studies to evaluate changes of the neurotransmitter amines after replacement therapy of the precursor amino acids in living patients with neurological disorders, such as PD.

The CEAS was used to study the metabolism of the precursor amino acids and their effects on the biogenic amines in the brain. The monoamines, their metabolites and precursor amino acids could be accurately quantitated in minimal sample volumes obtained by *in vivo* microdialysis of rat brains and in human tissues, such as CSF and IVF. The function and metabolism of the precursor amino acids, L-DOPA and L-*threo*-DOPS, were examined by the analysis of monoamines in microdialysis perfusates obtained from the rat striatum. These results were compared to those from CSF and IVF obtained from PD patients. The importance of the monitoring system for the drug therapy of neurological disorders is also discussed.

MICRODIALYSIS STUDY IN RAT STRIATUM

L-*threo*-DOPS and 3-O-methoxy-DOPS (3-O-MeDops) were kindly donated by Sumitomo Co. (Osaka, Japan). Protocatechualdehyde (PCAld) and protocatechuic acid (PA) were purchased from Nacalai tesque (Kyoto, Japan); the standard monoamines and their metabolites were from Sigma (St. Louis, MO); sodium octanesulfonate (SOS) was from Aldrich (Milwaukee, WI).

The procedures for rat brain microdialysis have been reported previously (Maruyama *et al.*, 1992a). The experimental protocols were approved by the Ethical Committee of The Department of Biosciences, Nagoya Institute of Technology. Adult male Wistar rats weighing 240-280g were anesthetized with sodium pentobarbital (50mg/kg i.p.) and guide cannulas were implanted above the striatum according to the stereotaxic atlas of Paxinos and Watson (1986). After recovery for 24 hours, a dialysis probe was inserted and the Ringer's solution was perfused at a flow rate of 2μL/min. Dialysate samples were collected every 20 minutes.

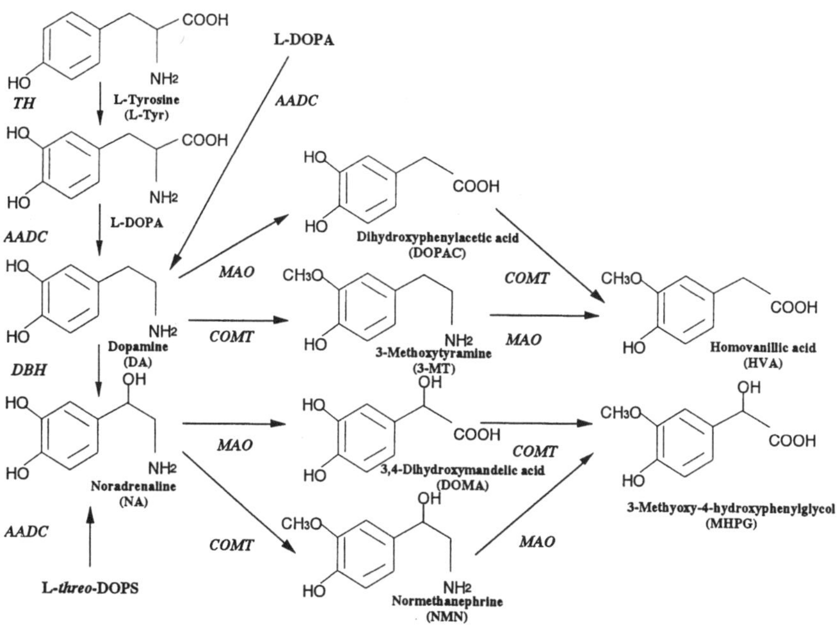

Figure 1. Metabolism of catecholamines in the human brain, and possible supplemental effects of L-DOPA and L-*threo*-DOPS. DA is synthesized from L-tyrosine (L-Tyr) through hydroxylation and decarboxylation, and ß-hydroxylation of DA produces NA. Catecholamines are catabolized by MAO and COMT. L-DOPA and L-*threo*-DOPS are decarboxylated to produce DA and NA, respectively.

and a 15μL aliquot of the 40μL sample was analyzed for monoamines, their metabolites and amino acids using the CEAS. The electrodes were set from 0 to +900mV in 60mV increments. Samples were eluted with an isocratic mobile phase at a flow rate of 0.6 mL/min and at a temperature of 25°C. The column was a prepacked reversed-phase column, Innertsil (GL Sciences, ODS-2, 4.6mm i.d. x 250mm). The mobile phase consisted of 90mM sodium acetate-35mM citric acid buffer, pH 4.35, containing 130μM disodium EDTA and 230μM SOS; methanol (10.5%, v/v). Correct identification of the peak was obtained from the retention time and the relative ratio of the peak height measured with two or three electrodes at different voltages (ratio accuracy). Quantitation of a compound was based on the peak height obtained for an external standard. After stabilization for 3 hours, the basal levels of monoamines and their metabolites were determined for

1 hour, during which time 3 samples were obtained. Then, 1mM of L-DOPA (n= 2) and L-*threo*- DOPS (n=4) dissolved in the Ringer's solution was perfused

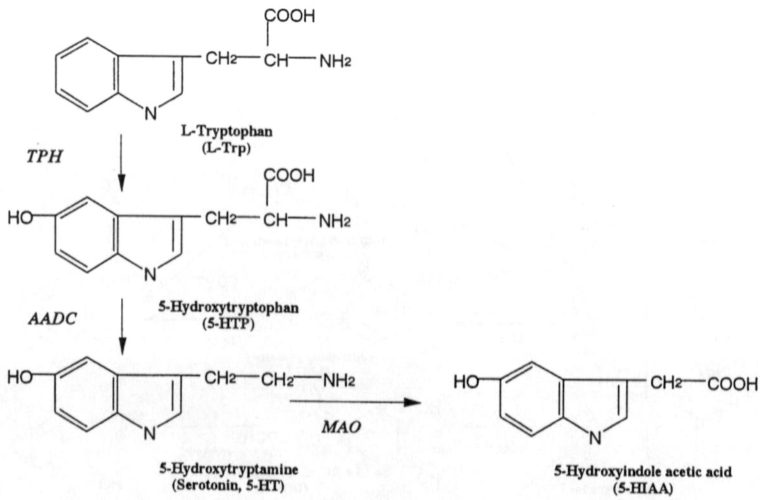

Figure 2. Metabolic map of 5-HT. 5-HT is synthesized from L-tryptophan (L-Trp) through hydroxylation and decarboxylation. 5-HT is catabolized by MAO producing 5-hydroxyindoleacetic acid (5-HIAA).

through the probe for 2 hours. The effects of L-DOPA and L-*threo*-DOPS perfusion on the number and amount of analytes in the striatal perfusate samples were studied and changes were statistically analyzed by one-sample runs test.

To convert the conjugated form of NA into the free form, the remainder of each dialysate sample was boiled for 30 minutes with 20mM perchloric acid containing 80μM sodium metabisulfite and 20μM disodium EDTA. Following centrifugation at 22,000g for 10 minutes the samples were then analyzed on the HPLC-ECD system. Using this procedure the amounts of conjugated and free forms were determined.

The metabolic pathway of catecholamines, DA and NA, and serotonin (5-hydroxytryptamine, 5-HT) in the human brain is shown in Figure 1 and Figure 2. Catecholamines are catabolized by monoamine oxidase [monoamine: oxygen oxidoreductase (deaminating), EC 1.4.3.4., MAO] and catechol-O-methyltransferase (S-adenosyl-L-methionine, catechol-O-methyltransferase, EC 2.1.1.6, COMT), while 5-HT is catabolized by MAO. The metabolism of administered L-DOPA into DA and further metabolism of DA and L-*threo*-DOPS into NA is also presented in Figure 1.

Figure 3 shows that metabolism of L-*threo*-DOPS in the brain can involve three different enzymatic pathways: by AADC into NA, which is metabolized further producing MHPG by COMT and MAO; by DOPS-aldolase (DOPS-ALD)

(Naoi *et al.*, 1987) producing protocatechualdehyde (PCald), which can be further oxidized producing protocatechuic acid (PA) by an aldehyde dehydrogenase

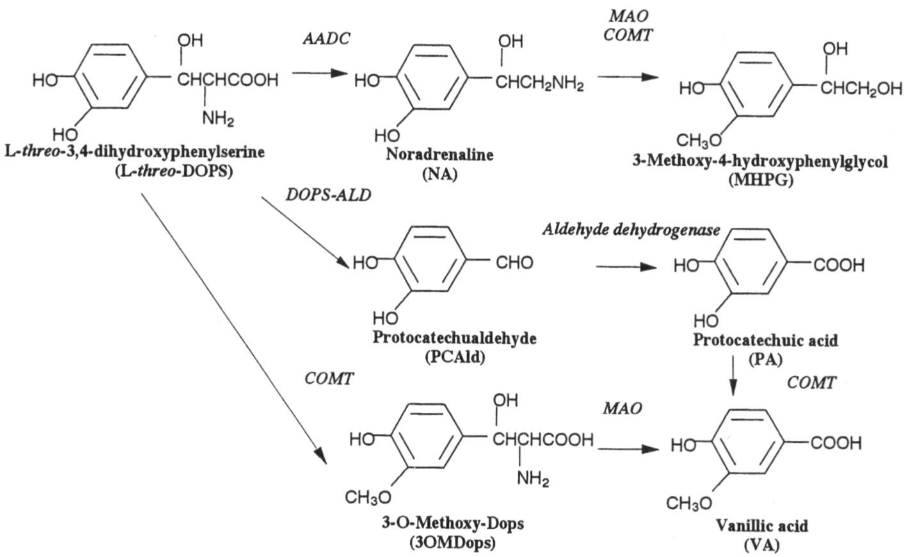

Figure 3. The metabolic pathway of L-*threo*-DOPS. L-*threo*-DOPS is catabolized by 3 enzymes, AADC to produce NA and ultimately MHPG, by DOPS-ALD, and by COMT.

[aldehyde: NAD^+ oxidoreductase, EC 1.2.1.3]; and by COMT producing 3-Me-DOPS, which can be further metabolized into vanillic acid (VA).

After 100 minutes perfusion of the precursor amino acids, there was significant changes in the chromatogram of the striatal dialysate compared to that before the perfusion (Figure 4). Before perfusion (Figure 4A), peaks corresponding to L-Tyr (peak 1), DOPAC (peak 2), 5-hydroxyindole acetic acid (5-HIAA) (peak 3), L-Trp (peak 4), and homovanillic acid (HVA) (peak 5) were identified. Following perfusion of L-DOPA into the striatum, the concentration of DA and its metabolites, DOPAC and HVA were markedly increased. In the case of L-DOPA perfusion only a few new peaks were detected. The major new peaks were 3OMDopa (peak 6), a methylated product of L-DOPA by COMT, and L-DOPA itself (Figure 4B). Conversely, after perfusion of L-*threo*-DOPS (Figure 4C), many new peaks were detected. Some of the novel peaks were identified as the metabolites of L-*threo*-DOPS by AADC [peak 8, NA and peak 9, 3-methoxy-4-phenylglycol (MHPG)], by DOPS-ALD [peak 10, PA; peak 12, PCAld] and by

COMT [peak 11, VA]. Others were identified as monoamines [peak 7, DA; peak 13, 5-HT].

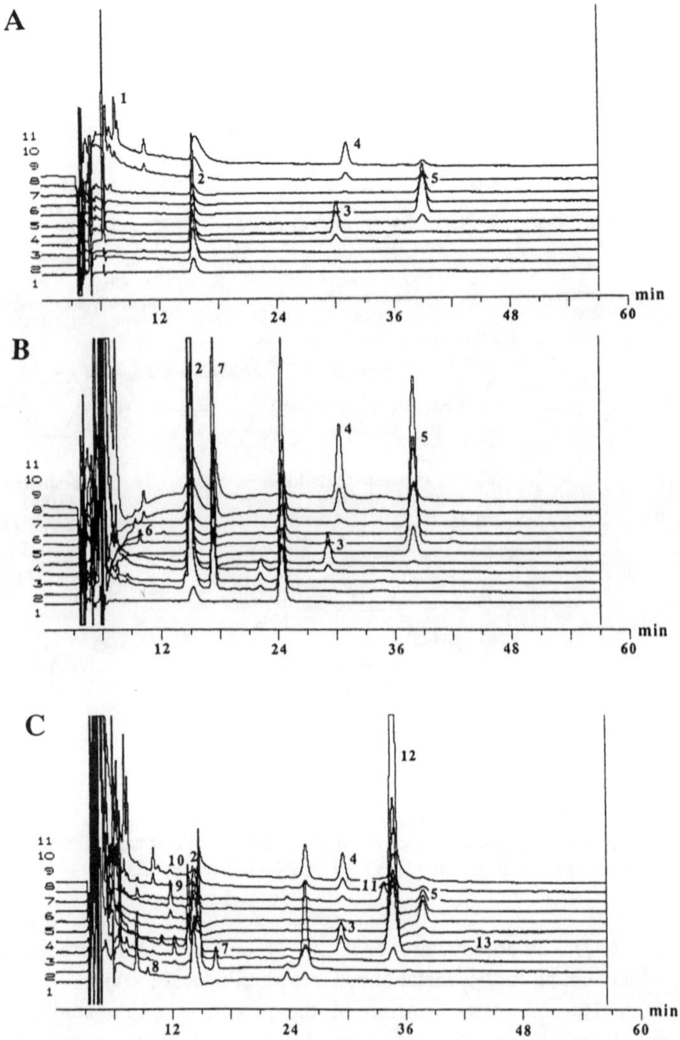

Figure 4. The HPLC-chromatograms of rat striatal dialysate before (A) and after perfusion of L-DOPA (B) and L-*threo*-DOPS (C) for 100 minutes. A 15μL dialysate sample was analyzed by the CEAS. The chromatograms are presented at a full scale current range of 200nA. The applied voltage to the electrode array was from 0 to +900mV in +60mV increments. Peak 1, L-Tyr; peak 2, DOPAC; peak 3, 5-HIAA; peak 4, L-Trp; peak 5, HVA; peak 6, 3OMDopa; peak 7, DA; peak 8, NA; peak 9, MHPG; peak 10, PA; peak 11, VA; peak 12, PCAld; and peak 13, 5-HT.

The conjugated form of NA was detected in the dialysate after L-*threo*-DOPS perfusion, and the relative concentration of the conjugated NA to the free form was found to be 1.06±0.30.

The metabolism of the two precursor amino acids, L-DOPA and L-*threo*-DOPS, was compared. The concentrations of the amino acids and their metabolites in the dialysate after 100 minutes perfusion are summarized in Table 1. L-DOPA was mainly metabolized by AADC into DA and its metabolites, DOPAC and HVA. In contrast, L-*threo*-DOPS was shown to be metabolized mainly by DOPS-ALD into PCAld and PA. The amounts of the metabolites by AADC was 0.4%, by COMT 2.1%, and by DOPS-ALD 97.5% of the total metabolites detected.

Table 1.
Concentration of L-*threo*- and L-*erythro*-DOPS and L-DOPA, and their metabolites in the rat striatal dialysate after 100 minutes perfusion.

	Precursor amino acids (μM)	Reaction Products of AADC (μM)		Reaction Products of COMT (μM)		Reaction Products of DOPS-ALD (μM)	
		NA	MHPG	3-Me-DOP	Vanillic acid	PCAld	PA
L-DOPA	236	Total (DA+DOPAC+HVA) = 31.5		Total (3-Me-DOPA) = 0.0420		-	
	261	20.8		0.0322		-	
L-*threo*-DOPS	191.5±29.7	0.026±0.01	0.085±0.01	0.401±0.51	0.208±0.02	26.9±17.2	1.5±0.5
		Total = 0.111±0.03		Total = 0.609±0.06		Total = 28.4±17.6	

The rat striatal dialysate was obtained after 100 minutes perfusion with 1mM solution of each drug. The compounds were quantitatively analyzed using HPLC-ECD. Each value represents the mean+SD (L-*threo*-DOPS, n=4; L-DOPA, n=2)

The effects of L-*threo*-DOPS on the monoamine levels in the striatum were examined. The concentrations of DA, 5-HT and their metabolites following 100 minutes perfusion of DOPS are summarized in Table 2. L-*threo*-DOPS increased extracellular levels of DA and 5-HT in the striatum. But the effect was more marked on DA than on 5-HT. On the other hand, the concentration of the monoamine metabolites, DOPAC, HVA and 5-HIAA were all decreased by L-*threo*-DOPS and the decrease of HVA was statistically significant ($p<0.05$ by analysis of variance followed by Scheffe F test).

ANALYSIS OF MONOAMINES IN CSF AND IVF

CSF samples were obtained from 5 untreated PD patients [DOPA(-) group], 5 PD patients treated with L-DOPA [DOPA (+) group], and 6 control patients without neurological diseases (Control group). Table 3 shows the clinical data of the patients whose CSF were analyzed. The purpose, possible risks and benefits were explained to the patients before the start of this study and their consent was obtained. The patients had been administered L-DOPA/carbidopa in three divided doses daily and the lumbar CSF was obtained in the morning 2 to 4 hours after receiving the last dose. CSF was frozen immediately and stored at -80°C until

Table 2.
The effects of L-*threo* and L-*erythro*-DOPS on extracellular levels of DA, 5HT and their metabolites in the striatum after 100 minutes perfusion.

	DA	5HT	DOPAC	HVA	5-HIAA
	(nM)		(% of the basal level)		
L-*threo*-DOPS	80.3±55.6	7.37±1.40	77.8±23.6	45.0±8.5	57.9±29.0
Basal level in striatal dialysate before perfusion	4.50±1.59	N.D.	1562±440	550±205	204±94.8

The rat striatal dialysate was obtained before and after 100 minutes perfusion of 1mM solution of each drug. The compounds in the dialysate were quantitatively analyzed using HPLC-ECD. Each value represents the mean±SD (n=4).

analysis. The assay system for analysis of CSF was the same as that used for analysis of microdialysate samples. The CSF was deproteinized by the addition of a 10% volume of perchloric acid. Following centrifugation the samples were filtered and the CSF levels of L-Tyr, L-Trp, monoamines and their metabolites in CSF were then measured on the CEAS.

Table 3.
Clinical data of the patients whose CSF was analyzed.

	Age (Years, mean±SD)	M/F	Duration of illness (years, mean±SD)	Dosage of L-DOPA (mg/day, mean±SD)
DOPA (+) Group (n=5)	69.5±2.6	4/1	7.7±4.0	363±160
DOPA (-) Group (n=5)	62.3±4.3	3/2	2.5±0.7	
Control Group (n=6)	50.3±15.6	4/2		

Twenty eight patients who were clinically diagnosed to be PD (including 1 juvenile Parkinsonism) were operated on stereotaxically (mostly thalamotomy) and 10 to 20mL samples of IVF were obtained during this procedure. In these Parkinsonian patients, 14 patients were treated with L-DOPA/carbidopa (DOPA group) and another 14 patients were treated with L-DOPA/carbidopa combined with L-*threo*-DOPS (DOPA/DOPS group). Although these patients had been treated with other anti-Parkinsonian drugs, the types and the doses were not significantly different between the 2 groups. Two patients with postural tremor, 2 patients with Huntington's disease, and 1 patient with dystonia also underwent the stereotaxic operation and IVF samples were obtained (Control group). These 5 patients received neither L-DOPA nor L-*threo*-DOPS. The IVF was stored at -70°C until analysis.

Table 4 summarizes the clinical profile of the patients who donated IVF samples. The average age of the DOPA/DOPS group was older than that of the DOPA group, and the Control group contained the youngest subjects. The average duration of the illness was longer and Hohen Yahr stage of the patients was more severe in the DOPA/DOPS group than the DOPA group.

Table 4.
Clinical data of the patients who donated IVF samples.

	Age (years mean+SD)	M/F	Duration of PD (years, mean±SD)	Yahr Stage	DOPA Dosage (mg/day, mean±SD)	L-threo-DOPS Dosage (mg/day, mean±sd)
DOPA group (n=14)	56.1±9.0	6/8	6.0±4.0	II:3 III:7 IV:4	358±164	-
DOPA/DOPS group (n=14)	62.1±8.8	9/5	8.1±4.7	III:5 IV:9	486±112	507±177
Control Group (n=5)	42.6±12.9	4/1			-	-

The IVF was analyzed by the CEAS using the same method which was used for the analysis of microdialysate samples. Conjugated forms of NA, DA, 5-HT and MHPG were also measured after 30 minutes boiling of the sample with 1/10 volume of 0.2M perchloric acid added as described above. DA, 5-HT, NMN, 3-MT, PCAld, and PA were found to be very low in IVF and were only analyzed after a concentration procedure. Each IVF sample was brought to pH 8.5 with 4mL of a solution prepared by dissolving 1M ammonium sulfate in 0.11M NaOH, and was loaded onto a phenylboronic acid cartridge column (Analytichem International, 100 mg/L). After washing with water (4mL) and methanol (4mL), the compounds were eluted with 4mL of 1M acetic acid in methanol, and the resulting eluent was evaporated to be dryness. The residue was dissolved in mobile phase and analyzed using the CEAS. L-*threo*-DOPS and 3,4-

dihydroxymandelic acid (DOMA) were also analyzed. In this case, the mobile phase consisted of 25mM potassium phosphate buffer, pH 4.0, containing 0.5mM tetra-*n*-butylammonium hydrogensulfate and 0.1mM EDTA; acetonitrile (4.8%, v/v). The flow rate was 0.6 mL/min, and the other analytical conditions were the same as described above.

The values were compared among the Control group, DOPA group, and DOPA/DOPS group. Statistical analysis was performed using analysis of variance (ANOVA) followed by Scheffe F-test. Table 5 summarizes the data of the quantitative analysis of L-Tyr, L-Trp, HVA and 5-HIAA in the lumbar CSF and

Table 5.
The concentration of the monoamines and their metabolites, L-Tyr and L-Trp in lumbar CSF and IVF.

	CSF			IVF	
	DOPA (-) (n=5)	DOPA (+) (n=5)	Control (n=6)	DOPA group (n=14)	Control Group (n=5)
L-Tyr (μM)	14.4±7.0	33.0±19.78	8.0±1.5	14.6±4.0	100.2±1.4
L-DOPA (nM)	N.D.	98.1±51.2	N.D.	55.0±80.5	N.D.
HVA (nM)	111±44	95.1±75.9	133±106	1300±590	1120±50*
L-Trp (μM)	13.8±3.0	17.6±3.8	14.2±2.0	3.00±1.15	3.13±0.21*
5-HIAA (nM)	64.5±30.9	21.4±11.5	62.7±39.8	315±113	306±78*

The concentration of monoamines and their metabolites, L-Tyr and L-Trp were measured quantitatively using the CEAS. Each value represents the mean+SD. N.D. - not determined. CSF and IVF were obtained from patients as described in text. * The difference between IVF and CSF of control group is statistically significant ($p<0.05$) by ANOVA.

IVF. The values were compared to the levels of analytes in CSF and IVF obtained from the Control group. There were significant increases in HVA and 5-HIAA levels but a significant decrease in L-Trp levels in IVF compared to CSF. The concentration of precursor amino acids, catecholamines, and their metabolites in the IVF from 3 groups of patients is shown in Figures 5 to 7. L-DOPA and L-*threo*-DOPS were detected in the IVF of DOPA or DOPA/DOPS group. The L-DOPA concentration was virtually the same between the two group; its concentration was 55.0±80.5nM in DOPA group and 56.3±48.9nM in DOPA/DOPS group. The concentration of L-*threo*-DOPS was 60.7±72.5nM in the DOPA/DOPS group.

Figure 5 shows the levels of DA and its metabolites in IVF. The total (conjugated+free) DA value was lower in PD patients, but the variance in the Control group was so large that the difference was not statistically significant between the PD and the Control groups. DA's principle metabolite, DOPAC, was increased in the DOPA group but even more markedly in the DOPA/DOPS group. The difference was significant between the DOPA/DOPS group and both Control and the DOPA group ($p<0.05$ by ANOVA). On the other hand, the level of a metabolite of DA by COMT, 3-MT, was higher in the DOPA group than in the Control group, but in the DOPA/DOPS group its concentration was lower than either the Control or DOPA groups. The difference was statistically significant between the DOPA group and the DOPA/DOPS group ($p<0.05$ by ANOVA). The concentration of HVA, a metabolite of DA by both MAO and COMT, was similar in all 3 groups.

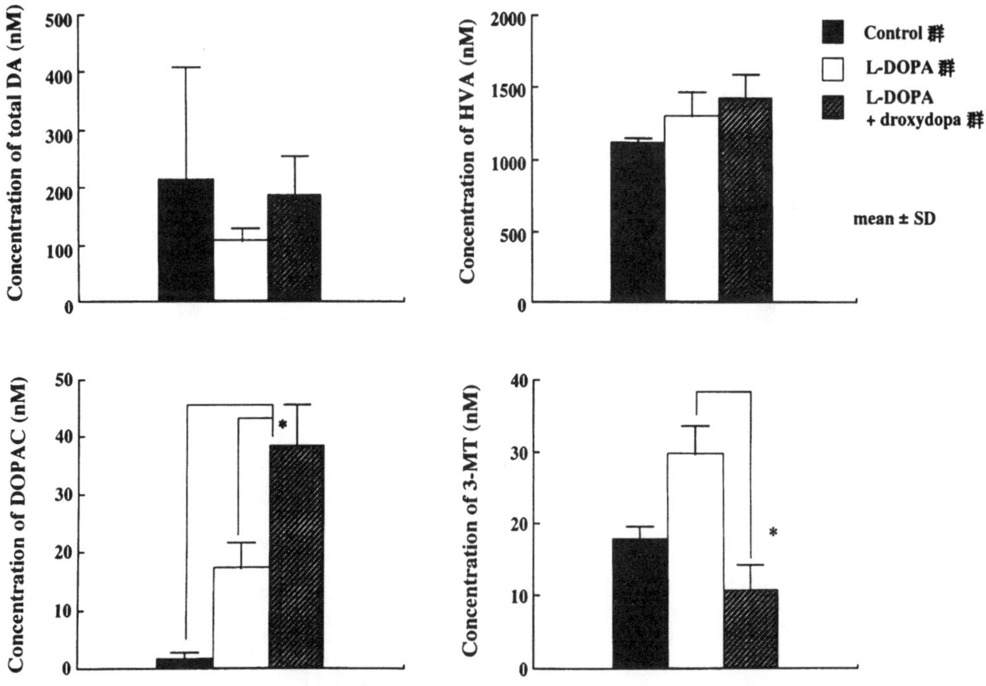

Figure 5. The levels of DA and its metabolites in IVF. Each column and bar represent mean±SD. Filled column - Control group. Empty column - L-DOPA group. Hatched column - DOPA/DOPS group. * Difference is statistically significant ($p<0.05$) by ANOVA.

The amount of total NA (free+conjugated) was lower in the PD group than the Control group, but the difference was not statistically significant (Figure 6). Although free NA could not be detected in the Control group, a significant amount was detected in the DOPA and the DOPA/DOPS groups (4.57±3.26nM in DOPA group and 5.67±3.38 in the DOPA/DOPS group). NA metabolite levels were also affected by L-DOPA and L-*threo*-DOPS administration. A metabolite of NA by MAO, DOMA, was increased in the DOPA group and the DOPA/DOPS group. A metabolite of NA by COMT, NMN, was decreased significantly in DOPA/DOPS group compared with the Control and L-DOPA groups. The concentration of MHPG, a metabolite of NA by MAO and COMT, was significantly decreased in both the L-DOPA and DOPA/DOPS groups.

The concentration of 5-HT and a metabolite of 5-HT by MAO, 5-HIAA, are summarized in Figure 7. There was no significant difference in their contents between the Control and the DOPA group. In the DOPA/DOPS group, the level of L-Trp and 5-HIAA increased significantly compared to the Control and DOPA groups. Free 5-HT could not be detected in all three groups. The conjugated 5-HT concentration was the highest in DOPA/DOPS group, but the variation among the patients was so large that the difference was not statistically significant.

To examine the effects of L-*threo*-DOPS administration on catecholamine metabolism in the brain, the ratio of the levels of the precursors and metabolites was calculated. As shown in Table 6, the 3-MT/DA ratio decreased, while HVA/3-MT and MHPG/NMN ratio increased markedly in the DOPA/DOPS group and the difference from those in the DOPA group was significant ($p < 0.05$ by ANOVA).

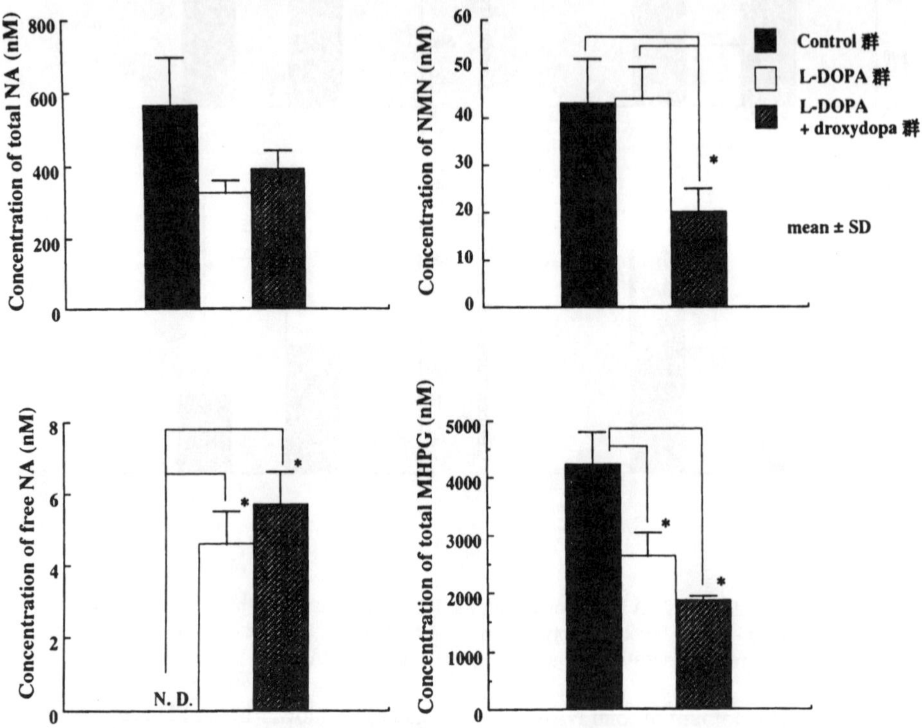

Figure 6. The levels of NA and its metabolites in IVF samples. Each column and bar represent mean±SD. Filled column - Control group. Empty column - L-DOPA group. Hatched column - DOPA/DOPS group. * Difference is statistically significant (p<0.05) by ANOVA.

Table 6.
Metabolic ratio of monoamines in the IVF.

	DOPA group	DOPA/DOPS group
3-MT/DA	0.39±0.33	0.07±0.08*
DOPAC/DA	0.24±0.09	0.26±0.09
HVA/3-MT	49.0±13.7	261±166*
HVA/DOPAC	122±105	51.3±18.2
NMN/NA	0.20±0.21	0.17±0.43
DOMA/NA	#	0.005±0.004
MHPG/NMN	60.9±41.8	328±421*
MHPG/DOMA	#	3533±5745

Each value represents mean ± SD. *The difference between the DOPA group and the DOPA/DOPS group is statistically significant ($p<0.05$) by ANOVA. # - Could not be calculated.

The correlation of the concentration of L-DOPA to the total DA metabolites (DA+DOPAC+3-MT+HVA), and that of L-*threo*-DOPS to the total NA metabolites (NA+MHPG+NMN+DOMA), were examined (Figure 8). The total DA metabolite concentration positively correlated to the L-DOPA concentration, both in the DOPA group ($r = 0.625$) and the DOPA/DOPS group ($r = 0.566$). On the other hand, the concentration of L-*threo*-DOPS had no correlation to the total NA metabolite concentration ($r = 0.221$).

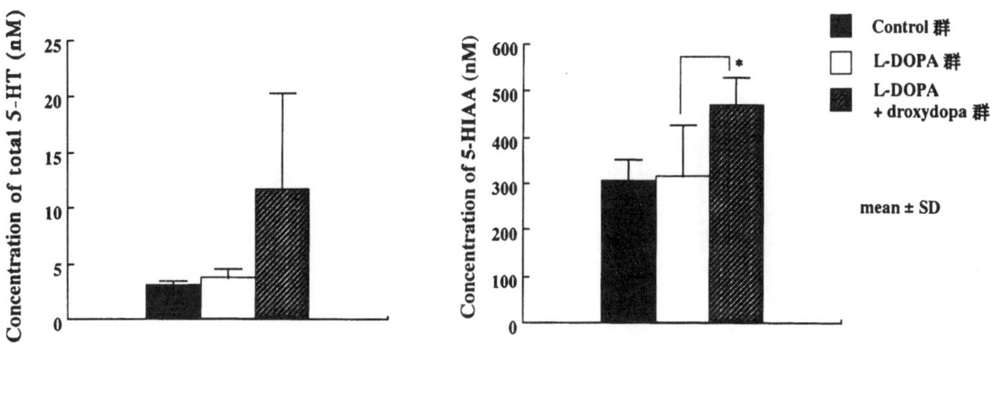

Figure 7. The levels of 5-HT and its metabolites in IVF samples. Each column and bar represent mean and SD. Filled column - Control group. Empty column - L-DOPA group. Hatched column - DOPA/DOPS group. * Difference is statistically significant ($p < 0.05$) by ANOVA.

The metabolites of L-*threo*-DOPS by DOPS-ALD and COMT were identified in the IVF of the DOPA/DOPS group. There were large varieties in the concentration of L-*threo*-DOPS metabolites; expressed as mean±SD in nM (range of the value), PAld; 20.4±35.4 (2.31-117), PA; 33.9±73.8 (0.28-236), 3OMDops; 386±379 (52.8-1210), VA; 32.7±29.9 (4.15-116). There was no correlation between the concentrations of the metabolite by DOPS-ALD and COMT. In most patients the COMT derived metabolites were predominant, but the patients with high concentrations of the metabolites by DOPS-ALD were shown to have low concentrations of the COMT derived metabolites.

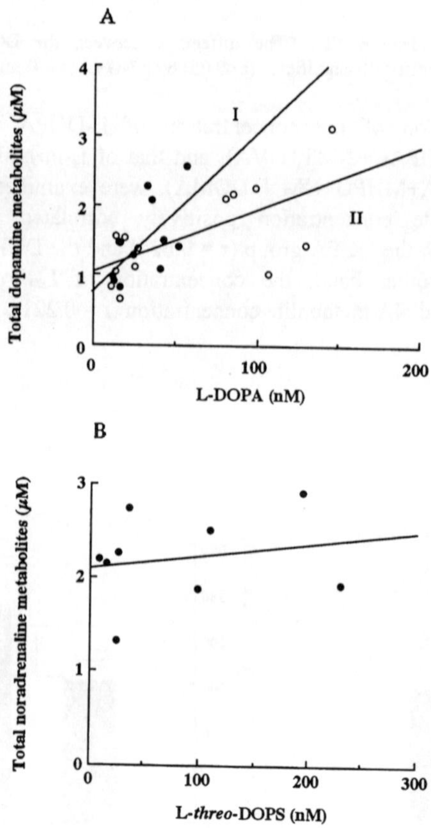

Figure 8. The effects of precursor amino acids on monoamine metabolism in the brain. A; Correlation of the concentration of L-DOPA and of total DA metabolites (DA+DOPAC+3-MT+HVA) in IVF. Filled circles (I) represent values in the DOPA group and open circles (II) represent those in the DOPA/DOPS group. B; Correlation of the concentration of L-*threo*-DOPS and total NA metabolites (NA+DOMA+NMN+MHPG) in the IVF samples obtained from the DOPA/DOPS group.

DISCUSSION

The levels of monoamines and related compounds were determined both in CSF and IVF. By using the CEAS, biological samples were analyzed with a minimum of sample preparation. Identification of the peaks corresponding to the monoamines, and their metabolites and precursor amino acids, was done precisely and easily in a single sample and was used to evaluate the complexities of monoamine metabolism (Maruyama *et al.,* 1992a). In the Control group, the concentration of HVA, a metabolite of DA, was significantly higher in IVF than that in CSF, but that of L-Tyr, the DA precursor amino acid, was not different (Table 2). On the other hand, the concentration of a 5-HT metabolite, 5-HIAA, was also higher in IVF, but the difference was not so marked as in the case of HVA, and the concentration in CSF was about 1/5 of that in IVF. On the contrary, the concentration of L-Trp, the precursor amino acid of 5-HT, was significantly higher in CSF than that in IVF. The CSF rostral-caudal gradient of HVA and 5-HIAA has previously been reported (Blennow *et al.*, 1993) and the gradient was also confirmed between CSF and IVF. In addition, our results indicate that the existence of a "negative" gradient of L-Trp. Since the concentration of L-Trp regulates 5-HT synthesis in the brain (Grahame-Smith, 1971), the cause and the importance of this gradient of L-Trp should be clarified further. Free NA and its metabolites could not be detected in CSF, suggesting that most of these compounds are in the conjugated form. In addition, it has previously been reported that DA and 5-HT metabolites are, at least partially, originating from the brain, whereas NA metabolites are not thought to be derived from the brain (Marrello *et al.*, 1994). In conclusion, IVF is more appropriate than CSF to evaluate monoamine metabolism in the brain.

This is the first report to analyze human IVF samples systematically and to evaluate the effects of the precursor amino acids on monoamine metabolism. In the DOPA group, increased levels of the DA metabolite by MAO, DOPAC and that by COMT, 3-MT, and the level of HVA, a DA metabolite by both MAO and COMT, were not different when compared with the Control group. This may indicate that L-DOPA administration enhances DA turnover in the brain and that the HVA level was normalized. On the contrary, in IVF samples from the patients treated with L-DOPA, free NA level increased, but the concentration of DOMA and NMN did not significantly change and the total MHPG level was lower than in the Control group. L-DOPA can synthesize NA by 2 enzymatic steps, AADC and dopamine ß-hydroxylase [3,4-dihydroxy-phenylethylamine, ascorbate: oxygen oxidoreductase (ß-hydroxylating), EC 1.14.17.2, DBH]. However, the results here indicate that L-DOPA is a poor precursor of NA and this may be due to the decreased activity of DBH in PD reported by Nagatsu *et al.* (1982). L-DOPA administration did not affect the 5-HT metabolism in the brain.

Perfusion of L-*threo*-DOPS in the rat striatum demonstrated that the effects and the metabolism of this artificial amino acid are quite different from those of L-DOPA (Maruyama *et al.*, 1994). Differences in the chromatograms obtained before and after the perfusion of L-DOPA and L-*threo*-DOPS will now be discussed. L-DOPA perfusion led to significant increases in the levels of DA and its metabolites. L-*threo*-DOPS perfusion produced many new peaks, some of which were monoamines. L-*threo*-DOPS has previously been suggested to act as a monoamine releaser (Nishino *et al.*, 1987) and indeed, in IVF samples from the DOPA/DOPS group, all the monoamines such as total DA, free and total NA, and total 5-HT levels were higher than DOPA group. This is significant as the DOPA/DOPS group is comprised of patients that are much older and PD has progressed further than the DOPA group and without L-*threo*-DOPS administration the monoamine levels were lower (Fig 3). L-*threo*-DOPS might indeed act as a monoamine releaser. In addition, the *in vivo* microdialysis study showed that L-*threo*-DOPS reduced the concentration of DA and 5-HT metabolites, indicating that L-*threo*-DOPS is a potent inhibitor of monoamine catabolizing enzymes, MAO (Naoi and Nagatsu, 1986) and COMT. To examine the effect of L-*threo*-DOPS on monoamine metabolism in the brain, the ratio of the compounds of subsequent enzymatic reaction in IVF was calculated. The ratio which indicate *in vivo* COMT activity, 3-MT/DA, HVA/DOPAC, and NMN/NA, decreased in DOPA/DOPS patients more than DOPA patients. On the other hand, the ratios which indicate *in vivo* MAO activity, HVA/3-MT and MHPG/NMN, both increased. L-*threo*-DOPS might be a potent COMT inhibitor in humans. COMT inhibitors are now used to treat PD patients and it has been reported to economize L-DOPA and decrease fluctuation of the symptoms in the patients treated with L-DOPA (Marello *et al.*, 1994; Roberts *et al.*, 1994). The COMT inhibitory properties of L-*threo*-DOPS may have some significance but further studies are still required.

In the rat striatal microdialysis perfusate, the production of free NA and MHPG was confirmed after L-*threo*-DOPS administration. However, the metabolites by DOPS-ALD were 97.5% and those by COMT 2.1%, while NA and its metabolites were only 0.4% after 100 minutes perfusion of L-*threo*-DOPS. In contrast, 99.9% of the metabolites of L-DOPA detected in the perfusate was by AADC, that is, DA and its metabolites. The metabolites by DOPS-ALD and COMT were confirmed in human IVF samples, indicating that these enzymes may play a role in the human brain. There was large variation in the concentration of the metabolites among the patients and it may be attributed to the variation in the activity of metabolizing enzymes. It is known that clinical response to L-*threo*-DOPS differs in each patient, and this might be attributed to the different enzyme activities capable of metabolizing this artificial precursor amino acid among the patients. To improve the effects of L-*threo*-DOPS as an NA precursor, inhibition of DOPS-ALD might be required.

The CEAS has proven to be a useful tool in evaluating the effects and metabolism of a new drug. By comparing the results of rat *in vivo* microdialysis and analysis of human IVF samples, the mechanism of the drug effects can be clarified. To improve the drug therapy of neurological disorders, systems capable of monitoring the neurotransmitters are essential. The routine use of the CEAS for monitoring drug therapy should be established.

Acknowledgment
This work was supported by Grant-In-Aid for Scientific Research on Priority Area and (C) from the Ministry of Education, Science and Culture, Japan.

REFERENCES

Beal, M. F., Matson, W.R., Swartz, K. J., Gamache, P. H. and Bird, E. D. (1990). Kynurenine pathway measurement in Huntington's disease striatum: evidence for reduced formation of kynurenic acid. J. Neurochem. **55,** 1327-1339.

Blennow, K., Wallin, A., Gottfries, C. G., Mansson, J. -E. and Svennerholm, L. (1993). Concentration gradients for monoamine metabolites in lumbar cerebrospinal fluid. J. Neural Transm. [P-D Sect], **5,** 5-15.

Grahame-Smith, D. G. (1971). Studies in vivo on the relationship between brain tryptophan, brain 5-HT synthesis and hyperactivity in rats treated with a monoamine oxidase inhibitor and L-tryptophan. J. Neurochem. **18,** 1053-1066.

Garelis, E., Young, S. N., Lal, S. and Sourkes, T. L. (1974). Monoamine metabolites in Lumbar CSF: The question of their origin in relation to clinical studies. Brain Res. **79**, 1-8.

Marello, M., Lees, A .J., Webster, R., Bovingdon, M. and Gordin, A. (1994) Effect of entacapone, a peripherally acting catechol-O-methyltransferase inhibitor, on the motor response to acute treatment with levodopa in patients with Parkinson's disease. J. Neurol. Neurosurg. Psychiatry **57**, 186-189.

Maruyama, W., Nakahara, D., Ota, M., Takahashi, T., Takahashi, A., Nagatsu, T. and Naoi, M. (1992a). N-Methylation of dopamine-derived 6,7-dihydroxy-1,2,3,4-tetrahydroisoquinoline, (R)-salsolinol, in rat brains: in vivo microdialysis study. J. Neurochem. **59**, 395-400.

Maruyama, W., Naoi, M., Takahashi, A., Watanabe, H., Konagaya, Y., Mokuno, K., Hasegawa, S. and Nakahara, D. (1992b). The mechanism of perturbation in monoamine metabolism by L-DOPA therapy: in vivo and in vitro studies. J. Neural Transm. [P-D Sect] **90**, 183-197.

Maruyama, W., Nakahara, D. and Naoi, M. (1994). A new metabolic pathway of L-*threo*-3,4-dihydroxyphenylserine, a precursor amino acid of norepinephrine, in the rat brain. Studies by *in vivo* microdialysis. J. Neural Transm. [Gen Sect] 7, 21-33.

Maruyama, W., Naoi, M. and Narabayashi, H. (1996a). The metabolism of L-DOPA and L-*threo*-dihydroxyphenylserine and their effects on monoamines in the human brain: analysis of the intraventricular fluid from Parkinsonian patients. J. Neurol. Sci.. in press.

Maruyama, W., Abe, T., Tohgi, H., Dostert, P. and Naoi, M. (1996b). A dopaminergic neurotoxin, (*R*)-N-methylsalsolinol, increases in Parkinsonian CSF. Ann. Neurol. in press.

Maruyama, W., Narabayashi, H., Dostert, P. and Naoi, M. (1996c). Stereospecific occurrence of a Parkinsonism-inducing catechol isoquinoline, N-methyl(*R*)salsolinol, in human intraventricular fluid. J. Neural Transm. in press.

Matson, W. R., Langlais, P., Volicer, L., Gamache, P. H., Bird, E. and Mark, K. A. (1984). n-Electrode three-dimensional liquid chromatography with electrochemical detection for determination of neurotransmitters. Clin. Chem. **30**, 1477-1488.

Nagatsu, T., Waku, Y., Kato, T., Kato, T., Fujita, K., Kondo, T., Yokochi, F. and Narabayashi, H. (1982). Dopamine beta-hydroxylase activity in cerebrospinal fluid of Parkinsonian patients. Biomed. Res. **3**, 95-98.

Naoi, M. and Nagatsu, T. (1986). Inhibition of monoamine oxidase by 3,4-dihydroxyphenylserine. J. Neurochem. **47**, 604-607.

Naoi, M., Takahashi, T., Kuno, N. and Nagatsu, T. (1987). L-*Threo*-3,4-dihydroxyphenylserine (DOPS) aldolase: a new enzyme cleaving DOPS into protocatechualdehyde and glycine. Biochem. Biophys. Res. Commun. **143**, 482-488.

Naoi, M., Maruyama, W., Acworth, I.N., Nakahara, D. and Parvez, H. (1993) Multi-electrode detection system for determination of neurotransmitters. In: Methods in Neurotransmitter and Neuropeptide Research, Parvez, S.H., Naoi, M., Nagatsu, T. and Parvez, S. (Eds.). Elsevier Science Publisher, Amsterdam, Vol. 1 pp. 1-39.

Narabayashi, H., Kondo, T., Hayashi, A., Suzuki, T. and Nagatsu, T. (1981). L-*Threo*-3,4-dihydroxyphenylserine treatment for akinesia and freezing of parkinsonism. Proc. Japan Acad. **57,** 351-354.

Nishino, N., Fuji, Y., Kondo, M., Shuntoh, H., Fujisawa, H. and Tanaka, C. (1987). Effects of L-threo-3,4-dihydroxyphenylserine on efflux of monoamine and acetylcholine in guinea pig brain. J. Pharmacol. Exp. Ther. **242,** 621-628.

Paxinos, G. and Watson, C. (1986). The Rat Brain in Stereotaxic Coordinates. Academic Press, San Diego, 2nd ed.

Roberts, J. W., Cora-Locatelli, G., Bravi, D., Amantea M. A., Mouradian M. M. and Chase T. N. (1994). Catechol-O-methyltransferase inhibitor tolcapone prolongs levodopa/carbidopa action in Parkinsonian patients. Neurology 44, 2685-2688.

Progress in HPLC-HPCE, Vol. 6, pp. 291–299
I.N. Acworth *et al.* (Eds)

DETERMINATION OF ILLICIT DRUGS AND RELATED SUBSTANCES BY USING HPLC WITH AN ELECTROCHEMICAL COULOMETRIC ARRAY DETECTOR

GUIDO ACHILLI and GIAN PIERO CELLERINO

EuroService srl, Piazza Maggiolini 3, 20015 Parabiago, Italy

INTRODUCTION

The identification and quantitation of the drugs of abuse in biological samples requires a large commitment of manpower and time within forensic science laboratories, especially when instrumental techniques are to be used for confirmation purposes after the usual screening immunoassays. These researchers have tried to develop methods of wide application, as well as high specificity, to confirm the presence of drugs of abuse and determine their concentrations. Chromatographic methods provide an inexpensive, yet powerful approach for the identification and quantitation of the controlled substances. Liquid chromatography is particularly well suited for the analysis of a number of the commonly controlled substances which are relatively non-volatile or thermally unstable. However, to our knowledge, most of the methods published in recent years describe the determination of only one or a few molecules. We have conducted a bibliographic reference search on this topic mainly concentrating on references published in the last four years (i.e., January 1st 1991 to January 1st 1995). A total of ninety references were found and these are summarized in Table 1.

In all of the methods described in Table 1, the researchers were concentrating on the determination of only one or a few molecules. The paper reporting the greatest number of drugs measured simultaneously (five molecules) was published by Peterson *et al.* (1980). In general, the majority of papers reported the measurement of illicit drugs in biological fluids, with very few determining their levels in hair. The presence of a drug or its metabolites in the biological fluids is often limited by the mechanism of elimination from the body. However, drugs of abuse and their metabolites remain in hair for a much longer period of time. This means that subjects can still be screened for illicit drug use well after the drug has been eliminated from other "normally" measured body fluids (Tagliaro *et al.*, 1993). Furthermore, hair is readily obtainable and does not require invasive sampling procedures. The purpose of this study was to develop a method capable of identifying the greatest number of drugs of abuse in a single run, to minimize the analysis time and to maximize specificity. In order to achieve this, the specificity of electrochemical array detection (Matson *et al.*, 1984) was combined with the flexibility of gradient HPLC. This approach was used to separate and quantify fifteen molecules of great interest to forensic scientists including cocaine, codeine, ethylmorphine, hydrocodone, lidocaine, morphine, naloxone, procaine and thebaine. The application of this method was examined using samples of human hair subjected to standard sample preparation techniques.

Table 1.
Reported methods for determination of drugs of abuse during the period Jan 1st 1991 to Jan 1st 1995.

Technique	Matrix	Total number of papers	Percentage of the total
GC or GC MS	hair	5	5.5
	biological fluids	13	14.4
CZE	hair	1	1.1
	biological fluids	3	3.3
HPLC UV	hair	6	6.6
	biological fluids	27	30.0
HPLC PDA	biological fluids	3	3.3
HPLC MS	biological fluids	1	1.1
HPLC FLUO	biological fluids	7	7.7
HPLC RIA	biological fluids	1	1.1
RIA	hair	2	2.2
	biological fluids	2	2.2
ID	biological fluids	1	1.1
HPLC EC	biological fluids	18	20.0
All methods	hair	14	15.6
	biological fluids	76	84.4
	total	90	100.0

Abbreviations: GC = gas chromatography; GC MS = GC with mass spectrometry; CZE = capillary zone electrophoresis; HPLC UV = high performance liquid chromatography with ultra violet detection; HPLC PDA = HPLC with photodiode array detection; HPLC MS = HPLC with MS detection; HPLC FLUO = HPLC with fluorescence detection; HPLC RIA = HPLC with radio immuno assay; RIA = radio immuno assay; ID = isotope dilution; HPLC EC = HPLC with electrochemical detection.

MATERIALS AND METHODS

Mobile phases

Compounds were eluted using gradient HPLC. Mobile phase A consisted of 20mM monobasic sodium phosphate, 10% acetonitrile (v/v) (pH7.00). Mobile phase B consisted of 20mM monobasic sodium phosphate, 50% acetonitrile (v/v) (pH7.00). The mobile phases were filtered through 0.2μm PTFE lyophilic filters (Millipore, Bedford, MA, USA) and degassed by sonication under vacuum for 10 minutes prior to use. The water used for this method was purified with a Milli-Q R/O water purification system (Millipore) and filtered through 0.2μm PTFE lyophilic filters prior to use.

Equipment

The HPLC system was based around a CoulArray® detecor (ESA, Inc., Chelmsford, MA, USA). The HPLC system consisted of a refrigerated autosampler which is capable of variable volume injections with a 100μL loop. Gradient operation was achieved using two HPLC pumps (Model 580 (ESA, Inc.)) capable of operating from 0.05 to 10mL/min. The two outputs from the pumps were connected to a dynamic gradient mixer. The analytical column (150 x 4.6 mm i.d.) consisted of a stainless steel column packed with 3μm particles composed of a silica-based C-18 material (NBS column, ESA, Inc.). The detection system consisted of two coulometric array modules, each containing four flow-through graphite working electrode. Each working electrode has its own individual palladium reference and counter electrodes. The two array modules were arranged in series and were placed after the analytical column. The column was housed in a thermal-chamber maintained at 37°C. The autosampler, pumps, detectors, temperature-controlled box and all associated electronic circuitry were controlled by the CoulArray software which resided on a Model 486 computer equipped with a 350Mbyte hard disk and a 1.4Mbyte floppy disk drive. This computer system also performed data acquisition, storage, analysis and report generation. An appropriate software package was used for summary reports of the final data (e.g., Lotus® 1-2-3, Lotus Corp., Cambridge, MA, USA).

Chromatographic method

A gradient chromatographic method providing complete resolution of all fifteen 15 analytes was developed. The gradient used in the separation, expressed as a percentage of phase B was: 7% isocratic until 11.5 minutes after the injection, then the percentage of phase B was increased up to 50% at 19.5 minutes. The gradient then remained isocratic (50% B) until 24.5 minutes after the injection and was then followed by an increase of phase B up to 90% at 34.5 minutes. It then remained isocratic for 45 minutes when it rapidly decreased to the initial isocratic conditions prior to injection. The total flow rate was 0.8 mL/min. The time-line of this gradient profile is shown in Figure 1.

Analytes were measured using an oxidative array: from +550 to +850mV in 100mV increments (electrodes 1 to 5) followed by a potential range of +900 to +1000mV in 50mV increments (electrodes 6 to 8). The potentials applied to the graphite working electrodes are given with reference to the solid-state palladium reference electrode built into the coulometric cells. The absolute applied potential is therefore about 250mV lower than the corresponding potential applied using a Ag/AgCl reference electrode. At the end of each analysis, all cell potentials were increased to 1200mV for 1 minute to prevent long-term adsorption to the electrode surface. The electrodes were then allowed a period of 7.5 minutes to restabilize before the next injection.

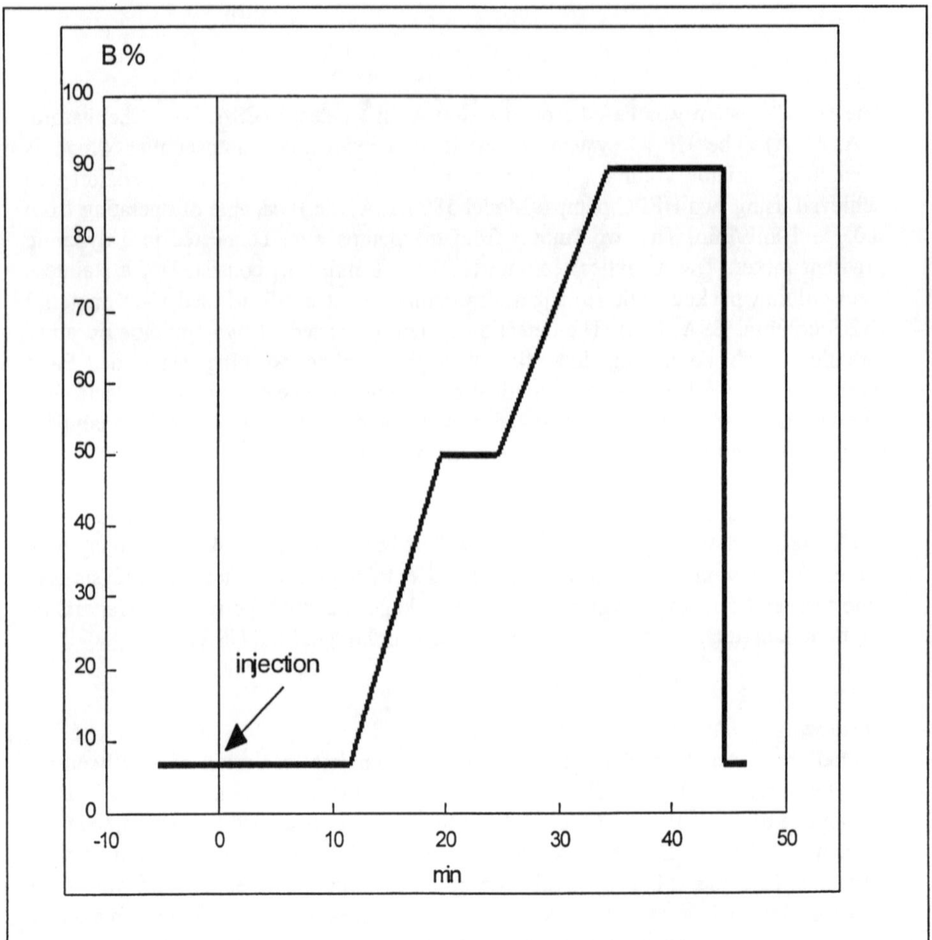

Figure 1. Time line showing the gradient profile used in the separation.

Although this time is more than sufficient to allow electrode re-equilibration, the system already required this time to allow re-equilibration of the column prior to the next analytical gradient analysis.

Standard and sample preparation

The pure standards of 3,4-methylenedioxymethamphetamine (MDMA), buprenorphin, cocaine, codeine, dihydroxycodeine, ethylmorphine, heroin, hydrocodone, lidocaine, methadone, morphine, naloxone, oxycodone, procaine and thebaine were obtained from Sigma Chimica (Milano, Italy). Stock standard solutions were prepared at a level of 1mg/mL in methanol and stored at -18°C. A 15 component working standard solution was prepared at various concentrations by combining aliquots of the individual stock standards and diluting them with mobile phase A. This working standard solution was prepared every day and kept refrigerated at 4°C. A 10µL volume was injected onto the system. Hair samples were prepared according to a previously described method (Tagliaro *et al.*, 1994). Briefly, hair samples (20-200mg) were incubated overnight in 2mL of 0.25 MHCl at 45°C and the resulting mixture was neutralized and extracted with a commercial liquid-liquid method (Toxi-Tubes A, Analytical Systems, Laguna Hills, CA, USA). The dried residue was reconstituted with mobile phase A and a 10µL sample was analyzed on the system.

Assay performance

Assay linearity and detection limit were carried out by adding known amounts of the drugs (from 1µg/L to 10mg/L) to hair extracts of subjects who had never consumed or were exposed to the compounds under investigation. These extracts were also tested prior to these assays to confirm the absence of drugs of abuse, thereby confirming their suitability as blanks. The within-run reproducibility was evaluated by replicated analysis (n = 10) of each standard solution added to the blank extracts and treated as samples. The between-run reproducibility was examined by preparing and analyzing standard extracts in duplicate on ten separate days.

RESULTS

Using the above method the fifteen illicit drug standards were readily resolved and quantified. A typical separation of a 10µL sample containing the 15 drug standards is shown in Figure 2. For graphic clarity traces from only six electrodes are shown. These traces correspond to electrodes 1 (+450mV), electrode 2 (+550mV), electrode 3 (+650mV), electrode 4 (+750mV), electrode 5 (+850mV) and electrode 7 (+950mV). The total analysis time was less than fifty minutes. Confirmation of peak identification was carried out in two steps. First, the samples were spiked with the relevant standard, thereby confirming correct peak identification based upon retention time. Second, the identity of individual analytes was confirmed based upon electrochemical behavior across the elctrode array (ratio accuracy - R). The use of the ratio accuracy to confirm peak purity and identity has been described in detail elsewhere in this volume. The detection characteristics of these 15 external standards are presented in Table 2.

Table 2.
Chromatographic and electrochemical behaviour of the tested drugs.

Key	Compound	Retention time (min)	Dominant potential (mV)	Detection limit (pg)	Within run R.S.D. %	Between run R.S.D. %	Concentration (ng/10μL)
A	morphine	4.66	350	17	3.2	3.8	17.00
B	procaine	8.75	750	31	4.3	4.9	10.00
C	codeine	15.53	750	18	2.8	3.5	25.00
D	mdma	15.95	450	107	5.2	6.1	10.00
E	dihydro-codeine	18.57	750	48	3.5	4.8	17.00
F	6-acetyl-morphine	18.98	650	261	4.8	5.8	10.00
G	ethyl-morphine	19.28	750	201	3.7	4.5	13.00
H	cocaine	21.58	750	264	3.2	4.8	29.00
I	naloxone	22.28	650	61	2.7	4.0	12.50
L	methadone	23.12	750	25	3.0	3.8	50.00
M	hydrocodone	28.08	750	15	1.8	3.8	20.00
N	thebaine	28.73	750	70	2.8	3.6	32.50
O	lidocaine	29.66	550	91	2.4	4.3	20.00
P	heroin	33.00	750	130	3.6	5.5	5.00
Q	buprenorphine	37.75	650	120	2.9	4.8	5.00

The precision (within-run and between-run), detection limits and dominant potentials are also reported (the dominant potential is that electrode potential where the maximum signal occurs). The retention time repeatability during the precision studies (carried out over a 10-day span) was found to be excellent (R.S.D. $< 2.0\%$) for all standards. This was due to the strict control of both the gradient profile and the column temperature. The curves of the standards for all drugs were linear throughout the range tested (better than 0.998, $p < 0.001$). The quantitation was therefore based on the use of external standards (Tagliaro *et al.*, 1994). The minimum detectable amount (at a signal to noise ratio of 3:1), as reported in the Table 2, was very low for all the analytes. In the worst case (cocaine) the sensitivity was less than 26.4µg/L, while in the best case (hydrocodone) it was less than 1.5µg/L. The within-run concentration variability (R.S.D.) ranged from 1.8% (hydrocodone) to 5.2 % (MDMA); the between-run concentration variability ranged from 3.5% (codeine) to 6.1% (MDMA). The analytical recovery for morphine from hair was found to be consistent with data previously published (Marigo *et al.*, 1986) .

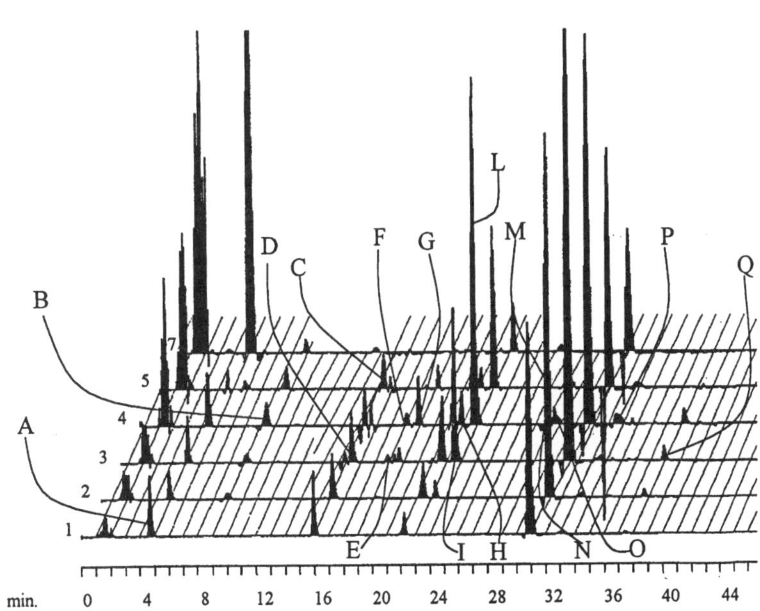

Figure 2. 6 channel chromatogram of the 15 external standards. The numbers represent the electrodes and the letters represent each compound. The applied potentials are: electrode 1: +450mV, electrode 2: +550mV, electrode 3: +650mV, electrode 4: +750mV, electrode 5: +850mV and electrode 7: +950mV. The external standards are: A morphine, B procaine, C codeine, D MDMA, E dihydro-codeine, F 6-acetyl-morphine, G ethyl-morphine, H cocaine, I naloxone, L methadone, M hydrocodone, N thebaine, O lidocaine, P heroin, Q buprenorphine.

Figure 3 shows the trace of the dominant channel (+450mV) for morphine in a sample of hair obtained from a morphine user. Also shown are traces obtained from a blank, from a hair blank and from the morphine standard. The combination of extraction and selectivity of the coulometric array detector ensures that individual drug peaks are all well resolved from each other and devoid of endogenous interfering substances. The morphine level determined by both RIA (Marigo *et al.*, 1986) and HPLC were comparable, being 133ng in the extracted sample.

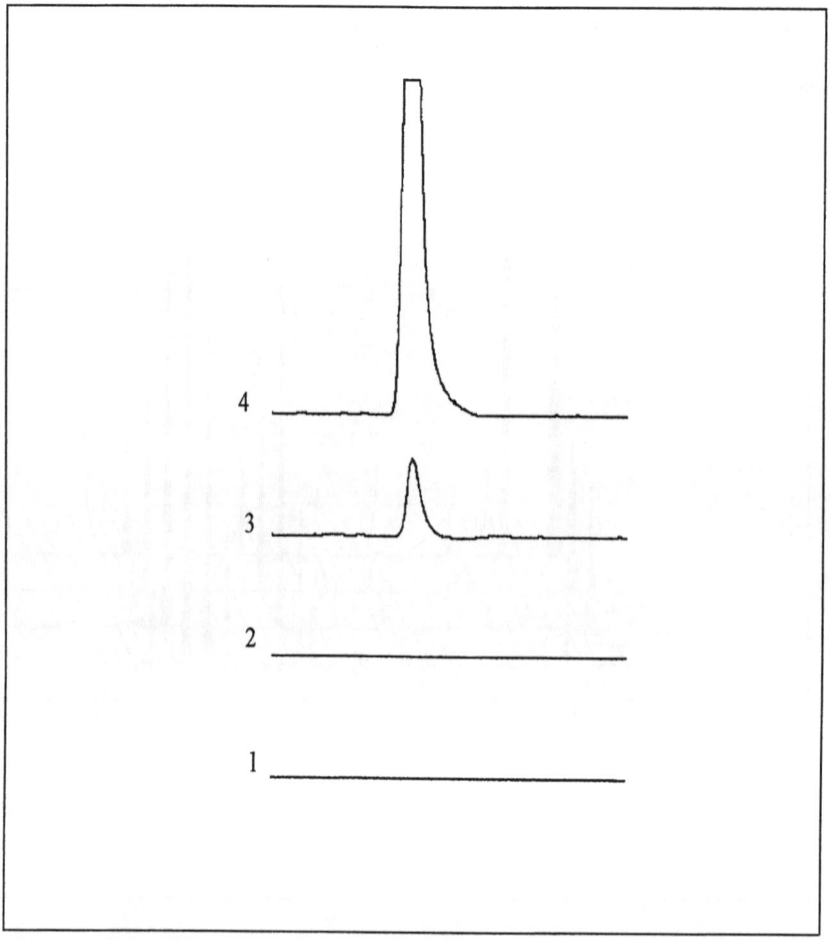

Figure 3. Comparison of the +450mV signals of four injected samples (full scale deflection is 1μA). Trace 1 represents a blank. Trace 2 is a normal hair extract. Trace 3 is a hair extract found positive for morphine (corresponding to 133 ng morphine). Trace 4 is an external standard of morphine standard extracted.

DISCUSSION AND CONCLUSION

The use of the coulometric array detection for the determination of neurochemicals in tissues and biological fluids has already been reported (Marigo *et al.*, 1986; Svendsen and Bird, 1989; Matson *et al.*, 1987; Gamache *et al.*, 1993). More recently, it has been used for the determination of 36 phenolic constituents in natural beverages and plant extracts (Achilli *et al.*, 1993) and for the measurements of phenols and herbicides in water (Achilli *et al.*, 1995). In this work we demonstrate that, with the combination of gradient reversed-phase chromatography with highly selective array coulometric detection, it is possible to separate and determine a large number of drugs in a single run of fifty minutes. This method yields a highly precise determination of drugs of different families in a single run. The coulometric efficiency of each element of the array allows a complete voltammetric resolution of analytes as a function of their reaction potential. Some peaks can be resolved by the detector even if they are unresolved when they leave the chromatographic column. For the samples analyzed here, we were able to easily differentiate morphine positive samples from controls.

In conclusion the proposed method for the simultaneous determination of several drugs in hair provides a fast and reproducible sampling procedure with good selectivity and sensitivity over a wide concentration range.

REFERENCES

Achilli, G., Cellerino, G. P., Gamache, P. H. and Melzi d'Eril, G. V. (1993). Identification and determination of phenolic constituents in natural beverages and plant extracts by means of a coulometric electrode array system. J. Chromatogr. **632**, 111-117.

Achilli, G., Cellerino, G. P., Melzi d'Eril, G. and Bird, S. (1995). Simultaneous determination of 27 phenols and herbicides in water by high-performance liquid chromatography with multi-electrode electrochemical detection. J. Chromatogr. **697**, 357-362.

Gamache, P. H., Kingery, M. and Acworth, I. N. (1993a). Urinary metanephrine and normetanephrine determined without extraction by using liquid chromatography and coulometric array detection. Clin. Chem. **93**, 1825-1830.

Marigo, M., Tagliaro, F. , Poiesi, C., Lafisca, S. and Neri, C. (1986). Determination of morphine in the hair of heroin addicts by high performance liquid chromatography with fluorimetric detection. J. Anal. Toxicol. **10**, 158-161.

Matson, W. R., Gamache, P. H, Beal, M. F. and Bird, E. D. (1987). EC array sensor concepts and data. Life Sci. **41**, 905-908.

Matson, W. R., Langlais, P., Volicer, L., Gamache, P. H., Bird, E. and Mark, K. A. (1984). n-Electrode three-dimensional liquid chromatography with electrochemical detection of neurotransmitters. Clin. Chem. **30**, 1477-1488.

Peterson, R. G., Rumack, B. H., Sullivan, J. B. and Makowski, A. (1980). Amperometric high-performance liquid chromatographic method for narcotic alkaloids. J. Chromatogr. **188**, 420-427.

Svendsen, C. N. (1993). Multi-electrode array detectors in high-performance liquid chromatography: A new dimension in electrochemical analysis. Analyst **118**, 123-129.

Tagliaro, F., Antonioli, C., De Battisti, Z., Ghielmi, S. and Marigo, M. (1994). Reversed-phase high-performance liquid chromatographic determination of cocaine in plasma and human hair with direct fluorometric detection. J. Chromatogr. **674**, 207-215.

Tagliaro, F., Antonioli, C., Moretto, S., Archetti, S., Ghielmi, S. and Marigo, M. (1993). High-sensitivity low-cost methods for determination of cocaine in hair: High-performance liquid chromatography and capillary electrophoresis. Forensic Sci. Int. **63**, 227-238.

Progress in HPLC-HPCE, Vol. 6, pp. 301–338
I.N. Acworth *et al.* (Eds)

ENANTIOSEPARATION OF BIOLOGICALLY ACTIVE COMPOUNDS BY HPLC

YULIN DENG[1], WAKAKO MARUYAMA[2], MASAO KAWAI[1], PHILIPPE DOSTERT[3] and MAKOTO NAOI[4]

[1]*Department of Applied Chemistry.* [2]*Laboratory of Biochemistry and Metabolism, Department of Basic Gerontology, National Institute for Longevity Sciences, Obe, Japan.* [3]*Pharmacia-Upjohn, Research and Development, Milan, Italy.* [4]*Department of Biosciences, Nagoya Institute of Technology, Nagoya 466, Japan*

INTRODUCTION

The development and application of chiral chromatographic methods have been gathering general interest, as reviewed previously (Armstrong, 1987a; Pirkle *et al.*, 1979; Allenmark *et al.*, 1994; Davankov, 1994; Okamoto *et al.*, 1994a). The resolution of enantiomers has been considered one of the more difficult problems in separation science. Naturally-occurring polymers, such as wool, lactose, cellulose, starch, and their derivatives, have been applied as sorbents to separate enantiomers, indicating the possible interactions between the chiral analytes and the structure of the sorbent (Dalgleish, 1952; Luettringhaus *et al.*, 1966; Davankov *et al.*, 1971; Hesse and Hagel, 1973). The use of these compounds proved to be effective also with modern HPLC equipment for the separation of enantiomers.

Optically active compounds have been attracting attention in various fields, especially in pharmacology, and therefore their preparation and analysis have been essentially required. At the molecular level, the biological environment is predominantly chiral, and biosynthesis and metabolism of biologically active compounds are severely controlled by chirality (Bentley, 1995). Many chiral drugs and agrochemicals show different biological characteristics with respect to their absolute configuration. A typical example was the thalidomide tragedy in the early 1960s. When racemic thalidomide was used as a sedative by pregnant women, many children were born with missing or abnormal limbs. Although the situation is still not completely clear, it is likely that only the *(R)*-form is a sedative, while the teratogenic activity is ascribed to the *(S)*-enantiomer. Similarly, 6,7-dihydroxy-1-methyl-1,2,3,4-tetrahydroisoquinoline (salsolinol, Sal) which was found to occur in human brain (Sjöquist *et al.*, 1982) and to be an endogenous neurotoxin (Naoi *et al.*, 1996), also possesses an asymmetric center and exists as both *(R)*- and *(S)*-enantiomers. *(R)*-Sal was found to inhibit type A of monoamine oxidase [monoamine: oxygen oxidoreductase (deaminating), EC 1.4.3.4] more potently than its enantiomeric counterpart (Minami *et al.*, 1992). The chirality of Sal and its *N*-methylated derivative (*N*MSal) was found to affect the affinity of tyrosine hydroxylase [tyrosine, tetrahydropteridine: oxygen oxidoreductase (3-hydroxylating), EC 1.14.16.2] to its substrate, L-tyrosine (Minami *et al.*, 1993). Using human dopaminergic neuroblastoma SH-SY5Y cells, only *(R)-N*MSal was transported by the dopamine uptake system (Takahashi *et al.*, 1994). These results indicate that the stereo-chemical structure of this alkaloid is involved in its biological activities.

For separation of enantiomers, HPLC is one of the most useful techniques but is dependent on high column efficiency and enantioselective recognition by a chiral selector. Two main strategies are generally used for chiral separation by HPLC; the indirect and the direct separation. Indirect chiral resolution requires the derivatization of the racemic mixture of analytes with an appropriate reagent of single enantiomeric form, to form a pair of diastereomers. The diastereomers have different chemical and physical properties and thus may be separated by conventional HPLC technique. For example, using a chiral derivatizing agent, *(S)*-1-(1-naphthyl) ethyl isothiocyanate, diastereomers of Sal enantiomers were obtained and separated by a conventional reversed-phase HPLC column (Pianezzola *et al.*, 1989). On the other hand, there are at least two general approaches for the direct separation of enantiomers. The first method uses a chiral stationary phase (CSP), and the second uses chiral additives in mobile phase with achiral stationary phases. These two approaches are essentially alternatives to the indirect one described previously, since these direct approaches are also based on the formation of diastereomers, but in this case it is only transient binding rather than the covalent modification. Although a chiral selector to a particular enantiomeric form is required for all these approaches, with the indirect separation approach the chiral selector can be used as the solid or mobile phase. In direct approaches, chiral resolution is affected by the difference in stability of transient diastereomeric complexes of the chiral selector with the enantiomers of analytes. For example, the enantiomer which can form the more stable adsorbate is preferentially retained on an CSP column. At present, a number of improved chiral stationary phases and additives have been introduced, and nearly 100 of the chiral stationary phases are commercially available. These include chiral π-acid and π-base stationary phase [Pirkle *et al.*, 1979 and 1981], chiral polymer (Okamoto *et al.*, 1981), protein-bonded phase (Allenmark *et al.*, 1982), and cyclodextrin-bonded phase (Fujimula *et al.*, 1983; Armstrong *et al.*, 1985). By development of successful chiral stationary phases, a number of theoretical and mechanistic studies on chiral recognition have been published.

In general, many chiral chromatographic systems were developed for relatively pure substance but they cannot be applied directly to crude biological samples. The identification and quantitation have been difficult because of the presence of contaminant peaks. As reviewed in this book, the introduction of multi-electrochemical detection system (Coulochem Electrode Array System (CEAS), ESA, Inc., Chelmsford, MA) can increase the accuracy of identifying and quantifying compounds in crude biological samples. In this system the identification of compounds is based on the comparison of an analyte's retention time and voltammetric behavior across the electrode array. In this article, we review the principles and methods for the direct enantioseparation using cyclodextrins.

CHIRAL RECOGNITION OF ENANTIOMERS WITH HPLC

For direct chiral resolution of enantiomers by HPLC, diastereomeric complexes must be formed from the chiral selector with at least one of the analyte enantiomers, and these must differ in their free energy of formation. These conditions will be discussed in this section.

As the enantiomers pass over the chiral stationary phase, each enantiomer forms a transient complex. Several factors influence the formation of these diastereomers including the attractive interaction between the CSP and the analyte enantiomers, the expulsion of the enantiomers from the mobile phase, and the passive diffusion of the analyte into the CSP. Even though no binding is taking place in the latter two instances, the enantiomers are in a chiral-environment. For ultimate separation of the enantiomers, formation of transient diastereomers is necessary but alone is not sufficient. If the difference in steric configuration is all that is required for chiral recognition of diastereomers, any chiral species can distinguish two enantiomers of any substance. This is obviously not the case. Dalgleish (1952) proposed a chiral recognition mechanism with a minimum of three point interactions, between the CSPs and either enantiomer, and that at least one of the interactions should be stereochemically dependent. The requirement for a minimum of three simultaneous interactions was verified by Baczuk (Baczuk *et al.*, 1971), as shown in Figure 1. A chiral stationary phase was prepared by bonding (*S*)-arginine to Sephadex beads through a cyanurate linkage. This CSP was used to separate the enantiomers of β-(3,4-dihydroxyphenyl) alanine (DOPA). In this work, three ionic interactions were involved, all of which stabilized the (*S*)-(*S*) transient diastereomer. Consequently, the (*S*)-enantiomer of DOPA was more strongly retained on the *(S)*-arginine CSP. When one of these interactions was removed, as in the case of phenylalanine (without the second phenolic hydroxyl group), no resolution was observed. It is thought that each of two enantiomers can interact with the chiral selector at two points if a third interaction is absent. However, the presence of an hydroxyl groups of the (*R*)-enantiomer actually inhibits binding. Thus, there must be at least three interaction sites between a selector and an analyte enantiomer that define a three-dimensional conformation adequate for chiral recognition. The selector distinguishes between the two analyte enantiomers by presence/absence or significant difference of at least one of three interactions.

Chiral recognition mechanisms of most chiral stationary phases are of the cooperative type, and depend on a series of interactions at the various points between the two particles. For instance, there are combinations of binding forces and chiral cavities in cyclodextrin-bonded stationary phases, polymeric or protein derived stationary phases.

Another condition for enantiomeric separation is the magnitude of the difference in the free energy to form transient diastereomers. Relative retention, α, is used to describe chromatographic separation of enantiomers, and defined as the relative ratio of the capacity ratios for the *(R)*- and *(S)*-enantiomers of the analytes on a given CSP and mobile phase at a given temperature.

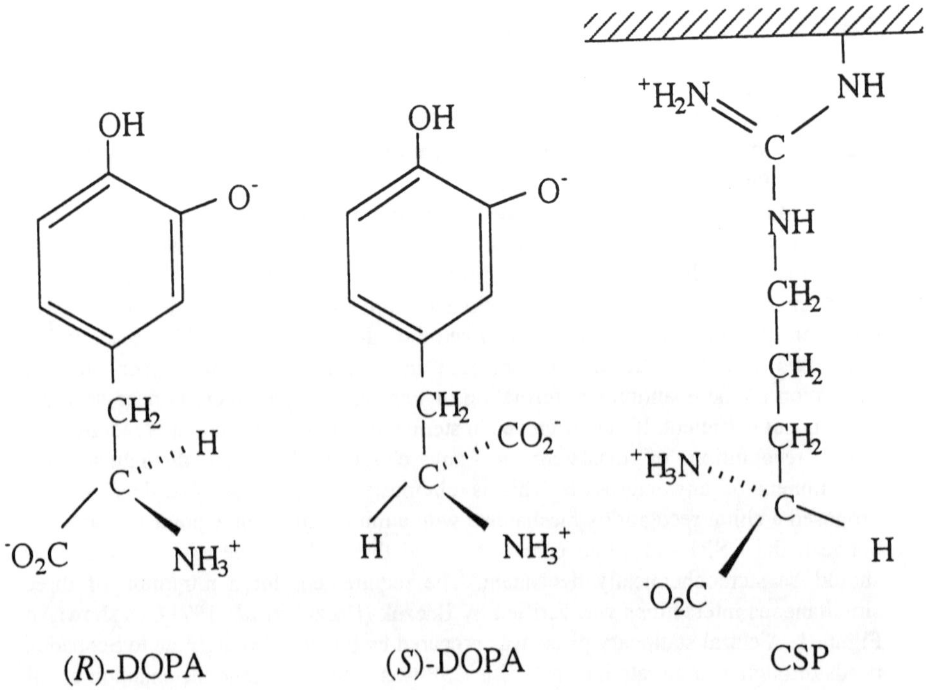

Figure 1. The comparison of the "three point interaction" between *(S)*- and *(R)*-DOPA with CSP.

$$\alpha_{(R/S)} = k_R / k_S \qquad (1)$$

Because capacity ratios are equilibrium constants of the analytes between the stationary phase and mobile phase, Equation 1 may be rewritten:

$$\alpha_{(R/S)} = \exp(-\Delta G_R/RT)/\exp(-\Delta G_S/RT) \qquad (2)$$

where ΔG_R and ΔG_S are the molar free energies of adsorption of *(R)*- and *(S)*-enantiomers, respectively. From statistical thermodynamic theory, an expression analogous to Equation 2 is derived as a function of the multitude of possible interactions (Boehm, 1988), and applied to models of chemically bonded CSPs which consist of chiral tetrahedral moieties or chiral molecular cavities. The enantioselectivity is the direct result of the interactions between the CSP and the analyte, assuming that the same number of solvent molecules are displaced from the stationary phase to the mobile phase by each analyte enantiomer. This assumption is experimentally valid based upon the fact that α is not so affected by temperature. The entropy is relatively unimportant to the magnitude of difference in the molecular free energy, ΔG. Rearrangement of Equation 2 gives the relationship between $\alpha_{(R/S)}$ and $\Delta(\Delta G)$, the difference in molar free energy of the formation of the transient complexes.

$$\Delta(\Delta G) = -RT \ln \alpha_{(R/S)} \quad (3)$$

In a highly efficient LC system, relatively small values of $\Delta(\Delta G)$ suffice for observable chromatographic separation. From Equation 3, if the plate requirement for the enantiomeric separation is about 10^4, $\Delta(\Delta G)$ should be only greater than about 40 cal/mol (equivalent to α =1.06 at 298 K). A variety of CSPs have been found to be effective in separating different enantiomers, although none are universally applicable to all types of chiral separations. This may be reasonable, since the molecular dimensions and/or structures of the chiral moieties of the CSP and the chiral analyte molecules must be suitable to form a minimum of three-point simultaneous interactions, and the energetic difference between the transiently formed diastereomers, $\Delta(\Delta G)$, must be sufficiently large according to Equation 3.

THE CHIRAL STATIONARY PHASE

There are two fundamental approaches for the preparation of chiral stationary phases (Table 1) (Francotte, 1994). In the first case, chiral organic polymers are used as CSPs in a pure form or in a diluted form where chiral polymers are coated or grafted on an achiral support. Polysaccharide derivatives, polyacrylamides, polyacrylic esters and protein-based phases belong to this type of CSPs. The second most commonly used approach is to attach optically active units to achiral carriers (mainly silica gel) by means of ionic or covalent bonding. A wide range of optically active moieties have already been applied, including amino acid derivatives, crown ethers, cinchona alkaloids, carbohydrates, amines, tartaric acid and derivatives, cyclodextrins and binaphthol. Some commercially available chiral stationary phases are summarized in Table 2.

Ligand exchange chiral stationary phases

Most successful works on ligand exchange chromatography were done by Davankov *et al.* (1971). As its name implies, this technique involves the reversible formation of a ternary complex between metal ion, analyte and CSP. In this case, both analytes and CSP have to carry functional groups able to interact as ligands. The mostly used complex-forming agents are α-amino acids possessing a carboxylic acid and an amino group. This type of CSP has been prepared by covalent bonding of amino acid on organic polymeric or on an inorganic silica support. As shown in Figure 2, if a chiral chelator such as (*S*)-proline is bound to a support, the CSP can form ternary diastereomeric complexes with a metal ion and a bidentate analyte enantiomer. The relative stability of the transient diastereomers, if different, leads to chromatographic separation of the analyte enantiomers. The relative stability of the diastereomeric complexes is controlled by several factors, the most critical of which is the choice of the bound chiral chelator. Proline and hydroxyproline are most often used, since their

Table 1
Two fundamental approaches for the preparation of chiral stationary phases

Chiral organic polymer	Achiral carrier modified with chiral moieties
1. Pure chiral polymer	Chiral molecules bonded on the surface of inorganic material (mainly silica gel)
2. Chiral polymer coating on inorganic support 	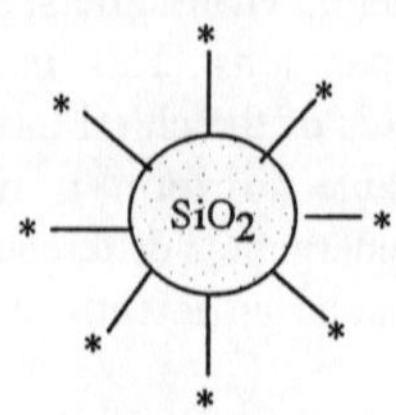
3. Grafted chiral polymer 	* Chiral moieties: such as amino acids (or derivatives), cyclodextrins and so on.

steric rigidity provides a high degree of enantioselectivity. A variety of studies have been done on this method, analyzing the effect of resin type (support), metal type (usually Cu^{2+} or Ni^{2+}), immobilized amino acid, and so on (Davankov *et al.*, 1983; Feibush *et al.*, 1983). Today varieties of this type phases are commercially available for liquid chromatography (Table 2).

Table 2
A list of some commercially available chiral stationary phases

Trade name	Chiral selector	Typical mobile phase used	Supplier
π-Acid and π-base phases:			
DNBPG	N-(3,5-Dinitrobenzoyl)-phenylglycine	Normal phase (hexane/2-propanol)	Baker
Sumipax OA-2000	N-(3,5-Dinitrobenzoyl)-phenylglycine	Normal phase (hexane/2-propanol)	Sumitomo
DNBleu	N-(3,5-Dinitrobenzoyl)-leucine	Normal phase (hexane/2-propanol)	Baker

Trade name	Chiral selector	Typical mobile phase used	Supplier
Supelcosil LC-Urea	1-(1-Naphthyl)ethylamine	Normal phase (hexane/2-propanol)	Supelco
Sumipax OA-1000	(*S*)-1-(1-Naphthyl)-ethylamine	Normal phase (hexane/2-propanol)	Sumitomo
Sumipax OA-2000	(*R*)-N-(3,5-dinitrobenzoyl)-phenylglycine	Normal phase (hexane/2-propanol)	Sumitomo
Sumipax OA-2100	(*S*)-2-(4-Chlorophenyl)-isovalelyl-(*R*)-phenylglycine	Normal phase (hexane/2-propanol)	Sumitomo
Sumipax OA-2200	N-[(1R,3R)-chrysanthemyl]-(*R*)-phenylglycine	Normal phase (hexane/2-propanol)	Sumitomo
Sumipax OA-2500	N-(3,5-Dinitrobenzoyl)-naphthylglycine	Normal phase (hexane/2-propanol)	Sumitomo
Sumipax OA-3000	N-(*t*-Butylaminocarbonyl)-(*S*)-valine	Normal phase (hexane/2-propanol)	Sumitomo
Sumipax OA-3100	N-Dinitrophenylamino carbonyl-(*S*)-valine	Normal phase (hexane/2-propanol)	Sumitomo
Cyclodextrin-bonded phases:			
Cyclobond I	β-Cyclodextrin	Reversed-phase (buffers/ acetonitrile or methanol)	AST
ChiraDex	β-Cyclodextrin	Reversed-phase (buffers/ acetonitrile or methanol)	Merck
Cyclobond II	γ-Cyclodextrin	Reversed-phase (buffers/ acetonitrile or methanol)	AST
Cyclobond III	α-Cyclodextrin	Reversed-phase (buffers/ acetonitrile or methanol)	AST
Cyclobond I AC	Acetyl- β-cyclodextrin	Reversed-phase (buffers/ acetonitrile or methanol)	AST
Cyclobond I RN	(*R*)-1-Naphthylethyl-carbamoyl-β-CD	Reversed-phase (buffers/ acetonitrile or methanol)	AST
Cyclobond I SN	(*S*)-Naphthylethyl-carbamoyl-β-CD	Reversed-phase (buffers/ acetonitrile or methanol)	AST
Cyclobond I RSP	*rac*-2-Hydroxypropyl-β-CD	Reversed-phase (buffers/ acetonitrile or methanol)	AST
Chiral polymer phases:			
Resolvosil	Bovine serum albumin	Aqueous buffer	Nagel
Enantiopac	α-Acid glycoprotein	Aqueous buffer	LKB
Chiralcel OA	Cellulose triacetate	Normal phase (hexane/2-propanol)	Daicel
Chiralcel OB	Cellulose tribenzoate	Normal phase (hexane/2-propanol)	Daicel
Chiralcel OJ	Cellulose tris(4-methyl-benzoate)	Normal phase (hexane/2-propanol)	Daicel
Chiralcel OC	Cellulose trisphenylcarbamate	Normal phase (hexane/2-propanol)	Daicel
Chiralcel OF	Cellulose tris(4-chloro-phenylcarbamate)	Normal phase (hexane/2-propanol)	Daicel
Chiralcel OG	Cellulose tris(4-methyl-phenylcarbamate)	Normal phase (hexane/2-propanol)	Daicel
Chiralcel OD	Cellulose tris(3,5-dimethyl-phenylcarbamate)	Normal phase (hexane/2-propanol)	Daicel
ChiralPak AD	Amylose tris(3,5-dimethyl-phenylcarbamate)	Normal phase (hexane/2-propanol)	Daicel

Trade name	Chiral selector	Typical mobile phase used	Supplier
Chiralcel OJ-R	Cellulose tris(4-methyl-benzoate)	Reversed-phase (water/ acetonitrile or methanol)	Daicel
Chiralcel OD-R	Cellulose tris(3,5-dimethyl-phenylcarbamate)	Reversed-phase (water/ acetonitrile or methanol)	Daicel
ChiraPak OT	(+)-Poly(triphenylmethyl methacrylate)	Normal phase or methanol	Daicel
ChiraPak OT	(+)-Poly(diphenyl-2-pyridilmethyl methacrylate)	Normal phase or methanol	Daicel
ChiraSpher	Poly[(S)-N-acryloyl-phenylalanine ethyl ester]	Methanol	Merck
Ligand exchange phases:			
ChiralPak WH, WA	Amino acid derivatives	Aqueous buffer/Cu^{2+}	Daicel
ChiralPak WE	2-Amino-1,2-diphenyl-ethanol	Aqueous buffer/Cu^{2+}	Daicel
Nucleosil Chiral-1	Hydroxyl-(*S*)-proline	Aqueous buffer/Cu^{2+}	Nagel
Chiral Hypro Cu	Hydroxy-(*S*)-proline	Aqueous buffer/Cu^{2+}	Serva
Chiral Pro Cu	(*S*)-Proline	Aqueous buffer/Cu^{2+}	Serva
Chiral Val Cu	(*S*)-Valine	Aqueous buffer/Cu^{2+}	Serva

Because ligand exchange chromatography requires no derivatization step and uses an aqueous mobile phase, it is one of the most effective means for the resolution of non-derivatized amino acid racemers. Detection of non-derivatized amino acid is difficult at low concentrations, and some derivatization methods are used to increase detection sensitivity. For example, after the separation on HPLC, enantiomers are reacted with a derivatizing agent such as *o*-phthalaldehyde to improve the sensitivity by fluorescence detection (Nimura *et al.*, 1981). Picomolar quantities of amino acids can be detected by such a method. Although ligand exchange chromatography is designed for bidentate analytes, this technique is applied also to other classes of compounds. Using a chiral (*S*)-proline-bonded phase, the biologically important β-hydroxyl-α-phenethylamines and catecholamines were separated after derivatization of these amines to a Schiff base with salicylaldehyde, as shown in Figure 3 (Gelber *et al.*, 1984). Recently, some novel chelating resins, which combine chelating behavior with chiral molecular cavity, were synthesized. For instance, a ligand exchange CSP which was prepared by bonding glycinocyclodextrin to silica gel can form an efficient complex with a metal ion (Cu^{2+}) and can be used to separate substituted phenol isomers (Liu *et al.*, 1994).

Chiral π-acid and π-base stationary phases

Pirkle (1979) developed the first commercially successful CSPs where optically active units were attached to an achiral silica gel support by means of ionic or covalent bonds. The enantioselective recognition can arise from hydrogen bonds, charge-transfer complexes, and dipole and steric interactions. These types of phases are called π-acid-base CSPs. The best known ones are the Pirkle-type CSPs. Pirkle-type phases (Figure 4) consist of a π-donor or π-acceptor-containing chiral selector attached to a support (commonly silica), in which the bonded chiral molecule can interact at three points with chiral analytes. Figure 5 shows a chiral recognition model for the separation of (*R*, *S*)-*N*-(2-naphthyl)-alanine undecenyl ester on the CSP prepared from *N*-(3,5-dinitrobenzoyl)-leucine (Pirkle *et al.*, 1985). Multipoint interactions are obtained by π-donor-acceptor overlap between the naphthyl ring of the analyte and the dinitrobenzoyl ring of the stationary phase, an hydrogen bond between the dinitrobenzamide NH of the phase and the carbonyl oxygen of analyte, and a second hydrogen bond between the amino NH of the analyte and the C-terminal carboxamide oxygen of the phase. The absolute configuration of the *(S)*-enantiomer of the analyte is the same as the chiral stationary phase. The *(S)*-enantiomer was very selectively retained with a separation factor of 10.2 (Pirkle *et al.*, 1985).

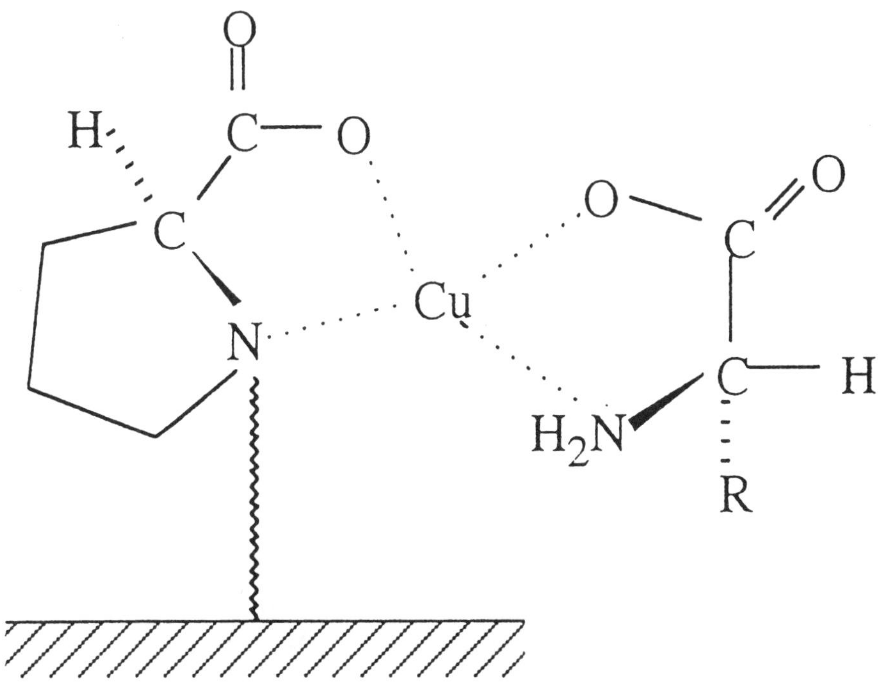

Figure 2. The formation of the ternary complex on the surface of a proline-bonded ligand exchange CSP.

By arrangement of hydrogen-bonding, π-donor-acceptor, dipole-stacking, and steric interactions between a CSP and an analyte, a number of highly selective CSPs were prepared by Pirkle and also by other research groups (Pirkle *et al.*, 1984; Oi *et al.*, 1986; Caude *et al.*, 1991). These CSPs can be synthesized with either absolute configuration. Changing the configuration of the CSP alters the elution order of the enantiomers, which is useful for the stereochemical study of chiral compounds. Although the covalently bonded phases can be used with relatively polar mobile phases, the enantioselectivity and stability are much greater when used in normal-phase. Recently, polysiloxane-based CSPs containing π-acidic or basic selectors were developed. The polysiloxane-based CSPs was found to be highly enantioselective under both normal phase and reversed-phase conditions (Schleimer *et al.*, 1994).

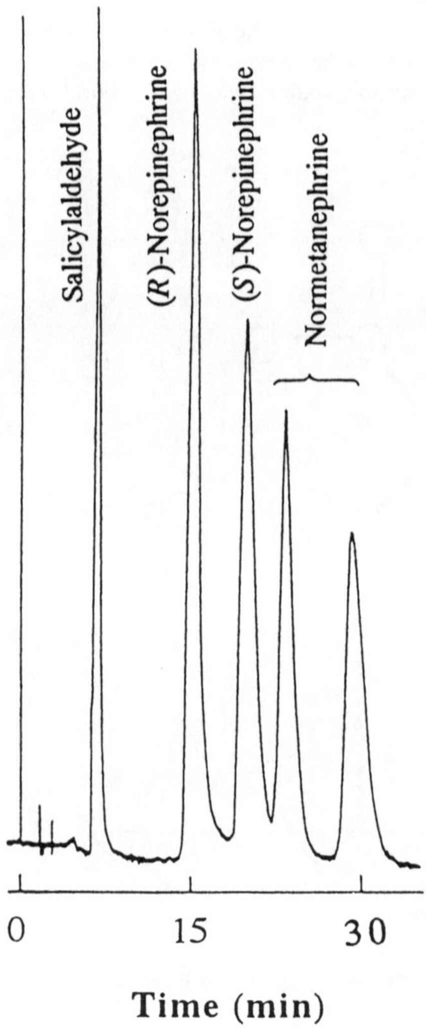

Figure 3. Resolution of enantiomeric norepinephrine and normetanephrine as salicylaldehyde Schiff bases. Column: C_{20}-Proline phase; Mobile phase: 5 x 10^{-4}M Cu(II), 0.125 M ammonium acetate, pH 5.0, 50% methanol; Flow, 1mL/min; UV Detector: 350nm. (Reproduced with permission from Gelber *et al.*, (1984). J. Am. Chem. Soc. **106**, 7729-7734).

(A)

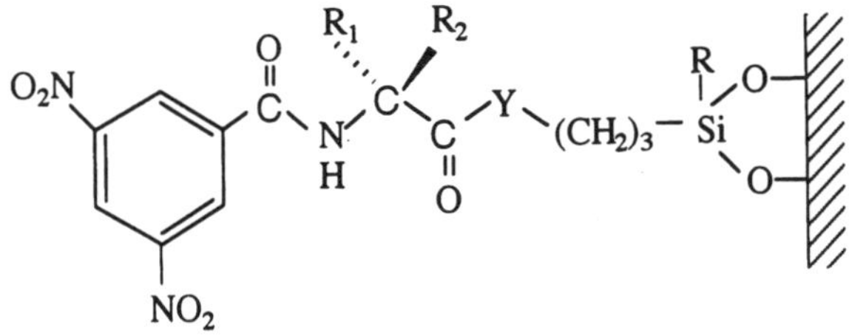

A1: $R_1 = C_6H_5$, $R_2 = H$; A2: $R_1 = H$, $R_2 = CH_2CH(CH_2)_3$

$Y = O^-NH_3^+$; Ionic phase

$Y = NH$; Covalent phase

(B)

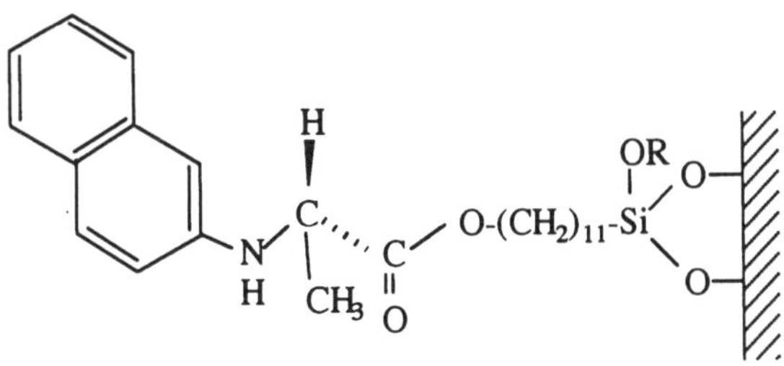

Figure 4. Structure of π-acid (A) and π-base (B) chiral stationary phases.

Chiral polymer stationary phases

Some classical CSPs have been improved and are now widely used as stationary phases. The CSPs use natural or synthetic chiral polymers, such as protein, cellulose and so on. Protein-bonded stationary phases are derived from serum transport proteins, such as bovine serum albumin (BSA), and are commonly used because of ease of preparation and commercial availability. BSA was the first protein to be used for enantiomeric separation by analytical HPLC (Allenmark *et al.*, 1982). BSA is composed of a single polypeptide chain, contains no carbohydrate and has an isoelectric point near 4.7. The 17 disulfide bridges give rise to 9 loops, grouped into three homologous domains, each containing three loops. BSA has a number of different types of binding sites for ionic compounds, and a CSP, where BSA is irreversibly bound to surface such as silica or agarose, can separate a variety of chiral compounds. Recently, in addition to BSA phase, several column types have been developed which were derived from α1-acid glycoprotein (AGP) (Hermansson, 1983), ovomucoid (Miwa *et al.*, 1987) and human serum albumin (Dominici *et al.*, 1990) by bonding to silica. These CSPs are applied mainly for separation of acidic and neutral hydrophobic compounds, but some basic compounds are also successfully resolved. One of the advantages of protein-based chiral sorbents is that the chromatography is performed in a reversed-phase model with aqueous mobile phase. This means that the retention and selectivity are controlled by varying pH and ionic strength of the buffer and by the addition of small amounts of organic modifiers such as propanol, diethyloctylamine or ion-pair reagents (Iredale *et al.*, 1991). However, these protein type sorbents have relatively low sample capacities, indicating limited availability of binding sites. In some cases, overloading results in reversal of the elution order due to the presence of multi types of binding site. In addition, the protein itself is not particularly durable due to degradation over time and is affected by extreme temperature and pH (Allenmark *et al.*, 1983).

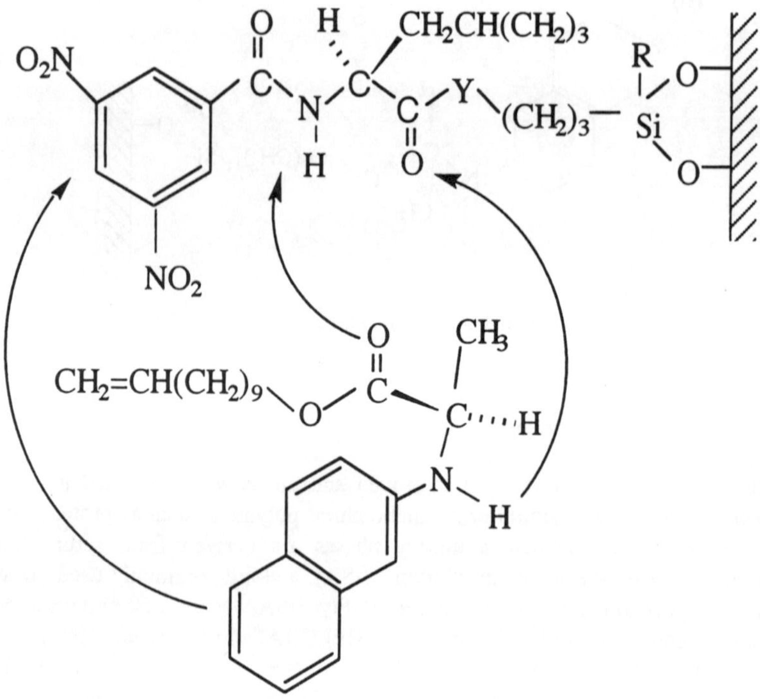

Figure 5. Chiral recognition between π-acid CSP and *N*-(2-naphthyl)alanine undecenyl ester.

Cellulose is a linear polymer of D-glucopyranose connected via a β-1,4-linkage. The β-glucose units have a chair conformation with the 2-OH, 3-OH and CH_2OH group all in equatorial positions. Native cellulose is rather fragile and its high polarity and porous structure give rise to unfavorable kinetic behavior. Cellulose triacetate (CTA-I) was prepared by acetylation of microcrystalline cellulose under heterogeneous conditions (Hesse and Hagel, 1976). The original structure of the polymer is partially reserved, which is important for enantioselectivity. The mechanism of the enantioselectivity on CTA-I may be by inclusion, namely by shape-selective adsorption into chiral cavities in the polymer network (Francotte *et al.*, 1985). However, if the process of acetylation is performed in such a way that the triacetate product dissolves during the acetylation, the product after re-precipitation has a completely different structure, which is defined as CTA-II. The mechanism of retention in the later phase is attractive rather than inclusive, with dipole-stacking interactions (Wainer, 1986).

In order to improve the chromatographic behaviors of cellulose-derived chiral polymer phases, Okamoto and co-workers developed carbamate chiral stationary phases by treatment of cellulose and amylose with aryl isocyanate (Okamoto *et al.*, 1986), as shown in Figure 6. Today, a number of different types of these cellulose-derived phases are available on the market (Table 2). In fact, a wide range of racemers with different chemical structures can be separated using these phases, for example, Chiralcel OD (3,5-dimethylphenylcarbamyl derivative of cellulose), Chiralcel OJ (p-methylbenzoyl derivative of cellulose), and Chiralcel AD (3,5-dimethylphenyl-carbamyl derivative of amylose). In this Okamoto-type packing, the chiral component is coated on macroporous γ-aminopropyl silica. Normal phase hexane-alcohol mixture is most commonly used with these types of stationary phases. Recently, a few reversed-phase columns of these types also were commercialized as Chiralcel OD-R and Chiralcel OJ-R, with which buffered water-methanol or water-acetonitrile mixtures can be used. For (*R*, *S*)-indapamide, the chromatogram obtained using reversed-phase conditions on a Chiralcel OD-R column is shown in Figure 7 (Ikeda *et al.*, 1989). Francotte *et al.* (1991) developed the spherical cellulose derivatives which are cellulose ester of aromatic acids in the pure form of almost spherical, partially crystalline particles with relatively good mechanical stability. These materials have characteristics which are comparable with those of the corresponding coated cellulose derivatives.

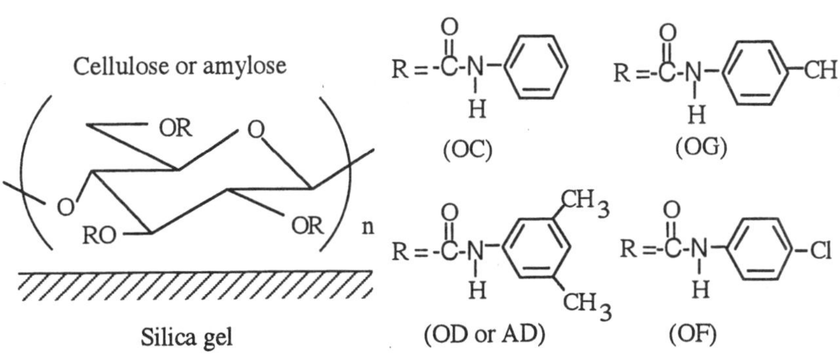

Figure 6. Structure of cellulose-derived chiral polymer phases.

Other commercially available polymeric CSPs are prepared from the asymmetric polymerization of triphenylmethyl methacrylate (or its derivatives), as shown in Figure 8. Okamoto *et al.* (1981) first used this methacrylate polymer as a stationary phase. The polymerization of this chiral polymer was initiated using chiral anions (e.g., lithium *(R)*-phenylethylanilide), whose chirality was obtained by a stable helical structure generated by bevel-gear interactions between the triphenylmethyl groups (Okamoto *et al.*, 1979). The helical polymers can be used, either as solid polymeric particles with high-molecular-weight or as soluble particles with lower-molecular-weight adsorbed on diphenyl silanized silica gel, with which methanol is

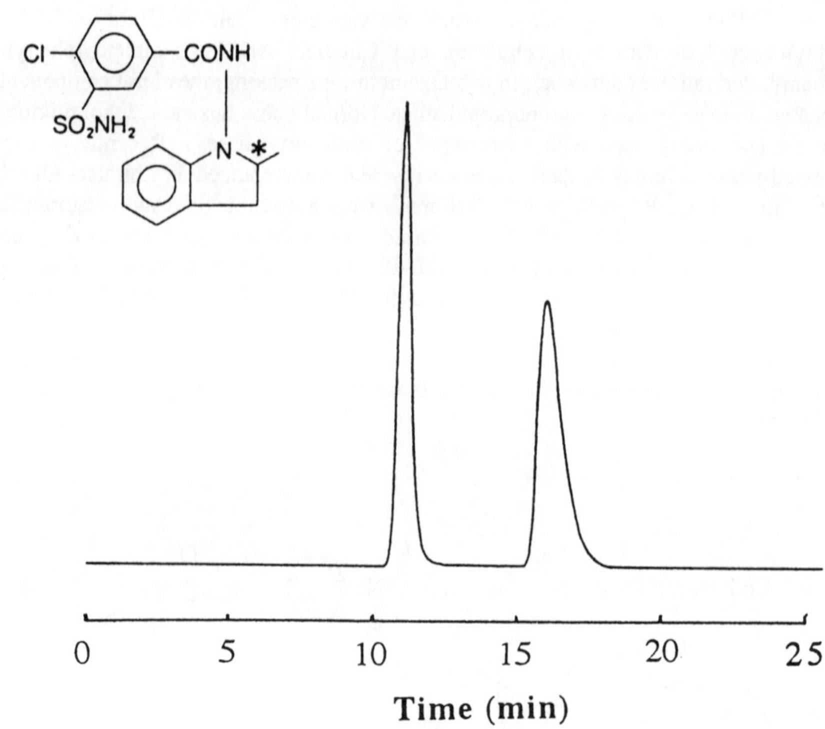

Figure 7. Resolution of indapamide enantiomers using a Chiralcel OD-R column. Mobile phase: 40% CH3CN aqueous solution; Flow: 1.0mL/min; UV Detector: 254nm. (From Ikeda *et al.*, (1989). Chem. Lett. **6**, 1089-1090).

most commonly used as the mobile phase. The enantioselectivity is often dependent on the state of the polymers on the silica support. The helical polymer phases can separate the enantiomers of compounds that are less polar or have an aromatic substituent at the chiral center. Stable cyclodextrin-bonded phases developed by Armstrong *et al.* (1984) were the first commercial chiral stationary phases, which can be used in a conventional reversed-phase mode. Cyclodextrin-bonded CSPs will be described in another section of this review.

The first chiral binaphthyl-based crown ether CSP was prepared by Cram's group (Dotsevi *et al.*, 1975). A substance with a primary amino group at the chiral center can be included into the center of the crown structure of the chiral 18-crown-6, and stability in the complex may depend on other groups around the chiral center. Therefore, chiral compounds containing primary amines can be bound into a complex and often resolved by these crown ethers. Currently, these phases have been commercialized, with which the reversed-phase mode can be used (Okamoto *et al.*, 1994b).

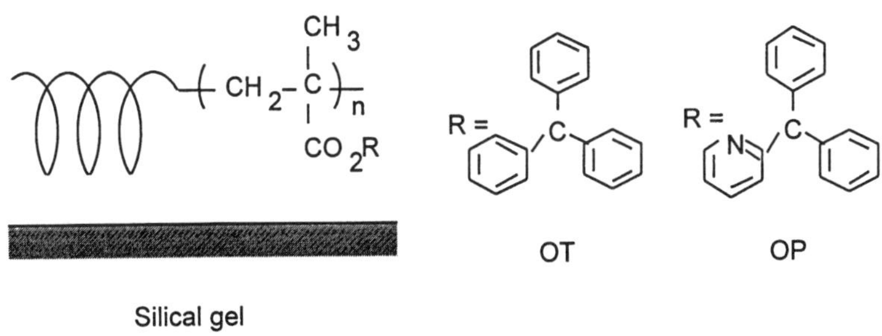

Figure 8. Structure of poly(triphenylmethyl methacrylate) CSPs.

CHIRAL MOBILE PHASE ADDITIVES

Many racemic mixtures can be separated on conventional achiral HPLC columns by inclusion of an appropriate chiral additive into the mobile phase. Although the CSP method has been able to separate enantiomers, there are some disadvantage. Chiral additives were used as the mobile phase as an alternative method. Advantages of this technique are that a less expensive conventional column can be used; the chiral additive can be eluted and changed; there are a wide variety of possible additives; and the selectivity is sometimes different from the corresponding CSP. However, many of the optically pure additives are not commercially available, and the subsequent removal of the additive from the analyte also is a disadvantage for application of this method in preparative-scale separations.

Ligand exchange additives

This useful variation of chiral ligand exchange chromatography was first reported by Gil-Av (Hare *et al.*, 1979) and Karger (LePage *et al.*, 1979), respectively. Chiral amines, (*S*)-2-alkyl-4-octyldiethylenetriamine with Zn (II) in the mobile phase, were used to resolve a series of dansyl-amino acid enantiomers (LePage *et al.*, 1979). In this case, the additive contains a hydrophobic moiety. The ligand associates strongly with a C-8 reversed-phase column by the hydrophobic binding. As in ligand exchange CSPs, a ternary complex is formed with the ligand exchange additive, Zn (II) ion, and appropriate racemic analytes (bidentate compounds, particularly amino acids). The resolution behavior is dependent on the hydrophobicity of the additive and the stationary phase as well as the polarity of the mobile phase (Lam *et al.*, 1980). An interesting example is the use of tartaric acid mono-*N*-octylamide with a metal ion in the mobile phase and C-18 packing material for the resolution of amino acids (Lindner *et al.*, 1986).

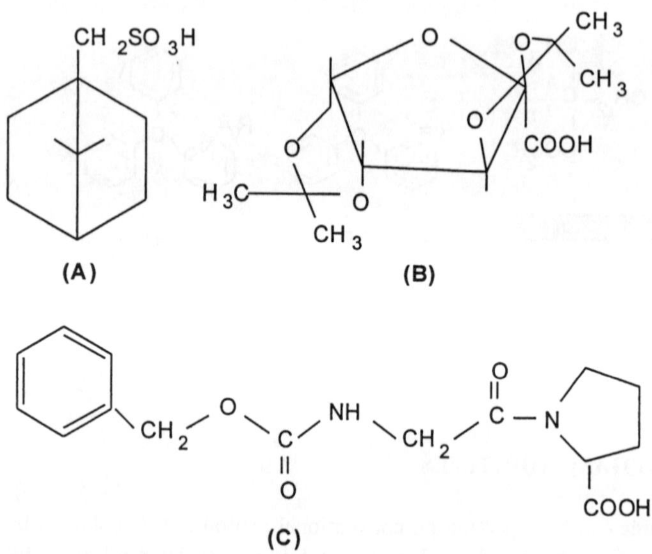

Figure 9. Structure of some chiral counter ions. (A), (+)-10-camphorsulphonic acid; (B), (*S*)-ZGP; (C), DIKGA.

Chiral counter ion additives

Chiral ion-pair chromatography is based on the formation of diastereomeric ion-pairs between the enantiomeric analytes and the chiral counter ion dissolved in the mobile phase. Most separation is carried out in solvents with low polarity (normal phase chromatography) in order to obtain the high stability of ion-pairs. The principle has also been applied in solvents of high polarity, where the counter ion was dissolved in water or water-methanol mixtures (reversed-phase chromatography). The counter ion interacts with the enantiomeric analytes yielding diastereomeric complexes (ion-pairs) that either differ in stability or distribution to an achiral stationary phase. Chiral ion-pair chromatography has successfully been utilized for separation of widely different kinds of enantiomers.

Different enantiomeric aminoalcohols (β-blockers) and eburnane alkaloids have been resolved with (+)-10-camphorsulphonic acid as a counter ion (Figure 9) (Pettersson *et al.*, 1981; Szepesi *et al.*, 1982). The counter ion has a rigid structure with acidic and hydrogen-accepting functional groups which can interact with the aminoalcohols at multi-points. Steric factors play important roles in the interaction between the counter ion and the analyte. Use of *N*-benzyloxycarbonylglycyl-(*S*)-proline [(*S*)-ZGP, see Figure 9] and its analogs as chiral counter ions was efficient to separate aminoalcohols (Pettersson *et al.*, 1986; Karlsson *et al.*, 1992). The *N*-protected dipeptides contain an acidic functional groups permitting electrostatic attraction and hydrogen bonding to the enantiomeric analytes. A separation of racemic metoprolol is presented in Figure 10 (Pettersson *et al.*, 1986).

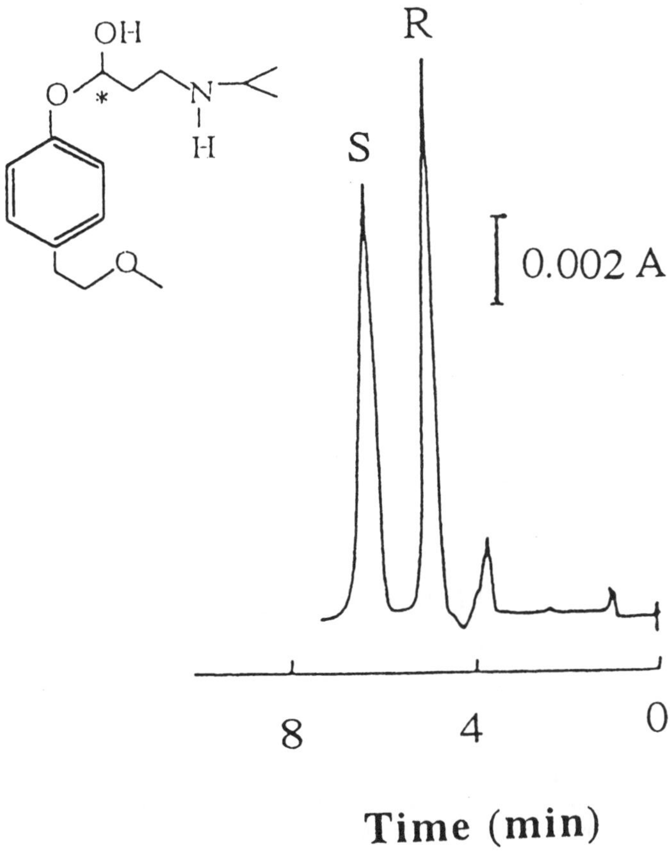

Figure 10, Chiral separation of metoprolol under normal phase conditions. Column: Lichrosorb DIOL; Mobile phase: 2.5mM L-ZGP and 1.0mM triethylamine in dichloromethane (80ppm H_2O). (Reproduced with permission from Pettersson and Josefsson. (1986). Chromatographia, **21**, 321-316).

On the other hand, (-)-2,3,4,6-di-O-isopropylidene-2-keto-(*S*)-gulonic acid (DIKGA, see Figure 9) has been used as a counter ion in the reversed-phase mode for direct separation of enantiomeric amines (Pettersson *et al.*, 1993). As shown in Figure 11, the complete resolution of remoxipride was obtained.

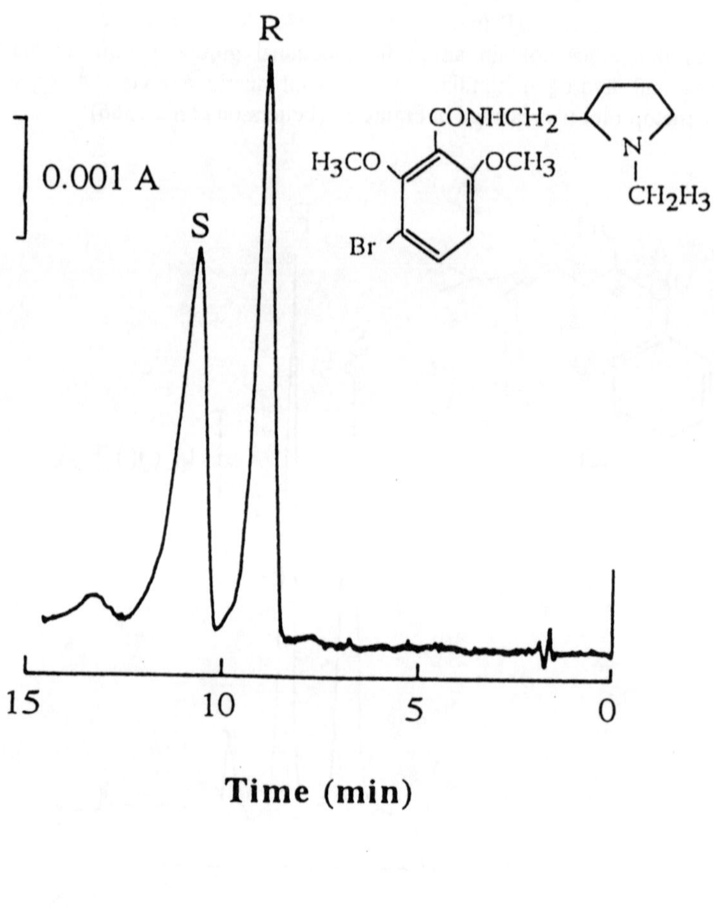

Figure 11, Separation of racemic remoxipride under reversed-phase conditions. Column: Hypercarb; Mobile phase: 5.0mM (-)-DIKGA and 2.5mM NaOH in water/methanol (20/80, v/v). (Reproduced with permission from Pettersson and Gioeli. (1993) Chirality, 5, 241-245).

The composition of the mobile phase, e.g., counter ion concentration and polar modifiers, can be changed to obtain optimal separation. The type of solid phase is also of importance for the separation. Porous graphite carbon (PGC) as the adsorbing phase seems to be superior over DIOL phase, a silica phase modified with vicinal hydroxylic groups, both with respect to equilibration time of the chromatographic system and to the enantioselectivity. The former can be used in either normal phase or reversed-phase conditions.

Chiral crown ethers have also been used as a mobile phase additive (Sousa *et al.*, 1974). However, when using this type of additives, the solvents have to be anhydrous and this potentially toxic chiral selectors should be handled carefully. For these reasons, the application of these additives in the mobile phase is limited and chiral crown ethers are mainly used in the immobilized form as CSPs. Throughout all this material on chiral mobile additives, cyclodextrins have played a major part, which will be described in detail in another section of this review. There are several advantages to use cyclodextrins in the mobile phase: 1) enantioselectivity due to fixed cavity sizes; 2) stability of cyclodextrins over a wide pH range; 3) solubility in water; 4) non-toxicity and non-volatility compared to organic modifiers in reversed-phase HPLC; and 5) reversibility of cyclodextrin inclusion processes.

CYCLODEXTRINS USED AS A CHIRAL SELECTOR IN HPLC

Structure and properties of cyclodextrins

Cyclodextrins (CD) are natural macrocyclic oligosaccharides composed of D-(+)-glucopyranose units bonded via α-1,4-linkages. Most common members of them are α-, β-, and γ-CDs possessing 6, 7, and 8 glucose units, respectively. All glucose units in CDs are in a C_1 (D) chair conformation, as shown in Figure 12. The CD conformations in solution are nearly identical with those observed in the crystalline state. The structural features of CDs are cone-shaped molecules with a relatively hydrophobic internal cavity. The exterior is relatively hydrophilic due to the primary and secondary hydroxyls. The hydroxymethyl groups at C-6 are free to rotate and can partially block the CD cavity from one end. The mouth of the opposite end of the CD cavity is encircled by the C-2 and C-3 secondary hydroxyl groups and has a larger circumference. The restricted conformational freedom and orientation of these secondary hydroxyl groups is thought to account for the chiral recognition by the CDs.

The naturally occurring α-, β-, and γ- CDs are commercially available. Table 3 summarizes some of the important physical properties of the three CDs.

The most remarkable property of CDs is their ability to include a wide variety of guest organic and inorganic molecules or ions selectively into their hydrophobic cavity. Complex formation between the CD host and guest substrate molecules, S, is a dynamic equilibrium process that can be represented by Equation 4. The stability of the inclusion complex, assuming that the 1:1 stoichiometry of CD·S is formed, can be quantitatively described in terms of a formation equilibrium constant, K_f, as defined

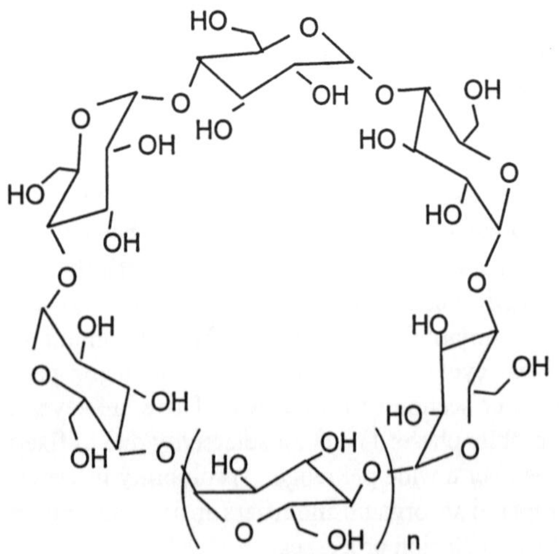

Figure 12. Structure and molecular shape of cyclodextrin.

$$CD + S \overset{K_f}{=} CD{\cdot}S \tag{4}$$

$$K_f = \frac{[CD{\cdot}S]}{[CD]\,[S]} \tag{5}$$

in Equation 5 where [CD·S] represents the concentration of the inclusion complex. The greater the K_f value, the more stable is the CD inclusion complex.

The most important factors in the formation of an inclusion complex are the relative size and geometry of the guest molecule in relation to the dimensions of the host CD cavity. The internal diameter of α-CD varies from about 4.5 to 6 Å, which is a proper size for binding to a single five or six-member aromatic ring. The 6 to 8 Å internal diameter of β-CD is suitable for the inclusion of molecules with the size of biphenyl or naphthalene. The internal diameter of γ-CD varies from about 8 to 10 Å, to which molecules as large as substituted pyrenes can be bound. A tight CD inclusion complex seems to be required for chiral recognition in many cases. In addition, there are several parameters which can affect the binding forces between the guest and the CD molecule in solution: 1) hydrophobic effect due to favorable dipole-dipole interactions; 2) hydrogen bonding; 3) release of high energy water molecules in CD cavity during complex formation; and 4) in the case of α-CD, decrease of the strain energy in the CD ring frame system. In most cases, a combination of these factors seems to be operative, with the first two usually dominant. Several researchers indicated that many CD complexes also satisfy the three point chiral recognition model (Armstrong *et al.*, 1986a; Boehm *et al.*, 1988).

Cyclodextrin-bonded phases

Several research groups have tried to attach different CDs to silica gel via an amide linkage (Fujimura *et al.*, 1983; Kawaguchi *et al.*, 1983), or via an ether linkage (Armstrong *et al.*, 1985a). CD-bonded phases were the first CSPs deliberately designed to use in the reversed-phase mode (Armstrong et al., 1985a). Recent research has shown that CD-bonded phases can be used in normal phase (Armstrong *et al.*, 1991a; Stalcup *et al.*, 1991a) and polar-organic mode (Armstrong *et al.*, 1992a; Zukowski *et al.*, 1993; Chang *et al.*, 1993). There are several different cavity sizes (such as α-, β-, γ- CD) which can be used to separate a variety of enantiomers with different size. Currently, several native CD and derivatized CD columns are commercially available (see Table 2).

Table 3
Physical properties of cyclodextrins (CDs)

Cyclodextrin	Number of glucose units	Molecular weight	Cavity dimension, Å			Melting point °C	Solubility (in water) M
			E. D.[a]	I. D.[b]	Depth		
α-CD	6	972.86	14.6	4.5~6.0	4.5	278 (dec)	0.114
β-CD	7	1135.01	15.4	6.0~8.0	~7.0	298 (dec)	0.016
γ-CD	8	1297.15	17.5	8.0~10.0	~7.0	267 (dec)	0.179

a: E. D., external diameter; b: I. D., internal diameter. (Data from Hinze, Separation and Purification Methods, 10, (1981) 159-237).

Most of the original applications of CD-bonded phases used native CD, particularly β-CD, as the chiral selector in the reversed-phase mode. The hydrophobic inclusion may account for the enantioselectivity of CD-bonded phases in the reversed-phase mode. As mentioned previously, enantiomeric discrimination seems to require inclusion of the guest molecules within the CD cavity, and the fit between the host and guest molecules must be fairly tight. This may partially explain why the native β-CD has proven to be more useful than the α- or γ-CD phases (Armstrong *et al.*, 1987a; Hinze *et al.*, 1985). CD-bonded phases are most often used with water-methanol or water-acetonitrile mobile phases (Armstrong *et al.*, 1984 and 1985a). The addition of buffers and/or salts to the mobile phase improves the efficiency significantly in many separations. Changing the pH of the mobile phase greatly affects the retention and selectivity of ionizable compounds. Retention and resolution are also controlled by altering the amount of organic modifier (e.g., methanol, acetonitrile, ethanol, etc.) in the mobile phase. Generally, retention and resolution decrease with an increase in modifier concentration under conventional reversed-phase conditions.

Enantiomeric pairs with aromatic functional groups at the chiral centers are well resolved using the native β-CD columns. Substitution on the chiral center of the analyte with polar or hydrogen-bonding functional groups also seems to enhance chiral recognition. The aromatic moiety is assumed to be included in the hydrophobic CD cavity, while the polar or hydrogen-bonding groups interact with the hydroxyl groups of CD. A three dimensional computer-generated projection of inclusion complexes of (*R*)- and (*S*)-propranolol with β-CD was made based on the X-ray crystallographic data (Armstrong *et al.*, 1986a). When the aromatic moiety of (*R*)- and (*S*)-propranolol were placed identically within the CD cavity, important differences were observed between the complexes of (*R*)- and (*S*)-enantiomers with respect to their secondary amine groups, as shown in Figure 13. For the (*S*)-enantiomer, the nitrogen was ideally placed for hydrogen bonding to both a 2- and 3-hydroxyl group of the CD, with hydrogen bond distances of 3.3 and 2.8Å, respectively. The nitrogen in (*R*)-propranolol was less favorable for hydrogen bonding, and the bond distances to the closest 2- and 3- hydroxyl groups of the CD are 3.8 and 4.5Å, respectively. This suggests that the (*S*)-enantiomer would preferentially interact with the β-CD and thus be retained more strongly.

CD-bonded CSPs were used for the separation of a broad variety of enantiomer analytes with the reversed-phase. Dansyl and naphthyl amino acids (Armstrong *et al.*, 1984; Hinze *et al.*, 1985), several aromatic drugs, steroids and alkaloids (Armstrong *et al.*, 1986a and 1987b), metallocenes (Armstrong *et al.* 1985b), and binaphthyl crown ethers (Armstrong *et al.*, 1985c) were resolved using β-CD bonded columns. Aromatic amino acids, such as tryptophan, phenylalanine, tyrosine, and their analogs were separated using α-CD-bonded columns (Armstrong *et al.*, 1987c).

Unfortunately, many chiral analytes cannot be resolved using native CD-bonded phases. Some derivatized CD-bonded CSPs have been developed in recent years. For instance, an acetyl-β-CD column improved separation for scopolamine better than the native β-CD phase under reversed-phase conditions (Stalcup *et al.*, 1991b). Hydroxylpropyl-β-CD CSPs, which *(S)*- or racemic 2-hydroxylpropyl groups were attached to the native β-CD, were shown to be effective for enantioseparation of *N-tert*-butoxycarbonyl (Boc)-protected amino acids, which are important building blocks in peptide synthesis (Chang *et al.*, 1992). The *N*-Boc derivatives of nonaromatic amino acids were separated on native β-CD-bonded CSP. All twenty five enantiomeric pairs of Boc amino acids were resolved using hydroxypropyl-β-CD CSP with the reversed-phase conditions in less than 30 minutes. The separation chromatogram of several Boc amino acids is shown in Figure 14. The hydroxypropyl groups seem to play several roles in the mechanism of the chiral recognition by these phases. For some analytes, the hydroxypropyl group appeared to provide additional but non-stereospecific binding sites that are unavailable with the native CD. For example, analytes such as 1-indanol, 1,2,3,4-tetrahydro-1-naphthol or 1-benzocyclobutene carboxylic acid, which have the small size of the hydrophobic moiety are not adequate to fill the CD cavity or separated on the native β-CD CSP. However, these enantiomeric analytes are resolved on the hydroxypropyl-β-CD CSP. The enantioselectivity was found to be independent of the optical purity of the hydroxypropyl substituent, i.e., the near equivalent selectivity was obtained on the *(S)*- and racemic 2-hydropropyl β-CD CSPs (Table 4) (Stalcup *et al.*, 1990). In other cases, the configuration of the hydroxypropyl substituents plays a definite role in chiral recognition and some analytes (such as 3,5-diiodothyronine) are separated on the *(S)*-hydropropyl-β-CD CSPs, but not on the racemic hydropropyl-β-CD CSPs (Stalcup *et al.*, 1990).

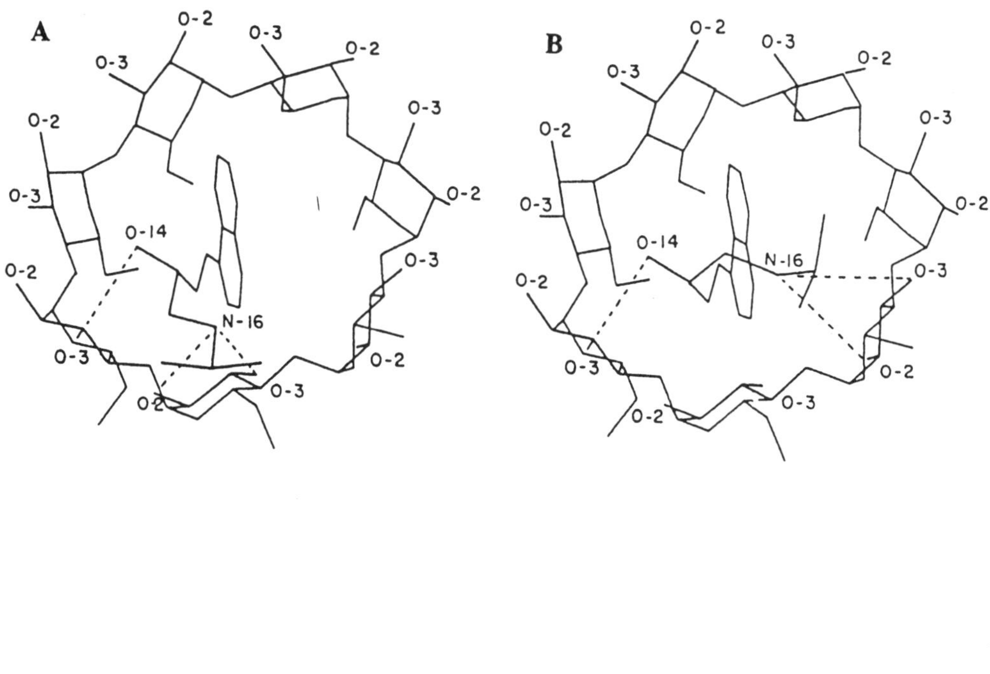

Figure 13. Computer projections of inclusion complexes of propranolol with β-CD taken from X-ray crystallographic data. (A), [*R*)-propranolol; (B), (*S*)-propranolol. (Reproduced with permission from Armstrong *et al.*, (1986). Science, **232**, 1132-1135.

As in the case of the hydroxypropyl-β-CD, incorporation of the naphthylethyl carbamate (NEC) moiety onto the CD introduces additional chiral centers. The NEC-β-CD CSP proved to be very effective in the normal phase as well as in the reversed-phase mode. In the normal phase mode, the enantioselectivity of these phases seems to be rather similar to that obtained on the Pirkle-type phases, the derivatized cellulose and amylose phases. The configuration of these additional stereogenic centers was shown to play an important role in the enantioselectivity of these NEC-β-CD CSPs. A total of 121 racemic compounds were separated in the normal phase mode on an *(S)*-NEC-β-CD CSP and 74 on the *(R)*-NEC-β-CD CSP (Berthod *et al.*, 1992). In many instances, elution order was shown to depend on the configuration of the naphthylethylcarbamate substituents, and to be the reversal on the *(S)*-column compared to the *(R)*-column (Stalcup *et al.*, 1991; Berthod *et al.*, 1992). For instance, Table 5 shows the difference in elution order of some enantiomeric 1-naphthylethylamine derivatives on the (*S*) and *(R)*-NEC-β-CD CSPs. With the reversed-phase conditions, many enantiomeric analytes were resolved on the NEC-CD CSPs, which were not resolved on the native β-CD or NEC-CD phase under normal phase conditions (Armstrong *et al.*, 1991b). Indeed, the NEC-CD CSP may prove to be the first effective multi-mode CSP for chiral liquid chromatography.

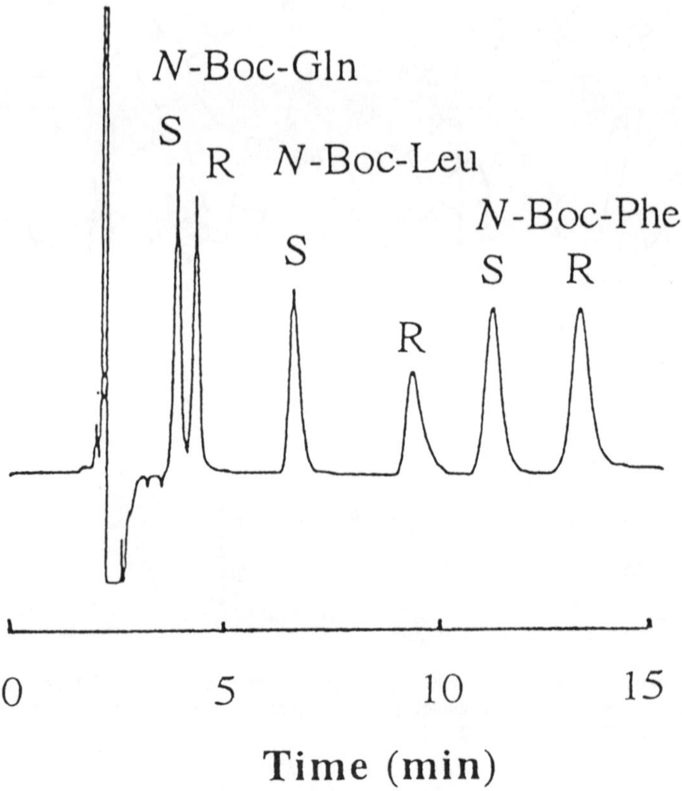

Figure 14. Resolution of three racemic *N*-Boc-amino acids using hydroxypropyl-β-CD. Column: Cyclobond I RSP; Mobile phase: 7% acetonitrile/93% buffer (1% triethylammonium acetate, pH7.1); Flow: 1.0mL/min; Detector: 225nm (UV). (Reproduced with permission from Chang *et al.*, (1992). J. Liquid. Chromatogr. **15**, 1411-1429).

Table 4.
The effect of the chiral center of hydropropyl substituent on enantioseparation[a]

Compound	CSP[b]	k'[c]	α	Mobile phase[d]
1-Indanol	S	1.27	1.07	100% buffer
	Rac	1.00	1.11	100% buffer
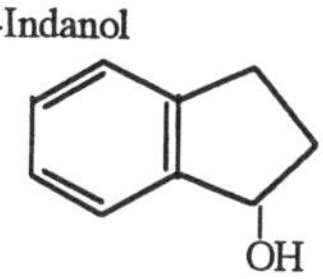				
1,2,3,4-Tetrahydro-1-naphthol	S	2.36	1.08	100% buffer
	Rac	1.92	1.08	100% buffer
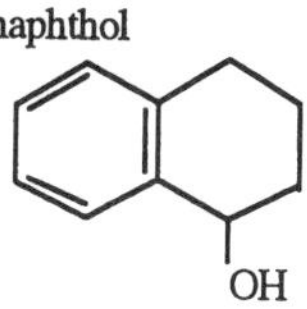				
(±)-1-Benzocyclobutene carboxylic acid	S	2.98	1.14	100% buffer
	Rac	1.60	1.10	95% buffer/MeCN
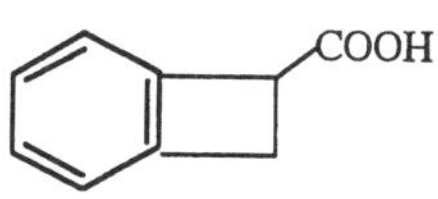				
3,3',5-Triiodo-D,L-thyronine	S	7.18	1.05	85% buffer/MeCN
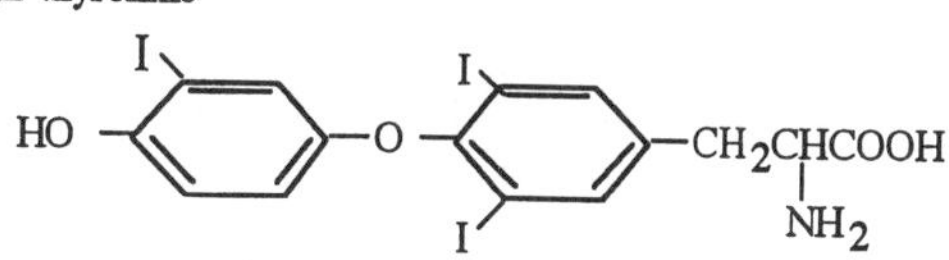				
3,5-Diiodo-D,L-thyronine	S	8.42	1.07	90% buffer/MeCN
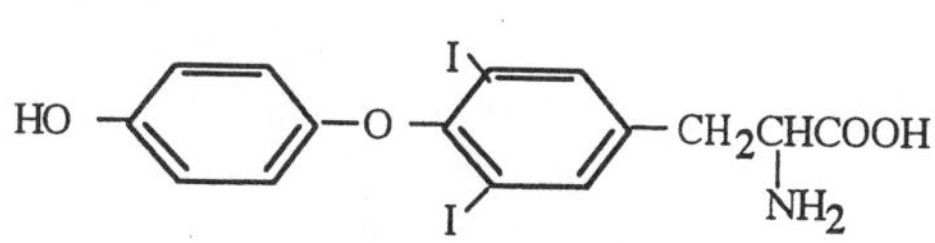				

a: Data selected from Ref. Stalcup, et al., 1990.
b: S, (*S*)-hydropropyl-β-CD CSP; Rac, *Rac*-hydropropyl-β-CD CSP.
c: Capacity ratio of the first eluting enantiomer.
d: 1% Triethylamine acetate, pH 4.1; MeCN, acetonitrile.

As mentioned previously, reversed-phase system on CD-bonded phases uses water (or aqueous buffers)-organic modifier (such as acetonitrile or methanol) as the mobile phases. Increases in the proportion of buffer in the mobile phase prolongs the retention time, whereas an increase in the concentration of the organic modifier reduces both retention and enantiomeric resolution. It was noted that in some cases a retention minimum was reached between 40 and 80% (by volume) of organic modifier in the mobile phase (Seeman *et al.*, 1988). Further elevation in the modifier concentration increased retention and enantioselectivity. The presence of a retention minimum suggests that the mechanism of retention at high content of aqueous buffer (conventional reversed-phase mode) may differ from that at high concentration of the

Table 5
The difference in elution order of enantiomeric 1-naphthylethylamine derivatives on the (*S*)- and (*R*)-NEC-β-CD CSPs [a]

Compound	R substituent	CSP	k'^{b}	α	Mobile phase[c]
1-(1-Naphthyl)-	Acetyl	(*R*)-NEC	6.35^{S}	1.04	10/90
ethylamine		(*S*)-NEC	7.82^{R}	1.04	10/90
CH_3-CH-NHR	Benzoyl	(*R*)-NEC	1.17^{S}	1.09	30/70
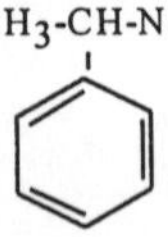		(*S*)-NEC	$1..28^{R}$	1.11	30/70
	DNB	(*R*)-NEC	3.33^{S}	3.05	30/70
		(*S*)-NEC	5.82^{R}	3.59	30/70
	Naphthoyl	(*R*)-NEC	1.67^{S}	1.14	30/70
		(*S*)-NEC	1.75^{R}	1.14	30/70

a: Data selected from Ref. Berthod, et al., 1992.
b: Capacity ratio of the first eluting enantiomer, configuration indicated as a superscript.
c: Mobile phase: 2-propanol/hexane.

organic modifier (polar-organic mode). The mechanism of separation in the polar-organic mode was thought to be ascribed to the sitting of the analyte atop the CD cavity like a "cover" rather than inclusion in the complex. The chiral recognition involves hydrogen bonding of the analyte with the secondary hydroxyl groups on the CD. Adequate dipole stacking and steric factors should also be required. For the separation with this mode, analyte enantiomers must have at least one polar group near to the chiral center with proper hydrogen bonding ability, such as a hydroxyl group, secondary amino group, amido group, etc. The molecular size of these separated analytes is usually larger than that of the CD cavity. The use of this new mobile phase allows the separation of enantiomers that cannot be resolved in the normal phase or reversed-phase mode. Both native CD and derivatized CD-bonded CSPs can be used in the polar-organic mode. Figure 15 shows the enantiomeric resolution of proglumide on five CD-bonded phases (native α- and β-CD, acetyl-β-CD, *(R)*- and *(S)*-NEC-β-CD-bonded phases) using the polar-organic mode (Chang *et al.*, 1993).

Several applications of CD-bonded CSPs in biological samples were reported. Unnatural D enantiomers of various amino acids in human urine, plasma, cerebrospinal (CSF) and amniotic fluid were determined by chromatographic techniques using coupled columns. A C-18 column was used to isolate the analytes from matrix interferences. This was coupled though a six-port switching valve to two

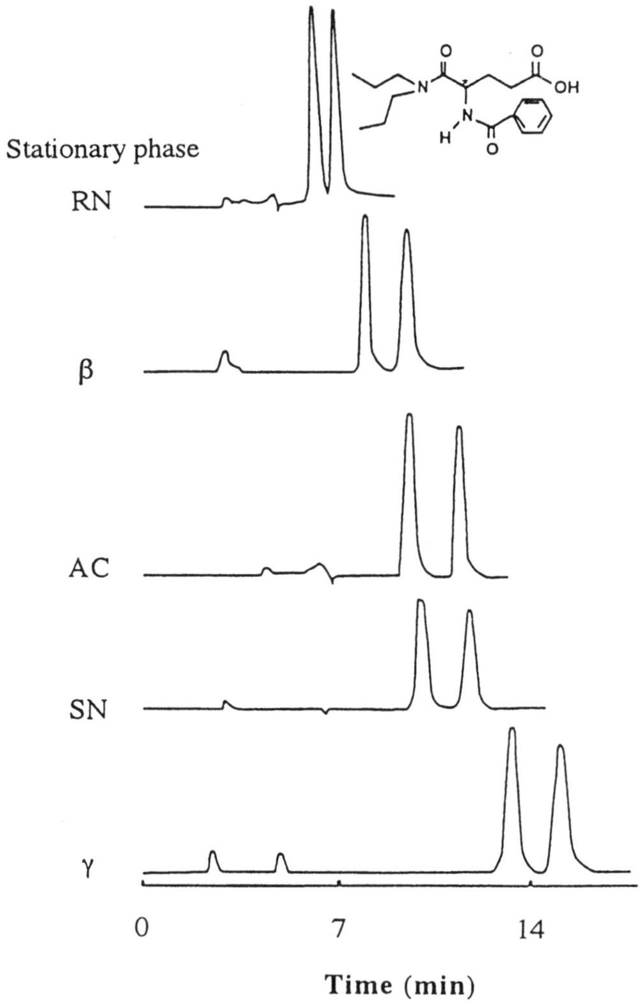

Figure 15. Chromatograms of the enantiomeric resolution of proglumide in the polar-organic mode on five CD-bonded phases. Column: Cyclobond I RN ((*R*)-1-naphthylethyl carbamoyl β-CD, RN), Cyclobond I (β-CD, β), Cyclobond I AC (acetyl β-CD, AC), Cyclobond I SN ((*S*)-1-naphthylethyl carbamoyl β-CD, SN), and Cyclobond II (γ-CD, γ); Detector: 254nm (UV). (Reproduced with permission from Chang *et al.*, (1993). Trends in Anal. Chem. **12**, 144-153).

CD-bonded columns in series to resolve the amino acid enantiomers in these biological samples (Armstrong *et al.*, 1993). It was not only found that D-amino acids exist in all physiological fluids tested, but that their levels also varied considerably.

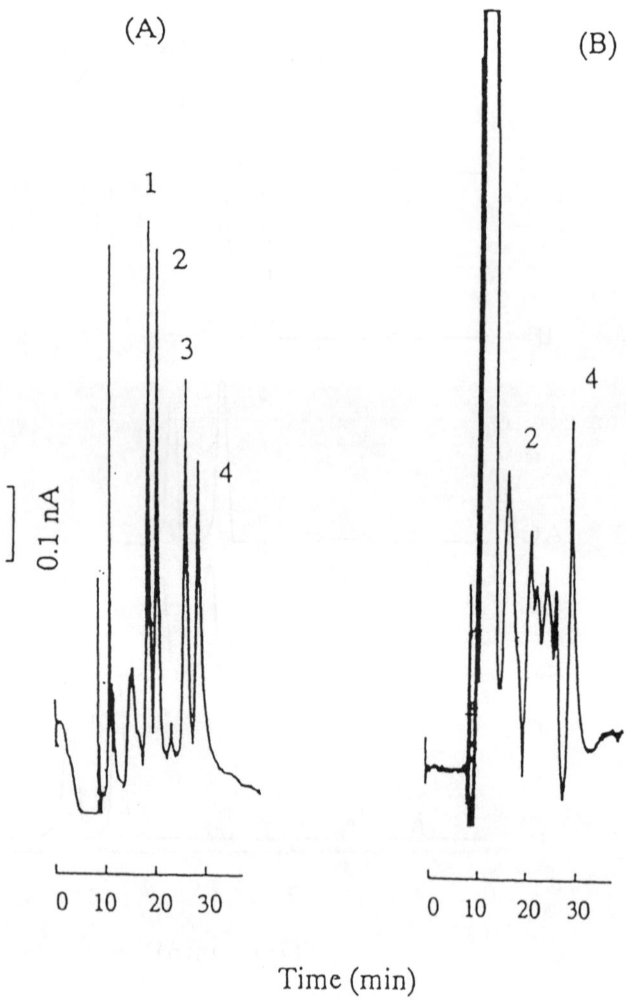

Figure 16. The resolution of the racemic mixture of salsolinol and their derivatives using an HPLC-ECD system. Column: LichroCART 250-4 ChiraDex; Mobile phase: 0.1M phosphate buffer (pH5.2), 2.5% methanol. Peak: 1; dopamine, 2=*(S)*-Sal, 3=*(R)*-Sal, and 4=*(S)*-NMSal.

The lowest levels of D-amino acids were usually found in amniotic fluid and CSF, and the highest in urine. Our laboratory developed a direct enantioseparation method for neurotoxic salsolinol, 6,7-dihydroy-1-methyl-1,2,3,4-tetrahydroisoquinoline, and the derivatives using β-CD-bonded column. The method was used to determine the enantiomeric composition of these neurotoxins in human brain samples (Deng *et al.*, 1995). Figure 16 shows that the racemic mixture of salsolinols and their *N*-methylated derivatives was completely separated from each other and from their catecholamine precursor dopamine using an HPLC-ECD system.

In addition, direct application of biological samples, such as human serum, onto the CD-bonded column has been explored (Haginaka *et al.*, 1990; Stalcup *et al.*, 1992). The combination of the chiral selectivity of CD with the hydrophilic diol functional group was examined. CDs and diol functional groups are bonded to internal and external surfaces of special silica supports with a pore size of 80Å. Macromolecules such as serum proteins were excluded without destructive accumulation on the hydrophilic diol surface, but small molecules such as drugs were retained and enantiospecifically separated on the CD surfaces (Haginaka *et al.*, 1990).

Cyclodextrins as chiral mobile phase additives

The use of CDs as mobile phase modifiers was first reported in thin-layer chromatography (TLC) for the separation of substituted aromatics (Hinze *et al.*; 1980). The success of CDs as mobile phase additives has led to their extensive use in reversed-phase HPLC. Debowski *et al.* (1982a, b) first used α- and β-CD as mobile phase additives for the enantioseparation of mandelic acid and its derivatives.

From the several equilibria achieved in a chromatographic system containing CD in the mobile phase, the retention behavior of analyte enantiomer is described by the following relationship, assuming that only one species of each enantiomer of the analyte is present and 1:1 stoichiometric complexes are formed (Walhagen and Edholm, 1991).

$$1/k' = 1/k'_0 + [CD]_m K_f / k'_0 \qquad (6)$$

where K_f is the apparent formation constant for the inclusion complex, and k' and k'_0 are the capacity ratio in the presence and absence of CD.

A similar equation (Equation 7) was developed for the ion-pair chromatographic system using CD and a counter ion in the mobile phase which was effective for the separation of hydrophilic catecholamines (Deng *et al.*, 1996).

$$1/k' = 1/k'_{0i} + [X]_m [CD]_m K_f / k'_{0i} \qquad (7)$$

where K_f is the apparent formation constant for ternary complex of CD with the analyte and counter ion, and $[X]_m$ is the concentration of the counter ion. k'_{0i} is the capacity ratio of analyte in the mobile phase containing the counter ion but no CD. Equation 6 and 7 show that the retention value is influenced by the capacity ratio of analyte in the absence of CD, which in turn is mainly determined by the type of stationary phase, and the type and concentration of the organic modifier and counter ion.

To evaluate the influence of CD on chiral recognition, the separation factor, $\alpha_{(2/1)}$, for (*R*)/(*S*)-enantiomers was obtained from Equation 7 and is represented as follows:

$$\alpha_{(2/1)} = \frac{K_{f1}}{K_{f2}} \left[1 - \frac{K_{f1} - K_{f2}}{K_{f1} + K_{f1} K_{f2} [CD]_m [X]_m}\right] \quad (8)$$

where K_{f1} and K_{f2} are the formation constants for two enantiomers and the subscripts "1" and "2" indicate the elution order. From Equation 8, it can be seen that the separation factor is dependent on the concentration of CD in the mobile phase and the values of the formation constants, K_{f1} and K_{f2}.

Generally, there are several factors required for chiral separation. The formation constants (K_f) of CD with guest analytes must be sufficiently large and markedly different between two enantiomers, and the retention also must be high enough. The formation of an inclusion complex is mainly determined by the relative size of the host CD cavity. The relationship of the size of CD to the guest analyte was examined with steroid norgesterol and α-, β-, and γ-CD additives (Gazdag *et al.*, 1986). It was found that the larger cavity was required for resolution of the large steroid molecule. The concentration of CD is essential for chiral separation. Unfortunately, the CDs are not readily soluble and a maximum concentration of 10-20mM is obtained in water, but this is dependent upon the amount of the organic modifier present. Solubility can be increased by the addition of urea up to 500mM for β-CD. Another possibility is to utilize derivatized CDs. Methylcyclodextrin, which has the secondary hydroxyl groups on the rim, is about 100 times more soluble than β-CD. The polymethylated derivatives of β-CD enantioselectively resolve barbiturates containing a chiral carbon in the heterocyclic ring (Zukowski *et al.*, 1986). This derivative is adsorbed strongly onto the C-18 stationary phase because of its high hydrophobicity. As a result, small amounts of this additive produce a significant increase in the retention time of the analytes. Other derivatized CDs are the hydroxypropyl- and hydroxyethyl-CDs, which actually increase the bulk of the cavity around the secondary hydroxyl groups. Another approach to enhance the retention of ionic analytes is the ion-pair chromatographic mode. Small amounts of the counter ion significantly improve the retention behavior of ionic compounds on the C-18 stationary phase. Ion-pair chromatography was found to be effective for the enantioseparation of hydrophilic catecholamines, such as the neurotoxin salsolinol (Deng *et al.*, 1996b). This novel procedure was applied to determine the enantiomeric composition of salsolinol and N-methylsalsolinol in human brain and some food and beverage samples (Deng *et al.*, 1996a). Figure 17 shows the typical chromatographic patterns of the assay for enantiomeric salsolinol in dried banana and wine samples using native β-CD as a chiral mobile phase additive under ion-pair reversed-phase conditions. There have been a few reports on the use of CD mixtures as mobile phase components. Gazdag *et al.* (1988) reported the separation of some racemic molecules (such as budesonides) into their enantiomers using a mixture of α-, β-, and γ-CDs in a methanol-water mobile phase.

Since inclusion complex formation is facilitated in aqueous media, an aqueous-organic solvent mixture is appropriate for this type of separation. The organic solvent should be selected with consideration to the solubility of CD, and the retention behavior of the analyte, as well as to the stability of the inclusion complex. A water-methanol mixture is the most suitable solvent for the separation using modified CD mobile phases. Warner's group also noted that 2-methyl-2-propanol, used as a co-modifier in water-methanol mobile phase, can stabilize the analyte-β-CD complex (Husain *et al.*, 1993). The pH is an important parameter for ionic compounds since the concentration of their ionic forms is determined by the pH of the mobile phase.

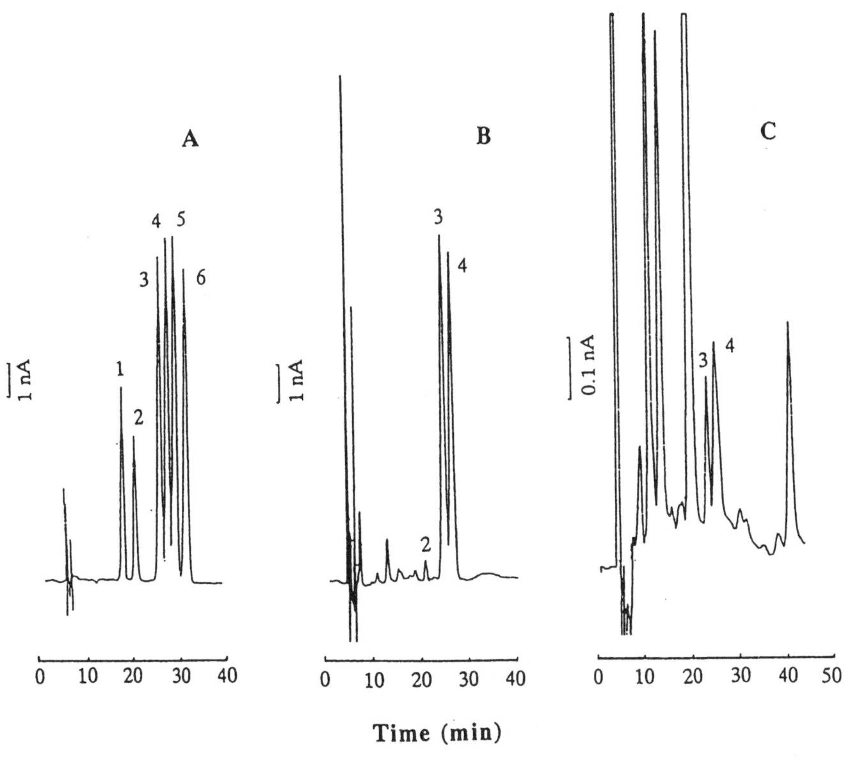

Figure 17. Chromatographic examples of the assay for the enantiomeric composition of salsolinol in dried banana and wine samples. (A): Standards (20pmol for each compound); (B): sample prepared from dried banana; (C) sample prepared from red German wine. Column: Inertosil ODS-3; Mobile phase: 25mM phosphate buffer, pH3.0, containing 12mM β-cyclodextrin, 1mM SHS and 3% acetonitrile. Peaks: 1, NMNorsal; 2, dopamine; 3, (*R*)-Sal; 4, (*S*)-Sal; 5, (*R*)-NMSal; and 6, (*S*)-NMSal.

It is interesting to note that even though vendors sell the same kind of reversed stationary phases, their products show very different behavior when CDs are used as chiral mobile phase additives. As shown in Figure 18, the retention times were markedly different with C-18 phases from three different sources, suggesting that stationary and mobile phases should be selected carefully (Walhagen *et al.*, 1991).

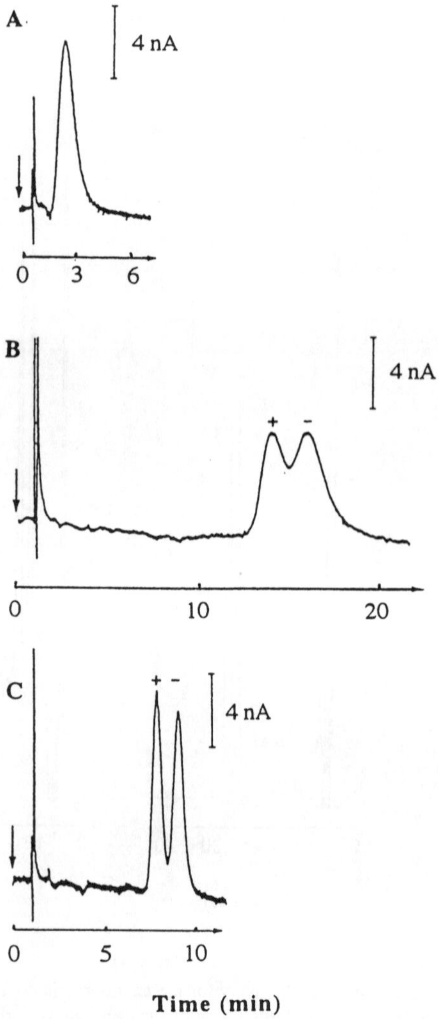

Figure 18. Chromatograms of racemic terbutaline on C-18 phases from different suppliers, (A) Kromasil™ C-18, (B) Spherisorb ODS2 C-18, (C) Nucleosil C-18. Mobile phase: 8.4mM β-CD in ammonium acetate 0.05M (pH6.0); Flow: 1 mL/min. (Reproduced with permission from Walhagen and Edholm, (1991). Chromatographia **32**, 215-223).

In addition to enantiomeric separation, CDs can be used as mobile phase additives in the separation of isomers. The ortho-, meta-, and para-isomers of nitrophenol and nitroaniline were separated using β-CD in the reversed-phase mobile phase, but the separation was incomplete in the absence of β-CD (Fujimura *et al.*, 1986). CD-modified mobile phases were also used for the analysis of several PAHs (polynuclear aromatic hydrocarbons), nitrogen heterocycles, and aromatic hydroxyl compounds (Mohseni *et al.*, 1990).

This approach by the addition of CDs to the mobile phase can also be used to determine the formation constants and stoichiometry for CD inclusion complexes. As seen in Equation 6, the reciprocal of the capacity factor versus the CD concentration is linear, and the slope represents K_f/k'_0 and the intercept $1/k'_0$. The ratio of the slope over the intercept directly provides a value for K_f. The complex equilibrium and equations involved in these methods were examined in details by Armstrong *et al.* and other groups (Armstrong *et al.*, 1986b; Deng *et al.*, 1996b).

CONCLUSION

Chiral HPLC is now a common analytical technique. The development of enantioselective HPLC has enabled the determination of many enantiomeric drugs and metabolites in biological samples. Enantioseparation has now become routine by the commercial availability of over 100 CSPs for HPLC and the introduction of a wide variety of pure enantiomeric reagents as chiral additive to the mobile phase.

For electrochemical detection, the use of an aqueous mobile phase system limits the types of enatiometric methods used. However, cyclodextrin-bonded CSPs are compatible with HPLC-ECD. Furthermore, they are relatively inexpensive and have wide applicability. Protein-bonded CSPs and ligand exchange CSPs are also performed in a reversed-phase mode, and reversed-phase columns of other types, such as cellulose-derived CSPs, are also commercialized. The addition of a chiral selector to the mobile phase increases further variations in enantioseparation method by HPLC. This approach has the advantage of reasonable cost and flexibility in selection of additives and achiral stationary phases, and sometimes gives enantiomeric resolutions which have not been achieved by the respective CSP.

Acknowledgment
This work was supported by a Grant-in-Aid for Scientific Research on Primary Area and (C) from the Ministry of Education, Science and Culture, Japan.

REFERENCES

Allenmark, S., Bomgren, B. and Boren, H. (1982). Direct resolution of enantiomers by liquid affinity chromatography on albumin-agarose under isocratic conditions. J. Chromatogr. **237**, 473-477.

Allenmark, S., Bomgren, B. and Boren, H. (1983). Direct ligand chromatographic separation of enantiomers on immobilized protein stationary phases. III. Optical resolution of a series of *N*-aroyl D,

L-amino acids by high-performance liquid chromatography on bovine serum albumin covalently bound to silica. J. Chromatogr. **264**, 63-68.

Allenmark, S. and Andersson, S. (1994). Proteins and peptides as chiral selectors in liquid chromatography. J. Chromatogr. A **666**, 167-179.

Armstrong, D. W. and DeMond, W. J. (1984). Cyclodextrin bonded phases for the liquid chromatographic separation of optical, geometrical, and structural isomers. J. Chromatogr. Sci. **22**, 411-415.

Armstrong, D. W. (1985a). Bonded phase material for chromatographic separations. US patent 4,539,399.

Armstrong, D. W., DeMond, W. J. and Czech, B. P. (1985b). Separation of metallocene enantiomers by liquid chromatography: Chiral recognition via cyclodextrin bonded phase. Anal. Chem. **57**, 481-484.

Armstrong, D. W., Ward, T. J., Czech, A., Czech, B. P. and Bartsch, R. A. (1985c). Synthesis, rapid resolution, and determination of absolute configuration of racemic 2,2'-binaphthyldiyl crown ethers and analogues via β-cyclodextrin complexation. J. Org. Chem. **50**, 5556-5559.

Armstrong, D. W., Ward, T. J., Armstrong, R. D. and Beesley, T. E. (1986a). Separation of drug stereoisomers by the formation of β-cyclodextrin inclusion complexes. Science **232**, 1132-1135.

Armstrong, D. W., Nome, F., Spino, L. A. and Golden, T. D. (1986b). Efficient detection and evaluation of cyclodextrin multiple complex formation. J. Am. Chem. Soc. **108**, 1418-1421.

Armstrong, D. W. (1987a). Optical isomer separation by liquid chromatography. Anal. Chem. **59**, 84A-91A.

Armstrong, D. W., Yang, X., Han, S. M. and Menges, R. A. (1987b). Direct liquid chromatographic separation of racemates with an α-cyclodextrin bonded phase. Anal. Chem. **59**, 2594-2596.

Armstrong, D. W., Spino, L. A., Han, S. M., Seeman, J. I. and Secor, H. V. (1987c). Enantiomeric resolution of racemic nicotine and nicotine analogs by microcolumn liquid chromatography with β-cyclodextrin inclusion complexes. J. Chromatogr. **411**, 490-493.

Armstrong, D. W., Hilton, M. and Coffin, L. (1991a). Multimodel chiral stationary phase for liquid chromatography: *(R)*- and *(S)*-naphthylethyl-carbamate-derivatized β-cyclodextrin. LC-GC. **9**,646-652.

Armstrong, D. W., Chang, C. D. and Lee, S. H. (1991b). *(R)*- and *(S)*-Naphthylethyl-carbamate-substituted β-cyclodextrin bonded stationary phases for the reversed-phase liquid chromatographic separation of enantiomers. J. Chromatogr. **539**, 83-90.

Armstrong, D. W., Chen, S., Chang, C. D. and Chang, S. (1992). A new approach for the direct resolution of racemic beta adrenergic blocking agents by HPLC. J. Liq. Chromatogr. **15**, 545-556.

Armstrong, D. W., Gasper, M., Lee, S. H., Zukowski, J. and Ercal, N. (1993), D-Amino acid levels in human physiological fluids. Chirality **5**, 375-378.

Baczuk, R. J., Landrom, G. K., Dubois, R. J. and Dehm, H. C. (1971). Liquid chromatographic resolution of racemic β-3,4-dihydroxyphenylalanine. J. Chromatogr. **60**, 351-361.

Bentley, R., (1995). Chirality in Biology. In: Molecular biology and biotechnology. Meyers, R. A. (Ed.), VCH, New York, pp. 168-173.

Berthod, A., Chang, S. C. and Armstrong, D. W. (1992). Empirical procedure that uses molecular structure to predict enantioselectivity of chiral stationary phases. Anal. Chem. **64**, 395-404.

Blaschke, G. (1971). Angew Chem., Int. Ed. Engl. **83**, 547.

Boehm, R. E., Martire, D. E. and Armstrong, D. W. (1988). Theoretical considerations concerning the separation of enantiomeric solutes by liquid chromatography. Anal. Chem. **60**, 522-528.

Caude, M., Tambute, A. and Siret, L. (1991). Chiral stationary phase derived from tyrosine. J. Chromatogr. **550**, 357-382.

Chang, S. C., Wang, L. R. and Armstrong, D. W. (1992). Facile resolution of *N-tert*-butoxycarbonyl amino acids: The importance of enantiomeric purity in peptide synthesis. J. Liq. Chromatogr. **15**, 1411-1429.

Chang, S. C., Reid, G. L., Chen, S., Chang, C. D. and Armstrong, D. W. (1993). Evaluation of a new polar-organic high-performance liquid chromatographic mobile phase for cyclodextrin-bonded chiral stationary phase. Trends in Anal. Chem. **12**, 144-153.

Dalgleish, C. E. (1952). Synthesis of *N'*-formyl-DL-kynurenine, N^a-acetyl-DL-kynurenine, and related compounds, and observation on the synthesis of kynurenine. J. Chem. Soc. **137**, 3940-3942.

Davankov, V. A. and Rogozhin, S. V. (1971). Liquid chromatography as a novel method for the investigation of mixed complexes. Stereoselective effects in α-amino acid-copper (II) complexes. J. Chromatogr. **60**, 280-283.

Davankov, V. A., Kurganov, A. A. and Bochkov, A. S. (1983). Resolution of racemates by high-performance liquid chromatography. Adv. Chromatogr. **22**, 71-116.
Davankov, V. A. (1994). Chiral selectors with chelating properties in liquid chromatography: fundamental reflections and selective review of recent developments. J. Chromatogr. A **666**, 55-76.
Debowski, J., Sybilska, D. and Jurczak, J. (1982a). β-Cyclodextrin as a chiral component of the mobile phase for separation of mandelic acid into enantiomers in reversed-phase systems of high-performance liquid chromatograpy. J. Chromatogr. **237**, 303-306.
Debowski, J., Sybilska, D. and Jurczak, J. (1982b). The resolution of some chiral compounds in reversed-phase high-performance liquid chromatograpy by means of β-cyclodextrin inclusion complexes. Chromatographia **16**, 198-200.
Deng, Y., Maruyama, W., Dostert, P., Takahashi, T., Kawai, M. and Naoi, M. (1995). Determination of the *(R)*- and *(S)*-enantiomers of salsolinol and *N*-methylsalsolinol by use of a chiral high-performance liquid chromatographic column. J. Chromatogr. B **670**, 47-54.
Deng, Y., Maruyama, W., Kawai, M., Dostert, P., Yamamura, H., Takahashi, T. and Naoi, M. (1996a). Assay for the (*R*)- and (*S*)-enantiomers of salsolinols in biological samples and foods with ion-pair HPLC using β-cyclodextrins as a chiral mobile additive. J. Chromatogr. B in press.
Deng, Y., Maruyama, W., Yamamura, H., Kawai, M., Dostert, P. and Naoi, M. (1996b). Mechanism of enantioseparation of salsolinols, endogenous neurotoxins in human brain, with ion-pair chromatography using β-cyclodextrin as a mobile phase additive. Anal. Chem. in press.
Domenici, E., Bertucci, C., Salvadori, P., Felix, G., Cahagne, I., Motellier, S. and Wainer, I. W. (1990). Synthesis and chromatographic properties of an HPLC chiral stationary phase upon human serum albumin. Chromatographia **29**, 170-176.
Dotsevi, G., Sogah, Y. and Cram, D. J. (1975). Chromatographic optical resolution through chiral complexation of amino ester salts by a host covalently bound to silica gel. J. Am. Chem. Soc. **97**, 1259-1261.
Feibush, B., Cohen, M. J. and Karger, B. L. (1983). The role of bonded phase composition on the ligand-exchange chromatography of dansyl-D, L-amino acids. J. Chromatogr. **282**, 3-26.
Francotte, E., Wolf, R. M., Lohmann, D. and Mueller, R. (1985). Chromatographic resolution of racemates on chiral stationary phases. I. Influence of the supramolecular structure of cellulose triacetate. J. Chromatogr. **347**, 25-37.
Francotte, E. and Wolf, R. M. (1991). Benzoyl cellulose beads in the pure polymeric form as a new powerful sorbent for the chromatographic resolution of racemates. Chirality **3**, 43-55.
Francotte, E. (1994). Contribution of preparative chromatographic resolution to the investigation of chiral phenomena. J. Chromatogr. A **666**, 565-601.
Fujimura, K., Ueda, T. and Ando, T. (1983). Retention behavior of some aromatic compounds on chemically bonded cyclodextrin silica stationary phase in liquid chromatography. Anal. Chem. **55**, 446-450.
Fujimura, K., Ueda, T., Kitagawa, M., Takayanigi, H. and Ando, T. (1986). Reversed-phase retention behavior of aromatic compounds involving β-cyclodextrin inclusion complex formation in the mobile phase. Anal. Chem. **58**, 2668-2272.
Gazdag, M., Szepesi, G. and Huszar, L. (1986). Separation of isomeric compounds as cyclodextrin inclusion complexes on a cyanopropylsilica stationary phase. J. Chromatogr. **371**, 227-234.
Gazdag, M., Szepesi, G. and Huszar, L. (1988). α-, β- and γ-Cyclodextrins as mobile phase additives in the high-performance liquid chromatographic separation of enantiomeric compounds: II. Optimization of the separation method by using α-, β- and γ-cyclodextrins in a mixture. J. Chromatogr. **436**, 31-38.
Gelber, L. R., Karger, B. L., Neumeyer, J. L. and Feibush, B. (1984). Ligand exchange chromatography of amino alcohols. Use of Schiff bases in enantiomer resolution. J. Am. Chem. Soc. **106**, 7729-7734.
Haginaka, J. and Wakai, J. (1990). β-Cyclodextrin bonded silica for direct injection analysis of drug enantiomers in serum by liquid chromatography. Anal. Chem. **62**, 997-1000.
Hare, P. E. and Gil-Av, D. (1979). Separation of D and L amino acids by liquid chromatography: Use of chiral eluants. Science **204**, 1226-1228.
Hesse, G. and Hagel, R. (1973). Complete separation of a racemic mixture by elution chromatography on cellulose triacetate. Chromatographia **6**, 277-280.
Hesse, G. and Hagel, R. (1976). Inclusion chromatography and a new retention mechanism for benzene derivatives. Chromatographia **9**, 62-68.
Hermansson, J. (1983). Direct ligand chromatographic resolution of racemic drugs using α1-acid glycoprotein as the chiral stationary phase. J. Chromatogr. **269**, 71-80.

Hinze, W. L. and Armstrong, D. W. (1980). Thin layer chromatographic separation of ortho, meta, and para-substituted benzoic acids and phenols with aqueous solution of α-cyclodextrin. Anal. Lett. **13**, 1093-1104.

Hinze, W. L. (1981). Applications of cyclodextrins in chromatographic separations and purification methods. Separation and Purification Methods **10**, 159-237.

Hinze, W. L., Riehl, T. E., Armstrong, D. W., DeMond, W., Alak, A. and Ward, T. (1985). Liquid chromatographic separation of enantiomers using a chiral β-cyclodextrin-bonded stationary phase and conventional aqueous-organic mobile phases. Anal. Chem. **57**, 237-242.

Husain, N., Anigbogu, V. C., Cohen, M. R. and Warner, I. M. (1993). Reversed-phase liquid chromatographic measurement of the influence of a co-modifier functional group on the retention behavior of the β-cyclodextrin-pyrene complex. J. Chromatogr. **635**, 211-219.

Ikeda, K., Hamasaki, T., Kohno, H., Ogawa, T., Matsumoto, T. and Sakai, J. (1989). Direct separation of enantiomers by reversed-phase high-performance liquid chromatography on cellulose tris (3,5-dimethylphenyl) carbamate. Chem. Lett. **6**, 1089-1090.

Iredale, J., Aubry, A. F. and Wainer, I. W. (1991). The effects of pH and alcoholic organic modifiers on the direct separation of some acidic, basic and neutral compounds on a commercially available ovomucoid column. Chromatographia **31**, 329-334.

Karlsson, A. and Pettersson, C. (1992). Separation of enantiomeric amines and acids using chiral ion-pair chromatography on porous graphitic carbon. Chirality **4**, 323-332.

Kawaguchi, Y., Tanaka, M., Nakae, M., Funazo, K. and Shono, T. (1983). Chemically bonded cyclodextrin stationary phases for liquid chromatographic separation of aromatic compounds. Anal. Chem. **55**, 1852-1857.

Lam, S., Chow, F. and Karmen, A. (1980). Reversed-phase high-performance liquid chromatographic resolution of D- and L-Dns-amino acids by mixed chelate complexation. J. Chromatogr. **199**, 295-305.

LePage, J., Lindner, W., Davies, G., Seitz, D. E. and Karger, B. L. (1979). Resolution of the optical isomers of dansyl amino acids by reversed-phase liquid chromatography with optically active metal chelate additives. Anal. Chem. **51**, 433-435.

Liu, C. Y., Chang, C. H., Chyueh, S. D. and Hu, C. C. (1994). Combination of chelating behavior with molecular recognition as interaction for ligand exchange chromatography. Fresenius J. Anal. Chem. **350**, 132-134.

Lindner, W. and Hirchback, I. (1986). Chromatographic resolution of amino acids using tartaric acid móno-n-octylamide as mobile phase additive. J. Liq. Chromatogr. **9**, 551-571.

Luettringhaus, A. and Peters, K. C. (1966). Conformational enantiomerism of a derivative of cyclotriveratrylene. Angew. Chem., Intern. Ed. Engl. **5**, 593-594.

Minami, M., Takahashi, T., Maruyama, W., Takahashi, A., Dostert, P., Nagatsu, T. and Naoi, M. (1992). Inhibition of tyrosine hydroxylase by *R* and *S* enantiomers of salsolinol, 1-methyl-6,7-dihydroxy-1,2,3,4-tetrahydroisoquinoline. J. Neurochem. **58**, 2097-2101.

Minami, M., Maruyama, W., Dostert, P., Nagatsu, T. and Naoi, M. (1993). Inhibition of type A and B monoamine oxidase by 6,7-dihydroxy-1,2,3,4-tetrahydroisoquinolines and their *N*-methylated derivatives. J. Neural Transm. **92**, 125-135.

Miwa, T., Ichikawa, M., Tsuno, M., Hattori, T., Miyakawa, T., Kayano, M. and Miyake, Y. (1987). Direct liquid chromatographic resolution of racemic compounds. Use of ovomucoid as a column ligand. Chem. Pharm. Bull. **35**, 682-686.

Mohseni, R. M. and Hurtubise R. J. (1990). Retention characteristics of several compound classes in reversed-phase liquid chromatography with β-cyclodextrin as a mobile phase modifier. J. Chromatogr. **499**, 395-410.

Naoi, M., Maruyama, W., Dostert, P., Hashizume, Y., Nakahara, D., Takahashi, T. and Ota, M. (1996). Dopamine-derived endogenous 1(*R*), 2(*N*)-dimethyl-6,7-dihydroxy-1,2,3,4-tetrodoisoquinoline, *N*-methyl-(*R*)-salsolinol, induced Parkinsonism in rat: Biochemical pathological and behavioral studies. Brain Res. **709**, 285-295.

Nimura, N., Suzuki, T., Kasahara, Y. and Kinoshita, T. (1981). Reversed-phase liquid chromatographic resolution of amino acid enantiomers by mixed chelate complexation. Anal. Chem. **53**, 1380-1383.

Oi, N. and Kitahara, H. (1986). Enantiomer separation by HPLC with some urea derivatives of L-valine as novel chiral stationary phases. J. Chromatogr. **257**, 111-117.

Okamoto, Y., Honda, S., Okamoto, I., Yuki, H., Murata, S., Noyori, R. and Takaya, H. (1981). Novel packing material for optical resolution: (+)-Poly(triphenyl methyl methacylate) coated on macroporous silica gel. J. Am. Chem. Soc. **103**, 6971-6973.

Okamoto, Y., Suzuki, K., Ohta, K., Hatada, K. and Yuki, H. (1979). Optically active poly(triphenylmethyl methacrylate) with one-handed helical conformation. J. Am. Chem. Soc. **101**, 4763-4765.

Okamoto, Y., Kawashima, M. and Hatada, K. (1986). Chromatographic resolution. XI. Controlled chiral recognition of cellulose triphenylcarbamate derivatives supported on silica gel. J. Chromatogr. **363**, 173-186.

Okamoto, Y. and Kaida, Y. (1994a). Resolution by high-performance liquid chromatography using polysaccharide carbamates and benzoates as chiral stationary phases. J. Chromatogr. A **666**, 403-419.

Okamoto, M., Takahashi, K. I. and Doi, T. (1994b). Direct stereochemical resolution of 3,4-dihydroxy phenylserine using a chiral crown ether stationary phase. J. Chromatogr., A **675**, 244-247.

Pianezzola, E., Bellotti, V., Fontana, E., Moro, E., Gal J. and Desai, D. M. (1989). Determination of the enantiomeric composition of salsolinol in biological samples by high-performance liquid chromatography with electrochemical detection. J. Chromatogr. B **495**, 205-214.

Pettersson, C. and Schill, G. (1981). Separation of enantiomeric amines by ion-pair chromatography. J. Chromatogr. **204**, 179-183.

Pettersson, C. and Josefsson, M. (1986). Chiral separation of amino alcohols by ion-pair chromatography. Chromatographia **21**, 321-316.

Pettersson, C. and Gioeli, C. (1993). Chiral separation of amines using reversed-phase ion-pair chromatography. Chirality **5**, 241-245.

Pirkle, W. H. and House, D. W. (1979). Chiral high-performance liquid chromatographic stationary phases. 1. Separation of the enantiomers of sulfoxides, amines, amino acids, alcohols, hydroxy acids, lactones, and mercaptans. J. Org. Chem. Chem. Sci. **44**, 1957-1960.

Pirkle, W. H., Finn, J. M., Schreiner, J. L. and Hamper, B. C. (1981). A widely useful chiral stationary phase for the high-performance liquid chromatography separation of enantiomers. J. Am. Chem. Soc. **103**, 3964-3966.

Pirkle, W. H., Hyun, M. H. and Bank, B. (1984). A rational approach to the design of highly effective chiral stationary phase. J. Chromatogr. **316**, 585-604.

Pirkle, W. H., Pochasky, T. C., Mahler, G. S. and Field, R. E. (1985). Chromatographic separation of enantiomers of 2-carboalkoxyindolines and *N*-aryl-α-amino esters on chiral stationary phases derived from *N*-(3,5-dinitrobenzoyl)-α-amino acids. J. Chromatogr. **348**, 89-96.

Pirkle, W. H. and Pochapsky, T. C. (1989). Considerations of chiral relevant to the liquid chromatographic separation of enantiomers. Chem. Rev. **89**, 347-362.

Seeman, J. I., Secor, H. V., Armstrong, D. W., Immons, K. D. and Ward, T. J. (1988). Enantiomeric resolution and chiral recognition of racemic nicotine and nicotine analogues by β-cyclodextrin complexation. Structure-enantiomeric resolution relationship in host-guest interactions. Anal. Chem. **60**, 2120-2127.

Sjoequist, B. Eriksson, A. and Winblad, B. (1982). Salsolinol and catecholamines in human brain and their relation to alcoholism. Prog. Clin. Biol. Res. **90**, 57-67.

Sousa, L. R., Hoffman, D. H., Kaplan, L. and Cram, D. J. (1974). Total optical resolution of amino esters by designed host-guest relationships in molecular complexation. J. Am. Chem. Soc., **96**, 7100-7101.

Stalcup, A. M., Chang, S. C., Armstrong, D. W. and Pitha, J. (1990). *(S)*-2-Hydroxypropyl-β-cyclodextrin, a new chiral stationary phase for reversed-phase liquid chromatography. J. Chromatogr. **513**, 181-194.

Stalcup, A. M., Chang, S. C. and Armstrong, D. W. (1991a). Effect of the configuration of the substituents of derivatized β-cyclodextrin. J. Chromatogr. **540**, 113-128.

Stalcup, A. M., Fanlker, J. R. Jr., Tang, Y., Armstrong, D. W., Levy, L. W. and Regalado, E. (1991b). Determination of the enantiomeric purity of scopolamine isolated from plant extracts using achiral/chiral coupled column chromatography. Biomed. Chromatogr. **5**, 3-7.

Stalcup, A. M. and Williams, K. L. (1992). Determination of enantiomers in human serum by direct injection onto a β-cyclodextrin HPLC bonded phase. J. Liq. Chromatogr. **15**, 29-37.

Szepesi, G., Gazdag, M. and Ivancsics, R. (1982). Normal-phase dynamic (solvent-generated) molecular complexation chromatography using anionic ion exchange. II. Separation of optical isomers. J. Chromatogr. **244**, 33-48.

Takahashi, T., Deng, Y., Maruyama, W., Dostert, P., Kawai, M. and Naoi, M. (1994). Uptake of a neurotoxin-candidate, *(R)*-1,2-dimethyl-6,7-dihydroxy-1,2,3,4-tetrahydroisoquinoline into human dopaminergic neuroblastoma SH-SY5Y cells by the dopamine transport system. J. Neural Transm. **98**, 107-118.

Wainer, I. W. and Alembik, M. C. (1986). Resolution of enantiomeric amides on a cellulose-based chiral stationary phase. Steric and electronic effects. J. Chromatogr. **358**, 85-93.

Walhagen, A. and Edholm, L. E. (1991). Chiral separation on achiral stationary phase with different functionalities using β-cyclodextrin in the mobile phase and application to bioanalysis and coupled columns. Chromatographia **32**, 215-223.

Zukowski, J., Pawlowska, M., Nagatkina, M. and Armstrong, D. W. (1993). High-performance liquid chromatographic enantioseparation of glycyl di- and tripeptides on native cyclodextrin bonded phase. Mechanistic considerations. J. Chromatogr. **629**, 169-179.

Zukowski, J., Sybilska, D. and Bojarski, J. (1986). Application of α- and β-cyclodextrin and heptakis (2,6-di-methyl)-β-cyclodextrin as mobile phase components for the separation of some chiral barbiturates into enantiomers by reversed-phase high-performance liquid chromatography. J. Chromatogr. **364**, 225-232.

Progress in HPLC-HPCE, Vol. 6, pp. 339–350
I.N. Acworth *et al.* (Eds)

ANALYSIS OF ENDOGENOUS ENANTIOMERS OF NEUROTOXINS IN CLINICAL SAMPLES BY A MULTI-ELECTRODE DETECTION SYSTEM WITH A CHIRAL COLUMN

WAKAKO MARUYAMA[1], YULIN DENG[2], PHILIPPE DOSTERT[3] and MAKOTO NAOI[4]

[1]Laboratory of Biochemistry and Metabolism, Department of Basic Gerontology, National Institute for Longevity Sciences, Obu, Japan. [2]Department of Applied Chemistry and 4 Department of Biosciences, Nagoya Institute of Technology, Nagoya, Japan. [3]Pharmacia-Upjohn, Milan, Italy

INTRODUCTION

Parkinson's disease (PD) is one of the most common neurodegenerative diseases in the aged and its prevalence was estimated to be about 1.5% in the population older than 65 years of age (de Rijk *et al.*, 1995). The major pathological findings of this disease are selective neuronal loss of dopamine (DA) neurons in the substantia nigra and occurrence of Lewy bodies. The etiology of PD has not yet been elucidated, but recently there have been increasing reports suggesting that endogenous or xenobiotic neurotoxins may be involved in the mechanism of cell death specific for DA neurons (see reviews: Collins, 1994; Naoi *et al.*, 1995a; Tipton *et al.*, 1996). Dopamine-derived endogenous 1(*R*),2(*N*)-dimethyl-6,7-dihydroxy-1,2,3,4-tetrahydroisoquinoline [*N*-methyl(*R*)salsolinol, *N*M(*R*)Sal] is now proposed as the most likely PD-inducing neurotoxin candidate. This catechol isoquinoline was found to induce Parkinsonism in rats after continuous injection into the striatum, where the terminals of DA neurons exist (Naoi *et al.*, 1996b). The rat showed behavioral abnormalities, such as bradykinesia, stiffness of the tail, twitching of the limbs, which are very similar to some symptoms of PD. In addition, the number of DA neurons, DA content and the activity of tyrosine hydroxylase [tyrosine, tetrahydropteridine: oxygen oxidoreductase (3-hydroxylating)] were reduced in the substantia nigra, where the cell bodies of DA neurons occur. Among derivatives of 1-methyl-6,7-dihydroxy-1,2,3,4-tetrahydro-isoquinoline (salsolinol, Sal), only the (*R*)-enantiomer of *N*-methylsalsolinol (*N*MSal) was found to induce Parkinsonism in rats, whereas the (*S*)-enantiomer of *N*MSal and other related catechol isoquinolines did not.

Sal was found to occur in the human brain (Sjöquist *et al.*, 1982) and is probably synthesized *in situ*, since Sal cannot pass through the blood-brain-barrier (Origitano *et al.*, 1981). Sal has been considered to be synthesized from DA and acetaldehyde in racemic form by a non-enzymatic Pictet-Spengler reaction. However, enantio-selective synthesis of (*R*)-Sal in humans was suggested by the fact that the (*R*)-enantiomer was the only Sal detected in human urine of healthy volunteers (Strolin Benedetti *et al.*, 1989a, b; Dostert *et al.*, 1990). On the other hand, (*S*)-Sal was detected in urine only after L-DOPA administration when free DA increases (Dostert *et al.*, 1989) or after drinking alcohol which contains racemic Sal (Strolin Benedetti *et al.*, 1989b; Dostert *et al.*, 1991). Unlike urine, it has never been shown whether racemic Sals can be produced in the brain by L-DOPA therapy.

These results indicate that (*R*)-enantiomer of Sal is synthesized enzymatically. The enzymatic biosynthesis pathway of (*R*)-Sal involves the condensation of DA with pyruvic acid, producing 1-(*R*)methyl, carboxyl-6,7-dihydroxy-1,2,3,4-tetrahydro-isoquinoline [salsolinol-1-carboxylic acid]. This is then decarboxylated into 1,2-dehydrosalsolinol followed by a reduction into salsolinol (Dostert *et al.*, 1990), as shown in Figure 1. The intermediates in the proposed biosynthesis, salsolinol-1-carboxylic acid (Sjöquist and Ljumgquist, 1985) and 1,2-dehydrosalsolinol (Ung-Chhun *et al.*, 1985) were also detected in the brain. However, this pathway, especially the final enantio-selective reduction into (*R*)-Sal, has never been proven *in vivo* or *in vitro*. More recently, we isolated a novel enzyme from human brains that catalyzes enantio-selective synthesis of (*R*)-Sal directly from DA and acetaldehyde (Naoi *et al.*, 1996c). *N*MSal is synthesized in the brain by *N*-methylation of (*R*)-Sal, as proven by *in vivo* microdialysis of the rat brain (Maruyama *et al.*, 1992). The *in vivo N*-methylation activity was found to be most prominent in the substantia nigra.

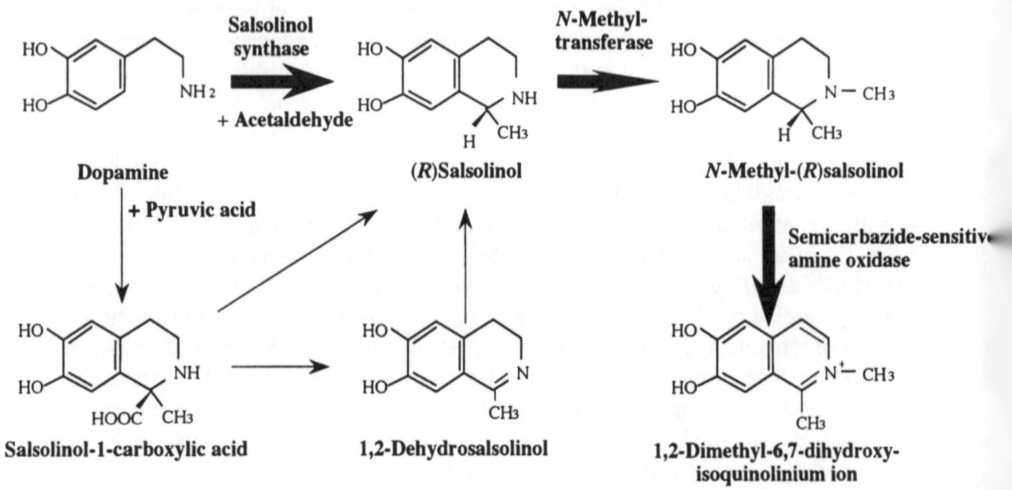

Figure 1. The metabolic pathway of (*R*)-Sal and *N*M(*R*)-Sal. From DA and acetaldehyde (*R*)-Sal is synthesized by a salsolinol synthase and *N*M(*R*)-Sal is produced by an *N*-methyltransferase. The non-enzymatic (Maruyama *et al.*, 1995a) and enzymatic oxidation (Naoi *et al.*, 1995b) generate the 1,2-dimethyl-6,7-dihydroxyisoquinolinium ion from *N*M(*R*)-Sal.

The quantitative analysis of the enantiomers of Sal and its derivatives in the brain and other biological samples has been very difficult, because they occur in very small quantities and they readily racemize by a Pictet-Spengler reaction during the extraction and preparation procedures. The two enantiomers of Sal were enantio-selectively identified by gas chromatography with nitrogen-phosphorous detection after methylation with diazomethane and then derivatization to diastereomers with N-trifluoroacetyl-L-propyl chloride (Dostert *et al.*, 1987). With the same derivatization; a high-performance liquid chromatography (HPLC) method was devised and applied to the analysis of banana, which contains equimolar (*R*)- and (*S*)-Sal (Strolin Benedetti *et al.*, 1989a). Later, this method was modified by use of a chiral derivatization agent, (S)-1-(1-naphthyl)ethylisothiocyanate, and the two enantiomers were well separated and identified (Strolin Benedetti *et al.*, 1989b). Using this method definite amounts of (*S*)-Sal, in addition to (*R*)-Sal, which may be derived from food were detected in human urine,. All of these methods require the derivatization into the diastereomers, and they also lack sufficient sensitivity for the analysis of clinical samples where sample volumes are limited, e.g., cerebrospinal fluid (CSF) and intraventricular fluid (IVF). A simpler method has been required for quite some time. Recently two papers reported new sensitive enantio-selective analyses using a cyclodextrin-bonded column and HPLC-electrochemical detection (ECD). Sällström Baum and Rommelspacher (1994) analyzed Sal in human plasma using a Nucleodex β-OH column (Macherey-Nagel, Düren, Germany), after solid-phase extraction of Sal. We used another β-cyclodextrin-bonded column, LichroCart 250-4 ChiraDex (Merck, Darmstadt, Germany), for analysis of Sal and *N*MSal in the brain (Deng *et al.,* 1995) without concentration of the sample on a solid extraction column. Unfortunately, the chiral column from Merck was so fragile that the crude biological samples could not be analyzed using this column.

This paper reviews our recent results on analysis of the isoquinolines in tissues. Enantiomers of Sal and *N*MSal in the CSF (Maruyama *et al.*, 1996a) and the IVF (Maruyama *et al.*, 1996b) were analyzed. The combination of a chiral Nucleodex β-OH column and HPLC with coulometric electrode array system (CEAS, ESA, Inc., Chelmsford, MA) increased the selectivity and sensitivity in identification and quantitation of Sal and *N*MSal in human materials. Using HPLC-CEAS the identification of a compound can be readily achieved based upon its retention time and also by its relative response across an array of ECD electrodes with different but fixed oxidation voltages (Naoi *et al.*, 1993). It was confirmed that (*R*)-enantiomers of *N*MSal and Sal are the only enantiomers detected in the human CSF and IVF. In CSF from untreated Parkinsonian patients, *N*M(*R*)Sal was found to increase significantly, and in IVF the enantio-specificity was not affected by L-DOPA therapy. These results are discussed in relation to the possible involvement of this catechol isoquinoline in the pathogenesis of PD.

IDENTIFICATION OF SALSOLINOLS IN IVF: EFFECT OF L-DOPA THERAPY

The (*R*)- and (*S*)-enantiomer of Sal and *N*MSal were synthesized according to Teitel *et al.* (1972). L-DOPA, DA, homovanillic acid (HVA) and other chemicals were purchased from Sigma (St. Louis, MO).

IVF was obtained from 13 patients with PD [Male/Female, 8/5; age, 56.2 ± 9.3 years (mean ± SD); L-DOPA (150-600 mg/day)/carbidopa administrated; duration of the disease, 6.0 ± 4.0 years (mean ± SD); Hoehn and Yahr stage, Stage II - 2; Stage III - 6; Stage IV - 2; Stage V - 2]. The patients underwent thalamotomy and about 10mL of IVF was obtained during the operation. All the patients were fully informed about the purpose of the analysis and their consent was confirmed.

IVF samples were stored at -70°C until analysis. To the sample 1/10 volume of 1M perchloric acid containing 1mM EDTA and 1mM sodium metabisulfite was added, mixed and centrifuged at 22,000g for 10 minutes. The supernatant was filtered through a Millipore filter (pore size, 0.45μm; filter diameter, 4mm) prior to analysis.

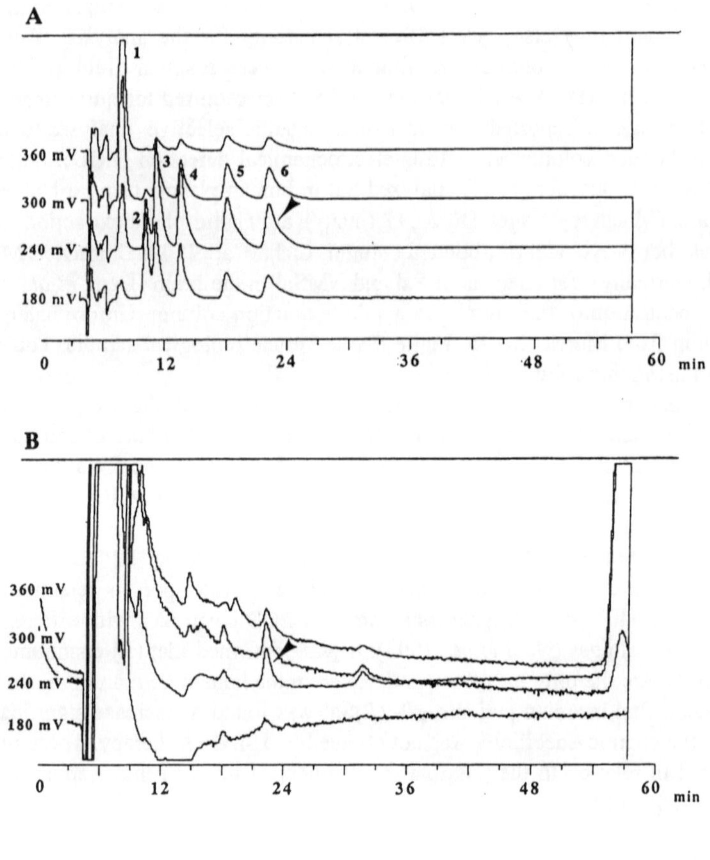

Figure. 2. HPLC patterns of IVF from Parkinsonian patients. Standards of salsolinol derivatives and the sample were applied to a ß-cyclodextrin-bonded column connected to the HPLC-CEAS. A; Standards (80 pmoles and full scale of 500nA), Peak 1, salsolinol-1-carboxylic acid; 2, DA; 3, (*S*)-Sal; 4, (*R*)-Sal; 5, *N*M(*S*)-Sal and 6, *N*M(*R*)-Sal. B; Sample prepared from Parkinsonian IVF (full scale, 20 nA).

An eighty μL sample was analyzed using the CEAS with 12 ECD electrodes. The conditions for the analysis of (*R*)- and (*S*)-enantiomer of Sal and *N*MSal, and HVA were a modification of the method reported by Sällström *et al.* (1994). The column used was a Nucleodex ß-OH column (200 x 4mm, Macherey-Nagel, Düren, Germany) and the mobile phase was 0.1M sodium phosphate buffer, pH3.6, containing 4.76% methanol. The flow rate was 0.4mL/min. The 12 electrodes were set from +60 to +720mV with a +60mV increments. Monoamines and the metabolites were analyzed by HPLC-CEAS as reported previously (Maruyama *et al.*, 1992, 1996a, b; Naoi *et al.*, 1993). The peaks were identified by comparison of the retention time and the redox (voltammetric) behavior across the electrodes (ratio accuracy) to those of authentic standards.

Figure 2 shows a typical HPLC chromatogram of standard (*R*)-Sal (peak 4), (*S*)-Sal (peak 3), *N*M(*R*)-Sal (peak 6), and *N*M(*S*)-Sal (peak 5), salsolinol-1-carboxylic acid (peak 1), and DA (peak 2) (A), and that of IVF sample (B). A peak corresponding to *N*M(*R*)-Sal was identified (shown by an arrow) in human IVF samples.

In all the samples *N*M(*R*)-Sal and (*R*)-Sal were identified, whereas (S)-enantiomers of *N*MSal, Sal, salsolinol-1-carboxylic acid and were not detected (the detection limit, <0.01nM). As summarized in Table 1, the identification of the peak was carried out by both the retention time and relative reactivity across the electrodes with graded voltage.

Table 1.
Retention time and relative accuracy of *N*M(*R*)-Sal in IVF sample compared with authentic standard.

	Retention time (mins.)	Ratio accuracy 240/180(mV)	300/240(mV)	360/420(mV)
Authentic standard of *N*M(*R*)Sal	22.03	2.69	0.667	0.400
IVF sample	21.90	2.80	0.682	0.455

IVF samples were analyzed by HPLC-CEAS. Ratio accuracy was calculated by the comparison of the voltammetric behavior across the electrode array of the standard and the peak in the IVF sample.

The effects of the duration and the progress of the disease on the level of *N*M(*R*)-Sal in IVF were examined (Figure 3). There were large difference in the level of *N*M(*R*)-Sal, but higher levels were observed in the patients with the disease duration around 5 years (Figure 3A) and with Hoehn and Yahr stage III and IV (Figure 3B). In relation to the progress of PD, *N*M(*R*)-Sal level increased at the earlier stage of the disease and decreased at the later stage, which may be due to the release of the catechol isoquinoline from dying neurons in the earlier stage and depletion of the neurons in the later stage.

In the IVF sample, the L-DOPA concentration was 50.8±78.7nM, the total DA (free and conjugated) and HVA were 109±64nM and 1.35±0.56μM, respectively. The correlation amongst the total DA and *N*M(*R*)-Sal and HVA was examined. As shown in Figure 5, there does not seem to be any correlation between *N*M(*R*)-Sal and total DA concentration, whereas HVA concentration is well correlated with the total DA concentration (Figure 4).

The stereo-specific occurrence of *N*M(*R*)-Sal was also confirmed in IVF even after L-DOPA therapy and the concentration was not correlated with DA level. This suggests that the synthesis of the catechol isoquinoline is not dependent on DA level, but on other factors such as the activity of the enzymes involved in its metabolism.

ENANTIOMERIC ANALYSIS OF *N*MSAL AND SAL IN PARKINSONIAN CSF

The lumbar CSF samples from 16 patients with newly diagnosed and untreated PD were also obtained [Male/Female, 6/10; age, 62.6±10.4 years (mean±SD), Hoehn and Yahr stage, Stage II - 1; Stage III - 14; Stage IV - 1; duration of the disease, 3.18±2.73 years]. Control subjects without neurological disorders (control, n=29) [Male/Female, 20/9; age, 52.7±15.8 years (mean±SD)] and patients with multiple system atrophy (MSA) (disease control, n=5) [Male/Female, 2/3; age, 54.6±4.1 years (mean±SD); duration of the disease, 2.50±1.67 years (mean±SD)] were analyzed. Differences in data from the 3 groups were analyzed for significance nonparametrically by the Kruskal-Wallis test. The differences between the 2 groups were examined by the Mann-Whitney U test separately. When $p<0.05$ data were considered to be significant. The methods used for the analysis of Sal derivatives using the HPLC-CEAS with a chiral column and for the analysis of the monoamines using the HPLC-CEAS with a reverse phase column are the same as those reported for IVF sample analysis above.

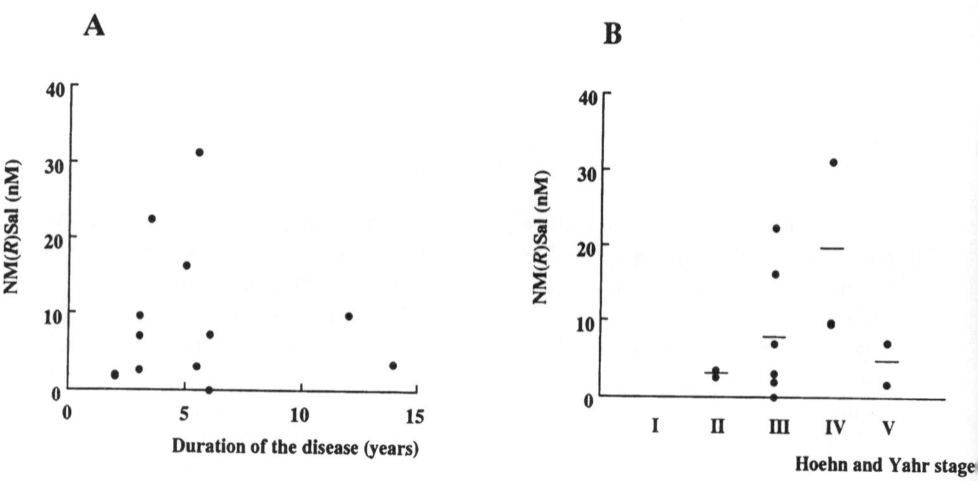

Figure 3. Effects of duration of the disease and the Hoehn and Yahr stage of the patients on the concentration of *N*M(*R*)-Sal in the IVF. **A**; Duration of the disease and **B**; Hoehn-Yahr stage. Bars denote mean levels.

*N*M(*R*)Sal was detected in all CSF samples from controls, patients with PD and MSA without L-DOPA therapy and IVF from PD patients treated with L-DOPA (Table 2). However, another enantiomer, *N*M(*S*)-Sal, was below the detection limit in all IVF and CSF samples (<0.01nM).

The CSF concentration of NM(R)-Sal obtained from normal control subjects was not affected by the age (22 to 76 years) (Figure 5). *N*M(*R*)Sal levels were >6nM in most Parkinsonian CSF (12 out of 16 patients), but <6nM in all control patients or those with MSA, except 2 cases. *N*M(*R*)-Sal level in the PD patients was significantly higher than that in either control subjects ($p<0.0001$) or patients with MSA ($p<0.0022$) (Table 2). To evaluate the biosynthesis rate of Sals from DA, the concentration of *N*M(*R*)-Sal was compared in each sample with that of HVA, a major metabolite of DA. HVA concentrations were lower in PD samples (141±75nM) and MSA samples (156±75nM) than in the control samples (180±92nM), but the difference was not statistically significant. The ratio [*N*M(*R*)-Sal/HVA] was significantly increased in PD group (0.086±0.067) than in the control group (0.034±0.030) ($p<0.001$) or MSA group (0.034±0.029) ($p<0.05$).

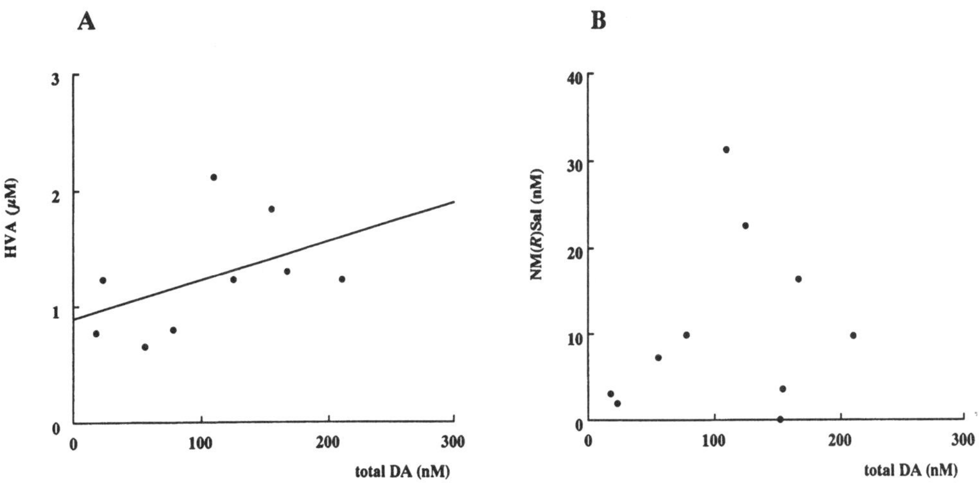

Figure 4. Relation of the level of total DA with HVA or *N*M(*R*)-Sal. The total DA concentration was plotted against HVA (**A**) or *N*M(*R*)-Sal (**B**).

Table 2.
The concentration of (*R*)-Sal and NM(*R*)-Sal in CSF and IVF sample.

	(*R*)-Sal (nM)	*N*M(*R*)-Sal (nM)
CSF from		
Normal control	n.d.	4.53 ± 2.08
PD patients	n.d.	8.32 ± 2.89*†
MSA patients	n.d.	3.59 ± 1.52
IVF	0.39 ± 0.21	9.15 ± 9.08

Each value represents the mean±SD. n.d. = not detected. * $p<0.0001$ compared with CSF of normal control patients, † $p<0.0022$ compared with CSF of MSA patients.

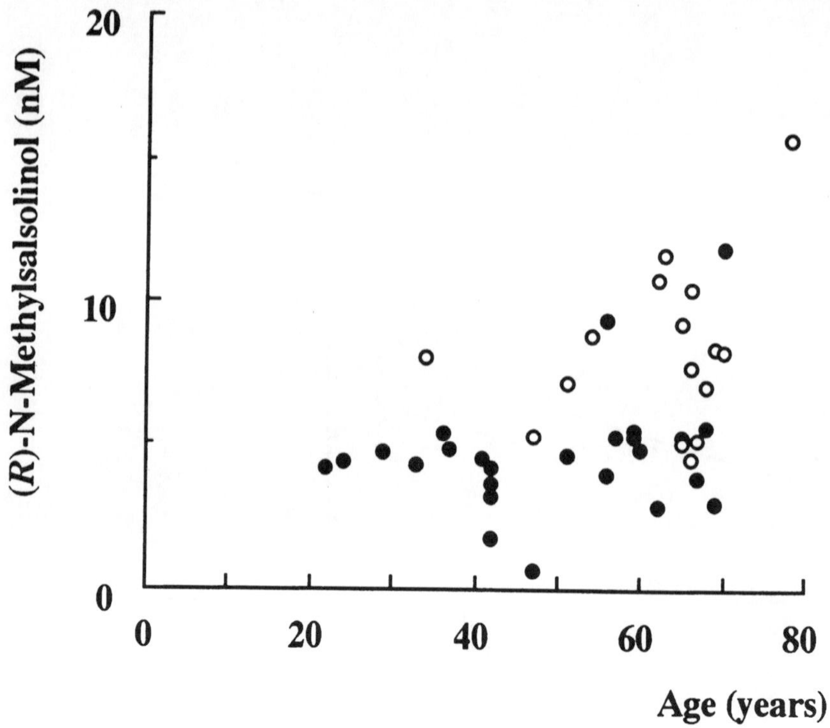

Figure 5. The effects of aging on *N*M(*R*)-Sal concentration in CSF. The *N*M(*R*)-Sal concentration was plotted against age. The open circles represents CSF from PD patients and the filled circles represent CSF from control patients.

These results clearly show the enantio-selective occurrence and increase of *N*M(*R*)-Sal in Parkinsonian CSF. Considering that salsolinols are produced from DA and DA synthesis is reduced in PD patients, the increase in the *N*M(*R*)-Sal level is noteworthy and may be due to its increased biosynthesis or reduced catabolism, or both. As described above, the two enzymes, a (*R*)-salsolinol synthase and an *N*-methyltransferase, are involved in the biosynthesis, and their activities may be altered in PD. The analysis of the enzymes may give us a clue to understanding the pathogenesis of this disease. The *N*M(*R*)Sal concentration in CSF might prove to be an effective preclinical and early marker of PD.

DISCUSSION

Up until now, three classes of endogenous or xenobiotic compounds have been proposed as neurotoxin candidates and have been implicated in the degeneration of DA neurons in PD: 6,7-dihydroxy-1,2,3,4-tetrahydroisoquinoline (Naoi *et al.*, 1995a), 1,2,3,4-tetrahydroisoquinoline (Nagatsu and Yoshida, 1988) and β-carbolines (Matsubara, 1996). In human brain *N*MSal, 2-methyl-6,7-dihydroxy-1,2,3,4-tetrahydroisoquinoline (N-methylnorsalsolinol) (Niwa *et al.*, 1991), 1,2,3,4-tetrahydroisoquinoline (Niwa *et al.*, 1987; Ohta *et al.*, 1987), and β-carbolines (Matsubara *et al.*, 1993) have all been detected. These neurotoxin candidates were identified by gas chromatography-mass spectrometry. Even though this method is accurate, it requires relative large amounts of samples (at least 10g of brain tissues) and involves complicated extraction and derivatization procedures. On the other hand, HPLC-ECD methods are sensitive enough to detect these compounds in only a few mL of sample. However, in brain and plasma samples there are often so many electrochemically reactive compounds that extraction, purification and concentration procedures are usually required for HPLC with a conventional electrochemical detector. The CEAS system is a very useful tool for the identification and quantitation of small amounts of compounds in human clinical materials (Naoi *et al.*, 1993). The CEAS can eliminate many electrochemically-reactive contaminants by oxidizing them on the upstream electrodes (lower oxidation voltages) thereby acting as a screen. Removal of such contaminants thus makes the identification of analytes much easier than with conventional ECD system. The volume of IVF and CSF used for the studies reported here commonly used <100μL for each measurement. The combination of the CEAS with a chiral column proved to be a useful approach for the enantiomeric separation and identification of bioactive compounds, such as neurotoxins, in human materials.

In CSF from Parkinsonian patients, another catechol isoquinoline, *N*-methylnorsalsolinol, was detected, but not in samples from control subjects (Moser *et al.*, 1995). Its concentration in PD patients was correlated with increased DA turnover, and negatively correlated with the disease duration, suggesting that the increase in the levels of this isoquinoline is a secondary change following DA depletion. In our animal experiment, *N*-methylnorsalsolinol did not induce Parkinsonism in rats (Naoi *et al.*, 1996a), and our newly-purified (*R*)-salsolinol synthase did not synthesize N-methylnorsalsolinol (Naoi *et al.*, 1996). Among β-carbolines, the 2,9-dimethyl-norharmanium cation and 2-methylated β-carbolinium cations were found to be increased in CSF from PD patients (Matsubara *et al.*, 1995). These cations may be derived from the environment and are probably not synthesized in the brain. Further evidence is required to confirm whether these alkaloids may be also involved in the pathogenesis of PD.

In this study increases in *N*M(*R*)Sal were clearly confirmed in the CSF obtained from PD patients compared to controls and patients with MSA (a disease control). These results suggest that levels of *N*M(*R*)-Sal may also increase in the central nervous system. In addition, (*S*)enantiomers of *N*MSal were not detected in CSF. The importance of the increase in NM(R)-Sal should be emphasized, since in PD brain DA neurons are degenerated and the level of DA and its metabolites, such as HVA, are decreased. This suggests that the DA level in the brain may not directly regulate the concentration of (*R*)-Sal and *N*M(*R*)-Sal. This assumption may be relevant with the fact that total DA level did not correlate with that of *N*M(*R*)-Sal in IVF, even though it correlated well with HVA levels, and that the (*S*)-enantiomers of *N*MSal and Sal were not produced from DA which was, in turn, increased following L-DOPA therapy. The true neurotoxin, *N*M(*R*)-Sal proved to be synthesized by two enzymes, a (*R*)-salsolinol synthase and an *N*-methyltransferase, and the enzyme activity of these enzymes may regulate the *N*M(*R*)-Sal level in PD patients, suggesting that the enzymes may be involved in the pathogenesis of PD as an endogenous factor. The polymorphism or mutation of these enzymes or enzymes catabolyzing *N*M(*R*)-Sal may account for the increase of *N*M(*R*)-Sal in PD. To clarify this point, analyses of clinical samples from Parkinsonian patients are now under way.

Acknowledgments
This work was supported by Grant-In-Aid for Scientific Research on Priority Area and from the Ministry of Education, Science and Culture, Japan.

REFERENCES

Collins, M. A. (1994). Potential Parkinsonian protoxicants within and without. Neurobiol. Aging **15**, 277-278.

Deng, Y., Maruyama, W., Dostert, P., Takahashi, T., Kawai, M. and Naoi, M. (1995). Determination of the (*R*)- and (*S*)-enantiomers of salsolinol and N-methylsalsolinol by use of a chiral high-performance liquid chromatographic column. J. Chromatogr. B. **670**, 47-54.

de Rijk, M. C., Breteler, M. M. B., Graveland, G. A., Ott, A., Grobbee, D. E., van der Meche F. G. A. and Hofman, A. (1995). Prevalence of Parkinson's disease in the elderly: The Rotterdam study. Neurology **45**, 2143-2146.

Dostert, P., Strolin Benedetti, M. and Dedieu, M. (1987). Ratio of enantiomers of salsolinol in human urine. Pharmacol. Toxicol. [Suppl. 1], **60**, 13.

Dostert, P., Strolin Benedetti, M., Dordain, G. and Vernay, D. (1989). Enantiomeric composition of urinary salsolinol in Parkinsonian patients after Madopar. *J. Neural Transm. [P-D Sect]* **1**, 269-278.

Dostert, P., Strolin Benedetti, M., Bellotti, V., Allievi, C. and Dordain, G. (1990). Biosynthesis of salsolinol, a tetrahydroisoquinoline alkaloid, in healthy subjects. J Neural Transm. **81**, 215-223

Dostert, P., Strolin Benedetti, M., Dordain, G. and Vernay, D. (1991). Urinary elimination of salsolinol enantiomers in alcoholics. J. Neural Transm. **85**, 51-59.

Maruyama, W., Nakahara, D., Ota, M., Takahashi, T., Takahashi, A., Nagatsu, T. and Naoi, M. (1992). N-Methylation of dopamine-derived 6,7-dihydroxy-1,2,3,4-tetrahydroisoquinoline, (*R*)-salsolinol, in rat brains: *In vivo* microdialysis study. J. Neurochem. **59**, 395-400.

Maruyama, W., Dostert, P., Matsubara, K. and Naoi, M. (1995a). N-Methyl(*R*)-salsolinol produces hydroxyl radicals: Involvement to neurotoxicity. Free Rad. Biol. Med. **19**, 67-75.

Maruyama, W., Abe, T., Tohgi, H., Dostert, P. and Naoi, M. (1996a). A dopaminergic neurotoxin, (*R*)-N-methylsalsolinol, increases in Parkinsonian CSF. Ann. Neurol. **40**. 119-122.

Maruyama, W., Narabayashi, H., Dostert, P. and Naoi, M. (1996b). Stereospecific occurrence of a Parkinsonism-inducing catechol isoquinoline, N-methyl(*R*)-salsolinol, in the human intraventricular fluid. J. Neural Transm. in press.

Matsubara, K., Collins, M., Akane, A., Ikebuchi, J., Neafsey, E., Kagawa, M. and Shiono, H. (1993). Potential bioactivated neurotoxicants, N-methylated β-carbolinium ions, are present in human brain. Brain Res. **610**, 90-96.

Matsubara, K., Kobayashi, S., Kobayashi, Y., Yamashita, K., Koide, H., Hatta, M., Iwashita, K., Tanaka, O. and Kimura, K. (1995). β-Carbolinium cations, endogenous MPP^+ analogs, in the lumbar cerebrospinal fluid of patients with Parkinson's disease. Neurology **45**, 2240-2245.

Matsubara, K. (1996). Occurrence of neurotoxic β-carbolinium cations in mammalian central nervous system. Biogenic Amines **12**, 161-169.

Moser, A., Scholz, J., Nobbe, F., Vieregge, P., Böhme, V. and Bamberg, H. (1995). Presence of N-methyl-norsalsolinol in the CSF: correlations with dopamine metabolites of patients with Parkinson's disease. J. Neurol. Sci. **131**, 183-189.

Nagatsu, T. and Yoshida, M. (1988). An endogenous substance of the brain, tetrahydroisoquinoline, produces Parkinsonism in primates with decreased dopamine, tyrosine hydroxylase and biopterin in the nigrostriatal regions. Neurosci. Lett. **87**, 178-182.

Naoi, M., Maruyama, W., Acworth, I. N., Nakahara, D. and Parvez, H. (1993). Multi-electrode detection systems for determination of neurotransmitters. In: Methods in Neurotransmitter and Neuropeptide Research, Part 1. **11**. Techniques in The Behavioral and Neural Sciences. Parvez, S. H., Naoi, M., Nagatsu, T., and Parvez, S. (Eds.), Elsevier Science Publishers BV, Amsterdam, pp. 1-39.

Naoi, M., Maruyama, W. and Dostert P. (1995a). Dopamine-derived 6,7-dihydroxy-1,2,3,4-tetrahydroisoquinoline; oxidation and neurotoxicity. Progr. Brain Res. **106**, 227-239.

Naoi, M., Maruyama, W., Zhang, J. H., Takahashi, T., Deng, Y. and Dostert, P. (1995b). Enzymatic oxidation of the dopaminergic neurotoxin, 1(*R*), 2(*N*)-dimethyl-6,7-dihydroxy-1,2,3,4-tetrahydroisoquinoline, into 1,2(N)-dimethyl-6,7-dihydroxyisoquinolinium ion. Life Sci. **57**, 1061-1066.

Naoi, M., Maruyama, W., Dostert, P. and Hashizume, Y. (1996a). Animal model of Parkinson's disease induced by naturally-occurring 1(*R*),2(*N*)-dimethyl-6,7-dihydroxy-1,2,3,4-tetrahydroisoquinoline. Biogenic Amines **12**, 135-147.

Naoi, M., Maruyama, W., Dostert, P., Hashizume, Y., Nakahara, D., Takahashi, T. and Ota, M. (1996b). Dopamine-derived endogenous 1(*R*),2(*N*)-dimethyl-6,7-dihydroxy-1,2,3,4-tetrahydroisoquinoline, *N*-methyl(*R*)-salsolinol, induces Parkinsonism in rats: Biochemical, pathological and behavioral studies. Brain Res. **709**, 285-295.

Naoi, M., Maruyama, W., Dostert, P., Kohda, K. and Kaiya, T. (1996c). A novel enzyme enantio-selectively synthesizes (*R*)salsolinol, a precursor of a dopaminergic neurotoxin, N-methyl(*R*)-salsolinol. Neurosci. Lett. **212**, 1-4.

Niwa, T., Takeda, N., Kaneda, N., Hashizume, Y. and Nagatsu, T. (1987). Presence of tetrahydroisoquinoline and 2-methyltetrahydroquinoline in Parkinsonian and normal human brains. Biochem. Biophys. Res. Acta **144**, 1084-1089.

Niwa, T., Takeda, N., Yoshizumi, H., Tatematsu, A., Yoshida, M., Dostert, P., Naoi, M. and Nagatsu, T. (1991). Presence of 2-methyl-6,7-dihydroxy-1,2,3,4-tetrahydroisoquinoline and 1,2-dimethyl-6,7-dihydro-1,2,3,4-tetrahydroisoquinoline, endogenous amines, in Parkinsonian and normal human brains. Biochem. Biophys. Res. Commun. **177**, 603-609.

Origitano, T., Hannigen, J. and Collins, M. A. (1981). Rat brain salsolinol and blood-brain barrier. Brain Res. **274**, 446-451.

Ohta, S., Kohno, M., Makino, Y., Tachikawa, O. and Hirobe, M. (1987). Tetrahydroisoquinoline and 1-methyl-tetrahydroisoquinoline are present in the human brain: Relation to Parkinson's disease. Biomed. Res. **8**, 453-456.

Sällström Baum, S. and Rommelspacher, H. (1994). Determination of total dopamine, R- and S-salsolinol in human plasma by cyclodextrin bonded-phase liquid chromatography with electrochemical detection. J. Chromatogr. B. **660**, 235-241.

Sjöquist, B., Eriksson, A., and Winblad, B. (1982). Brain salsolinol levels in alcoholism. Lancet, i 675-676.

Sjöquist, B. and Ljungquist, C. (1985). Identification and quantification of 1-carboxysalsolinol and salsolinol in biological samples by gas chromatography-mass spectrometry. J. Chromatogr. **343**, 1-8.

Strolin Benedetti, M., Bellotti, V., Pianezola, E., Moro, E., Carminati, P. and Dostert, P. (1989a). Ratio of the R and S enantiomers of salsolinol in food and human urine. J. Neural Transm. **77**, 47-53.

Strolin Benedetti, M., Dostert, P. and Carminati, P. (1989b). Influence of food intake on the enantiomeric composition of urinary salsolinol in man. J.Neural Transm. **78**, 43-51.

Teitel, S., O'Brien, J., and Brossi, A. (1972). Alkaloids in mammalian tissue. II. Synthesis of (+) and (-) substituted-6,7-dihydroxy-1,2,3,4-tetrahydroisoquinolines. J. Med. Chem. **15**, 845-846.

Tipton, K. F., Naoi, M., Parvez, S. H. and Nagatsu, T. (1996). Naturally occurring and xenobiotic toxins and neurodegenerative disorders. Biogenic Amines **12**, 83-190.

Ung-Chhun, N., Cheng, B. Y., Proger, D. A., Serrano, P., Chavez, B., Fernandez Perrez, R., Morales, J. and Collins, M. A. (1985). Alkaloid adducts in human brain: Coexistence of 1-carboxylated and noncarboxylated isoquinolines and β-carbolines in alcoholics and nonalcoholics. Prog. Clin. Biol. Res. **183**, 125-136.

Progress in HPLC-HPCE, Vol. 6, pp. 351–376
I.N. Acworth *et al.* (Eds)

ELECTROCHEMICAL MEASUREMENT OF ELECTROCHEMICALLY UNREACTIVE COMPOUNDS: AN EXAMINATION OF THE USE OF DERIVATIVES AND PHOTOLYSIS WITH HPLC AND COULOMETRIC DETECTION.

IAN N. ACWORTH and JOHN WARASKA

ESA, Inc., 22 Alpha Road, Chelmsford, MA 01824-4171

INTRODUCTION

High-performance liquid chromatography with electrochemical detection (HPLC-ECD) is the technique of choice when substances must be measured at trace or ultra-trace levels. HPLC-ECD offers superior sensitivity and selectivity over HPLC-spectrophotometric techniques, however, sometimes the selectivity is found to be too restrictive because not all compounds are electrochemically active.[1] Additionally, some compounds are only electrochemically detectable at high potentials, where the signal-to-noise (s/n) ratio is poor and the resulting decreased selectivity increases the chance for co-elution or analyte misidentification. Several approaches have been used to broaden the applicability of HPLC-ECD to electrochemically unreactive analytes while still maintaining sensitivity. These techniques also often lower the oxidation potential required for detection thereby improving the s/n ratio. Such methods include pre-column derivatization, post-column photo-activation, post-column chemical reactions and post-column enzymatic reactions (Bratin *et al.*, 1984; ; Krull *et al.*, 1984, 1985, 1994; Bachman and Stewart, 1989; Bovington and Webster, 1994; Gorog and Gazdag, 1994; Shimada and Mitamura, 1994; Yasaka and Tanaka, 1994).

In this chapter we examine the use of pre-column derivatization and post - column photo-activation with coulometric array detection of example compounds and the application of these techniques for the measurement of analytes that are otherwise undetectable by HPLC-ECD.

[1] Many compounds are electrochemically active and include aromatic alcohols, aromatic amines, nitro-aromatics, heterocyclics, aliphatic amines and aliphatic polyenes (Musch *et al.*, 1985; Jane *et al.*, 1985; Smyth, 1987). Other compounds either require too high a potential to be measured satisfactorily or are electrochemically inert (see Acworth and Bowers, this volume).

PRE-COLUMN DERIVATIZATION

Why derivatize?

Electrochemical detectors for HPLC are often thought of as exquisitely sensitive and selective for the measurement of catecholamines. Although it is true that they have been used extensively and indeed have become the detector of choice for this analysis, their applicability is much broader than just biogenic amines. Jane *et al.* (1985) and Musch *et al.* (1985) demonstrated that a significant number of biologically active compounds can be detected using HPLC-ECD. Over the last several years the technique has been used in a wide variety of applications ranging from environmental analysis of phenolics to detection of food adulteration to drug analysis. A common characteristic of all these analyses is the need for low levels of detection with a high level of selectivity.

Unfortunately, not all compounds lend themselves to analysis by HPLC-ECD or any other high sensitivity technique in their native state. Over the years a variety of techniques have been used to enhance the detectability of these compounds. One approach is the use of derivatizing reagents. Derivatives for the enhancement of detection in HPLC were originally adapted from specific reactions for spectroscopy. Hence many of the methods were for use with HPLC-UV/vis. Both pre-column and post-column procedures have been reported. Later, again in an effort to improve detection limits, specific derivatives for fluorescence were developed. Today a wide variety of derivatives are available for a number of functional groups and HPLC detectors.

Why electrochemical array detection with derivatization?

With so many potential derivatives and detectors to choose from, it is fair to ask why select electrochemical detection? With older ECDs the issues of stability and equilibration times made this a reasonable question. While this may have been true for older amperometric electrochemical detectors, the current flow-through coulometric detectors have been refined to the point of eliminating these concerns from consideration. Today's coulometric detectors are stable, relatively maintenance free and equilibrate rapidly.

The reasons for choosing an ECD for use with derivatives are much the same as the reasons for choosing an ECD for any compound. First, electrochemical detectors deliver extremely low levels of detection. Derivatives are used to enhance sensitivity by improving the response over that of the native analyte or to make a compound responsive that would normally not respond. For example, serotonin can be detected by its native fluorescence but the signal is not adequate to measure it in tissue. When the appropriate derivative is made, its levels in tissue can be measured. In the case of derivatization with ECD, the derivatizing agent is used to add an electroactive group that will oxidize (or reduce) under reasonable conditions. (Thermodynamically all compounds can be oxidized or reduced, but there are practical limits to the potential ranges that can be used to carry out these processes in a detector for HPLC.) Once an electroactive group is added, the inherent sensitivity provided by ECD is the primary reason for its use. In other cases the addition of the electroactive group allows detection at a lower potential than would normally be required for the native compound. Hence the improvement in detection limit is a result of decreased noise.

A second characteristic of ECD is its selectivity. Although a derivatizing agent is chosen for its specificity and selectivity for a particular functional group on the compound of interest, often other compounds are present in the sample which do not derivatize but do produce a signal. This is a common problem when using UV detectors. The selectivity of ECD often eliminates these extraneous peaks, thus providing a "cleaner" chromatogram, resulting in reduced sample preparation and/or chromatographic run time.

The sensitivity and selectivity advantages of ECD are extended by use of the coulometric electrode array. The principles of coulometric array detection have been described in detail elsewhere (see Acworth and Bowers, this volume). In this detector a series of high efficiency constrained flow electrodes are set at varying potentials (typically incrementally increasing potentials for oxidation or decreasing for reduction). As a sample moves from the HPLC column through the detector, compounds react at specific electrodes depending upon their redox potential. A compound that oxidizes at a low potential will be oxidized at an early electrode. Compounds that oxidizes at high potentials will react (and produce a signal) at latter electrodes. What makes the coulometric electrode array unique is that, because of the high surface area and reduced diffusion distance resulting from the constrained flow, 100% of a compound can be reacted at a given electrode. The subsequent electrodes, therefore, are left to react with compounds that oxidize at high potentials free from signal produced by compounds that oxidize at lower potentials.

This means that an array can be used to "screen-out" compounds or to focus only on compounds that react in a specific potential window. Potentials can also be set such that a compound can be oxidized and then measured reductively or *vice versa* (if the compound is readily reversible). This concept can be extended to redox cycling to enhance s/n ratios. An array configured in this manner can be used to enhance selectivity, to simplify a chromatogram, to eliminate sample preparation, or to enhance sensitivity.

The array can also be selected with potentials adjusted to extend across the oxidation range for a compound. This results in data similar to a hydrodynamic voltammogram or "electrochemical spectrum" that can be used to confirm the identity of a compound, much the same way as spectral data from photodiode array UV detectors are used.

Another often overlooked capability of EC array detection is the ability to operate using gradient elution. This is a result of specific instrument capability rather than an inherent character of coulometric electrodes but is a highly useful ability of the system.

Considerations in choosing a derivative

A wide variety of EC-active derivatizing agents exist as presented in Table 1. There are several factors that can affect the choice of a derivatizing agent:

1. Availability and stability of the reagent.
2. Ease of use and automation, specificity, unwanted side-reactions, stability of the derivative, reaction time, reaction conditions and adaptability to pre- and post-column use.

3. Compatibility with reversed-phase HPLC-ECD mobile phase, sensitivity of the method, ease of oxidation, chromatographic resolution, linear range, and quantitation.

4. Effect of excess reagent.

Table 1.
Derivatizing agents used with HPLC-ECD.

Derivatizing agent	Abbreviation	Analyte	Reference
O-Acetylsalicyloyl chloride	OAS-Cl	Primary amines and amino acids	Smith *et al.*, 1988.
p-Aminophenol	PAP	Isocyanates	Meyer and Tallman, 1983.
Biuret		Dipeptides, dipeptide amides, and peptides	Warner and Weber, 1989; Tsai and Weber, 1990, 1991.
2-Carboxy-4,6-dinitrophthalic anhydride	CDNPA	Peptides	Frei and Lawrence, 1981.
7-Chloro-4-nitrobenzo-2-oxa-1,3-diazole	NBD-Cl	Secondary amino acids	Sessa *et al.*, 1986; Welch *et al.*, 1993.
2-Cyano-acetamide	2-CA	Sugars	Honda *et al.*, 1984.
1-(2,4-Dihydroxyphenyl)-2-bromoethanone	2,4-DBE	Fatty acids, metabolites and drugs with free carboxyl groups	Shimada *et al.*, 1987a; Munns *et al.*, 1988.
1-(2,5-Dihydroxyphenyl)-2-bromoethanone	2,5-DBE	Fatty acids, metabolites and drugs with free carboxyl groups	Shimada *et al.*, 1987a; Munns *et al.*, 1988.
Dimethoxyaniline	DMA	Prostaglandins	Knopse *et al.*, 1988.
2,4-Dinitro-fluoro-benzene, 2-chloro-3,5-dinitropyridine, 2,4,6 trinitrobenzene sulfonic acid	DNFB, DNCP, TNBS	Amines and amino acids	Jacobs and Kissinger, 1982.
3,6-Dinitrophthalic anhydride	DNPA	Peptides	Frei and Lawrence, 1981.
1-Ferrocenyl ethylamine or 1-ferrocenyl-propylamine, other ferrocenyl derivatives	1-FE, 1-FP	N-acetyl amino acids, primary amines and amino acids, ibuprofen, naproxen, peptides, proteins and thiols	Shimada *et al.*, 1987b,c, 1989; Eckert and Koller, 1990.

7-Fluoro-4-nitrobenzo-2-oxa-1,3-diazole	NBD-F	Secondary amino acids	Imai and Watanabe, 1981; Watanabe and Imai, 1981, 1982, 1984.
1-(4-Hydroxy-phenyl)-2-bromoethanone	4-HBE	Fatty acids, metabolites and drugs with free carboxyl groups	Munns, *et al.*, 1988.
Maleimides - Anilinonaphthyl-maleimide, dimethyl-amino-phenylmaleimide, N-(ferrocenyl)-maleimide	APM, DAM, NFM	N-acetylcysteine, captopril and reduced thiols	Shimada *et al.*, 1982, 1983, 1984b, 1987c.
1-(2-Methoxyphenyl)-piperazine	MPP	aromatic and aliphatic isocyanates	Warwick *et al.*, 1981.
Naphthalene dialdehyde (naphthalene-2,3-dicarboxaldehyde)/ cyanide	NDA/CN	Primary amino acids, primary amines, some peptides	Lunte *et al.*, 1989; Lunte and Wong, 1990; Nussbaum *et al.*, 1992.
ß-Naphthoquinone-4-sulfonate	BNQS	Primary and secondary amines and amino acids	Nakahara *et al.*, 1989.
(S) or (R)-1-(1-Naphthyl) ethyl isocyanate	NEI	Salsolinols	Pianezzola *et al.*, 1989.
Nitrophenylhydrazine, dinitrophenylhydrazine	NPH, DNPH	Aldehydes, ketones, ketosteroids and other carbonyls	Kissinger *et al.*, 1979; Shimada *et al.*, 1980; Jacob and Kissinger, 1982; Shimada *et al.*, 1984a; Chiavari and Bergamini, 1985; Goldring *et al.*, 1993.
α-Phenyl-*tert*-butylnitrone	PBN	Alkyl radicals, hydroxyl radicals	Stronk *et al.*, 1984.
Phenylisothiocyanate, p-N,N-dimethylamino phenylisothiocyanate	PITC, DMA-PITC	Primary and secondary amino acids	Musson and Sternson, 1980; Granberg, 1984; Mahachi *et al.*, 1984; Sherwood *et al.*, 1990.
o-Phthalaldehyde/ *ß*-mercaptoethanol	OPA/ßME	Primary amines, primary amino acids and some peptides	Joseph and Davis, 1983; Allison *et al.*, 1984; Donzanti and Yamamoto, 1988.
o-Phthalaldehyde/ sulfite	OPA/S	Primary amino acids, diamines, some peptides	Jacobs, 1987; Smith and Sharp, 1994; Turiak and Volicer, 1994.

o-Phthalaldehyde/ *tert*-butylthiol	OPA/TBT	Primary amines, primary amino acids and some peptides	Allison *et al.*, 1984; Zielke, 1985; Jacobs *et al.*, 1986. Hikal *et al.*, 1988; Shea and Jacobs, 1989; Globus *et al.*, 1991.
o-Phthalaldehyde/ N-acetylcysteine (or 1-thio-*ß*-D-glucose, N-acetyl-D-penicillamine, 2,3,4,6-tetra-O-1-thio-*ß*-D-glucopyranoside, thiolactic acid, mercaptopropanoic acid	OPA/X	Enantiomeric determination of primary amino acids, amphetamine, rimantidine, tocainide, tranylcypromine and some peptides	Einarsson *et al.*, 1987; Desai and Gal, 1993.
2,2,6,6-Tetramethyl-4-piperidone	TMP	Singlet oxygen	Motohashi and Mori, 1989.

For example, the reaction of dehydroepiandrosterone sulfate with NPH requires heating to 60°C for 20 minutes in 0.3% trichloroacetic acid/benzene solution before preparation for HPLC-ECD (Shimada *et al.*, 1984a). This method is not very practical as the removal of excessive reagent involves the disconnection of the ECD from the HPLC for 5 minutes following injection. Another example is the use of DNFB, DNCP or TNBS to measure amino acids. HPLC-ECD analysis of the alanine adduct is hampered by the hydrolysis of the derivatizing agents with the production of phenolic by-products capable of chromatographic interference (Jacobs and Kissinger, 1982). Also the use of PAP to measure isocyanates yields complicated chromatograms, as isocyanates react with both the amino and phenol groups of the derivatizing reagent (Meyer and Tallman, 1983). In each of these cases, a derivative of the compound of interest can be formed but other factors limit their utility for routine analytical procedures.

Methods for amino acid analysis

Perhaps the most widely studied and analyzed group of compounds that require derivatization are the amino acids. The first amino acid analyzers used ion-exchange chromatography with ninhydrin as a post-column derivatizing reagent. Later, in an effort to improve sensitivity, the amino acid adduct of *o*-phthalaldehyde (OPA) was utilized as a post-column fluorescent derivative. Pre-column methods were introduced to simplify the analytical instrumentation. Some of these have also used OPA, others have incorporated reagents such as phenylisothiocyanate (PITC) and its variants. Because amino acid analysis is so widespread and because the principles embodied in this analysis are typical and illustrative of the process and problems, the remainder of this discussion will focus on amino acid determination with electrochemistry.

The derivatization of amino acids, alkylamines and small peptides with OPA and a suitable nucleophile (typically a thiol reducing agent) for fluorometric detection was first reported by Roth (1971). These fluorescent products were later shown to be 1-(alkylthio)-2-alkylisoindoles (Simons and Johnson, 1976). The isoindoles formed using OPA and ß-mercaptoethanol (ßME) were also found to be electrochemically active (Joseph and Davies, 1983).

Although the OPA/ßME precolumn-derivatization method is now widely used and permits the sensitive measurement of primary amino acids, the instability of the isoindole derivative limits the utility of this method. To improve the analytical precision, careful timing of the reaction or even the use of instrumental automation is required (Furst *et al.*, 1990).

The stability of the OPA/ßME isoindole derivative varies and is dependent upon the amino acid studied. Of the several neurotransmitter amino acids measured, the isoindole derivatives of glutamate and aspartate remained stable while those of GABA, glycine and taurine had half-lives of 4-7 minutes (Lindroth and Mopper, 1979; Joseph and Davis, 1983; Allison *et al.*, 1984; Donzanti and Yamamoto, 1988). This is of great concern when trying to measure these amino acids at low concentrations, as for example in microdialysate perfusates obtained from the brain of living organisms. Here, GABA levels are exceptionally low, typically <0.1 pmol/μL, so that any loss of signal (due to derivative instability) may render GABA levels unmeasurable (Bourdelais and Kalivas, 1991, 1992).

The use of several thiol and thiol derivatives have produced more stable primary amine derivatives (Allison *et al.*, 1984; Garcia Alvarez-Coque *et al.*, 1989). The use of OPA/*tert*-butylthiol produced a GABA derivative which was stable for more than 30 minutes. Other thiols such as thiolactic acid produced multiple peaks making chromatographic interpretation difficult (Allison *et al.*, 1994).

Recently sulfite has successfully replaced the toxic, volatile and malodorous organic thiols. Jacobs used OPA/sulfite to measure primary amines (1987). This method was modified to measure GABA in *post-mortem* human brain tissue (Pearson *et al.*, 1991). Smith and Sharp (1994) reported that the OPA/sulfite GABA derivative was stable for up to one hour and used their method to measure <0.02pmole/μL in striatal microdialysis perfusates. Bossi *et al.*, (1996) measured the OPA/sulfite derivative of GABA in insulinoma cell secretions, brain tissue homogenates and human CSF. They reported <2% degradation of this derivatize when stored at 4°C for 7 hours and a lower detection limit of 20pg.

The use of the coulometric electrode array with pre-column derivatization

Achilli *et al.* (1994) developed a gradient coulometric array method using a CEAS (Coulochem® Electrode Array System; Model 5500, ESA, Inc.) capable of measuring a variety of aromatic amines (tryptamine, tyramine, phenethylamine, and histamine) and aliphatic diamines (putrescine, cadaverine and 1,6-diaminohexane). These compounds are commonly found in wine where they are not only responsible for generating distinct flavors but also for causing undesirable effects such as headache, vomiting and diarrhea. Their technique used OPA/ßME derivatization and was sensitive, selective and rapid. The method was used to accurately measure trace levels of amines and diamines in filtered wine and port samples. Compounds were identified both by retention time and by characteristic voltammetric behavior across the array and could be measured readily even in these complex biological matrices. Although Borolo wine and port showed similar levels in all compounds measured, Gewurztraminer was exceptionally high in histamine, tyramine and putrescene.

Gamache *et al.* (1993) developed a CEAS method capable of analyzing monoamines, their metabolites and OPA/ßME derivatized amino acids concurrently. This method was used to analyze both striatal tissue homogenates as well as striatal microdialysis perfusates. An aliquot of sample was initially analyzed for monoamines and metabolites by isocratic elution and detection on a serial electrode array of eight coulometric flow-through graphite electrodes (0 to +490mV in 70mV increments). The remaining sample was then derivatized with OPA/ßME (based on the method of Donzanti and Yamamoto, 1988) and, after column switching, was analyzed for amino acids using a second isocratic system with electrochemical detection on four electrodes (250, 450, 550 and 650mV). The assay had a limit of detection of 1.4ng/g wet weight for the monoamines and 8.25ng/g wet weight for the amino acids in striatal tissue, and 2.5pg/20µL sample for the monoamines and 15pg/20µL sample for amino acids in striatal perfusates. The assay was completed in under 25 minutes and was capable of measuring a wide variety of compounds as shown in Figure 1. Compounds were identified both by retention time and by their voltammetric behavior across the array (Table 2)[2]. This dual approach to compound identification is crucial, especially with the analysis of microdialysis perfusates where analyte levels are often extremely low. Unfortunately, compounds in microdialysis samples are normally identified solely by the comparison of their retention time to that of an external standard with no ability to measure peak purity. Recently, the method of Gamache *et al.* was refined to include the measurement of drugs (Acworth *et al.*, 1994). Acworth *et al.* reported the capability of measuring monoamines, their metabolites, amino acids and drugs (hydralazine and morphine administered peripherally) in striatal and hippocampal microdialysis perfusates. Using this approach, it is now possible to examine the interplay between different neuronal pathways in the awake freely-moving animal, to examine the effects of drugs on those pathways and to monitor changes in behavior at the same time as obtaining a drug's pharmacokinetic data on-line.

Recently the original method of Gamache *et al.* (1993) has been adapted to simultaneously measure hydrophilic and hydrophobic amino acids without gradient elution. Following derivatization with OPA/ßME, an aliquot of sample was injected onto the first isocratic system for measurement of glutamate, aspartate and other hydrophilic amino acids (100mM sodium phosphate, 25% methanol, 2.5% acetonitrile, pH 6.3, at 1.0mL/min; 3µm C18, 4.6 x 80mm column). After switching to the second isocratic system using a ten-port valve, hydrophobic amino acids (e.g. leucine, isoleucine, tryptophan valine) were then resolved (50mM sodium phosphate, 18.5% acetonitrile, 10% methanol, pH 6.8, at 1.2mL/min; 5µm C18, 4.6 x 150mm column). Using this method twenty-eight amino acids could be well resolved in under 40 minutes.

As discussed above, some amino acids such as GABA form unstable derivatives with OPA/ßME. One way to overcome this problem is to use sulfite as the nucleophile. Turiak and Volicer used a CEAS to measure a variety of amino acids and their methyl esters, and compared the use of sulfite and ßME (1994). Using gradient chromatography with a sixteen coulometric electrode array they reported good linearity (0.1-10µM) and a detection limit of 100-200 fmol (on column). It is also interesting to note that the OPA/sulfite derivative was not only more hydrophilic, eluting much earlier than the equivalent OPA/ßME derivative, but also required a higher oxidation potential (+180mV). Although the method developed by Turiak and Volicer showed excellent sensitivity and selectivity, they failed to measure GABA.

[2] For an in depth explanation of voltammetric behavior and the concept of ratio accuracy see Cullison and Gamache, this volume.

Based on the method of Smith and Sharp (1994) a CEAS method was recently developed capable of measuring GABA as an OPA/sulfite derivative (Ryan *et al.*, 1995). The resulting isoindole-sulfonate is well resolved in under ten minutes using a 5μm C18 4.6 x 150mm column and 0.1M sodium phosphate (pH 6.5) containing 15% methanol, and has been quantitated in rat striatal microdialysis perfusates and in human CSF samples.

A). B).

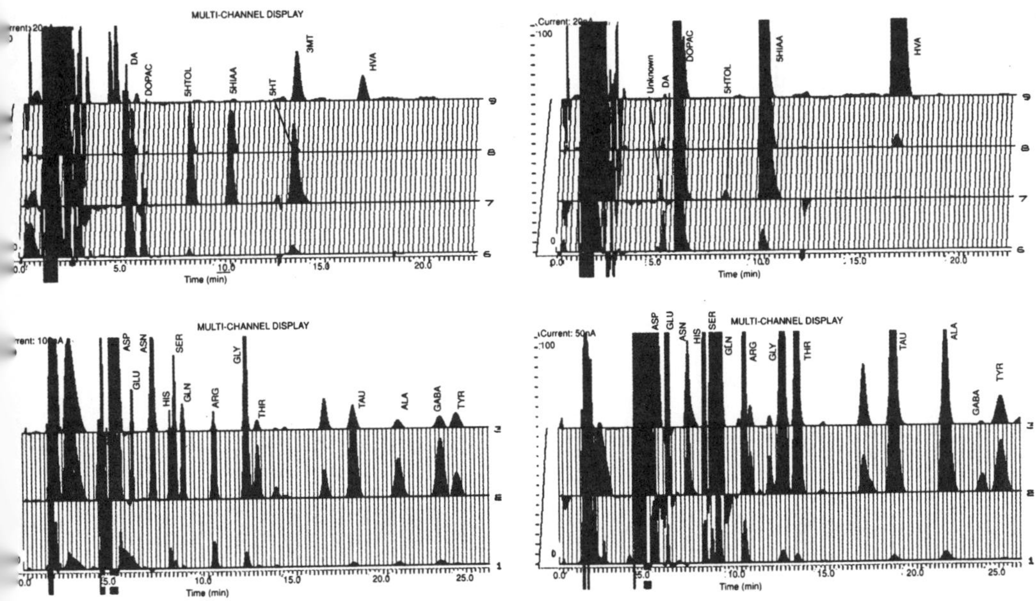

Figure 1. (A) Chromatogram for external standard mixture for microdialysis perfusates showing separation of monoamines and metabolites (upper figure: 20nA sensitivity; 100pg on column) and OPA-ßME derivatized amino acids (lower figure: 100nA sensitivity; 10ng on column). Electrochemical resolution of co-eluting compounds is illustrated by 5HT and 3MT. For clarity only channels 1, 2 and 3 (+250, +450 and +550mV) for amino acids and channels 6,7,8 and 9 (+70, +140, +210 and +280mV) for monoamines and amino acids are shown. (B) Chromatogram of a basal striatal microdialysis perfusate showing sensitive measurement of DA and GABA (Note that an unidentified metabolite elutes very closely to DA but is resolved electrochemically). Conditions as in A except that the lower chromatogram for amino acid measurement is 50nA. Taken from Gamache *et al* (1993) with permission from Elsevier Science Publishers.

POST-COLUMN PHOTO-ACTIVATION

The use of high-energy UV light to transiently activate and make inert compounds detectable has been known since the 1970s (see Birks and Frei [1982] for a review on photochemical reaction detection). Such photochemical devices extend the sensitivity and specificity of detection while minimizing sample preparation for HPLC. The

Table 2.
Typical levels and ratio accuracies for metabolites found in homogenates and microdialysis perfusate samples.

	Retention time	Striatal tissue homogenate concentration (µg/g) (n=2)	Striatal tissue homogenate ratio accuracy	Striatal microdialysis perfusate (pg/collection) (n=10-14)	Striatal microdialysis perfusate ratio accuracy	Hippocampal microdialysis perfusate (pg/collection) (n=3)	Hippocampal microdialysis perfusate ratio accuracy
5-Hydroxy-indoleacetic acid	10.78	0.46±0.001	0.97	2000±240	0.99	1530±63	0.99
Serotonin	14.13	0.64±0.03	0.99	ND		19.4±11	0
5-Hydroxy- tryptophol	8.75	ND		16.2	0	ND	
Dopamine	5.57	6.42±0.01	0.81	41.0±3.6	0.99	6.3±0.5	0.96
Dihydroxyphenyl-acetic acid	6.34	2.59±0.36	0.89	7160±1240	0.7	259±47	0.95
Homovanillic acid	17.58	1.31±0.03	1.00	8160±1280	0.79	117±14	0.95
3-Methoxy-tyramine	14.12	0.43±0.02	0.67	ND		ND	
Alanine	21.69	53.60±3.3	0.97	3400	0.95	7695±1215	0.98
Aspartic acid	5.56	233.00±8.8	0.88	2320±360	0	8730±90	0.23
γ-aminobutyric acid	24.12	296.90±26.40	0.86	321±46	0	293±45	0
Glutamine	9.22	853.67±47.8	0.95	239160±25480	0.91	21105±4680	0.62
Glutamate	6.28	1356.20±88.7	0.99	28200±5680	0.93	4860±270	0.94
Glycine	12.75	165.51±13.11	0.97	6080±800	0.93	5850±765	0.75
Serine	8.76	87.60±4.35	0.99	11636±2080	0.99	10395±225	0.96
Taurine	18.99	873.00±86.00	0.82	12440±1400	0.94	8973±270	0.77
Threonine	13.40	80.62±10.28	0.96	21360±3520	0.96	16470±405	0.99
Tyrosine	25.02	54.01±2.21	0.98	4328±464	0.98	5625±270	0.98

ND- not determined.
Taken from Gamache *et al.* (1993) and Acworth *et al.* (1994).

photochemical reactor generally consists of an UV source and a tubular irradiation coil, and usually is placed after the analytical column and before the detector in the HPLC system. The reactor coil is typically made from PTFE (because it is UV transparent and its flexibility permits different geometrical configurations) and can be crocheted (Poulsen *et al.*, 1986), knotted (Selavka *et al.*, 1987) or knitted (Engelhardt and Neue, 1982). These designs overcome the band-broadening that is associated with simple coiled reactors (see Bachman and Stewart [1989] and references therein; Joshua, 1992). The reactor coils can be either air or water cooled.

Photolytic reactors have been used with HPLC-fluorescence techniques to measure a wide variety of compounds including aflatoxins, antibacterial agents, antipsychotics, aromatic pesticides, diuretics, and vitamin K homologs (Bachman and Stewart, 1989; Joshua, 1993; Aoki *et al.*, 1994; Hansen and Hansen, 1994). Furthermore, this technique is not limited only to analytes that can form suitable fluorophores. Photosensitizers such as anthraquinone-2,6-disulfonate (Gandelman and Birks, 1982) or titanium dioxide (Patel and Moye, 1991) can be used to permit measurement of non-fluorogenic analytes such as aliphatic alcohols, aldehydes, ethers and nitrogenous pesticides.

Photochemical reactors have also been used with HPLC-ECDs to measure a wide variety of compounds (see Table 3). Much of the preliminary work was done by groups headed by Ira Krull (Bratin *et al.*, 1984; Krull *et al.*, 1984b, 1985; Chang *et al.*, 1987), Dennis Johnson (Snider and Johnson, 1979) and Steve Weber (Elbicki *et al.*, 1985). This approach has also been used to measure electrochemically active species at a much lower potential resulting in better limits of detection and fewer interferences (Bachman and Stewart, 1990).

Table 3.
Analytes measured using HPLC-UVEC.

Analyte	Reference
Amino acids	Dou and Krull, 1990.
Amino acids as their 2,4-dinitrofluorobenzene derivatives	Chang *et al.*, 1987.
Amino alcohols as their 2,4-dinitrofluorobenzene derivatives	Chang *et al.*, 1987.
Anions (e.g carbonate, chromate iodate etc.)	LaCourse *et al.*, 1985.
Aspartame	Galletti and Bocchini, 1996
Barbiturates	Selavka *et al.*, 1985.
Benzodiazepines	Selavka *et al.*, 1985.
Cholesterol hydroperoxides	Korytowski *et al.*, 1991.
Cocaine	Selavka *et al.*, 1986.

Analyte	Reference
Diuretics	Macher and Wintersteiger, 1995; Bachman and Stewart, 1990.
Explosives	Krull *et al.*, 1984; Acworth and Waraska, 1996.
Narcotics	Selavka *et al.*, 1985a.
Nitrosamines	Righezza *et al.*, 1987.
Organohalides	Selavka and Krull, 1987a,b; Selavka *et al.*, 1988.
Penicillins	Selavka *et al.*, 1986; Yamazaki *et al.*, 1993. Lihl *et al.*, 1996
Peptides	Dou and Krull, 1990.
Pesticides	Ding and Krull, 1984; Acworth, 1996.
Proteins	Dou and Krull, 1990; Dou *et al.*, 1991.
Triazines	Pacakova *et al.*, 1988.
Vomitoxin	Childress *et al.*, 1990.

The use of HPLC-photolysis array ECD for the measurement of explosives

Many compounds can be made to be EC active using photolysis. The principle will be discussed using the example of the analysis of explosives. There are a wide variety of chemicals used as explosives, but most contain nitro groups attached to carbon (aromatic and aliphatic nitro compounds), oxygen (nitrate esters) or nitrogen (nitramines). Explosive residues are toxic and carcinogenic (Burrows *et al.*, 1989). Of great environmental concern is the corruption of the water-table by explosive residues resulting from contaminated soil. Two of the most widely used explosives, 2,4,6-trinitrotoluene (TNT) and hexahydro-1,3,5-trinitro-1,3,5-triazine (RDX), unlike many other organic compounds, are fairly mobile in soil and readily pollute the water-table. This contamination has occurred at many military installations and most munitions plants throughout the last century due to uncontrolled discharge of waste water during the manufacture of explosives, from residues produced during the destruction of out-of-date bombs and from the use of munitions during training. Based on the studies of Burrows *et al.* (1989), the Environmental Protection Agency and Oak Ridge National Laboratory have set permissible levels for drinking water contamination which range from 0.0068µg/L (2,6-DNT -2,6-dinitrotoluene) to 400µg/L (HMX - octahydro-1,3,5,7-tetranitro-1,3,5,7-tetrazocine) (see Walsh *et al.*, 1993 and references therein).

TNT and RDX are two commonly used military explosives but their measurement in soil and drinking water is far more complex than just analyzing for these two compounds. Biotransformation and metabolism (Walsh and Jenkins, 1992), photolysis, and the impurity of the ingredients (Walsh *et al.*, 1993) requires an analytical technique capable of discriminating between many chemically related species.

A variety of approaches have been used to measure explosive residues, including thin-layer chromatography, gas chromatography, supercritical fluid chromatography and high performance liquid chromatography with UV, photodiode array, electron capture, thermal energy analysis, off-line chemical ionization mass spectrometry, electrochemical and photolytic-electrochemical detectors (Vouros *et al.*, 1977; Walsh *et al.*, 1993 and references therein). However, not all techniques can be used for the routine measurement of trace levels of explosive residues due to lack of sensitivity, poor selectivity or expense. Many of the methods suffer from the need for extensive sample preparation.

HPLC-ECD offers both sensitivity and selectivity. Although, theoretically, aromatic and aliphatic nitro-compounds should be measured reductively, practically this approach is limited due to problems associated with dissolved oxygen, metal impurities and unstable working electrodes (Bratin *et al.*, 1981). The mechanisms by which explosives react electrochemically has been studied using cyclic voltammetry on a gold/mercury amalgam electrode and this information was then used to measure a mixture of explosives on either a glassy carbon electrode (at -1000mV *vs.* Ag/AgCl) or gold amalgam (-850mV *vs.* Ag/AgCl) (Bratin *et al.*, 1981). Krull *et al.* (1984a) measured the nitrite generated photolytically from explosives on dual glassy-carbon amperometric electrodes (set at +800mV and +1200mV *vs.* Ag/AgCl). They used the ratio of responses at these two electrodes as an indicator of peak purity and reported lower detection limits for the explosives from 500pg to 4.0ng (on column).

The use of the coulometric array in the measurement of explosives

Achilli and Cellerino (1992a) have developed a gradient CEAS method and used it to examine the electrochemical behavior (both oxidative and reductive) of dinitronaftalene, hexogene, pentrite and octogene. They also used this method, with the collaboration of The French Gendarmerie, to fingerprint explosive residues. Using this approach it was possible to identify the explosive used (e.g C3, N40 or bentolite). Furthermore from the residue patterns obtained from spent cartridges and hand washings, it was possible to identify who had fired the gun and what cartridges had been used (Achilli and Cellerino, 1992b).

Recently we have combined UV photolysis with coulometric array detection (CoulArray™; Model 5600, ESA, Inc.) and UV detection to measure explosives and metabolites (Acworth and Waraska, 1996). The schematic of the hardware setup is shown in Figure 2. Analytes leaving the column are first subjected to electrochemical treatment using four coulometric electrodes (to measure EC active metabolites) before photolytic activation. Compounds are then measured electrochemically using a second set of four coulometric electrodes where ratioing of the response across this array is used to help positively identify the reacting species. Finally, the eluent passes through an UV detector for further validation. Separation of the analytes is based on the method of Jenkins and Golden (1993). Analytes were resolved either on a Meta-250 column (4.6 x 250mm, 5µm, C18; ESA, Inc.) using a 50mM sodium phosphate (pH 3.5) mobile phase containing 50% methanol at a flow rate of 0.7mL/min or a Keystone NA column (4.6 x 250mm, 5µm) using a mobile phase composed of 66%

(0.1M sodium phosphate)/34% methanol, pH 3.5 at 1.0mL/min. To keep the system back pressure within tolerable limits the column was maintained at 35°C in a column

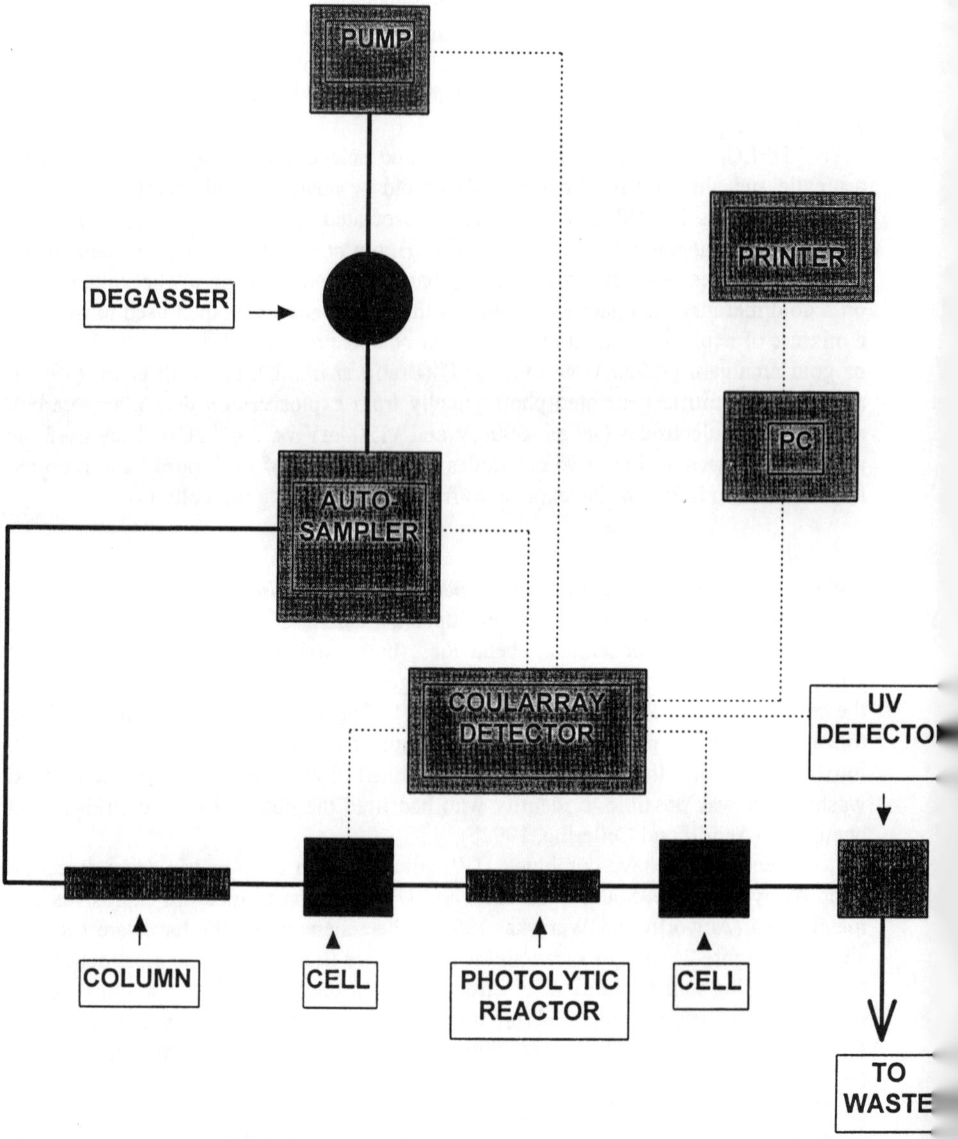

Figure 2. Schematic of analytical system used for the measurement of explosives. The figure highlights the position of the photolytic reactor between the up-stream and down-stream electrode arrays and the placement of the UV detector.

oven. All equipment was from ESA, Inc. except the photolytic unit which was obtained from Aura Industries, Inc. ("Phred" consisting of an high energy UV lamp and 15m x 0.25mm i.d. knitted reactor coil). The UV detector (Model 520, ESA, Inc.) wavelength was 254nm (with the signal processed through the CoulArray™ system).

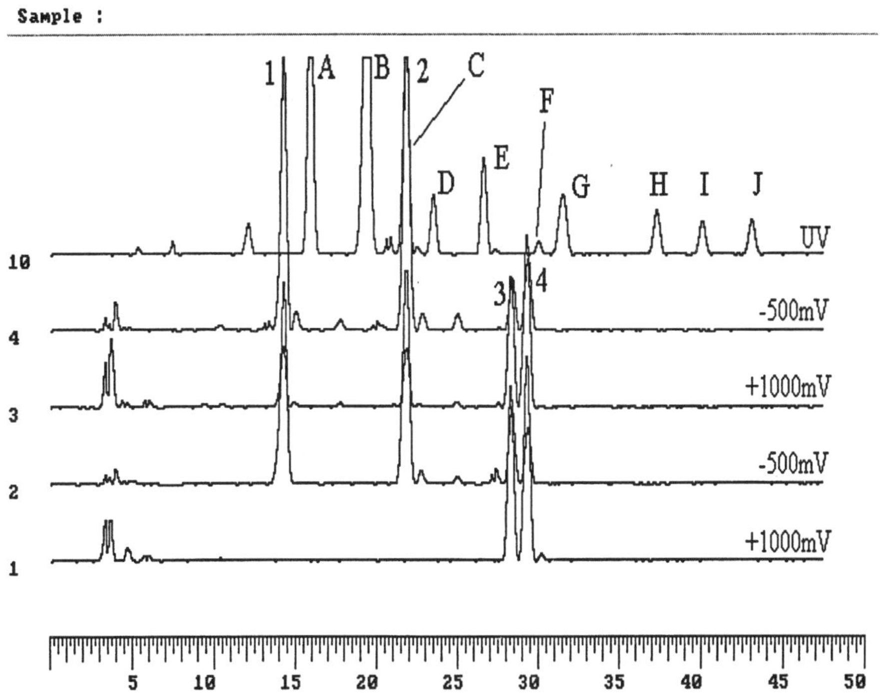

Figure 3. Chromatogram (200ng on column) showing that not all explosives are inherently EC active. TNB (1) and TETRYL (2) readily reduced while 2A-4,6-DNT [2-amino-4,6-dinitrotoluene] (4) and 4A-2,6-DNT [4-amino-2,6-dinitrotoluene] (3) were readily oxidized on the upstream electrode array. Most explosives could be measured using UV including TNB [1,3,5-trinitrobenzene] (A), DNB [1,3-dinitrobenzene] (B), NB [nitrobenzene; hidden behind peak 2] (C), TETRYL (D), TNT (E), 4A-2,6-DNT (F), 2A-4,6-DNT and 2,6-DNT (G), 2-NT [2-nitrotoluene] (H), 4NT [4-nitrotoluene] (I) and 3-NT [3-nitrotoluene] (J). In general the EC response was much larger than the UV. All data were obtained with the photolytic unit turned off and are presented at a gain of 1.5μA. Note that there was approximately a 1.5 minute delay from EC measurement to UV response (a consequence of the size of reactor coil and flow rate). Mixed standards were obtained from Accustandard, Inc. New Haven, CT. At the sensitivity shown, the signal from the UV detector appears to be equivalent to the signal from the ECD. However, ECD offers superior detection limits than UV and can function at much greater sensitivities (see Table 4).

Initial experiments examined the effects of pre-photolysis electrochemistry. Repetitive redox (where potentials of -500mV, +1000mV, -500mV, and +1000mV were applied to the four working electrodes) or repetitive oxred (where potentials of +1000mV, -500mV, +1000mV and -500mV were applied to the same four working electrodes) showed that only a few of the fifteen explosives were electrochemically reactive (see Figure 3 and Table 4). 1,3,5-trinitrobenzene (TNB) and tetryl were readily reduced but only weakly oxidized, whereas both 2-amino-4,6-dinitrotoluene and 4-amino-2,6-dinitrotoluene were readily oxidized and reduced. RDX presented only a weak oxidation response, whereas 2,4-dinitrotoluene and 2,6-dinitrotoluene presented weak reductive responses. Similar results were found for the downstream electrode array when the photolysis unit was turned off. When the photolytical unit was turned on all fifteen explosives could be electrochemically cycled on the downstream coulometric array. However, the oxidation response was typically much larger than the reduction response. In agreement with Bratin *et al.* (1981) we found the sensitive measurement of explosives reductively was limited by noise. This situation was somewhat improved by the use of an on-line degassing unit. Rather than use the reductive electrodes for analyte quantitation, they were used only to generate reduced species capable of reacting at lower potentials at subsequent oxidation electrodes.

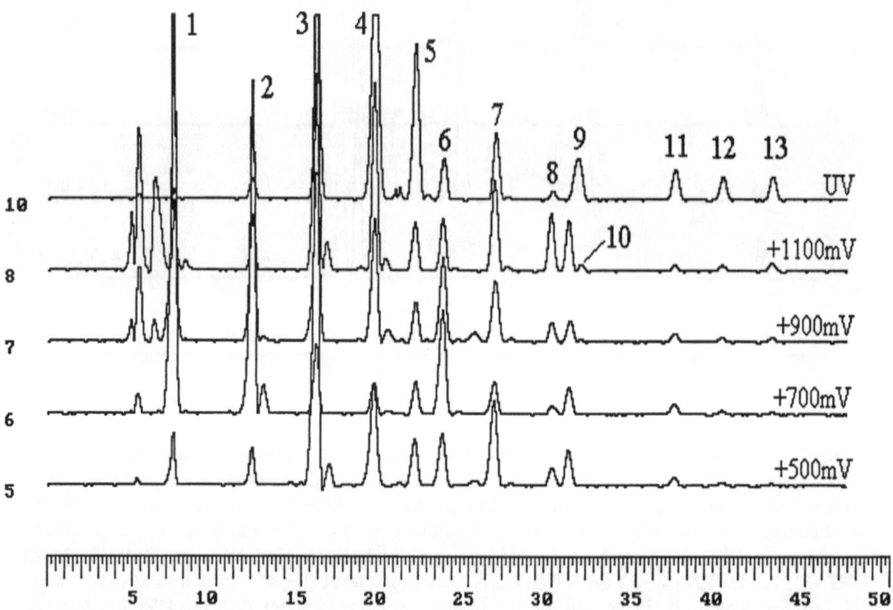

Figure 4. Photolysis permits electrochemical measurement of all explosives in the mixed standard (200ng on column) at the downstream electrode array. HMX (1), RDX (2), TNB (3), DNB (4), NB (5), TETRYL (6), TNT (7), 4A-2,6-DNT (8), 2A-4,6-DNT and 2,6-DNT (9), 2,4-DNT [2,4-dinitrotoluene] (10), 2NT (11), 4NT (12) and 3NT (13). Note that there is no delay between the EC and UV responses. At the sensitivity shown, the signal from the UV detector appears to be equivalent to the signal from the ECD. However, ECD offers superior detection limits than UV and can function at much greater sensitivities (see Table 4).

A second approach kept the redox cycling on the upstream array but utilized an oxidation array downstream (+500mV to +1100mV in +200mV increments). Again, few analytes could be measured on the downstream array when the photolysis unit was off, but all analytes responded when it was on (Figure 4 and Table 4). The limits of detection using only the downstream dominant channel for oxidation

Table 4.
Comparison between the UV signal and ECD signal obtained from the analysis of 200ng (on column) of explosives and residues.

COMPOUND	UV[1] (nA)	DOMINANT[2] REDOX (DR) (nA)	CUMULATIVE[3] REDOX (CR) (ΣnA)	DR/UV	CR/UV	DOMINANT OXIDATIVE (DO) (nA)	CUMULATIVE OXIDATIVE (CO) (ΣnA)	DO/UV	CO/UV
HMX	147	-	-	-	-	2230	4544	15	31
RDX	171	-	-	-	-	1680	3268	10	19
1,3,5-TNB	300	922	1661	3	6	1790	4216	6	14
1,3-DNB	432	34	46	0.1	0.1	1060	2194	3	5
NB	261	-	-	-	-	220	579	1	2
TETRYL	131	1070	2815	8	22	624	1806	5	14
TNT	159	92	161	1	1	671	1774	4	11
4-Amino 2,6-DNT	114	803	1548	7	14	512	1199	5	11
2-Amino 4,6-DNT	141	914	2421	7	17	521	1358	4	10
2,6-DNT	190	23	26	0.1	0.1	127	311	1	2
2,4-DNT	393	31	37	0.1	0.1	336	835	1	2
2-NT	79	-	-	-	-	118	399	2	5
4-NT	58	-	-	-	-	72	227	1	4
3-NT	85	-	-	-	-	74	194	1	2

[1] Obtained using a high-pressure flow cell with 2mm path length. $\lambda = 254$nm. Data were obtained with both photolytic unit and electrode arrays turned off.
[2] Dominant channel is the electrode presenting greatest signal.
[3] Cumulative response is obtained by summation of the signals obtained from up to four electrodes in the array (either redox [pre-photolytic unit] or oxidative [post-photolytic unit]).

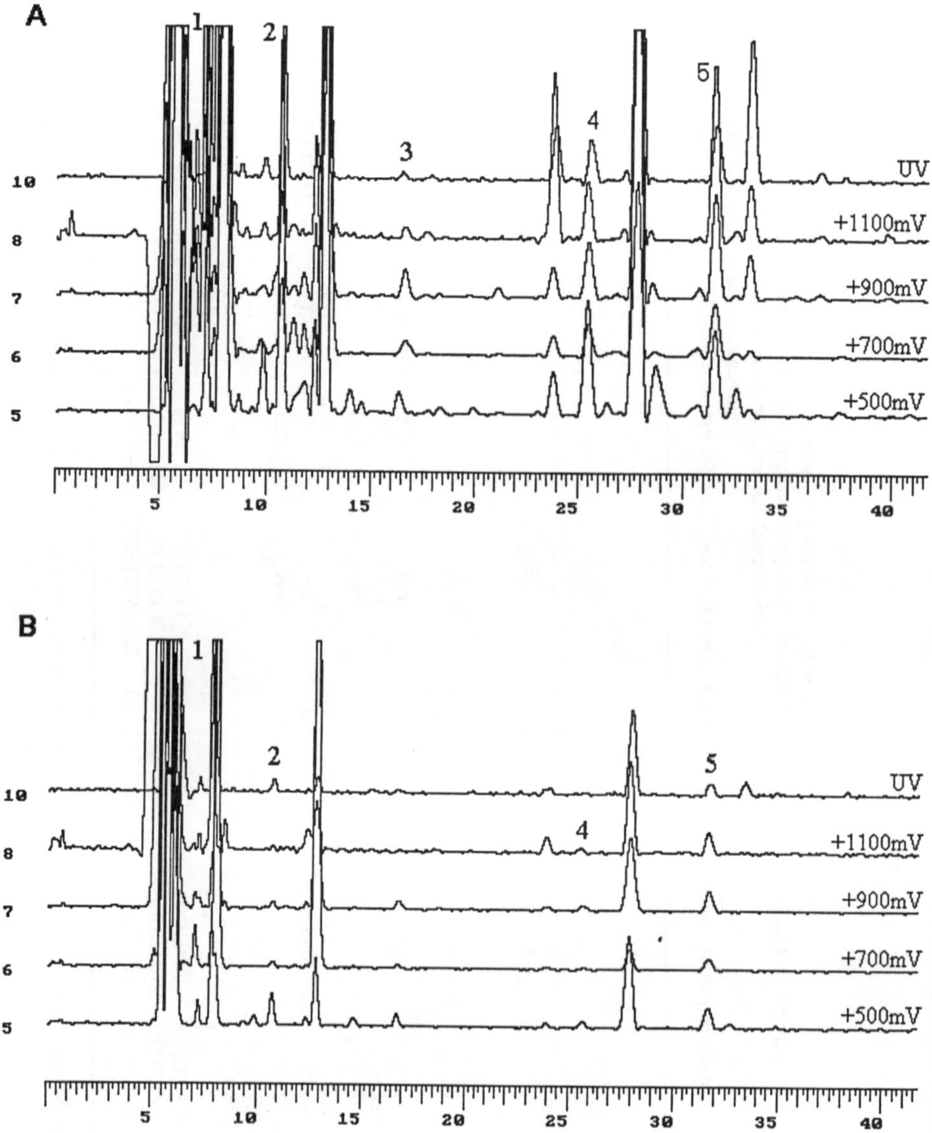

Figure 5. Analysis of contaminated waste water using the coulometric array method outlined above. Water samples were analyzed directly without sample pre-treatment. (A) Sample showing heavy contamination with HMX (1), RDX (2), TNB (3), TNT (4) and DNT (5). (B) Water sample from same site showing much less contamination. The authors are grateful to Dr. Thomas Jenkins (CRREL, Hanover, NH) for providing samples for analysis. See Acworth and Waraska (1996) for further details.

measurement were found to be 20-100pg for HMX, RDX and tetryl, 0.5-2ng for the nitrotoluenes and 200-500pg for the other explosives and residues (on column with s/n=3:1). This approach was also used to study contaminants in unextracted well water (Figure 5).

We wondered whether the use of photolysis and ECD had an effect on each compound's native UV absorbance and whether it would be possible to compare limits of detection for the UV and EC detectors. With the ECD arrays off, most compounds' UV signal were little effected by activation of the photolysis unit with three exceptions (Table 5). The signals from HMX and RDX disappeared (possibly a result of a shift in λ-max due to structural alteration induced by photolysis) while the signal from TNB increased slightly. Activation of the ECD arrays had little effect on the UV signal. Overall we found between a 1 to 31 fold improvement in signal for EC over UV detection (Table 4).

It is interesting to note that the signal across the downstream array due to photolysis is far broader than has been experienced with other compounds. This is probably a result of the formation of multiple compounds as a result of photolysis. Preliminary data indicate that this pattern and peak response can be altered by manipulation of a variety of variables including the composition of the mobile phase, wavelength and intensity of the incident UV light, and length and composition of the reactor coil. These may prove to be important variables in developing or duplicating a method.

Table 5.
The effects of photolysis and electrochemical detection on the UV signal of explosives and residues.

COMPOUND	% UV SIGNAL[1] (ECD OFF PHOTOLYSIS OFF)	% UV SIGNAL (ECD OFF PHOTOLYSIS ON)	% UV SIGNAL (ECD ON PHOTOLYSIS ON)
HMX	100	0	14
RDX	100	0	7
1,3,5-TNB	100	129	121
1,3-DNB	100	97	97
NB	100	92	96
TNT	100	101	82
2-NT	100	89	96
4-NT	100	95	95
3-NT	100	100	102

[1]The UV signal obtained when both the photolytic unit and ECD arrays were turned off was used to normalize data obtained when either the photolytic unit was turned on or when both the photolytic unit and ECD arrays were turned on.

CONCLUSION

The exquisite sensitivity provided by ECD can be extended to the analysis of compounds that are not inherently electroactive through the use of a number of techniques including derivatization and UV photolysis. Additional utility is provided by the use of the coulometric electrode array. One of the most useful of these is the ability to use gradient elution.

When derivatives are used, it is important to select a reagent that is compatible with the requirements of HPLC-ECD. This typically means reversed-phase chromatography compatibility. When amino acids are derivatized, limits in the low pg range are typically achieved. Because of the ability of coulometric electrochemical array technology to be used with gradient chromatography it is now possible to measure many more amino acids than is possible isocratically. An alternative approach is to use the array as "separate systems". With this method both monoamines and derivatized amino acids can be measured in the same sample. Coulometric array technology also has a wide dynamic response permitting the measurement of low and high levels of analytes concurrently. In this way it is possible, for example, to measure drug levels and their effects on metabolism at the same time.

Another approach for enhancing the detection of undetectable species is the use of UV photolysis. By irradiating the sample with high intensity UV light it is possible to create species that now react electrochemically. For explosives limits of detection as low as 20pg have been achieved. A number of parameters may effect detectability using this technique and they may need to be optimized.

By using the array in non-standard configurations it is possible to perform a number of experiments that would be difficult if not impossible using traditional ECD.

Coulometric EC arrays can be used for analysis of compounds with high sensitivity and selectivity and with the added dimension of virtually on-line hydrodynamic voltammetry for compound identification/verification. By using techniques such as derivatization and UV photolysis, these capabilities can be extended to compounds that are generally believed not to be candidates for EC detection.

REFERENCES

Acworth, I. N. (1996). The use of UV photolysis and coulometric array detection for the measurement of pesticides. In Preparation.

Acworth, I. N., and Waraska, J. (1996). The measurement of explosive residues by UV photolysis and coulometric array detection. In Preparation.

Acworth, I. N., Yu, J., Ryan, E., Gamache, P., Hull, K. and Maher, T. (1994). Simultaneous measurement of monoamine, amino acid, and drug levels, using high-performance liquid chromatography and coulometric array detection. J. Liq. Chromatogr. **17**, 685-705.

Achilli, G. and Cellerino, G. P. (1992a). Determination of explosives using the coulometric electrode array system. ESA report #1037.

Achilli, G. and Cellerino, G. P. (1992b). Determination of explosives and residues using the coulochem electrode array system. ESA report #1039.

Achilli, G, Cellerino, G. P., and Melzi d-Eril, G. V. (1994). Determination of amines in wines by high-performance liquid chromatography with electrochemical detection after precolumn derivatization. J. Chromatogr. **661**, 201-205.

Allison, L. A., Mayer, G. S. and Shoup, R. E. (1984). *o*-Phthalaldehyde derivatives of amines for high-speed liquid chromatography/ electrochemistry. Anal. Chem. **56**, 1089-1096.

Aoki, H., Ohshima, Y., Tanaka, M., Okazaki, O. and Hakusui, H. (1994). High-performance liquid chromatographic determination of the quinoline antibacterial agent DU-6859a in human serum and urine using solid-phase extraction with photolysis-fluorescence detection. J. Chromatogr., **660**, 365-374.

Bachman, W. J. and Stewart, J. T. (1989). HPLC postcolumn derivatization techniques. LC-GC. **7**, 38-48.

Bachman, W. J. and Stewart, J. T. (1990). HPLC-photolysis-electrochemical detection in pharmaceutical analysis: application to the determination of spironolactone and hydrochlorothiazide in tablets. J. Chromatogr. Sci. **28**, 123-128.

Birks, J. W. and Frei, R. W. (1982). Trends Anal. Chem. **1**, 361-364.

Bossi, S. R., Morrison, S. A. and Maher, T. J. (1996). Stable OPA/sulfite derivative of GABA in cellular incubation matrix, CSF and tissue homogenates via HPLC with electrochemical detection. 26th Annual Meeting of The Society for Neuroscience, Washington D.C.

Bourdelais, A. J. and Kalivas, P. W. (1991). High sensitivity HPLC assay for GABA in brain dialysis studies. J. Neurosci. Methods **39**, 115-121.

Bourdelais, A. J. and Kalivas, P. W. (1992). Modulation of extracellular γ-aminobutyric acid in the ventral pallidum using *in vivo* microdialysis. J. Neurochem. **58**, 2311-2320.

Bovington, M. E. and Webster, R. A. (1994). Derivatization reactions for neurotransmitters and their automation. J. Chromatogr. **659**, 157-183.

Bratin, K., Blank, C. L., Krull, I. S., Lunte, C. E. and Shoup, R. E. (1981). Recent advances in LCEC and voltammetry. Am. Lab. 33-62.

Burrows, E. P., Rosenblatt, D. H., Mitchell, W. R. and Parmer, D. L. (1989). Organic explosives and related compounds: Environmental and health considerations. USA Biomedical Research and Development Laboratory, Fort Detrick, Maryland. TR-8901.

Chang, M-Y., Chen, L-R., Ding, X-D., Selavka, C. M., Krull, I. S. and Bratin, K. (1987). Precolumn derivatization for improved detection in liquid chromatography-photolysis-electrochemistry. J. Chromatogr. Sci. **25**, 460-467.

Chiavari, G. and Bergamini, C. (1985). High-performance liquid chromatography of carbonyl compounds as 2,4-dinitrophenylhydrazones with electrochemical detection. J. Chromatogr. **318**, 427-432.

Childress, W. L., Krull, I. S. and Selavka, C. M. (1990). Determination of deoxynivalenol (DON, vomitoxin) in wheat by high-performance liquid chromatography with photolysis and electrochemical detection (HPLC-hν-EC). J. Chromatogr. Sci., **28**, 76-82.

Desai, D. M. and Gal. J. (1993). Enantiospecific drug analysis via the ortho-phthalaldehyde/homochiral thiol derivatization method. J. Chromatogr. **629**, 215-228.

Ding, X -D. and Krull, I. S. (1984). Trace analysis of organothiophosphate agriculture chemicals by high-performance liquid chromatography-photolysis-electrochemical detection. J. Agric. Food Chem., **32**, 622-628.

Donzanti, B. A. and Yamamoto, B. K. (1988). An improved an rapid HPLC-EC method for the isocratic separation of amino acid neurotransmitters from brain tissue and microdialysis perfusates. Life Sci. **43**, 913-922.

Dou, L. and Krull, I. S. (1990). Determination of aromatic and sulfur-containing amino acids, peptides and proteins using high-performance liquid chromatography with photolytic electrochemical detection. Anal. Chem. **62**, 2599-2606.

Dou, L., Holmberg, A. and Krull, I. S. (1991). Electrochemical detection of proteins in high-performance liquid chromatography using on-line, postcolumn photolysis. Anal. Biochem. **197**, 377-383.

Eckert, H. and Koller, M. (1990). Derivatizing reagents based on ferrocene for HPLC-ECD determination of peptides and proteins. J. Liq. Chromatogr. **13**, 3399-3414.

Einarsson, S., Folestad, S. and Josefsson, B. (1987). Separation of amino acid enantiomers using precolumn derivatization with o-phthalaldehyde and 2,3,4,6 tetra-O-acetyl-1-thio-β-glucopyranoside. J. Liq. Chromatogr. **10**, 1589-1601.

Einarsson, S., Josefsson, B. and Lagerkvist, S. (1983). Determination of amino acids with 9-fluorenylmethyl chloroformate and reversed-phase high-performance liquid chromatography. J. Chromatogr. **282**, 609-618.

Elbicki, J. M., Morgan, D. M. and Weber, S. G. (1985). Photoelectroanalytical chemistry: Electrochemical detection of a photochemically active species, tris(2,2'-bipyridine)ruthenium(II). Anal. Chem. **57**, 1746-1751.

Engelhardt, H. and Neue, U. D. (1982). Reaction detector with three dimensional open tubes in HPLC. Chromatographia **15**, 403-408.

Frei, R. W. and Lawrence, J. F. (Eds.). (1981) Chemical Derivatives in Analytical Chemistry. Vol. 1. Plenum Press, New York.

Furst, P., Pollack, T. A., Graser, T. A., Godel, H. and Stehle, P. (1990). Appraisal of four pre-column derivatization methods for the high-performance liquid chromatographic determination of free amino acids in biological materials. J. Chromatogr. **499**, 557-569.

Galletti, G. C. and Bocchini, P. (1996). High-performance liquid chromatography with electrochemical detection of aspartame with post-column photochemical reactor. J. Chromatogr. **729**, 393-398.

Gamache, P. H., Ryan, E., Svendsen, C. N., Murayama, K. and Acworth, I. N. (1993). Simultaneous measurement of monoamines, metabolites and amino acids in brain tissue and microdialysis perfusates. J. Chromatogr. **614**, 213-220.

Gandelman, M. S. and Birks, J. W. (1982). Anal. Chem. **54**, 2131-2135.

Garcia Alvarez-Coque, M. C., Medina Hernandez, M. J., Villaneuva Camanas, R. M. and Mongay Fernandez, C. (1989). Formation and instability of o-phthalaldehyde derivatives of amino acids. Anal. Biochem. **178**, 1-7.

Globus, M. Y. -T., Busto, R., Martinez, E., Valdes, I., Dietrich, W. D. and Ginsberg, M. D. (1991). Comparative effect of transient global ischemia on extracellular levels of glutamate, glycine and γ-aminobutyric acid in vulnerable and non-vulnerable brain regions in the rat. J. Neurochem. **57**, 470-478.

Goldring, C., Casini, A., Maellero, E., Bello, B. D. and Comporti, M. (1993). Determination of 4-hydroxynonenal by high-performance liquid chromatography with electrochemical detection. Lipids **28**, 141-145.

Gorog, S. and Gazdag, M. (1994). Enantiometric derivatization for biomedical chromatography. J. Chromatogr. **659**, 51-84.

Granberg, R. R. (1984). High resolution analysis of PITC-derivatized amino acids with UV and CE detection. LC. **2**, 776-781.

Hansen, B. B. and Hansen, S. H. (1994). Determination of zuclopenthixol and its main N-dealkylated metabolite in biological fluids using high-performance liquid chromatography with post-column photochemical derivatization and fluorescence detection. J. Chromatogr., **658**, 319-325.

Hikal, A. H., Lipe, G. W., Slikker, W., Scallet, A. C., Ali, S. F. and Newport, G. D. (1988). Determination of amino acids in different brain regions of the rat brain. Application to the acute effects of tetrahydrocannabinol (THC) and trimethyltin (TMT). Life Sci. **42**, 2029-2035.

Honda, S., Konishi, T. and Suzuki, S. (1984). Electrochemical detection of reducing carbohydrates in high-performance liquid chromatography after post-column derivatization with 2-cyanoacetamide. J. Chromatogr. **299**, 245-251.

Imai, K. and Watanabe, Y. (1981). Fluorimetric determination of secondary amino acids by 7-fluoro-4-nitrobenzo-2-oxa-1,3-diazole. Anal. Chim. Acta **130**, 377-383.

Jacobs, W. A. (1987). *o*-Phthalaldehyde-sulfite derivatization of primary amines for liquid chromatography-electrochemistry. J. Chromatogr. **392**, 435-441.

Jacobs, W. A. and Kissinger, P. T. (1982). Nitroaromatic reagents for the determination of amines and amino acids by liquid chromatography/ electrochemistry. J. Liq. Chromatogr. **5**, 881-895.

Jacobs, W. A., Leburg, M. W. and Madaj, E. J. (1986). Stability of *o*-phthalaldehyde-derived isoindoles. Anal. Biochem. **156**, 334-340.

Jane, I., McKinnon, A. and Flanagan, R. J. (1985). High-performance liquid chromatographic analysis of basic drugs on silica columns using non-aqueous ionic eluents. II. Application of UV, fluorescence and electrochemical oxidation detection. J. Chromatogr. **323**, 191-225.

Jenkins, T. F. and Golden, S.M. (1993). Development of an improved confirmation separation suitable for use with SW846 method 8330. CRREL Report #93-14.

Joseph, M. H. and Davies, P. (1983). Electrochemical activity of *o*-phthalaldehyde-mercaptoethanol derivatives of amino acids; Application to high-performance liquid chromatographic determination of amino acids in plasma and other biological fluids. J. Chromatogr. **277**, 125-136.

Joshua, H. (1992). The effect of experimental parameters on peak broadening in knitted open tubular reactors. Sixteenth International Symposium on Column Liquid Chromatography, Baltimore, MD. June 14-19th.

Joshua, H. (1993). Determination of aflatoxins by reversed-phase high-performance liquid chromatography with post-column in-line photochemical derivatization and fluorescence detection. J. Chromatogr. **654**, 247-254.

Kester, P. E. and Danielson, N. D. (1984). Determination of hydrazine and 1,1-dimethylhydrazine as salicaldehyde derivatives by liquid chromatography with electrochemical detection. Chromatographia **18**, 125-128.

Kissinger, P. T., Bratin, K., Davis, G. C. and Pachla, L. A. (1979). The potential utility of pre- and post-column chemical reactions with electrochemical detection in liquid chromatography. J. Chromatogr. Sci. **17**, 137-146.

Knopse, J., Steinhilber, D., Hermann, T. and Roth, H. (1988). Picomole determination of 2,4-dimethoxyanilides of prostaglandins by high-performance liquid chromatography with electrochemical detection. J. Chromatogr. **442**, 444-450.

Korytowski, W., Bachowski, G. J. and Girotti, A. W. (1991). Chromatographic separation and electrochemical determination of cholesterol hydroperoxides generated by photodynamic action. Anal. Biochem. **197**, 149-156.

Krull, I. S., Selavka, C. M., Duda, C. and Jacobs, W. (1985). Derivatization and post-column reactions for improved detection in liquid chromatography/ electrochemistry. J. Liq. Chromatogr. **8**, 2845-2870.

Krull, I. S., Deyl, Z. and Lingeman, H. (1994). General strategies and selection of derivatizing reactions for liquid chromatography and capillary electrophoresis. J. Chromatogr. **659**, 1-17.

Krull, I. S., Ding, X. -D., Selavka, C. M., Bratin, K. and Forcier, G. (1984a). The trace analysis of explosives and related compounds via high performance liquid chromatography-photolysis-electrochemical detection. J. Forensic Sci. **29**, 449-463.

Krull, I. S., Ding, X. -D., Selavka, C. M. and Nelson, R. (1984b). Electrochemical detection in HPLC. LC, **2**, 214-221.

LaCourse, W. R., Krull, I. S. and Bratin, K. (1985). Photoelectrochemical detector for high-performance liquid chromatography and flow injection analysis. Anal. Chem. **57**, 1810-1814.

Lihl, S., Rehorek, A. and Petz, M. (1996). High-performance liquid chromatographic determination of penicillins by means of automated solid-phase extraction and photochemical degradation with electrochemical detection. J. Chromatogr. **729**, 229-235.

Lindroth, P. and Mopper, K. (1979). High performance liquid chromatographic determination of subpicomole amounts of amino acids by precolumn fluorescence derivatization with o-phthaldialdehyde. Anal. Chem. **51**, 1667-1674.

Lunte, S. M., Mohabbat, T., Wong, O. S. and Kuwana, T. (1989). Determination of desmosine, isodesmosine, and other amino acids by liquid chromatography with electrochemical detection following precolumn derivatization with naphthalenedialdehyde/cyanide. Anal. Biochem. **178**, 202-207.

Lunte, S. M. and Wong, O. S. (1990). Precolumn derivatization with naphthalenedialdehyde/cyanide for fluorescence, chemiluminescence or electrochemical detection of primary amines and peptides. Curr. Sep. **10**, 19-25.

Macher, M. and Wintersteiger, R. (1995). Improved electrochemical detection of diuretics in high-performance liquid chromatography analysis by postcolumn on-line photolysis. J. Chromatogr. **709**, 257-264.

Mahachi, T. J., Carlson, R. M. and Poe, D. P. (1984). N,N-dimethylamino phenylisothiocyanate as an electrochemical label for high-performance liquid chromatographic determination of amino acids. J. Chromatogr. **298**, 279-288.

Meyer, S. D. and Tallman, D. E. (1983). The determination of toluene diisocyanate in air by high-performance liquid chromatography with electrochemical detection. Anal. Chim. Acta **146**, 227-236.

Motohashi, N. and Mori, I. (1989). High-performance liquid chromatography-electrochemical detection of singlet oxygen by reaction with 2,2,6,6-tetramethyl-4-piperidone. J. Chromatogr. **465**, 417-421.

Munns, R. K., Roybal, J. E., Shimoda, W. and Hurlbut, J. A. (1988). 1-(4-hydroxyphenyl)-, 1-(2,4-dihydroxyphenyl)- and 1-(2,5-dihydroxyphenyl)-2-bromoethanones: new labels for determination of carboxylic acids high-performance liquid chromatography with electrochemical and ultraviolet detection. J. Chromatogr. **442**, 209-218.

Musch, G., De Smet, M. and Massart, D. L. (1985). Expert system for pharmaceutical analysis. I. Selection of the detection system in high-performance liquid chromatographic analysis: UV *versus* amperometric detection. J. Chromatogr. **348**, 97-110.

Musson, D. G. and Sternson, L. A. (1980). Conversion of arylhydroxylamines to electrochemically active derivatives suitable for high-performance liquid chromatographic analysis with amperometric detection. J. Chromatogr. **188**, 159-167.

Nakahara, Y., Ishigami, A. and Takeda, Y. (1989). Electrochemical label for high-performance liquid chromatography. I. β-Naphthoquinone-4-sulfonate as an electrochemical detection labeling reagent of amines. J. Chromatogr. **489**, 371-376.

Nussbaum, M. A., Przedwiecki, J. E., Staerk, D. U., Lunte, S. M. and Riley, C. M. (1992). Electrochemical characteristics of amino acids and peptides derivatized with naphthalene-2,3-dicarboxaldehyde: pH effects and differences in oxidation potentials. Anal. Chem. **64**, 1259-1263.

Pacakova, V., Stulik, K. and Prihoda, M. (1988). High-performance liquid chromatography of s-triazines and their degradation products using ultraviolet photometric and amperometric detection. J. Chromatogr. **442**, 147-156.

Patel, B. M. and Moye, H. A. (1991). Postcolumn formation of fluorophores from nitrogenous pesticides by UV photolysis. Talanta **8**, 913-922.

Pearson, S. J., Czudek, C., Mercer, K. and Reynolds, G. P. (1989). Electrochemical detection of human brain transmitter amino acids by high-performance liquid chromatography of stable o-phthalaldehyde-sulphite derivatives. J. Neural. Transm. **86**, 151-157.

Peterkofsky, B. and Prockop, D. J. (1962). A method for the simultaneous measurement of the radioactivity of proline-C^{14} in biological materials. Anal. Biochem. **4**, 400-406.

Pianezzola, E., Bellotti, V., Fontana, E. and Moro, E. (1989). Determination of the enantiomeric composition of salsolinol in biological samples by high-performance liquid chromatography with electrochemical detection. J. Chromatogr. **495**, 205-214.

Poulsen, J. R. Birks, K. S., Gandelman, M. S. and Birks, J. W. (1986). Crocheted PTFE reactors for post-column photochemistry in HPLC. Chromatographia **22**, 231-234.

Prockop, D. J. and Udenfriend, S. (1960). A method for the simultaneous measurement of the radioactivity of proline-C^{14} and hydroxyproline C^{14} in biological materials. Anal. Biochem. **1**, 228-239.

Righezza, M., Murello, M. H. and Siouffi, A. M. (1987). Determination of nitrosamines by liquid chromatography with post-column photolysis and electrochemical detection. J. Chromatogr. **410**, 145-155.

Roth, M (1976). Fluorescence reaction for amino acids. Anal. Chem. **43**, 880-882.

Ryan, E. R., Gamache, P. H. and Acworth, I. N. (1995). A rapid method for the measurement of GABA: The use of OPA/sulfite derivatization and high-performance liquid chromatography with coulometric array detection. In Preparation.

Selavka, C. M. Jiao, K-S., Krull, I. S., Sheih, P., Yu, W. and Wolf, M. (1988). Liquid chromatography-photolysis-electrochemical detection of organobromides and organochlorides. Anal. Chem. **60**, 250-254.

Selavka, C. M. and Krull., I. S. (1987a). Liquid chromatography-photolysis-electrochemical detection for organoiodides. 1. Optimization and application. Anal. Chem. **59**, 2699-2703.

Selavka, C. M. and Krull., I. S. (1987b). Liquid chromatography-photolysis-electrochemical detection for organoiodides. 2. Operative mechanisms. Anal. Chem. **59**, 2704-2709.

Selavka, C. M., Krull, I. S. and Laurie, I. S. (1985). Photolysis derivatization for improved LCEC determination of pharmaceuticals in biological fluids. J. Chromatogr. Sci. **23**, 499-508.

Selavka, C. M., Krull, I. S. and Laurie, I. S. (1986). An improved method for the rapid screening of illicit cocaine preparations using high performance liquid chromatography with electrochemical detection. Forensic Sci. Int. **31**, 103-117.

Selavka, C. M., Nelson, R. J., Krull, I. S. and Bratin, K. (1986). Analysis of penicillins and cefoperazone by HPLC-photolysis-electrochemical detection (HPLC-hν-EC). J. Pharm. Biomed. Anal. **4**. 83-93.

Sessa, W. C., Rodgers, R. L. and Chichester, C. O. (1986). Quantification of collagen synthesis utilizing 4-chloro-7-nitrobenzofurazan derivatives. J. Chromatogr. **382**, 258-263.

Shea, P. A. and Jacobs, W. A. (1989). Isocratic LCEC determination of GABA in dialysates. Curr. Sep. **9**, 57-58.

Sherwood, R. A., Titheradge, A. C. and Richards, D. A. (1990). Measurement of plasma and urine amino acids by high-performance liquid chromatography with electrochemical detection using phenylisothiocyanate derivatization. J. Chromatogr. **528**, 293-303.

Shimada, K., Haniuda, E., Oe, T. and Nambara, T. (1987b). Ferrocene derivatization reagents for optical resolution of carboxylic acids by high-performance liquid chromatography with electrochemical detection. J. Liq. Chromatogr. **10**, 3161-3172.

Shimada, K. and Mitamura, K. (1994). Derivatization of thiol-containing compounds. J. Chromatogr. **659**, 227-241.

Shimada, K., Oe, T. and Nambara, T. (1987c). Sensitive ferrocene reagents for derivatization of thiol compounds in high-performance liquid chromatography with dual-electrode coulometric detection. J. Chromatogr. **419**, 17-25.

Shimada, K., Oe, T., Tanaka, M. and Nambara, T. (1989). Sensitive ferrocene reagents for derivatization of amines for high-performance liquid chromatography with electrochemical detection. J. Chromatogr. **487**, 247-255.

Shimada, K., Sakayori, C. and Nambara, T. (1987a). Determination of fatty acids by high performance liquid chromatography with electrochemical detection using ferrocene derivatization reagent. J. Liq. Chromatogr. **10**, 2177-2187.

Shimada, K., Tanaka, M. and Nambara, T. (1980). Derivatization of ketosteroids for high-performance liquid chromatography with electrochemical detection. Anal. Lett. **13** (B13), 1129-1136.

Shimada, K., Tanaka, M., Nambara, T., Imai, Y., Abe, K. and Yoshinaga, K. (1982). Determination of captopril in human blood by high-performance liquid chromatography with electrochemical detection. J. Chromatogr. **227**, 445-451.

Shimada, K., Tanaka, M. and Nambara, T. (1983). Sensitive derivatization reagent for thiol compounds in high-performance liquid chromatography with electrochemical detection. Anal. Chim. Acta **147**, 375-380.

Shimada, K., Tanaka, M. and Nambara, T. (1984a). Studies on steroids. CC. Determination of 17-ketosteroid sulfates in serum by high-performance liquid chromatography with electrochemical detection using pre-column derivatization. J. Chromatogr. **307**, 23-28.

Shimada, K., Tanaka, M., Nambara, T. and Imai, Y. (1984b). Determination of a new antihypertensive agent (2R,4R)-2-(2-hydroxyphenyl)-3-(3-mercaptopropionyl)-4-thiazolidinecarboxylic acid in blood by high-performance liquid chromatography with electrochemical detection. J. Pharm. Sci. **73**, 119-121.

Simons, S. S. Jr. and Johnson, D. F. (1976). The structure of the fluorescent adduct formed in the reaction of *O*-phthalaldehyde and thiols with amines. J. Am. Chem. Soc. **98**, 7098-7099.

Smith, R. M., Abdul Ghani, A., Haverty, D. G., Bament, G. S., Chasmi, A. Y. and Fogg, A. G. (1988). Derivatization of amines with O-acetylsalicyloyl chloride to enhance electrochemical detection in high-performance liquid chromatography. J. Chromatogr. **455**, 349-354.

Smith, S. and Sharp, T. (1994). Measurement of GABA in rat brain microdialysates using o-phthaldialdehyde-sulfite derivatization and high-performance liquid chromatography with electrochemical detection. J. Chromatogr. **652**, 228-233.

Smyth, W. F. (1987). Voltammetric determination of molecules of biological, environmental, and pharmaceutical importance. CRC Crit. Rev. Anal. Chem. **18**, 155-208.

Snider, B. G. and Johnson, D. C. (1979). A photo-electroanalyzer for the determination of volatile nitrosamines. Anal. Chim. Acta **106**, 1-13.

Stronk, H. J., Janzen, E. G. and Weber, J. R. (1984). Electrochemical detection of PBN spin adduct aminoxyls (nitroxides) separated by high performance liquid chromatography. Anal. Letts. **17**, 321-328.

Tsai, H. and Weber, S. G. (1990). Electrochemical detection of dipeptides and dipeptide amides. J. Chromatogr. **515**, 451-457.

Tsai, H. and Weber, S. G. (1991). Electrochemical detection of oligopeptides through the precolumn formation of biuret complexes. J. Chromatogr. **542**, 345-350.

Turiak, G. and Volicer, L. (1994). Stability of *o*-phthalaldehyde-sulfite derivatives of amino acids and their methyl esters: electrochemical and chromatographic properties. J. Chromatogr. **668**, 323-329.

Vouros, P., Petersen, B. A., Colwell, L., Karger, N. L. and Harris, H. (1977). Analysis of explosives by high-performance liquid chromatography and chemical ionization mass spectrometry. Anal. Chem. **49**, 1039-1044.

Walsh, M. E. and Jenkins, T. F. (1992). Identification of TNT products in soil. CRREL Report #92-16.

Walsh, M. E., Jenkins, T. F., Schnitker, P. S., Elwell, J. W. and Stutz, M. H. (1993). Evaluation of SW846 method 8330 for characterization of sites contaminated with residues of high explosives. CRREL Report #93-5.

Warner, A. M. and Weber, S. G. (1989). Electrochemical detection of peptides. Anal. Chem. **61**, 2664-2668.

Warwick, C. J., Bagon, D. A. and Purnell, C. J. (1981). Application of electrochemical detection to the measurement of free monomeric aromatic and aliphatic isocyanates in air by high-performance liquid chromatography. Analyst **106**, 676-685.

Watanabe, Y. and Imai, K. (1981). High-performance liquid chromatography and sensitive detection of amino acids derivatized with 7-fluoro-4-nitrobenzo-2-oxa-1,3-diazole. Anal. Biochem. **116**, 471-472.

Watanabe, Y. and Imai, K. (1982). Pre-column labeling for high-performance liquid chromatography of amino acids with 7-fluoro-4-nitrobenzo-2-oxa-1,3-diazole and its application to protein hydrolysates. J. Chromatogr. **239**, 723-732.

Watanabe, Y. and Imai, K. (1984). Sensitive detection of amino acids in human serum and dried blood disc of 3mm diameter for diagnosis of inborn errors of metabolism. J. Chromatogr. **309**, 279-286.

Welch, R. W., Acworth, I. N. and Levine, M. (1993). Coulometric electrochemical detection of hydroxyproline using 7-chloro-4-nitrobenzo-2-oxa-1,3-diazole. Anal. Biochem. **210**, 199-205.

Yamazaki, T., Ishikawa, T., Nakai, H., Miyai, M., Tsubota, T. and Asano, K. (1993). Determination of aspoxicillin in broncho-alveolar lavage fluid by high-performance liquid chromatography with photolysis and electrochemical detection. J. Chromatogr. **615**, 180-185.

Yasaka, Y. and Tanaka, M. (1994). Labeling of free carboxyl groups. J. Chromatogr. **659**, 139-155.

Zielke, H. R. (1985). Determination of amino acids in the brain by high-performance liquid chromatography with isocratic elution and electrochemical detection. J. Chromatogr. **347**, 320-324.

Subject Index